Visit classzone.com and get connected

Online resources provide instruction, practice, and learning support correlated to your text.

S0-BZM-924

- **Misconceptions database** provides solutions for common student misconceptions about science and their world.
- **Professional development links**, including SciLinks, offer additional teaching resources.
- **Animations and visualizations** help improve comprehension.
- **Math Tutorial** helps strengthen students' math skills.
- **Flashcards** help students review vocabulary.
- **State test practice** prepares students for assessments.

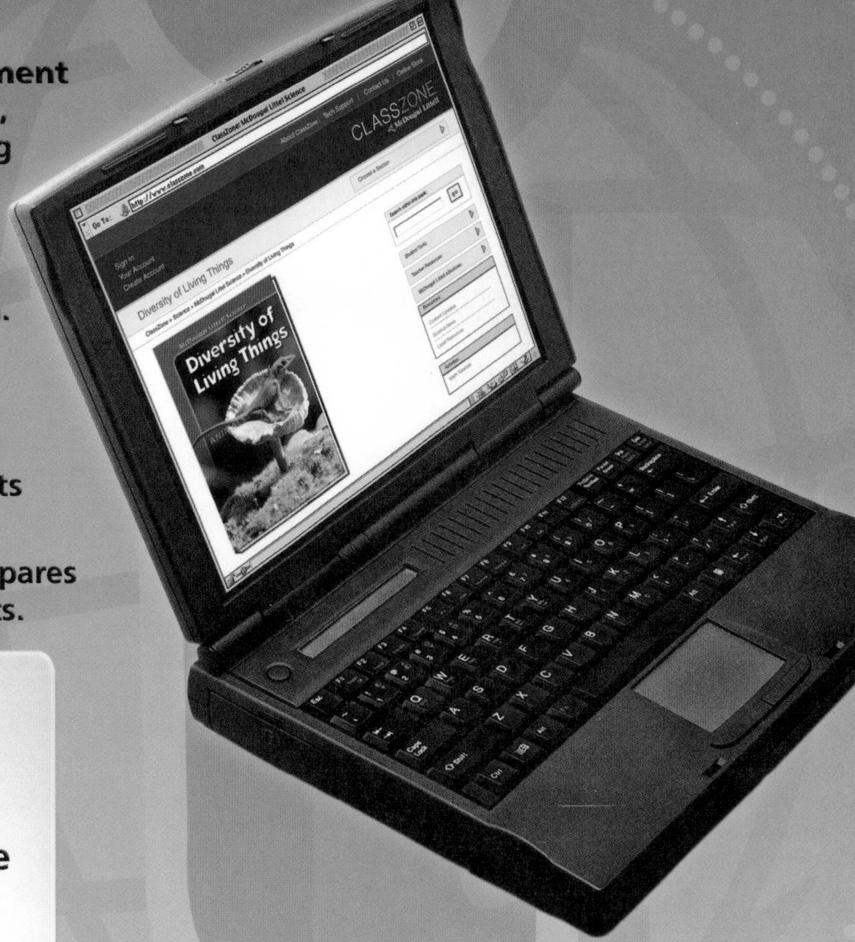

You have immediate access to *ClassZone's* teacher resources.

MCDTCOWDMSSZ

Use this code to create your own username and password.

Also visit *ClassZone* to learn more about these innovative and updated online resources.

- eEdition Plus Online
- eTest Plus Online
- EasyPlanner Plus Online
- Content Review Online

Now it all clicks!™

CLASSZONE.COM

McDougal Littell

TEACHER'S EDITION

McDougal Littell Science

Diversity of Living Things

Credits
5B Illustration by Steve Oh/KO Studios; **5C** *left* Illustration by Steve Oh/KO Studios; *right* Illustration by Peter Bull/Wildlife Art Ltd.; **39B** Illustration by Robin Storesund; **81B** Illustration by Debbie Maizels; **81C** Illustration by Debbie Maizels; © Ed Reschke; **119B** Illustration by Laurie O'Keefe; **119C** Illustration by Ian Jackson/Wildlife Art, Inc.; **153B** Illustration by Laurie O'Keefe **153C** *left* © Fritz Polking/Visuals Unlimited; *right* © Comstock.

Acknowledgements
Excerpts and adaptations from National Science Education Standards by the National Academy of Sciences. Copyright © 1996 by the National Academy of Sciences. Reprinted with permission from the National Academies Press, Washington, D.C.

Copyright © 2005 McDougal Littell, a division of Houghton Mifflin Company.

All rights reserved.

Warning: No part of this work may be reproduced or transmitted in any form or by any means, electronic or mechanical, including photocopying and recording, or by any information storage or retrieval system without the prior written permission of McDougal Littell unless such copying is expressly permitted by federal copyright law. Address inquiries to Supervisor, Rights and Permissions, McDougal Littell, P.O. Box 1667, Evanston, IL 60204.

ISBN: 0-618-33435-1 1 2 3 4 5 6 7 8 VJM 08 07 06 05 04

Internet Web Site: http://www.mcdougallittell.com

McDougal Littell Science

Effective Science Instruction Tailored for Middle School Learners

Diversity of Living Things Teacher's Edition Contents

Consultants and Reviewers

Science Consultants

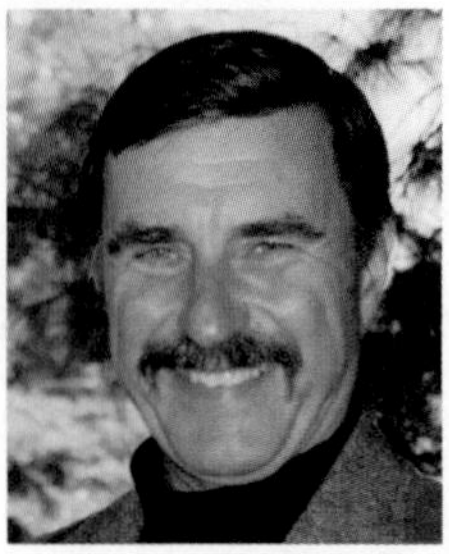

Chief Science Consultant

James Trefil, Ph.D. is the Clarence J. Robinson Professor of Physics at George Mason University. He is the author or co-author of more than 25 books, including *Science Matters* and *The Nature of Science.* Dr. Trefil is a member of the American Association for the Advancement of Science's Committee on the Public Understanding of Science and Technology. He is also a fellow of the World Economic Forum and a frequent contributor to *Smithsonian* magazine.

Rita Ann Calvo, Ph.D. is Senior Lecturer in Molecular Biology and Genetics at Cornell University, where for 12 years she also directed the Cornell Institute for Biology Teachers. Dr. Calvo is the 1999 recipient of the College and University Teaching Award from the National Association of Biology Teachers.

Kenneth Cutler, M.S. is the Education Coordinator for the Julius L. Chambers Biomedical Biotechnology Research Institute at North Carolina Central University. A former middle school and high school science teacher, he received a 1999 Presidential Award for Excellence in Science Teaching.

Instructional Design Consultants

Douglas Carnine, Ph.D. is Professor of Education and Director of the National Center for Improving the Tools of Educators at the University of Oregon. He is the author of seven books and over 100 other scholarly publications, primarily in the areas of instructional design and effective instructional strategies and tools for diverse learners. Dr. Carnine also serves as a member of the National Institute for Literacy Advisory Board.

Linda Carnine, Ph.D. consults with school districts on curriculum development and effective instruction for students struggling academically. A former teacher and school administrator, Dr. Carnine also co-authored a popular remedial reading program.

Donald Steely, Ph.D. serves as principal investigator at the Oregon Center for Applied Science (ORCAS) on federal grants for science and language arts programs. His background also includes teaching and authoring of print and multimedia programs in science, mathematics, history, and spelling.

Sam Miller, Ph.D. is a middle school science teacher and the Teacher Development Liaison for the Eugene, Oregon, Public Schools. He is the author of curricula for teaching science, mathematics, computer skills, and language arts.

Vicky Vachon, Ph.D. consults with school districts throughout the United States and Canada on improving overall academic achievement with a focus on literacy. She is also co-author of a widely used program for remedial readers.

Content Reviewers

John Beaver, Ph.D.
Ecology
Professor, Director of Science Education Center
College of Education and Human Services
Western Illinois University
Macomb, IL

Donald J. DeCoste, Ph.D.
Matter and Energy, Chemical Interactions
Chemistry Instructor
University of Illinois
Urbana-Champaign, IL

Dorothy Ann Fallows, Ph.D., MSc
Diversity of Living Things, Microbiology
Partners in Health
Boston, MA

Michael Foote, Ph.D.
The Changing Earth, Life Over Time
Associate Professor
Department of the Geophysical Sciences
The University of Chicago
Chicago, IL

Lucy Fortson, Ph.D.
Space Science
Director of Astronomy
Adler Planetarium and Astronomy Museum
Chicago, IL

Elizabeth Godrick, Ph.D.
Human Biology
Professor, CAS Biology
Boston University
Boston, MA

Isabelle Sacramento Grilo, M.S.
The Changing Earth
Lecturer, Department of the Geological Sciences
Montana State University
Bozeman, MT

David Harbster, MSc
Diversity of Living Things
Professor of Biology
Paradise Valley Community College
Phoenix, AZ

Richard D. Norris, Ph.D.
Earth's Waters
Professor of Paleobiology
Scripps Institution of Oceanography
University of California, San Diego
La Jolla, CA

Donald B. Peck, M.S.
Motion and Forces; Waves, Sound, and Light; Electricity and Magnetism
Director of the Center for Science Education (retired)
Fairleigh Dickinson University
Madison, NJ

Javier Penalosa, Ph.D.
Diversity of Living Things, Plants
Associate Professor, Biology Department
Buffalo State College
Buffalo, NY

Raymond T. Pierrehumbert, Ph.D.
Earth's Atmosphere
Professor in Geophysical Sciences (Atmospheric Science)
The University of Chicago
Chicago, IL

Brian J. Skinner, Ph.D.
Earth's Surface
Eugene Higgins Professor of Geology and Geophysics
Yale University
New Haven, CT

Nancy E. Spaulding, M.S.
Earth's Surface, The Changing Earth, Earth's Waters
Earth Science Teacher (retired)
Elmira Free Academy
Elmira, NY

Steven S. Zumdahl, Ph.D.
Matter and Energy, Chemical Interactions
Professor Emeritus of Chemistry
University of Illinois
Urbana-Champaign, IL

Susan L. Zumdahl, M.S.
Matter and Energy, Chemical Interactions
Chemistry Education Specialist
University of Illinois
Urbana-Champaign, IL

Safety Consultant

Juliana Texley, Ph.D.
Former K–12 Science Teacher and School Superintendent
Boca Raton, FL

English Language Advisor

Judy Lewis, M.A.
Director, State and Federal Programs for reading proficiency and high risk populations
Rancho Cordova, CA

Research-Based Solutions for Your Classroom

The distinguished program consultant team and a thorough, research-based planning and development process assure that *McDougal Littell Science* supports all students in learning science concepts, acquiring inquiry skills, and thinking scientifically.

Standards-Based Instruction

Concepts and skills were selected based on careful analysis of national and state standards.

- National Science Education Standards
- Project 2061 Benchmarks for Science Literacy
- Comprehensive database of state science standards

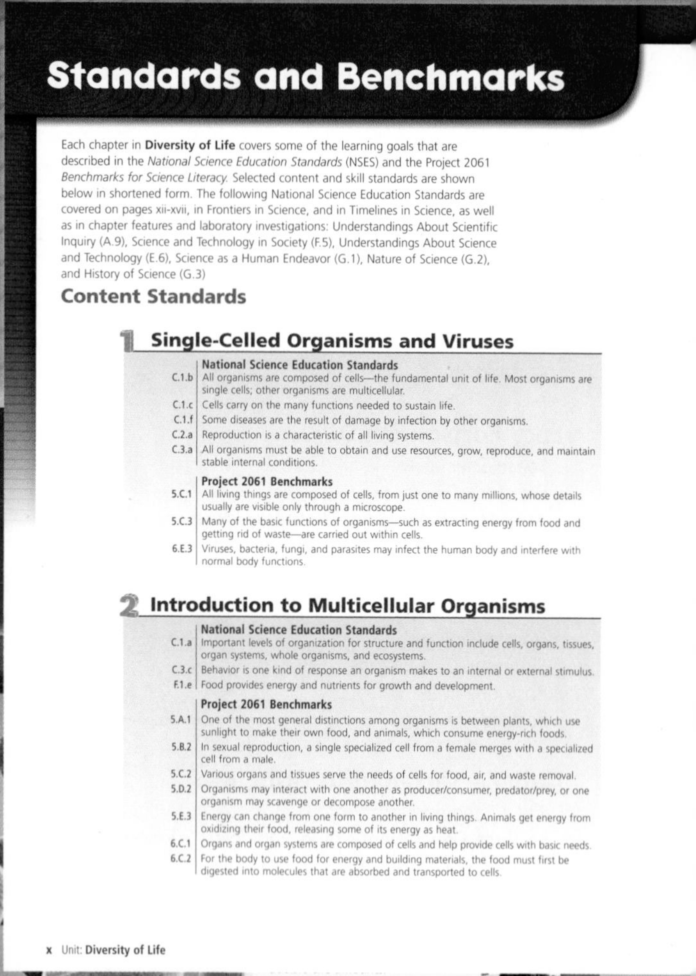

Standards and Benchmarks

Each chapter in **Diversity of Life** covers some of the learning goals that are described in the *National Science Education Standards* (NSES) and the Project 2061 *Benchmarks for Science Literacy.* Selected content and skill standards are shown below in shortened form. The following National Science Education Standards are covered on pages xii-xvii, in Frontiers in Science, and in Timelines in Science, as well as in chapter features and laboratory investigations: Understandings About Scientific Inquiry (A.9), Science and Technology in Society (F.5), Understandings About Science and Technology (E.6), Science as a Human Endeavor (G.1), Nature of Science (G.2), and History of Science (G.3)

Content Standards

1 Single-Celled Organisms and Viruses

National Science Education Standards

C.1.b All organisms are composed of cells—the fundamental unit of life. Most organisms are single cells; other organisms are multicellular.

C.1.c Cells carry on the many functions needed to sustain life.

C.1.f Some diseases are the result of damage by infection by other organisms.

C.2.a Reproduction is a characteristic of all living systems.

C.3.a All organisms must be able to obtain and use resources, grow, reproduce, and maintain stable internal conditions.

Project 2061 Benchmarks

5.C.1 All living things are composed of cells, from just one to many millions, whose details usually are visible only through a microscope.

5.C.3 Many of the basic functions of organisms—such as extracting energy from food and getting rid of waste—are carried out within cells.

6.E.3 Viruses, bacteria, fungi, and parasites may infect the human body and interfere with normal body functions.

2 Introduction to Multicellular Organisms

National Science Education Standards

C.1.a Important levels of organization for structure and function include cells, organs, tissues, organ systems, whole organisms, and ecosystems.

C.3.c Behavior is one kind of response an organism makes to an internal or external stimulus.

F.1.e Food provides energy and nutrients for growth and development.

Project 2061 Benchmarks

5.A.1 One of the most general distinctions among organisms is between plants, which use sunlight to make their own food, and animals, which consume energy-rich foods.

5.B.2 In sexual reproduction, a single specialized cell from a female merges with a specialized cell from a male.

5.C.2 Various organs and tissues serve the needs of cells for food, air, and waste removal.

5.D.2 Organisms may interact with one another as producer/consumer, predator/prey, or one organism may scavenge or decompose another.

5.E.3 Energy can change from one form to another in living things. Animals get energy from oxidizing their food, releasing some of its energy as heat.

6.C.1 Organs and organ systems are composed of cells and help provide cells with basic needs.

6.C.2 For the body to use food for energy and building materials, the food must first be digested into molecules that are absorbed and transported to cells.

x Unit: Diversity of Life

CHAPTER 1
Single-Celled Organisms and Viruses
the BIG idea
Bacteria and protists have the characteristics of living things, while viruses are not alive.
How can you tell if these structures, magnified 2800×, are alive?
Key Concepts
SECTION 1.1 Single-celled organisms have all the characteristics of living things. Learn about characteristics shared by all living things.
SECTION 1.2 Bacteria are single-celled organisms without nuclei. Learn about the characteristics of bacteria and archaea.
SECTION 1.3 Viruses are not alive but affect other living things. Learn about virus structure and how they affect other cells.
SECTION 1.4 Protists are a diverse group of organisms. Learn about protists and how they affect the environment.
Internet Preview
CLASSZONE.COM
Chapter 1 online resources: Content Review, two Visualizations, three Resource Centers, Math Tutorial, Test Practice.
Observe and Think How long did it take to fill the bottle?
Internet Activity: Microscopic Life and You
Go to ClassZone.com to learn about the single-celled organisms.
Observe and Think What types of organism live in the human body?
NSTA scilinks.org SCILINKS
Kingdom Protista Code: MDL039
C 6 Unit: Diversity of Living Things
Chapter 1: Single-Celled Organisms and Viruses 7 C

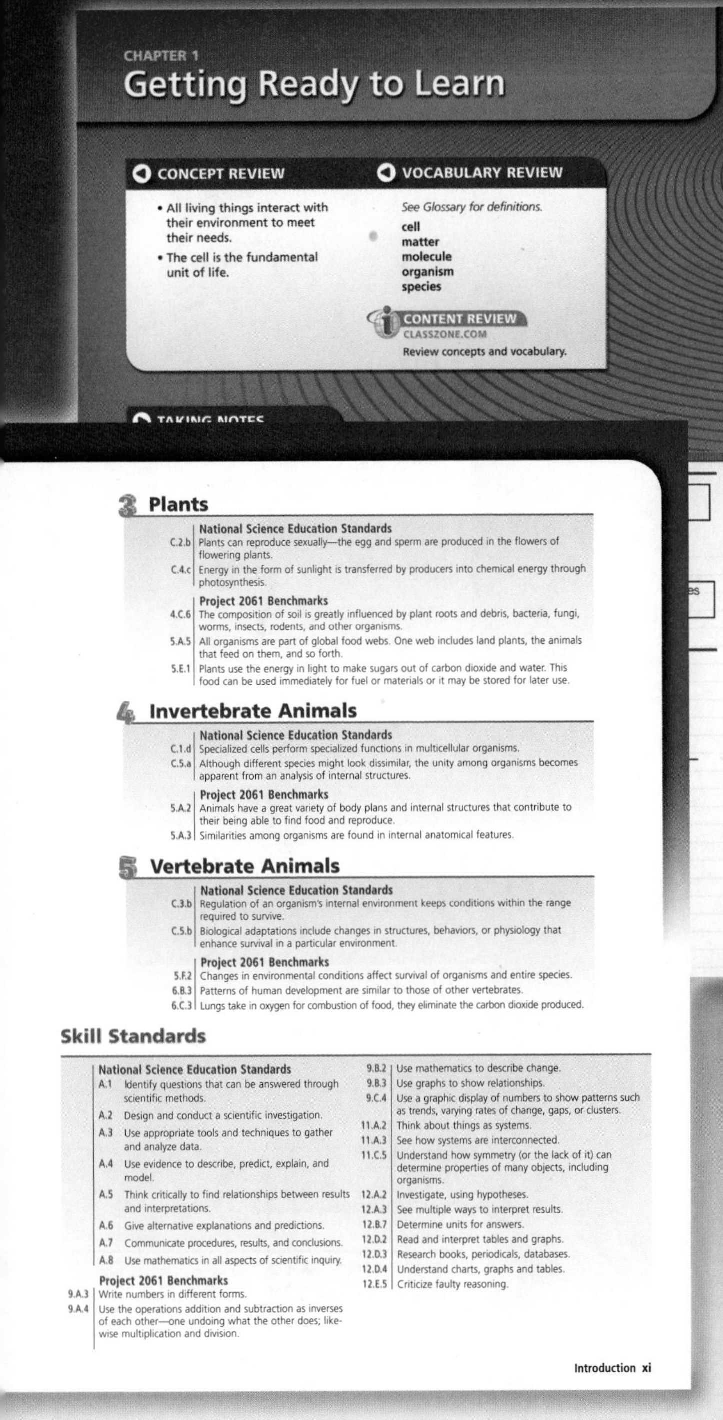

CHAPTER 1

Getting Ready to Learn

CONCEPT REVIEW

- All living things interact with their environment to meet their needs.
- The cell is the fundamental unit of life.

VOCABULARY REVIEW

See Glossary for definitions.

cell
matter
molecule
organism
species

CONTENT REVIEW
CLASSZONE.COM
Review concepts and vocabulary.

TAKING NOTES

3 Plants

National Science Education Standards

C.2.b Plants can reproduce sexually—the egg and sperm are produced in the flowers of flowering plants.

C.4.c Energy in the form of sunlight is transferred by producers into chemical energy through photosynthesis.

Project 2061 Benchmarks

4.C.6 The composition of soil is greatly influenced by plant roots and debris, bacteria, fungi, worms, insects, rodents, and other organisms.

5.A.5 All organisms are part of global food webs. One web includes land plants, the animals that feed on them, and so forth.

5.E.1 Plants use the energy in light to make sugars out of carbon dioxide and water. This food can be used immediately for fuel or materials or it may be stored for later use.

4 Invertebrate Animals

National Science Education Standards

C.1.d Specialized cells perform specialized functions in multicellular organisms.

C.5.a Although different species might look dissimilar, the unity among organisms becomes apparent from an analysis of internal structures.

Project 2061 Benchmarks

5.A.2 Animals have a great variety of body plans and internal structures that contribute to their being able to find food and reproduce.

5.A.3 Similarities among organisms are found in internal anatomical features.

5 Vertebrate Animals

National Science Education Standards

C.3.b Regulation of an organism's internal environment keeps conditions within the range required to survive.

C.5.b Biological adaptations include changes in structures, behaviors, or physiology that enhance survival in a particular environment.

Project 2061 Benchmarks

5.F.2 Changes in environmental conditions affect survival of organisms and entire species.

6.B.3 Patterns of human development are similar to those of other vertebrates.

6.C.3 Lungs take in oxygen for combustion of food, they eliminate the carbon dioxide produced.

Skill Standards

National Science Education Standards

A.1 Identify questions that can be answered through scientific methods.

A.2 Design and conduct a scientific investigation.

A.3 Use appropriate tools and techniques to gather and analyze data.

A.4 Use evidence to describe, predict, explain, and model.

A.5 Think critically to find relationships between results and interpretations.

A.6 Give alternative explanations and predictions.

A.7 Communicate procedures, results, and conclusions.

A.8 Use mathematics in all aspects of scientific inquiry.

Project 2061 Benchmarks

9.A.3 Write numbers in different forms.

9.A.4 Use the operations addition and subtraction as inverses of each other—one undoing what the other does; likewise multiplication and division.

9.B.2 Use mathematics to describe change.

9.B.3 Use graphs to show relationships.

9.C.4 Use a graphic display of numbers to show patterns such as trends, varying rates of change, gaps, or clusters.

11.A.2 Think about things as systems.

11.A.3 See how systems are interconnected.

11.C.5 Understand how symmetry (or the lack of it) can determine properties of many objects, including organisms.

12.A.2 Investigate, using hypotheses.

12.A.3 See multiple ways to interpret results.

12.B.7 Determine units for answers.

12.D.2 Read and interpret tables and graphs.

12.D.3 Research books, periodicals, databases.

12.D.4 Understand charts, graphs and tables.

12.E.5 Criticize faulty reasoning.

Introduction xi

Effective Instructional Strategies

McDougal Littell Science incorporates strategies that research shows are effective in improving student achievement. These strategies include

- Notetaking and nonlinguistic representations (Marzano, Pickering, and Pollock)
- A focus on big ideas (Kameenui and Carnine)
- Background knowledge and active involvement (Project CRISS)

Robert J. Marzano, Debra J. Pickering, and Jane E. Pollock, *Classroom Instruction that Works; Research-Based Strategies for Increasing Student Achievement* (ASCD, 2001)

Edward J. Kameenui and Douglas Carnine, *Effective Teaching Strategies that Accommodate Diverse Learners* (Pearson, 2002)

Project CRISS (Creating Independence through Student Owned Strategies)

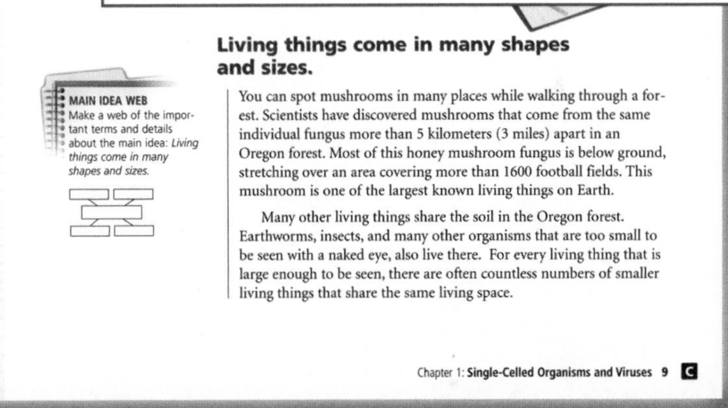

MAIN IDEA WEB
Make a web of the important terms and details about the main idea: *Living things come in many shapes and sizes.*

Living things come in many shapes and sizes.

You can spot mushrooms in many places while walking through a forest. Scientists have discovered mushrooms that come from the same individual fungus more than 5 kilometers (3 miles) apart in an Oregon forest. Most of this honey mushroom fungus is below ground, stretching over an area covering more than 1600 football fields. This mushroom is one of the largest known living things on Earth.

Many other living things share the soil in the Oregon forest. Earthworms, insects, and many other organisms that are too small to be seen with a naked eye, also live there. For every living thing that is large enough to be seen, there are often countless numbers of smaller living things that share the same living space.

Chapter 1: **Single-Celled Organisms and Viruses** 9 C

Comprehensive Research, Review, and Field Testing

An ongoing program of research and review guided the development of *McDougal Littell Science.*

- Program plans based on extensive data from classroom visits, research surveys, teacher panels, and focus groups
- All pupil edition activities and labs classroom-tested by middle school teachers and students
- All chapters reviewed for clarity and scientific accuracy by the Content Reviewers listed on page T5
- Selected chapters field-tested in the classroom to assess student learning, ease of use, and student interest

Content Organized Around Big Ideas

Each chapter develops a big idea of science, helping students to place key concepts in context.

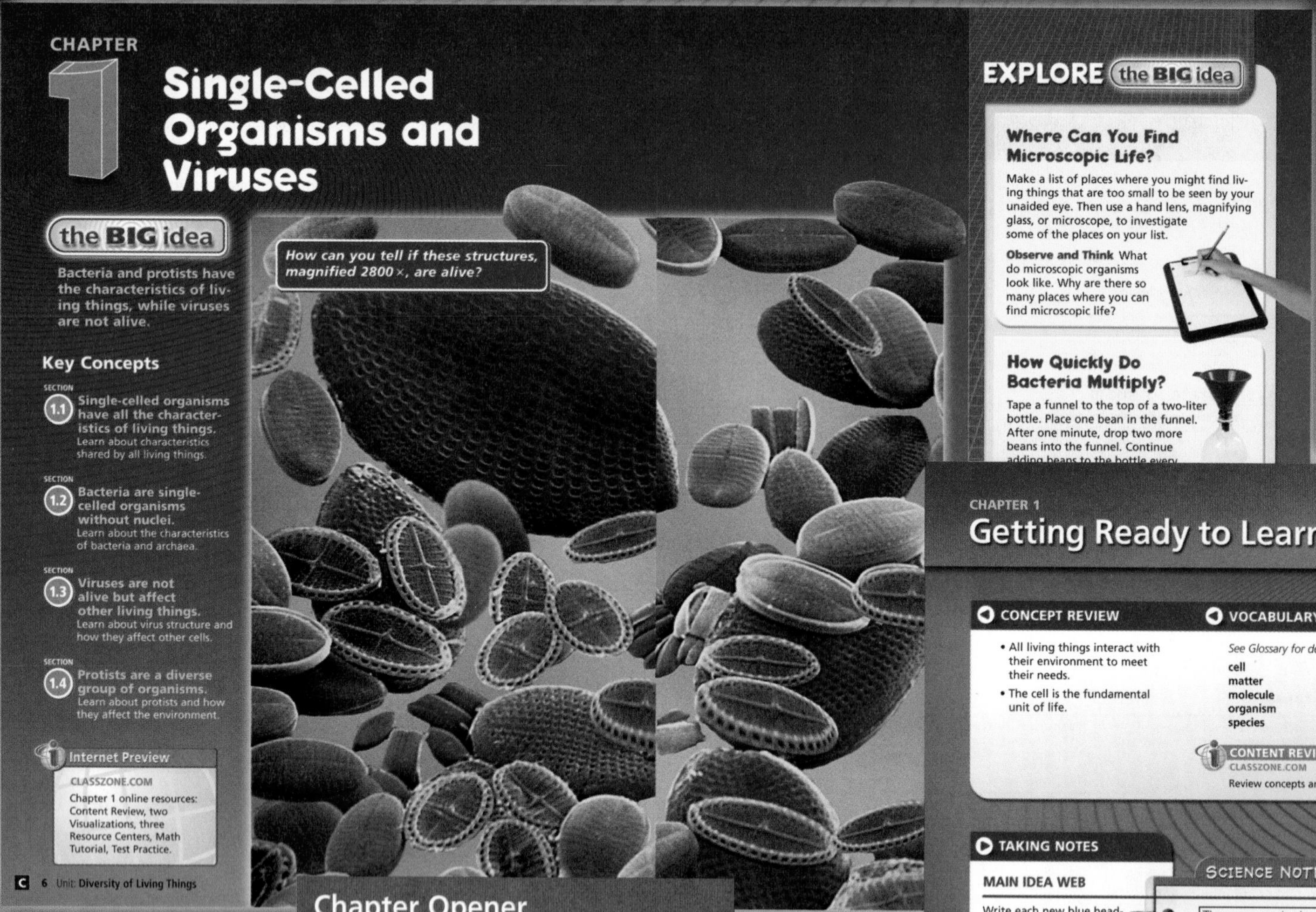

CHAPTER 1
Single-Celled Organisms and Viruses
the BIG idea
Bacteria and protists have the characteristics of living things, while viruses are not alive.
Key Concepts
SECTION 1.1 Single-celled organisms have all the characteristics of living things. Learn about characteristics shared by all living things.
SECTION 1.2 Bacteria are single-celled organisms without nuclei. Learn about the characteristics of bacteria and archaea.
SECTION 1.3 Viruses are not alive but affect other living things. Learn about virus structure and how they affect other cells.
SECTION 1.4 Protists are a diverse group of organisms. Learn about protists and how they affect the environment.
Internet Preview
CLASSZONE.COM
Chapter 1 online resources: Content Review, two Visualizations, three Resource Centers, Math Tutorial, Test Practice.
C 6 Unit: Diversity of Living Things
How can you tell if these structures, magnified 2800×, are alive?
EXPLORE the BIG idea
Where Can You Find Microscopic Life?
Make a list of places where you might find living things that are too small to be seen by your unaided eye. Then use a hand lens, magnifying glass, or microscope, to investigate some of the places on your list.
Observe and Think What do microscopic organisms look like. Why are there so many places where you can find microscopic life?
How Quickly Do Bacteria Multiply?
Tape a funnel to the top of a two-liter bottle. Place one bean in the funnel. After one minute, drop two more beans into the funnel. Continue

Chapter Opener

- Provides an advance organizer of the chapter Big Idea and Key Concepts
- Connects the Big Idea to the real world through an engaging photo and related question

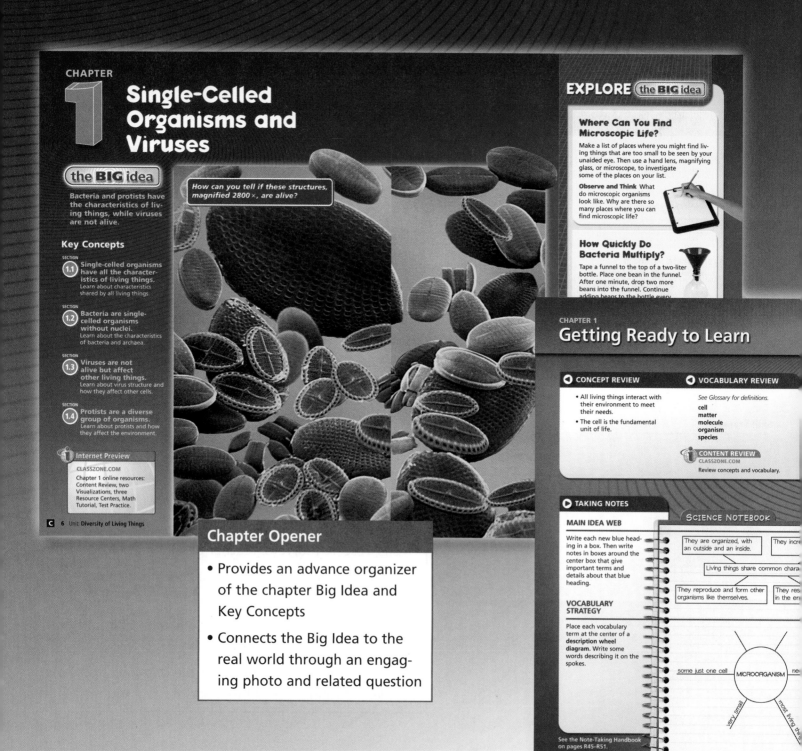

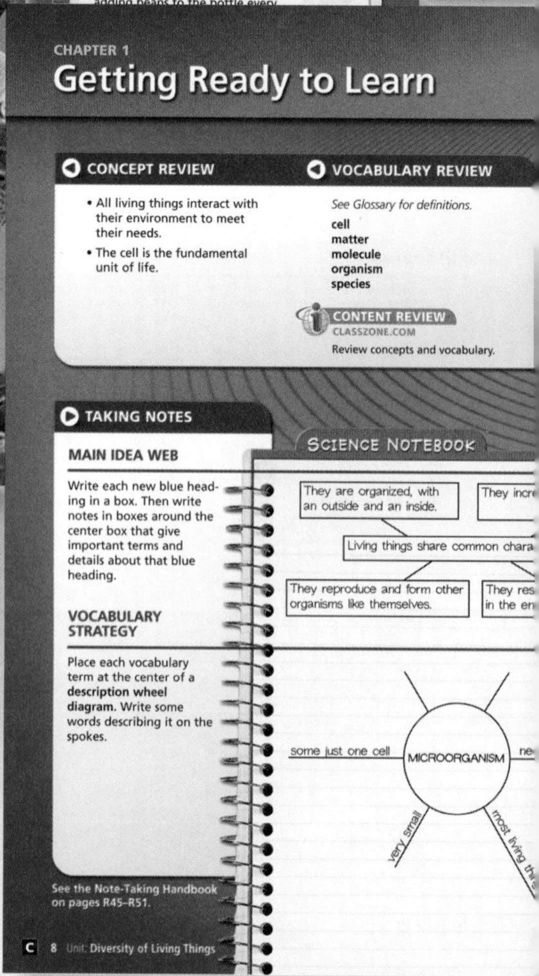

CHAPTER 1
Getting Ready to Learn
CONCEPT REVIEW
• All living things interact with their environment to meet their needs.
• The cell is the fundamental unit of life.
VOCABULARY REVIEW
See Glossary for definitions.
cell
matter
molecule
organism
species
CONTENT REVIEW
CLASSZONE.COM
Review concepts and vocabulary.
TAKING NOTES
MAIN IDEA WEB
Write each new blue heading in a box. Then write notes in boxes around the center box that give important terms and details about that blue heading.
VOCABULARY STRATEGY
Place each vocabulary term at the center of a description wheel diagram. Write some words describing it on the spokes.
See the Note-Taking Handbook on pages R45–R51.
SCIENCE NOTEBOOK
They are organized, with an outside and an inside.
Living things share common chara
They reproduce and form other organisms like themselves.
MICROORGANISM
some just one cell
very small
C 8 Unit: Diversity of Living Things

Visual Summary

- Summarizes Key Concepts using both text and visuals
- Reinforces the connection of Key Concepts to the Big Idea

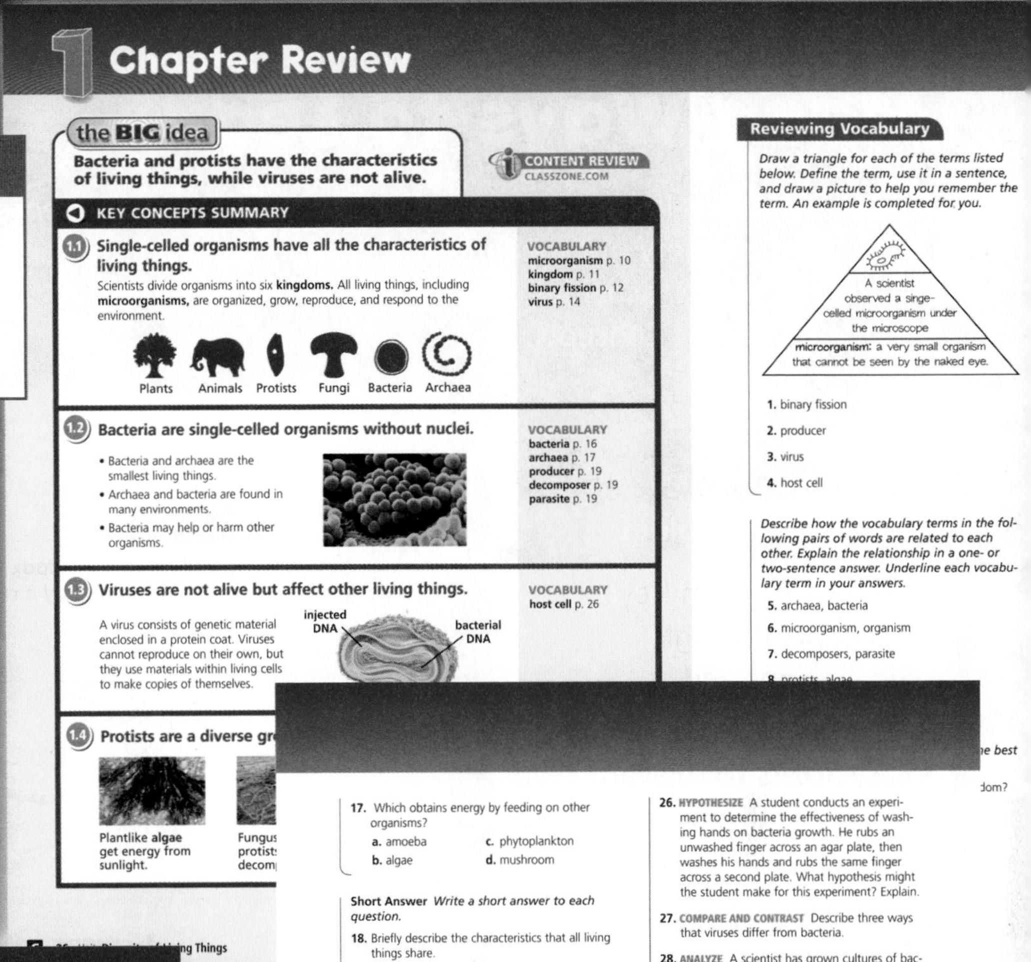

1 Chapter Review
the BIG idea
Bacteria and protists have the characteristics of living things, while viruses are not alive.
CONTENT REVIEW CLASSZONE.COM
KEY CONCEPTS SUMMARY
1.1 Single-celled organisms have all the characteristics of living things.
Scientists divide organisms into six kingdoms. All living things, including microorganisms, are organized, grow, reproduce, and respond to the environment.
Plants Animals Protists Fungi Bacteria Archaea
VOCABULARY microorganism p. 10 kingdom p. 11 binary fission p. 12 virus p. 14
1.2 Bacteria are single-celled organisms without nuclei.
• Bacteria and archaea are the smallest living things.
• Archaea and bacteria are found in many environments.
• Bacteria may help or harm other organisms.
VOCABULARY bacteria p. 16 archaea p. 17 producer p. 19 decomposer p. 19 parasite p. 19
1.3 Viruses are not alive but affect other living things.
A virus consists of genetic material enclosed in a protein coat. Viruses cannot reproduce on their own, but they use materials within living cells to make copies of themselves.
injected DNA
bacterial DNA
VOCABULARY host cell p. 26
1.4 Protists are a diverse gr
Plantlike algae get energy from sunlight.
Fungus protist decom
ing Things
Reviewing Vocabulary
Draw a triangle for each of the terms listed below. Define the term, use it in a sentence, and draw a picture to help you remember the term. An example is completed for you.
A scientist observed a single-celled microorganism under the microscope
microorganism: a very small organism that cannot be seen by the naked eye.
1. binary fission
2. producer
3. virus
4. host cell
Describe how the vocabulary terms in the following pairs of words are related to each other. Explain the relationship in a one- or two-sentence answer. Underline each vocabulary term in your answers.
5. archaea, bacteria
6. microorganism, organism
7. decomposers, parasite

Section Opener

- Highlights the Key Concept
- Connects new learning to prior knowledge
- Previews important vocabulary

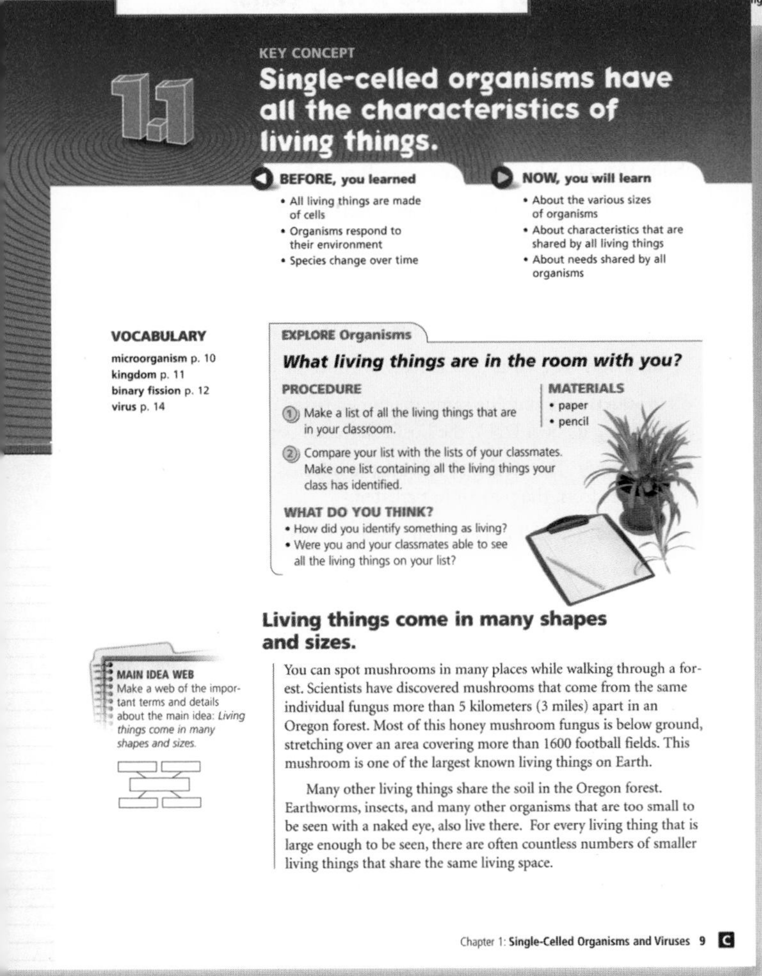

KEY CONCEPT
1.1 Single-celled organisms have all the characteristics of living things.
BEFORE, you learned
• All living things are made of cells
• Organisms respond to their environment
• Species change over time
NOW, you will learn
• About the various sizes of organisms
• About characteristics that are shared by all living things
• About needs shared by all organisms
VOCABULARY
microorganism p. 10
kingdom p. 11
binary fission p. 12
virus p. 14
EXPLORE Organisms
What living things are in the room with you?
PROCEDURE
1 Make a list of all the living things that are in your classroom.
2 Compare your list with the lists of your classmates. Make one list containing all the living things your class has identified.
MATERIALS
• paper
• pencil
WHAT DO YOU THINK?
• How did you identify something as living?
• Were you and your classmates able to see all the living things on your list?
MAIN IDEA WEB
Make a web of the important terms and details about the main idea: Living things come in many shapes and sizes.
Living things come in many shapes and sizes.
You can spot mushrooms in many places while walking through a forest. Scientists have discovered mushrooms that come from the same individual fungus more than 5 kilometers (3 miles) apart in an Oregon forest. Most of this honey mushroom fungus is below ground, stretching over an area covering more than 1600 football fields. This mushroom is one of the largest known living things on Earth.
Many other living things share the soil in the Oregon forest. Earthworms, insects, and many other organisms that are too small to be seen with a naked eye, also live there. For every living thing that is large enough to be seen, there are often countless numbers of smaller living things that share the same living space.
Chapter 1: Single-Celled Organisms and Viruses 9 C

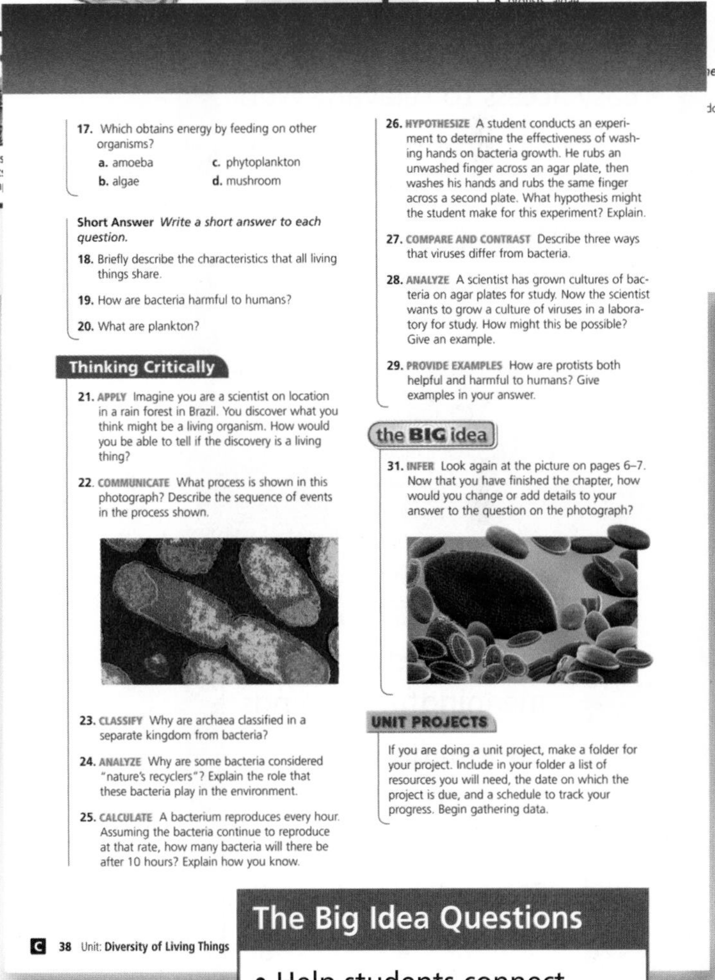

17. Which obtains energy by feeding on other organisms?
a. amoeba c. phytoplankton
b. algae d. mushroom
Short Answer Write a short answer to each question.
18. Briefly describe the characteristics that all living things share.
19. How are bacteria harmful to humans?
20. What are plankton?
Thinking Critically
21. APPLY Imagine you are a scientist on location in a rain forest in Brazil. You discover what you think might be a living organism. How would you be able to tell if the discovery is a living thing?
22. COMMUNICATE What process is shown in this photograph? Describe the sequence of events in the process shown.
23. CLASSIFY Why are archaea classified in a separate kingdom from bacteria?
24. ANALYZE Why are some bacteria considered "nature's recyclers"? Explain the role that these bacteria play in the environment.
25. CALCULATE A bacterium reproduces every hour. Assuming the bacteria continue to reproduce at that rate, how many bacteria will there be after 10 hours? Explain how you know.
26. HYPOTHESIZE A student conducts an experiment to determine the effectiveness of washing hands on bacteria growth. He rubs an unwashed finger across an agar plate, then washes his hands and rubs the same finger across a second plate. What hypothesis might the student make for this experiment? Explain.
27. COMPARE AND CONTRAST Describe three ways that viruses differ from bacteria.
28. ANALYZE A scientist has grown cultures of bacteria on agar plates for study. Now the scientist wants to grow a culture of viruses in a laboratory for study. How might this be possible? Give an example.
29. PROVIDE EXAMPLES How are protists both helpful and harmful to humans? Give examples in your answer.
the BIG idea
31. INFER Look again at the picture on pages 6–7. Now that you have finished the chapter, how would you change or add details to your answer to the question on the photograph?
UNIT PROJECTS
If you are doing a unit project, make a folder for your project. Include in your folder a list of resources you will need, the date on which the project is due, and a schedule to track your progress. Begin gathering data.
C 38 Unit: Diversity of Living Things

The Big Idea Questions

- Help students connect their new learning back to the Big Idea
- Prompt students to synthesize and apply the Big Idea and Key Concepts

Many Ways to Learn

Because students learn in so many ways, *McDougal Littell Science* gives them a variety of experiences with important concepts and skills. Text, visuals, activities, and technology all focus on Big Ideas and Key Concepts.

Integrated Technology

- Interaction with Key Concepts through Simulations and Visualizations
- Easy access to relevant Web resources through Resource Centers and SciLinks
- Opportunities for review through Content Review and Math Tutorials

Considerate Text

- Clear structure of meaningful headings
- Information clearly connected to main ideas
- Student-friendly writing style

Virus

All viruses, including this smallpox virus, contain genetic material surrounded by a capsid.

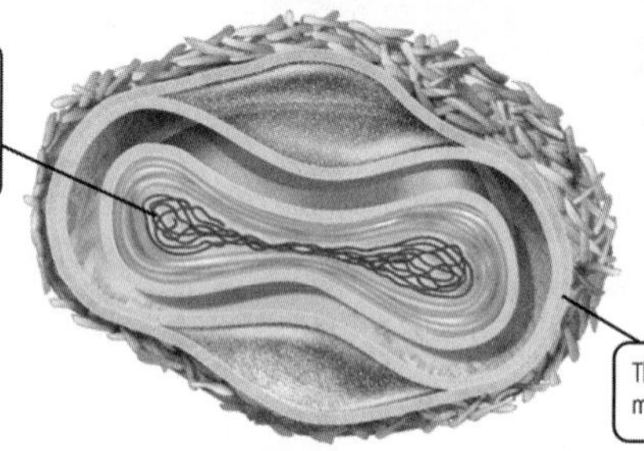

The **genetic material** stores information the virus needs to make copies of itself.
The **capsid** protects the genetic material and gives shape to the virus.

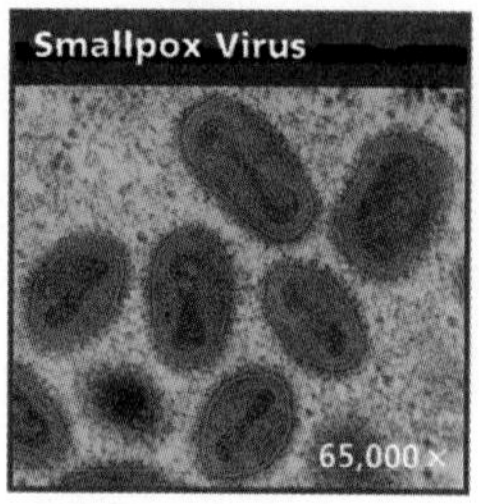

Smallpox Virus
65,000×

Viruses multiply inside living cells.

VISUALIZATION CLASSZONE.COM

See how viruses infect and multiply within bacteria.

Remember that all living things reproduce. Viruses cannot reproduce by themselves, which is one of the ways they are different from living things. However, viruses can use materials within living cells to make copies of themselves. The cells that viruses infect in order to make copies are called **host cells.** Despite their tiny size, viruses have the ability to cause a lot of damage to cells of other organisms.

One of the best studied viruses infects bacteria. It's called a bacteriophage (bak-TEER-ee-uh-FAYJ), which comes from the Latin for "bacteria eater." Some of the steps that a bacteriophage goes through to multiply are shown in the illustration.

1. **Attachment** The virus attaches to the surface of a bacterium.
2. **Injection** The virus injects its DNA into the bacterium.
3. **Production** Using the same machinery used by the host cell for copying its own DNA, the host cell makes copies of the viral DNA.
4. **Assembly** The viral DNA forces the infected cell to assemble new viruses from the parts it has created.
5. **Release** The cell bursts open, releasing 100 or more new viruses.

Viruses have proteins on their surfaces that look like the proteins that the host cell normally needs. The virus attaches itself to special sites on the host that are usually reserved for these proteins.

Not every virus makes copies in exactly the same way as the bacteriophage. Some viruses are inside host cells. Others use the host cell as a factory that produces new viruses one at a time. These viruses may not be as harmful to the infected organism because the host cell is not destroyed.

C 26 Unit: **Diversity of Living Things**

Visuals that Teach

- Information-rich visuals directly connected to the text
- Thoughtful pairing of diagrams and real-world photos
- Reading Visuals questions to support student learning

Making New Viruses

Viruses, such as this bacteriophage, use other cells to make new viruses.

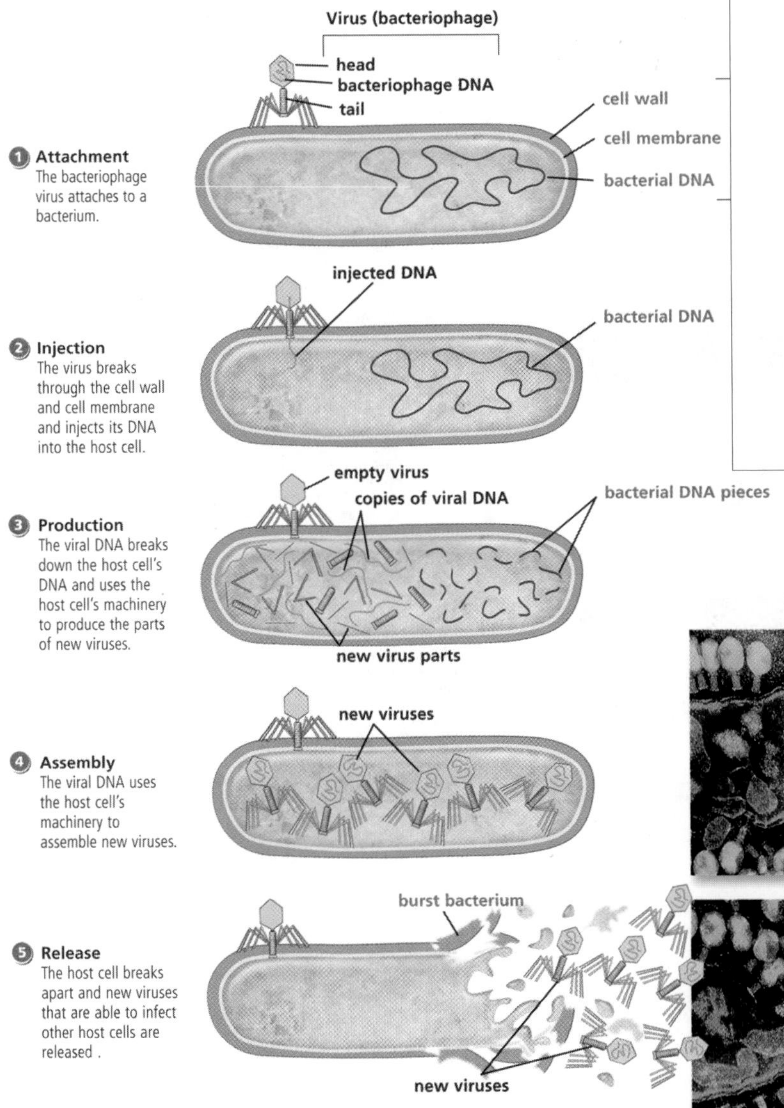

Virus (bacteriophage)
head
bacteriophage DNA
tail
cell wall
cell membrane
bacterial DNA
1 Attachment
The bacteriophage virus attaches to a bacterium.
injected DNA
bacterial DNA
2 Injection
The virus breaks through the cell wall and cell membrane and injects its DNA into the host cell.
empty virus
copies of viral DNA
bacterial DNA pieces
3 Production
The viral DNA breaks down the host cell's DNA and uses the host cell's machinery to produce the parts of new viruses.
new virus parts
new viruses
4 Assembly
The viral DNA uses the host cell's machinery to assemble new viruses.
burst bacterium
5 Release
The host cell breaks apart and new viruses that are able to infect other host cells are released .
new viruses

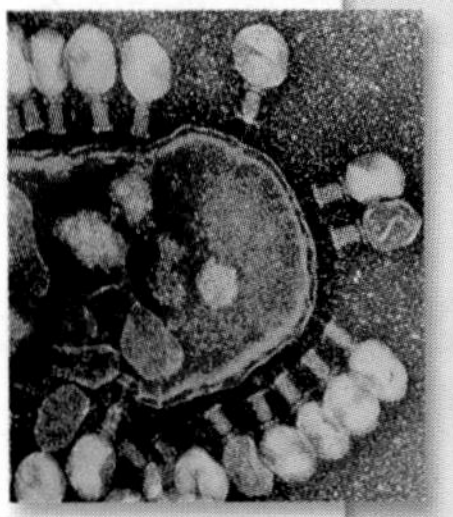

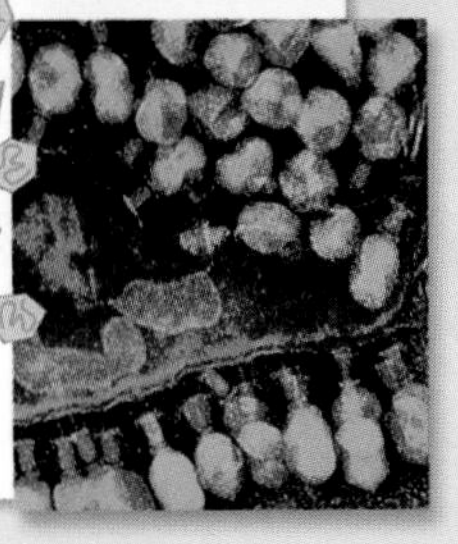

INVESTIGATE Viruses

SKILL FOCUS
Analyzing

MATERIALS
- paper cup
- sample liquid
- solution A
- solution B

TIME
30 minutes

How do infections spread?

PROCEDURE

1. Get a cup of sample liquid from your teacher. Pour half the liquid from your cup into the cup of a classmate, then pour the same amount back into the original cup. Your cup should then contain a mixture of the liquids from both cups.
2. Repeat step 1 with at least two other classmates.
3. Drop one drop of solution A into your paper cup. If it changes color, you are "infected." If you were "infected," add drops of solution B until your liquid turns clear again. Count how many drops it takes to "cure" you.

WHAT DO YOU THINK?

- If you were "infected," can you figure out who "infected" you?
- If you were not "infected," is it possible for anyone who poured liquid into your cup to be "infected"?

CHALLENGE Only one person in your class started out with an "infection." Try to figure out who it was.

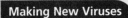

When a filter had removed all of the harmful organisms from a liquid, the liquid no longer caused any illnesses. This method worked when there was only bacteria in the liquid. Sometimes filtering did not prevent disease. Something much smaller than bacteria was in the liquid. Scientists called these disease-causing particles viruses, from the Latin word for "slimy liquid" or "poison."

RESOURCE CENTER
CLASSZONE.COM
Learn more about viruses.

CHECK YOUR READING How does the size of viruses compare with the size of bacteria?

Scientists have learned much about viruses, and can even make images of them with specialized microscopes. Viruses consist of genetic material contained inside a protective protein coat called a capsid. The protein coat may be a simple tube, such as the coat of an ebola virus, or have many layers, such as the smallpox virus shown on page 26.

Viruses may come in many shapes and sizes, but all viruses consist of a capsid and genetic material. The ability of viruses to make copies of their genetic material is one way that viruses are similar to living things. Also the protein coat is similar to a cell's outer membrane. But viruses do not grow, and viruses do not respond to changes in their environment. Therefore, viruses are not living organisms.

Hands-on Learning

- Activities that reinforce Key Concepts
- Skill Focus for important inquiry and process skills
- Multiple activities in every chapter, from quick Explores to full-period Chapter Investigations

Differentiated Instruction

A full spectrum of resources for differentiating instruction supports you in reaching the wide range of learners in your classroom.

1.1 INSTRUCT

Teach from Visuals

To help students interpret the visual of organisms in a mangrove swamp, ask:

- What types of organisms in the photograph are able to be seen? *plants and animals*
- What types of organisms in the photograph are unable to be seen? *Archaea, bacteria, protists, and fungi; the amoeba is a protist.*

EXPLORE the BIG idea

Revisit "Where Can You Find Microscopic Life?" on p. 7. Have students explain why there might be many microscopic organisms in the environment shown in the photograph on p. 10.

Arts Connection

Flemish artists of the 17th century often painted pictures of the living world as it was known at that time. Many of these paintings can be found on the Internet and printed. Show students a copy of one of these paintings, such as *Paradise* by Jan Brueghel the Younger. Point out that all of the organisms shown are either plants or animals.

Ongoing Assessment

Recognize that some organisms are too small to see.

Ask: How can you be sure that microscopic organisms are present in an environment such as a swamp? *You must use a microscope.*

CHECK YOUR READING *Answer: An organism is any living thing. Microorganisms are living things, but they are too small to be seen without a microscope.*

The honey mushroom fungus is one example of an organism. You, too, are an organism, and tiny bacteria living inside your body are also organisms. In fact, any living thing can be called an organism.

When you identify living things, you probably begin with those you can observe—plants, animals, and fungi such as mushrooms. However, most living things are too small to observe without a microscope. Even the tiniest organisms are made of cells. Very small organisms are called **microorganisms.** Some microorganisms are made of just one cell.

READING TIP: The prefix *micro-* means "very small." Therefore, *microscope* means "very small scope" and *microorganism* means "very small organism."

CHECK YOUR READING Compare and contrast the words *microorganism* and *organism.*

A visitor to a mangrove swamp forest can find an amazing variety of organisms. The mangrove trees themselves are the most obvious organisms. Roots from these trees grow above and below the muddy bottom of the forest. Other organisms live in almost every part of the mangrove tree.

Six Kingdoms of Life

All organisms are divided into six groups called kingdoms.

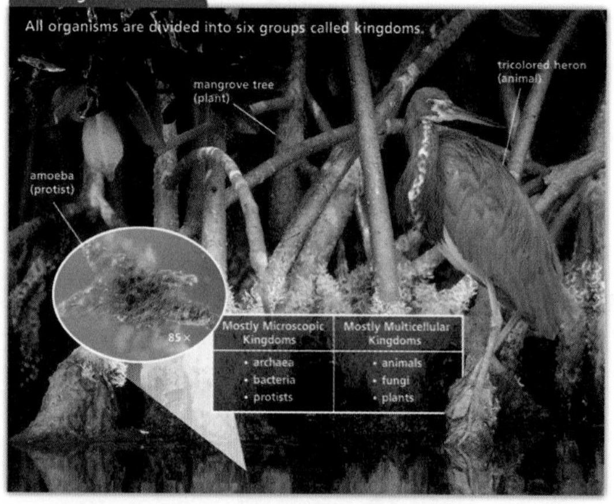

A single drop of water from a mangrove swamp may be living space for many microorganisms. The circled photograph on page 10 was taken using a microscope, and shows an amoeba that may be found in the water of the swamp. Larger organisms, such as manatees and fish, swim around the roots of mangrove trees. Birds, such as tricolored herons and roseate spoonbills, live on branches.

Scientists divide the organisms they identify into groups called **kingdoms.** This unit will cover all of the kingdoms of life, listed in the table on page 10. You are already familiar with plants and animals. Fungi are another kingdom. Fungi include mushrooms found in a forest. The other three kingdoms are composed of mostly microscopic life. You will learn more about microscopic organisms later in this chapter.

VOCABULARY: Add a description wheel for *kingdom* to your notebook. The spokes of your wheel should include examples from the six kingdoms.

Living things share common characteristics.

All living things—from the microorganisms living in a mangrove swamp to the giant organisms living in the open ocean—share similar characteristics. Living things are organized, grow, reproduce, and respond to the environment.

READING TIP: As you read about the four characteristics of all living things, note the examples of how single-celled organisms meet these four standards.

Organization

Cells, like all living things, have an inside and an outside. The boundary separating the inside from the outside of an individual cell is called the cell membrane. Within some cells, another structure called the nucleus is also surrounded by a membrane. Cells perform one or more functions that the organism needs to survive

In this chapter, you will read about organisms made of a single cell. Some types of single-celled organisms contain a nucleus and some do not. All single-celled organisms contain every structure they need to survive within their one cell. They have structures to get energy from complex molecules, structures to help them move, and structures to help them sense their environment. All of the structures are part of their organizations.

Growth

Living things increase in size. Organisms made of one cell do not grow as large as organisms made of many cells. But all living things consume food or other materials to get energy. These materials are also used to build new structures inside cells or replace worn-out structures. As a result, individual cells grow larger over time.

DIFFERENTIATE INSTRUCTION

More Reading Support

A What is an organism? *any living thing*

B What do you call an organism that cannot be seen with the naked eye? *a microorganism*

English Learners Have English learners write the definitions for *microorganism, kingdom, binary fission,* and *virus* in their Science Word Dictionaries. Have students look up the definition for the prefix *micro-* in the dictionary. Then have them list other words that begin with *micro-* (microscope, microwave, microphone), and explain the meaning of the prefix. *very small*

DIFFERENTIATE INSTRUCTION

More Reading Support

C What 3 kingdoms have organisms large enough to see? *fungi, plants, animals*

D How can an organism survive with only one cell? *It has all the structures needed in the cell.*

Below Level Demonstrate classification by having students separate a group of everyday items into categories according to their characteristics. Use items such as shoes or writing implements. Ask volunteers to explain why classifying is helpful when thinking about living things.

Teacher's Edition

- More Reading Support for below-level readers
- Strategies for below-level and advanced learners, English learners, and inclusion students

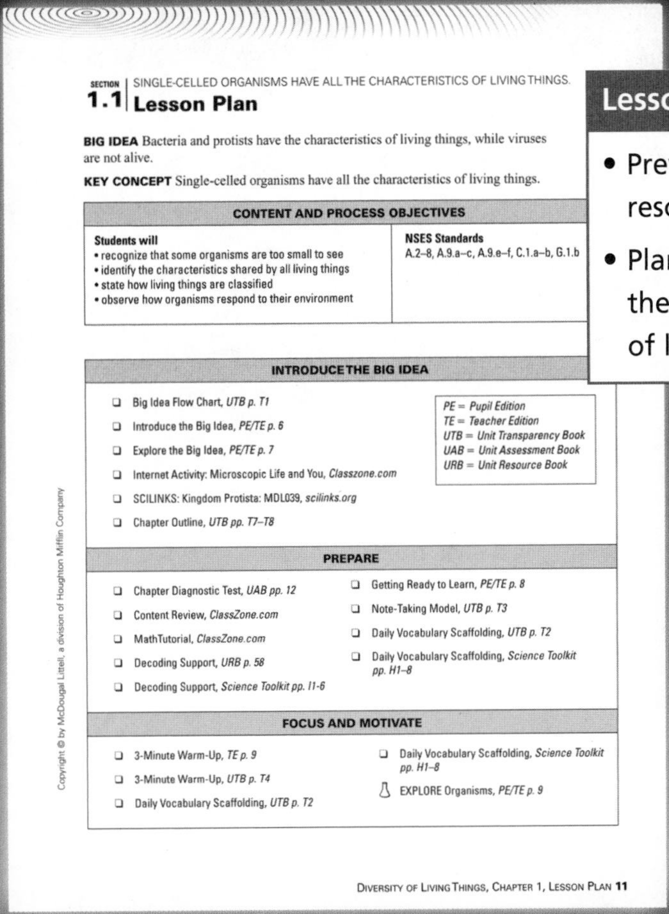

SECTION 1.1 | SINGLE-CELLED ORGANISMS HAVE ALL THE CHARACTERISTICS OF LIVING THINGS.

Lesson Plan

BIG IDEA Bacteria and protists have the characteristics of living things, while viruses are not alive.

KEY CONCEPT Single-celled organisms have all the characteristics of living things.

CONTENT AND PROCESS OBJECTIVES	
Students will • recognize that some organisms are too small to see • identify the characteristics shared by all living things • state how living things are classified • observe how organisms respond to their environment	**NSES Standards** A.2–8, A.9.a–c, A.9.e–f, C.1.a–b, G.1.b

INTRODUCE THE BIG IDEA

- ❑ Big Idea Flow Chart, *UTB p. T1*
- ❑ Introduce the Big Idea, *PE/TE p. 6*
- ❑ Explore the Big Idea, *PE/TE p. 7*
- ❑ Internet Activity: Microscopic Life and You, *Classzone.com*
- ❑ SCILINKS: Kingdom Protista: MDL039, *scilinks.org*
- ❑ Chapter Outline, *UTB pp. T7–T8*

PE = Pupil Edition
TE = Teacher Edition
UTB = Unit Transparency Book
UAB = Unit Assessment Book
URB = Unit Resource Book

PREPARE

- ❑ Chapter Diagnostic Test, *UAB pp. 12*
- ❑ Content Review, *ClassZone.com*
- ❑ MathTutorial, *ClassZone.com*
- ❑ Decoding Support, *URB p. 58*
- ❑ Decoding Support, *Science Toolkit pp. I1-6*
- ❑ Getting Ready to Learn, *PE/TE p. 8*
- ❑ Note-Taking Model, *UTB p. T3*
- ❑ Daily Vocabulary Scaffolding, *UTB p. T2*
- ❑ Daily Vocabulary Scaffolding, *Science Toolkit pp. H1–8*

FOCUS AND MOTIVATE

- ❑ 3-Minute Warm-Up, *TE p. 9*
- ❑ 3-Minute Warm-Up, *UTB p. T4*
- ❑ Daily Vocabulary Scaffolding, *UTB p. T2*
- ❑ Daily Vocabulary Scaffolding, *Science Toolkit pp. H1–8*
- EXPLORE Organisms, *PE/TE p. 9*

Copyright © by McDougal Littell, a division of Houghton Mifflin Company

DIVERSITY OF LIVING THINGS, CHAPTER 1, LESSON PLAN 11

Lesson Plans

- Preview differentiated resources
- Plan your path through the lesson for each type of learner

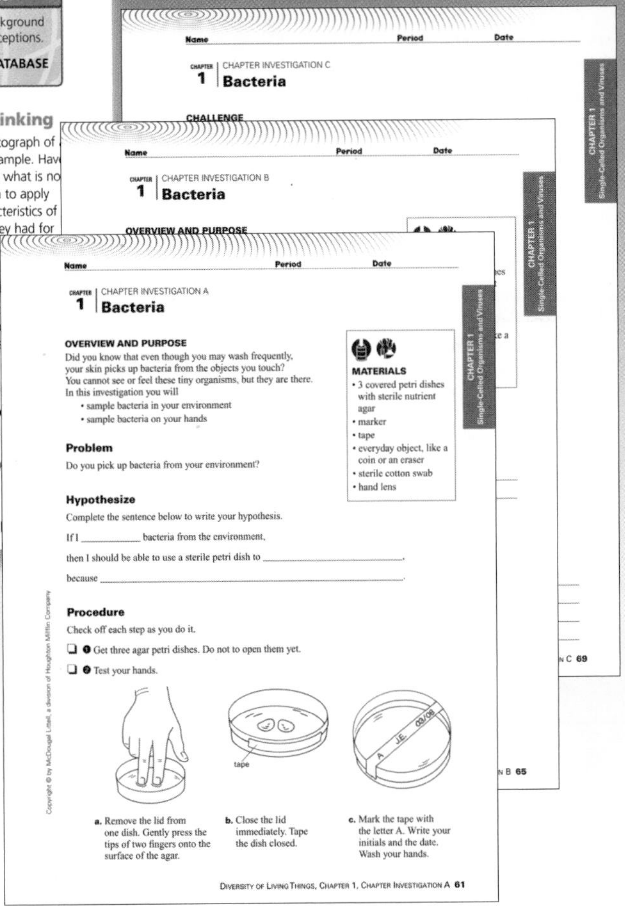

Name Period Date

CHAPTER 1 | CHAPTER INVESTIGATION A

Bacteria

OVERVIEW AND PURPOSE

Did you know that even though you may wash frequently, your skin picks up bacteria from the objects you touch? You cannot see or feel these tiny organisms, but they are there. In this investigation you will
- sample bacteria in your environment
- sample bacteria on your hands

Problem

Do you pick up bacteria from your environment?

Hypothesize

Complete the sentence below to write your hypothesis.

If I __________ bacteria from the environment,

then I should be able to use a sterile petri dish to ______________________.

because ______________________.

MATERIALS
- 3 covered petri dishes with sterile nutrient agar
- marker
- tape
- everyday object, like a coin or an eraser
- sterile cotton swab
- hand lens

Procedure

Check off each step as you do it.

❑ ❶ Get three agar petri dishes. Do not to open them yet.

❑ ❷ Test your hands.

a. Remove the lid from one dish. Gently press the tips of two fingers onto the surface of the agar.

b. Close the lid immediately. Tape the dish closed.

c. Mark the tape with the letter A. Write your initials and the date. Wash your hands.

DIVERSITY OF LIVING THINGS, CHAPTER 1, CHAPTER INVESTIGATION A 61

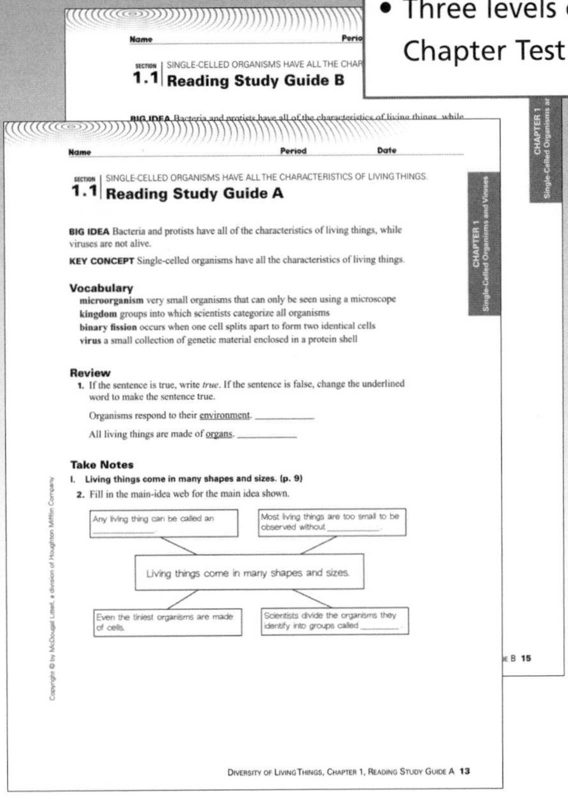

Name Period Date

SECTION 1.1 | SINGLE-CELLED ORGANISMS HAVE ALL THE CHARACTERISTICS OF LIVING THINGS.

Reading Study Guide A

BIG IDEA Bacteria and protists have all of the characteristics of living things, while viruses are not alive.

KEY CONCEPT Single-celled organisms have all the characteristics of living things.

Vocabulary

microorganism very small organisms that can only be seen using a microscope
kingdom groups into which scientists categorize all organisms
binary fission occurs when one cell splits apart to form two identical cells
virus a small collection of genetic material enclosed in a protein shell

Review

1. If the sentence is true, write *true*. If the sentence is false, change the underlined word to make the sentence true.

Organisms respond to their environment. __________

All living things are made of organs. __________

Take Notes

I. **Living things come in many shapes and sizes. (p. 9)**

2. Fill in the main-idea web for the main idea shown.

DIVERSITY OF LIVING THINGS, CHAPTER 1, READING STUDY GUIDE A 13

Leveled Resources

- Three levels of every Investigation (below level, on level, advanced)
- Below-level and on-level Reading Study Guides plus Challenge Readings for advanced students
- Three levels of every Chapter Test and Unit Test

Effective Assessment

McDougal Littell Science incorporates a comprehensive set of resources for assessing student knowledge and performance before, during, and after instruction.

Diagnostic Tests

- Assessment of students' prior knowledge
- Readiness check for concepts and skills in the upcoming chapter

Teach from Visuals

To help students compare the photographs of the rotavirus, and the animal cell, ask them to make a table listing the two items on the left side and the four characteristics of life across the top. Have students check each characteristic of life that applies to the items.

Ongoing Assessment

Identify the characteristics shared by all living things.

Ask students to explain the difference between growth and reproduction. *Growth occurs when an organism increases in size. Reproduction makes more organisms.*

Reinforce the BIG idea

Have students relate the section to the Big Idea.

R Reinforcing Key Concepts, p. 21

1.1 ASSESS & RETEACH

Assess

A Section 1.1 Quiz, p. 3

Reteach

Stage a "characteristics of life" scavenger hunt for students. Have the students work in teams with checklists and/or clipboards. The teams should list, within 5 minutes for each "hunt":

- As many examples as possible of organisms growing.
- As many examples as they can find of organisms with structures.
- . . . of organisms reproducing.
- . . . of organisms responding to the environment.

Technology Resources

Have students visit **ClassZone.com** for reteaching of Key Concepts.

CONTENT REVIEW

CONTENT REVIEW CD-ROM

C 14 Unit: **Diversity of Living Things**

Viruses are not alive.

Sometimes it's not easy to tell the difference between a living and a nonliving thing. A **virus** is a small collection of genetic material enclosed in a protein shell. Viruses have many of the characteristics of living things, including DNA. However, a virus is not nearly as complex as an animal cell and is not considered a living thing.

Rotavirus

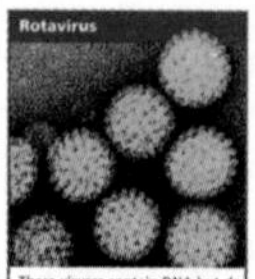

These viruses contain DNA but do not grow or respond to their environment. 570,000 ×

Animal Cell

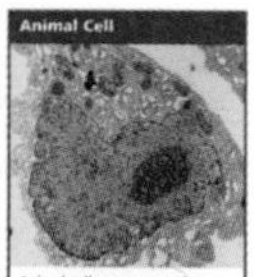

Animal cells grow, reproduce, and respond to external conditions. 4800 ×

Animal cells have structures that allow them to get materials or energy from their environment. Viruses do not grow once they have formed, and they do not take in any energy. Animal cells can make copies of their genetic material and reproduce by dividing in two. Viruses are able to reproduce only by "taking over" another cell and using that cell to make new viruses. Animal cells also have many more internal structures than viruses. Viruses usually contain nothing more than their DNA.

1.1 Review

KEY CONCEPTS

1. Give examples of organisms that are very large and organisms that are very small.
2. Name four characteristics that all living things share.
3. Name three things that living things must obtain to survive.

CRITICAL THINKING

4. **Synthesize** Give examples of how a common animal, such as a dog, is organized, grows, responds, and reproduces.
5. **Predict** In a certain lake, would you expect there to be more organisms that are large enough to see or more organisms that are too small for you to see? Why?

CHALLENGE

6. **Design** Try to imagine the different structures that a single-celled organism needs to survive in pond water. Then use your ideas to design your own single-celled organism.

C 14 Unit: **Diversity of Living Things**

ANSWERS

1. very large organism: huge fungus; very small organism: single-celled bacteria

2. growth, reproduction, organization, and response to environment

3. energy, materials, and living space

4. A dog is made of cells that perform specific functions. It starts as a puppy. As its cells reproduce, it grows to a full-sized dog. Male and female dogs reproduce sexually to create more puppies. Dogs bark at, run at, or run from different things they sense.

5. There is room in a lake for many billions of microscopic organisms but not for nearly as many larger organisms.

6. Examine students' designs for creativity and understanding of the characteristics and needs of life.

Ongoing Assessment

- Check Your Reading questions for student self-check of comprehension
- Consistent Teacher Edition prompts for assessing understanding of Key Concepts

Reviewing Vocabulary

Draw a triangle for each of the terms listed below. Define the term, use it in a sentence, and draw a picture to help you remember the term. An example is completed for you.

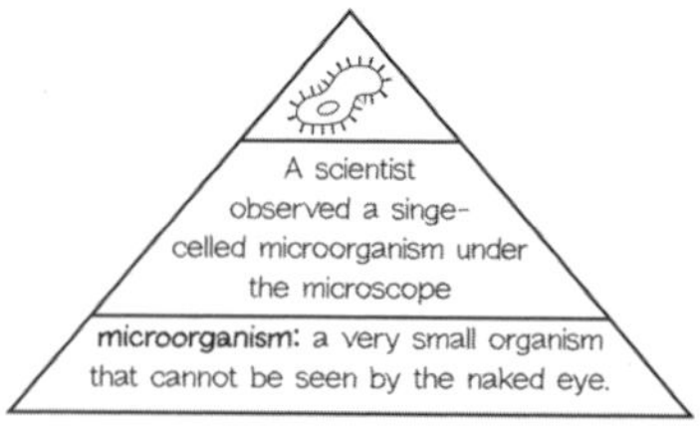
A scientist observed a singe-celled microorganism under the microscope
microorganism: a very small organism that cannot be seen by the naked eye.

1. binary fission
2. producer
3. virus
4. host cell

Describe how the vocabulary terms in the following pairs of words are related to each other. Explain the relationship in a one- or two-sentence answer. Underline each vocabulary term in your answers.

5. archaea, bacteria
6. microorganism, organism
7. decomposers, parasite
8. protists, algae

Reviewing Key Concepts

Multiple Choice *Choose the letter of the best answer.*

9. Which group is *not* a microscopic kingdom?
   a. fungi
   b. bacteria
   c. archaea
   d. protists
10. What happens in binary fission?
    a. DNA is combined into one cell.
    b. The daughter cells differ from the parent cell.
    c. Material from one cell is broken into two cells.
    d. One cell divides into four exact cells.
11. Which is a characteristic of a virus?
    a. obtains energy from sunlight
    b. responds to light and temperature
    c. doesn't contain DNA
    d. reproduces only within other cells
12. Which is the simplest type of organism on Earth?
    a. protists
    b. bacteria
    c. viruses
    d. parasites
13. Which statement about bacteria is *not* true?
    a. Bacteria reproduce using binary fission.
    b. Bacteria do not have a nucleus.
    c. Bacteria do not contain genetic material.
    d. Bacteria are either rod-, cone-, or spiral-shaped.
14. Archaea that can survive only in extreme temperatures are the
    a. methanogens
    b. halophiles
    c. thermophiles
    d. bacteria
15. A weakened viral or bacterial disease that is injected into the body is
    a. a filter
    b. a diatom
    c. a bacteriophage
    d. a vaccine
16. Which group of protists absorbs food from their environment?
    a. diatoms
    b. molds
    c. protozoa
    d. plankton

...nisms and Viruses 37 C

Section and Chapter Reviews

- Focus on Key Concepts and critical thinking skills
- A full range of question types and levels of thinking

Leveled Chapter and Unit Tests

- Three levels of test for every chapter and unit
- Same Big Ideas, Key Concepts, and essential skills assessed on all levels

Name Period Date

CHAPTER 1 | SINGLE-CELLED ORGANISMS AND VIRUSES
Chapter Test C

CHAPTER 1 Single-Celled Organisms and Viruses

15

Name Period Date

CHAPTER 1 | SINGLE-CELLED ORGANISMS AND VIRUSES
Chapter Test B

CHAPTER 1 Single-Celled Organisms and Viruses

11

Name Period Date

CHAPTER 1 | SINGLE-CELLED ORGANISMS AND VIRUSES
Chapter Test A

CHAPTER 1 Single-Celled Organisms and Viruses

Key Concepts

Choose the letter of the best answer. (5 points each)

____ 1. Very small organisms that you can see only by using a microscope are called
a. viruses
b. nuclei
c. single cells
d. microorganisms

____ 2. We know that a balloon is not alive because it
a. does not have an inside and an outside
b. never increases in size
c. cannot reproduce itself
d. does not change with changes in the environment

____ 3. Round, spiral, and rod are the three basic shapes of
a. bacteria
b. viruses
c. protozoa
d. protists

____ 4. The single-celled organisms without nuclei are
a. bacteria and archaea
b. viruses and protists
c. bacteria and protists
d. archaea and viruses

____ 5. A chickenpox virus uses its host cell for
a. energy
b. reproduction
c. shelter
d. food

Copyright © by McDougal Littell, a division of Houghton Mifflin Company

17. Which obtains energy by feeding on other organisms?
a. amoeba c. phytoplankton
b. algae d. mushroom

Short Answer *Write a short answer to each question.*

18. Briefly describe the characteristics that all living things share.
19. How are bacteria harmful to humans?
20. What are plankton?

Thinking Critically

21. **APPLY** Imagine you are a scientist on location in a rain forest in Brazil. You discover what you think might be a living organism. How would you be able to tell if the discovery is a living thing?

22. **COMMUNICATE** What process is shown in this photograph? Describe the sequence of events in the process shown.

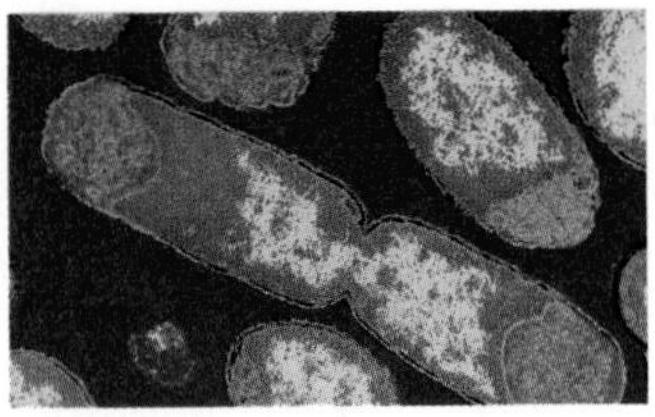

23. **CLASSIFY** Why are archaea classified in a separate kingdom from bacteria?

24. **ANALYZE** Why are some bacteria considered "nature's recyclers"? Explain the role that these bacteria play in the environment.

25. **CALCULATE** A bacterium reproduces every hour. Assuming the bacteria continue to reproduce at that rate, how many bacteria will there be after 10 hours? Explain how you know.

26. **HYPOTHESIZE** A student conducts an experiment to determine the effectiveness of washing hands on bacteria growth. He rubs an unwashed finger across an agar plate, then washes his hands and rubs the same finger across a second plate. What hypothesis might the student make for this experiment? Explain.

27. **COMPARE AND CONTRAST** Describe three ways that viruses differ from bacteria.

28. **ANALYZE** A scientist has grown cultures of bacteria on agar plates for study. Now the scientist wants to grow a culture of viruses in a laboratory for study. How might this be possible? Give an example.

29. **PROVIDE EXAMPLES** How are protists both helpful and harmful to humans? Give examples in your answer.

the BIG idea

31. **INFER** Look again at the picture on pages 6–7. Now that you have finished the chapter, how would you change or add details to your answer to the question on the photograph?

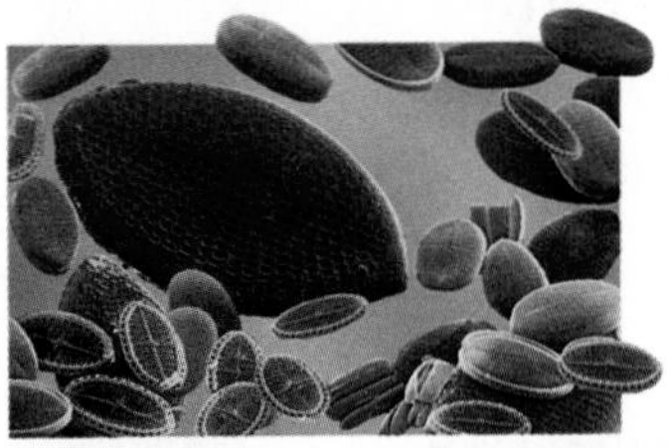

UNIT PROJECTS

If you are doing a unit project, make a folder for

Rubrics

- Rubrics in Teacher Edition for all extended response questions
- Rubrics for all Unit Projects
- Alternative Assessment with rubric for each chapter
- A wide range of additional rubrics in the Science Toolkit

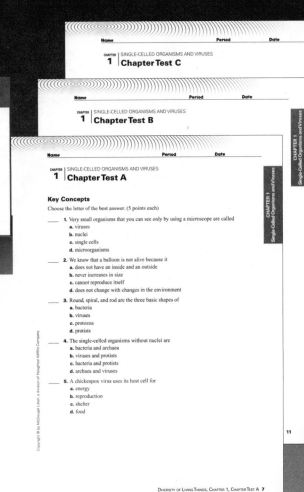

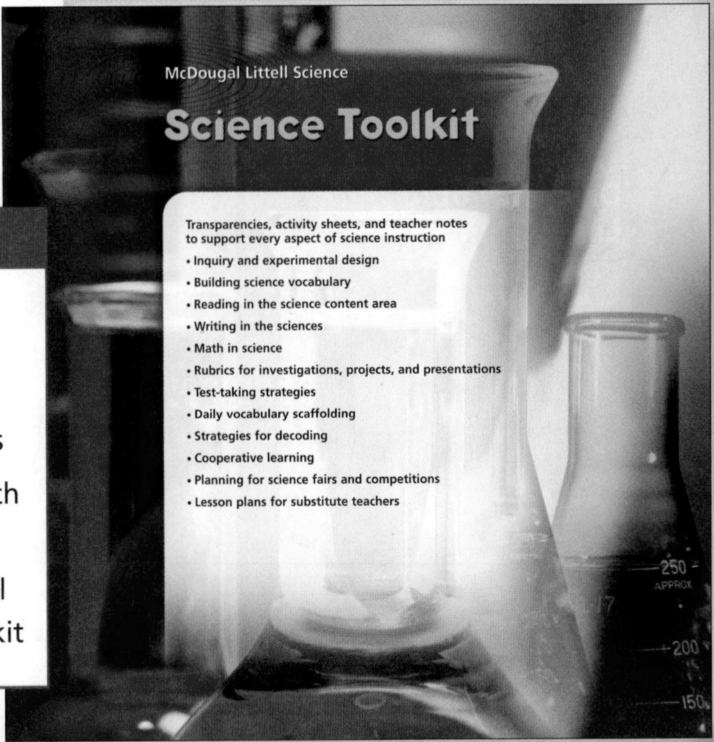
McDougal Littell Science
Science Toolkit
Transparencies, activity sheets, and teacher notes to support every aspect of science instruction
• Inquiry and experimental design
• Building science vocabulary
• Reading in the science content area
• Writing in the sciences
• Math in science
• Rubrics for investigations, projects, and presentations
• Test-taking strategies
• Daily vocabulary scaffolding
• Strategies for decoding
• Cooperative learning
• Planning for science fairs and competitions
• Lesson plans for substitute teachers
250
APPROX
200
150

McDougal Littell Science Modular Series

McDougal Littell Science lets you choose the titles that match your curriculum. Each module in this flexible 15-book series takes an in-depth look at a specific area of life, earth, or physical science.

- Flexibility to match your curriculum
- Convenience of smaller books
- Complete Student Resource Handbooks in every module

Life Science Titles

A ▶ Cells and Heredity
1. The Cell
2. How Cells Function
3. Cell Division
4. Patterns of Heredity
5. DNA and Modern Genetics

B ▶ Life Over Time
1. The History of Life on Earth
2. Classification of Living Things
3. Population Dynamics

C ▶ Diversity of Living Things
1. Single-Celled Organisms and Viruses
2. Introduction to Multicellular Organisms
3. Plants
4. Invertebrate Animals
5. Vertebrate Animals

D ▶ Ecology
1. Ecosystems and Biomes
2. Interactions Within Ecosystems
3. Human Impact on Ecosystems

E ▶ Human Biology
1. Systems, Support, and Movement
2. Absorption, Digestion, and Exchange
3. Transport and Protection
4. Control and Reproduction
5. Growth, Development, and Health

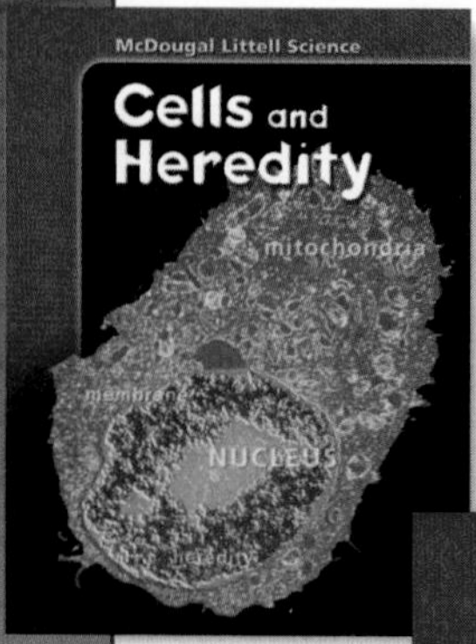

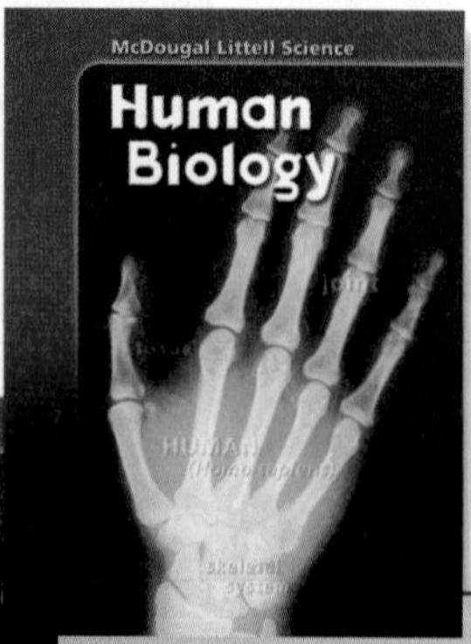

Earth Science Titles

A ▶ Earth's Surface
1. Views of Earth Today
2. Minerals
3. Rocks
4. Weathering and Soil Formation
5. Erosion and Deposition

B ▶ The Changing Earth
1. Plate Tectonics
2. Earthquakes
3. Mountains and Volcanoes
4. Views of Earth's Past
5. Natural Resources

C ▶ Earth's Waters
1. The Water Planet
2. Freshwater Resources
3. Ocean Systems
4. Ocean Environments

D ▶ Earth's Atmosphere
1. Earth's Changing Atmosphere
2. Weather Patterns
3. Weather Fronts and Storms
4. Climate and Climate Change

E ▶ Space Science
1. Exploring Space
2. Earth, Moon, and Sun
3. Our Solar System
4. Stars, Galaxies, and the Universe

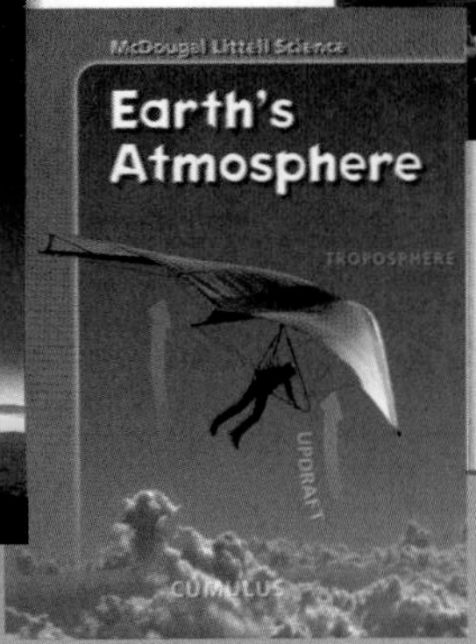

Physical Science Titles

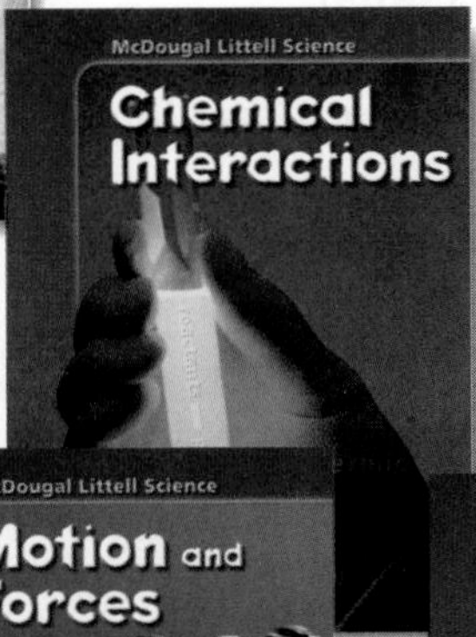

A ▶ Matter and Energy
1. Introduction to Matter
2. Properties of Matter
3. Energy
4. Temperature and Heat

B ▶ Chemical Interactions
1. Atomic Structure and the Periodic Table
2. Chemical Bonds and Compounds
3. Chemical Reactions
4. Solutions
5. Carbon in Life and Materials

C ▶ Motion and Forces
1. Motion
2. Forces
3. Gravity, Friction, and Pressure
4. Work and Energy
5. Machines

D ▶ Waves, Sound, and Light
1. Waves
2. Sound
3. Electromagnetic Waves
4. Light and Optics

E ▶ Electricity and Magnetism
1. Electricity
2. Circuits and Electronics
3. Magnetism

Teaching Resources

A wealth of print and technology resources help you adapt the program to your teaching style and to the specific needs of your students.

Book-Specific Print Resources

Unit Resource Book provides all of the teaching resources for the unit organized by chapter and section.

- Family Letters
- *Scientific American Frontiers* Video Guide
- Unit Projects
- Lesson Plans
- Reading Study Guides (Levels A and B)
- Spanish Reading Study Guides
- Challenge Readings
- Challenge and Extension Activities
- Reinforcing Key Concepts
- Vocabulary Practice
- Math Support and Practice
- Investigation Datasheets
- Chapter Investigations (Levels A, B, and C)
- Additional Investigations (Levels A, B, and C)
- Summarizing the Chapter

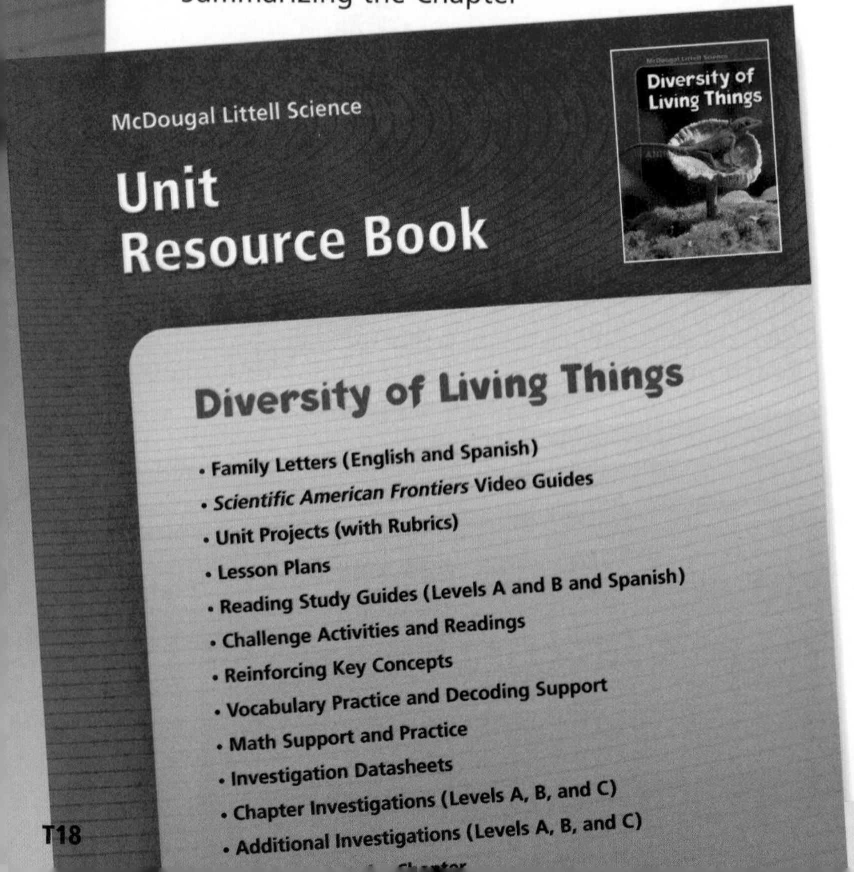

Unit Assessment Book contains complete resources for assessing student knowledge and performance.

- Chapter Diagnostic Tests
- Section Quizzes
- Chapter Tests (Levels A, B, and C)
- Alternative Assessments
- Unit Tests (Levels A, B, and C)

Unit Transparency Book includes instructional visuals for each chapter.

- Three-Minute Warm-Ups
- Note-Taking Models
- Daily Vocabulary Scaffolding
- Chapter Outlines
- Big Idea Flow Charts
- Chapter Teaching Visuals

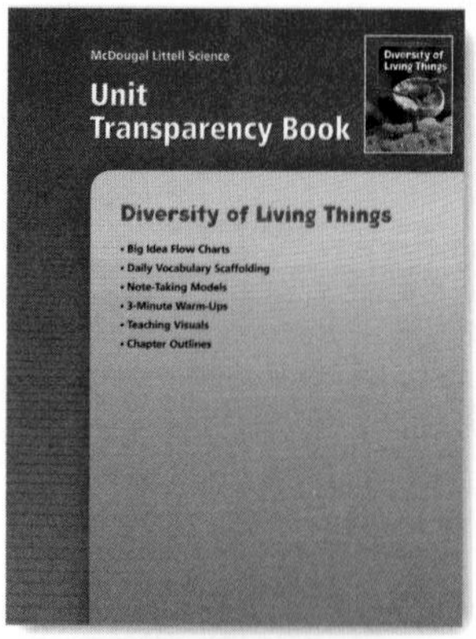

Unit Lab Manual

Unit Note-Taking/Reading Study Guide

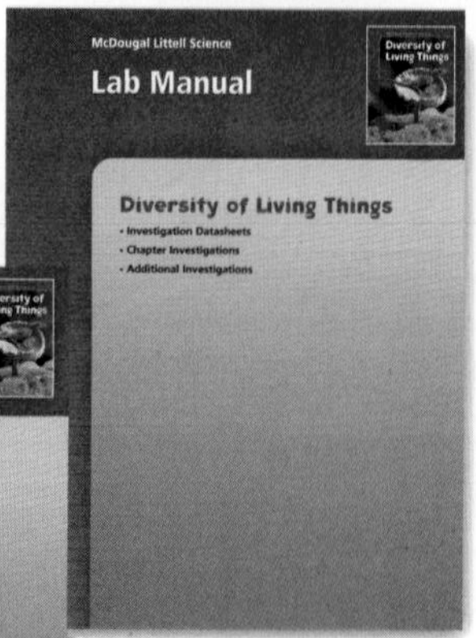

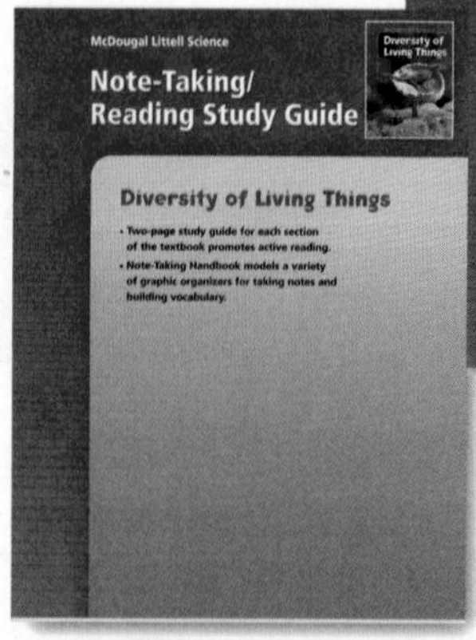

Program-Wide Print Resources

Process and Lab Skills

Problem Solving and Critical Thinking

Standardized Test Practice

Science Toolkit

City Science

Visual Glossary

Multi-Language Glossary

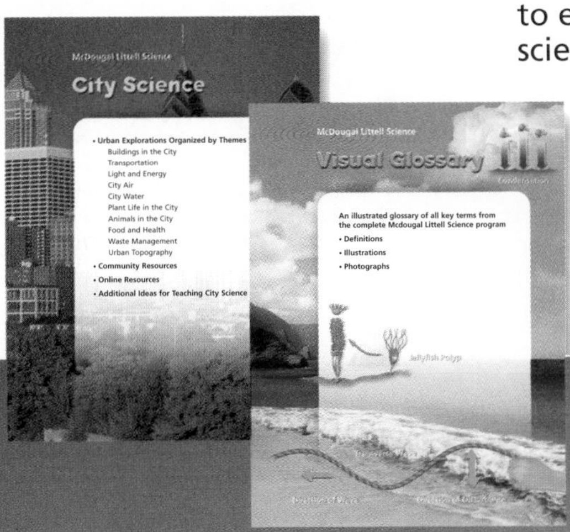

English Learners Package

***Scientific American Frontiers* Video Guide**

How Stuff Works Express
This quarterly magazine offers opportunities to explore current science topics.

Technology Resources

***Scientific American Frontiers* Video Program**
Each specially-tailored segment from this award-winning PBS series correlates to a unit; available on VHS and DVD

Audio CDs Complete chapter texts read in both English and Spanish

Lab Generator CD-ROM
A searchable database of all activities from the program plus additional labs for each unit; edit and print your own version of labs

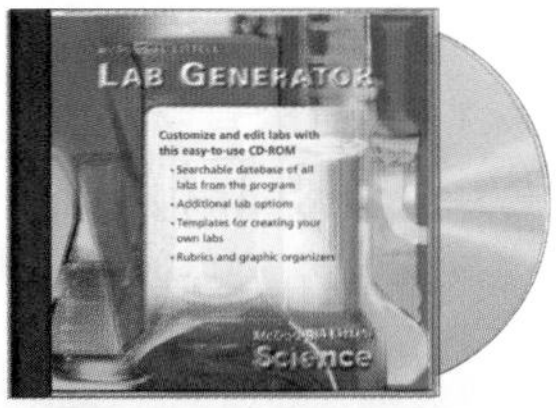

Test Generator CD-ROM

eEdition CD-ROM

EasyPlanner CD-ROM

Content Review CD-ROM

Power Presentations CD-ROM

Online Resources

- ClassZone.com
- Content Review Online
- eEdition Plus Online
- EasyPlanner Plus Online
- eTest Plus Online

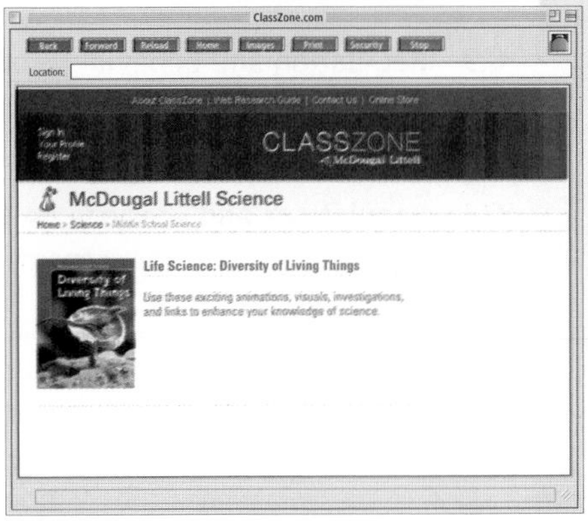

Correlation to National Science Education Standards

This chart provides an overview of how the five Life Science modules of *McDougal Littell Science* address the National Science Education Standards.

A Cells and Heredity
B Life Over Time
C Diversity of Living Things
D Ecology
E Human Biology

A. Science as Inquiry	Book, Chapter, and Section
A.1– A.8 **Abilities necessary to do scientific inquiry** Identify questions for investigation; design and conduct investigations; use evidence; think critically and logically; analyze alternative explanations; communicate; use mathematics.	All books (pp. R2–R44), All Chapter Investigations All Think Science features
A.9 **Understandings about scientific inquiry** Different kinds of investigations for different questions; investigations guided by current scientific knowledge; importance of mathematics and technology for data gathering and analysis; importance of evidence, logical argument, principles, models, and theories; role of legitimate skepticism; scientific investigations lead to new investigations.	All books (pp. xxii–xxv), A4.3, B1.3, C4.3, D2.2, E1.1

B. Physical Science	Book, Chapter, and Section
B.1 **Properties and changes of properties in matter** Physical properties; substances, elements, and compounds; chemical reactions.	A2.1, A3.1 (Connecting Sciences)
B.2 **Motions and forces** Position, speed, direction of motion; balanced and unbalanced forces.	C5.2 (Connecting Sciences)
B.3 **Transfer of energy** Energy transfer; forms of energy; heat and light; electrical circuits; sun as source of Earth's energy.	A2.2. D1.3, E4.2 (Connecting Sciences)

C. Life Science	Book, Chapter, and Section
C.1 **Structure and function in living systems** Systems; structure and function; levels of organization; cells and cell activities; specialization; human body systems; disease.	A1.1, A1.2, A1.3, A2.1, A2.2, A2.3, A3.1, B1, C1.1, C2, C4, D2.1, E1.1, E1.2, E1.3, E2.1, E2.2, E2.3, E3.1, E3.2, E.3.3, E4.1, E4.2, E4.3, E5.1, E5.2, E5.3
C.2 **Reproduction and heredity** Sexual and asexual reproduction; heredity and genes; traits determined by heredity and environment.	A3.2, A3.3, A4.1, A4.2, A4.3,, A5.1, A5.2, B2.1, C1.1, C1.2, C2.4, C3.3, C3.4, C4.2, C4.3, C4.4, C5.2, C5.3, C5.4, E4.3
C.3 **Regulation and behavior** Growth, reproduction, and maintenance of stable internal environment; regulation; behavior; evolution of behavior through adaptation to environment.	C1.1, C2.1, C4.1, C5.1, C5.2, C5.3, C5.4, E1.1, E4.1, E4.2, E5.1, E5.2
C.4 **Populations and ecosystems** Populations; ecosystems; producers, consumers, and decomposers; food webs; energy flow; population size and resource availability; population growth.	B3.1, B3.2, B3.3, C3.1, D1.1, D1.2, D1.3, D2.1, D2.2, D2.3, D3.1, D3.2, D3.3
C.5 **Diversity and adaptations of organisms** Unity and diversity; similarities in internal structures, chemical processes, and evidence of common ancestry; adaptation and biological evolution; extinction and fossil evidence.	B1.1, B1.1, B1.2, B1.3, B2.1, B2.2, B2.3, C4, C5

D. Earth and Space Science	Book, Chapter, and Section
D.1 Structure of the earth system Lithosphere, mantle, and core; plate movement and earthquakes, volcanoes, and mountain building; constructive and destructive forces on landforms; soil, weathering, and erosion; water and water cycle; atmosphere, weather, and climate; living organisms in earth system.	B3.3, D1.1
D.2 Earth's history Continuity of earth processes; impact of occasional catastrophes; fossil evidence.	B1.1, B (pp. R52–R57)
D.3 Earth in the solar system Sun, planets, asteroids, comets; regular and predictable motion and day, year, phases of the moon, and eclipses; gravity and orbits; sun as source of energy for earth; cause of seasons.	D1.3

E. Science and Technology	Book, Chapter, and Section
E.1– E.5 Abilities of technological design Identify problems; design a solution or product; implement a proposed design; evaluate completed designs or products; communicate the process of technological design.	All books (pp.xxvi–xxvii) B (p. 5)
E.6 Understandings about science and technology Similarities and differences between scientific inquiry and technological design; contributions of people in different cultures; reciprocal nature of science and technology; nonexistence of perfectly designed solutions; constraints, benefits, and unintended consequences of technological designs.	All books (pp. xxvi–xxvii) All books (Frontiers in Science, Timelines in Science) A5.3, D3.1, D3.2

F. Science in Personal and Social Perspectives	Book, Chapter, and Section
F.1 Personal health Exercise; fitness; hazards and safety; tobacco, alcohol, and other drugs; nutrition; STDs; environmental health.	C2.4, D3.2, E2.2, E5.2, E5.3
F.2 Populations, resources, and environments Overpopulation and resource depletion; environmental degradation.	B3.2, B3.3, D3.1, D3.2, D3.3
F.3 Natural hazards Earthquakes, landslides, wildfires, volcanic eruptions, floods, storms; hazards from human activity; personal and societal challenges.	D3.2
F.4 Risks and benefits Risk analysis; natural, chemical, biological, social, and personal hazards; decisions based on risks and benefits.	A5.3, E5.2, D3.3
F.5 Science and technology in society Science's influence on knowledge and world view; societal challenges and scientific research; technological influences on society; contributions from people of different cultures and times; work of scientists and engineers; ethical codes; limitations of science and technology.	A 5.3 All books (Timelines in Science)

G. History and Nature of Science	Book, Chapter, and Section
G.1 Science as a human endeavor Diversity of people working in science, technology, and related fields; abilities required by science .	All books (pp. xxii–xxv; Frontiers in Science)
G.2 Nature of science Observations, experiments, models; tentative nature of scientific ideas; differences in interpretation of evidence; evaluation of results of investigations, experiments, observations, theoretical models, and explanations; importance of questioning, response to criticism, and communication.	B1.2, B1.3, B2.3
G.3 History of science Historical examples of inquiry and relationships between science and society; scientists and engineers as valued contributors to culture; challenges of breaking through accepted ideas.	B1.2, E5.3 All books (Frontiers in Science; Timelines in Science)

Correlation to Benchmarks

This chart provides an overview of how the five Life Science modules of *McDougal Littell Science* address the Project 2061 Benchmarks for Science Literacy.

A Cells and Heredity
B Life Over Time
C Diversity of Living Things
D Ecology
E Human Biology

1. The Nature of Science	Book, Chapter, and Section
	The Nature of Science (pp. xxii–xxv); Scientific Thinking Handbook (pp. R2–R9); Lab Handbook (pp. R10–R35); Think Science Features: A1.3, A4.3, B1.3, C4.3, D2.2, E1.1

3. The Nature of Technology	Book, Chapter, and Section
	The Nature of Technology (pp. xxvi–xxvii); A5.3, B3.3, D3.3; Timelines in Science Features

4. The Physical Setting	Book, Chapter, and Section
4.B THE EARTH	B3.3, D1.1, D1.2, D3
4.C PROCESSES THAT SHAPE THE EARTH	B3.3, C3.4, D3.1, D3.2
4.D STRUCTURE OF MATTER	A2.1, D1.2
4.E ENERGY TRANSFORMATIONS	B1.3, C2.2

5. The Living Environment	Book, Chapter, and Section
5.A DIVERSITY OF LIFE	
5.A.1 Differences among plants, animals, and other organisms	C1.1, C2.2, C2.3, C2.4, D1.3
5.A.2 Body plans and internal structures for food and reproduction.	B2, C2.2, C2.3, C3, C4, C5
5.A.3 Similarities among organisms in internal and external structures used to determine relatedness	B2.1, B2.2, B2.3, C4.1, C4.2, C4.3, C4.4
5.A.4 For sexually reproducing organisms, a species comprises all organisms that can mate with one another to produce fertile offspring.	D2.1
5.A.5 Interconnected global food webs and cycles	C3.4, D1.3
5.B HEREDITY	
5.B.1 In some species, all the genes come from a single parent; whereas in organisms that have sexes, typically half of the genes come from each parent.	A3.3, A4
5.B.2 Sexual reproduction and the transmission of genetic information	A4.1, A4.3, A5.1, C2.1, E4.3
5.B.3 New varieties of plants and domestic animals from selective breeding	A5.2, A5.3, B1.2
5.C CELLS	
5.C.1 All living things are composed of cells; tissues and organs	A1, C1, C2.1
5.C.2 Cells divide to make more cells for growth and repair. Various organs and tissues function to serve the needs of cells for food, air, and waste removal.	A3.1, A3.2, A3.3, C2.1
5.C.3 Within cells, many basic functions of organisms are carried out; way in which cells function is similar in all living organisms.	A1.2, A2, C1, C2.1
5.C.4 About two thirds of the weight of cells is accounted for by water, which gives cells many of their properties.	A2.1

5.D2 INTERDEPENDENCE OF LIFE	
5.D.1 In all environments, organisms with similar needs may compete for resources; growth and survival of organisms depend on the physical conditions	B3.1, B3.2, D1.1, D2.2
5.D.2 Producer/consumer, predator/prey, or parasite/host relationships; scavengers and decomposers; competitive or mutually beneficial relationships.	C2.1, C2.2, C2.3, C2.4, D1.3, D2.2
5.E FLOW OF MATTER AND ENERGY	
5.E.1 Food molecules as fuel and building material for all organisms; photosynthesis and producers; consumers.	A2.1, A2.2, C2.2, C2.3, C3.1, D1.3
5.E.2 Flow of energy though living systems; amount of matter remains constant, even though its form and location change.	D1.2, D1.3
5.E.3 Energy can change from one form to another in living things; almost all food energy comes originally from sunlight.	A1.2, A2.2, C2.2, C2.3, D1.3
5.F EVOLUTION OF LIFE	
5.F.1 Small differences between parents and offspring can accumulate in successive generations so that descendants are very different from their ancestors.	A5.2, A5.3, B1.2, B1.3
5.F.2 Individual organisms with certain traits are more likely than others to survive and have offspring. Changes in environmental conditions can affect the survival of individual organisms and entire species.	B1.1, B1.2, C5
5.F.3 Sedimentary rock layers as evidence for the long history of the earth and of changing life forms whose remains are found in the rocks.	B1.1, B pp. R52–R57

6. The Human Organism	Book, Chapter, and Section
6.A HUMAN IDENTITY	E1, E2, E3, E4
6.B HUMAN DEVELOPMENT	C2.1, C5.4, E4, E5
6.C BASIC FUNCTIONS	C2.1, C5.4, E1, E2, E3, E4
6.E PHYSICAL HEALTH	C1.2, C1.3, C1.4, D3.2, E3, E5

8. The Designed World	A5.3, B3.3, D1.3, D3.2, D3.39.

9. The Mathematical World	All Math in Science Features

10. Historical Perspectives	A1.1, A1.2

12. Habits of Mind	Book, Chapter, and Section
12.A VALUES AND ATTITUDES	Think Science Features: A1.3, A4.3, B1.3, C4.3, D2.2, E1.1
12.B COMPUTATION AND ESTIMATION	All Math in Science Features, Lab Handbook (pp. R10–R35)
12.C MANIPULATION AND OBSERVATION	All Investigates and Chapter Investigations
12.D COMMUNICATION SKILLS	All Chapter Investigations, Lab Handbook (pp. R10–R35)
12.E CRITICAL-RESPONSE SKILLS	Think Science Features: A1.3, A4.3, B1.3, C4.3, D2.2, E1.1; Scientific Thinking Handbook (pp. R2–R9)

Planning the Unit

The Pacing Guide provides suggested pacing for all chapters in the unit as well as the two unit features shown below.

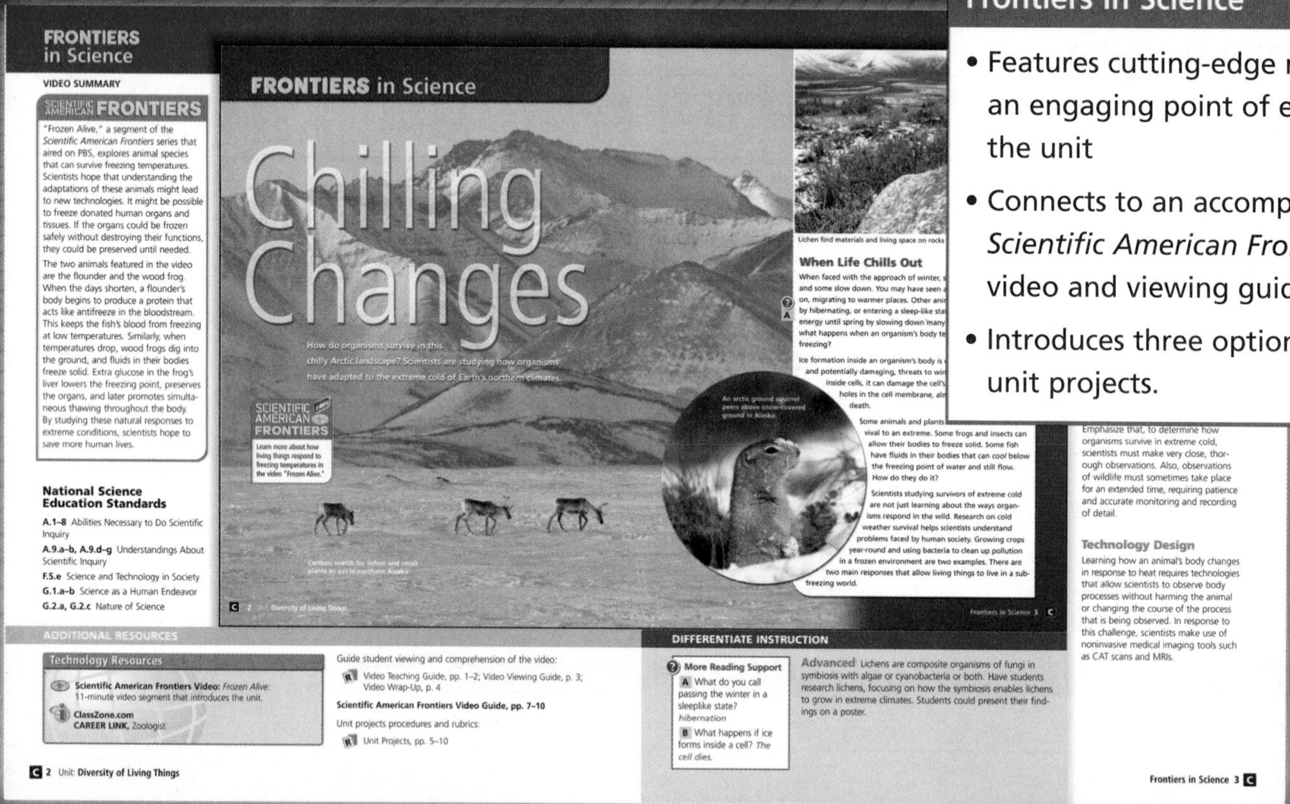

Frontiers in Science

- Features cutting-edge research as an engaging point of entry into the unit
- Connects to an accompanying *Scientific American Frontiers* video and viewing guide
- Introduces three options for unit projects.

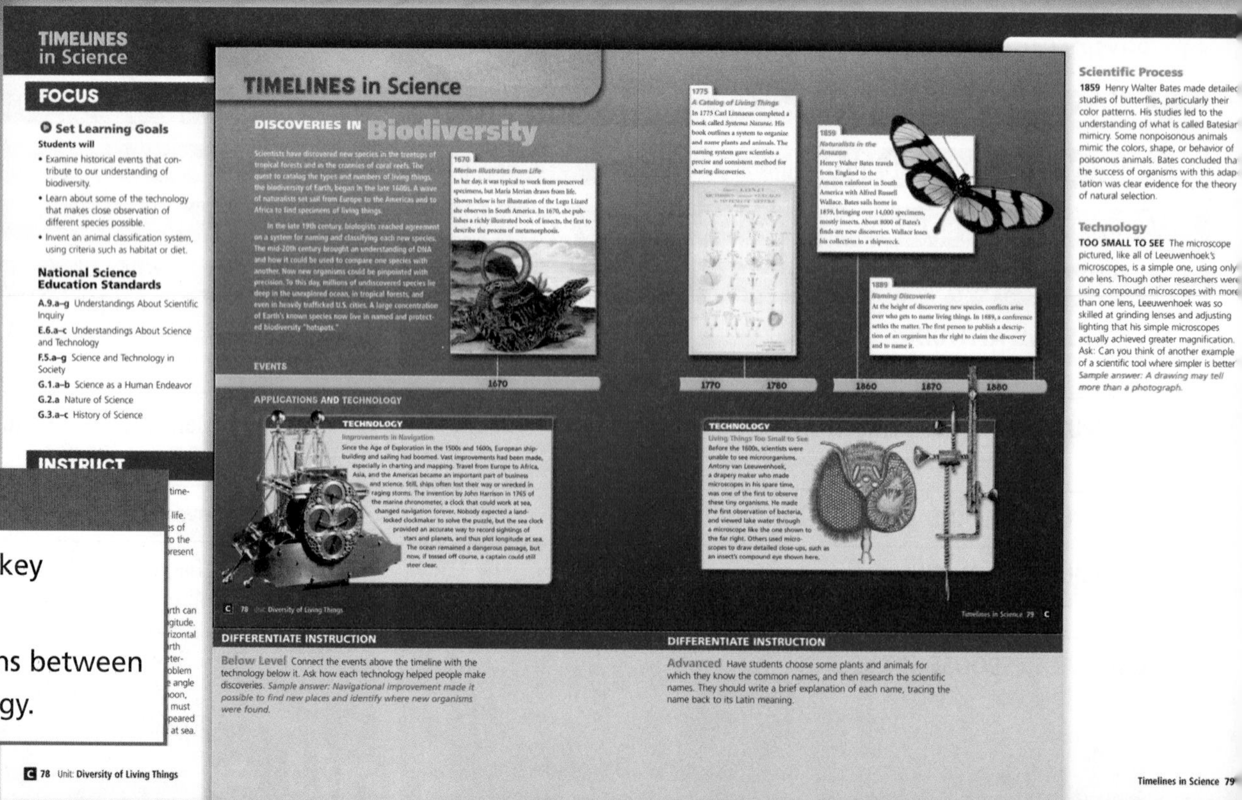

Timelines in Science

- Traces the history of key scientific discoveries
- Highlights interactions between science and technology.

Diversity of Living Things Pacing Guide

The following pacing guide shows how the chapters in ***Diversity of Living Things*** can be adapted to fit your specific course needs.

	TRADITIONAL SCHEDULE (DAYS)	BLOCK SCHEDULE (DAYS)
Frontiers in Science: Chilling Changes	1	0.5
Chapter 1 Single-Celled Organisms and Viruses		
1.1 Single-celled organisms have all the characteristics of living things.	2	1
1.2 Bacteria are single-celled organisms without nuclei.	2	1
1.3 Viruses are not alive but affect all living things.	2	1
1.4 Protists are a diverse group of organisms.	3	1.5
Chapter Investigation	1	0.5
Chapter 2 Introduction to Multicellular Organisms		
2.1 Multicellular organisms meet their needs in different ways.	2	1
2.2 Plants are producers.	2	1
2.3 Animals are consumers.	2	1
2.4 Most fungi are decomposers.	3	1.5
Chapter Investigation	1	0.5
Chapter 3 Plants		
3.1 Plants are adapted to living on land.	2	1
3.2 Most mosses and ferns live in moist environments.	2	1
3.3 Seeds and pollen are reproductive adaptations.	2	1
3.4 Many plants reproduce with flowers and fruit.	3	1.5
Chapter Investigation	1	0.5
Timelines in Science: Discoveries in Biodiversity	1	0.5
Chapter 4 Invertebrate Animals		
4.1 Most animals are invertebrates.	2	1
4.2 Cnidarians and worms have different body plans.	2	1
4.3 Most mollusks have shells and echinoderms have spiny skeletons.	2	1
4.4 Arthropods have exoskeletons and joints.	3	1.5
Chapter Investigation	1	0.5
Chapter 5 Vertebrate Animals		
5.1 Vertebrates are animals with endoskeletons.	2	1
5.2 Amphibians and reptiles are adapted for life on land.	2	1
5.3 Birds meet their needs on land, in water, and in the air.	2	1
5.4 Mammals live in many environments.	3	1.5
Chapter Investigation	1	0.5
Total Days for Module	**52**	**26**

Planning the Chapter

Complete planning support precedes each chapter.

Previewing Content

- Section-by-section science background notes
- Common Misconceptions notes

CHAPTER 1

Single-Celled Organisms and Viruses

Life Science
UNIFYING PRINCIPLES

PRINCIPLE 1	PRINCIPLE 2	PRINCIPLE 3	PRINCIPLE 4
All living things share common characteristics.	All living things share common needs.	Living things r needs through with the envir	

Unit: Diversity of Living Things
BIG IDEAS

CHAPTER 1 Single-Celled Organisms and Viruses	CHAPTER 2 Introduction to Multicellular Organisms	CHAPTER 3 Plants
Bacteria and protists have the characteristics of living things, while viruses are not alive.	Multicellular organisms live in and get energy from a variety of environments.	Plants are a diverse group of organisms that live in many land environments.

CHAPTER 1
KEY CONCEPTS

SECTION 1.1	SECTION 1.2	SECTION
Single-celled organisms have all the characteristics of living things. 1. Living things come in many shapes and sizes. 2. Living things share common characteristics. 3. Living things need energy, materials, and living space. 4. Viruses are not alive.	**Bacteria are single-celled organisms without nuclei.** 1. Bacteria and archaea are the smallest living things. 2. Archaea and bacteria are found in many environments. 3. Bacteria may help or harm other organisms.	**Viruses are affect living** 1. Viruses sha characteris things. 2. Viruses mu living cells. 3. Viruses ma cells.

The Big Idea Flow Chart is available on p. T1 in the **UNIT TRANSPARENCY BOOK**

Previewing Content

SECTION

1.1 Single-celled organisms have all the characteristics of living things. pp. 9–15

1. Living things come in many shapes and sizes.
Living things range in size from the enormous honey mushroom fungus that is more than 5 kilometers wide to microorganisms that can be seen only with a microscope. This diversity of life is

SECT 1.2

1. B
B

Previewing Content

SECTION

1.3 Viruses are not alive but affect living things. pp. 24–29

1. Viruses share some characteristics with living things.
Viruses are tiny infectious particles that consist of genetic material surrounded by a protein coat called a **capsid.** Viruses are not alive and do not grow or respond to the environment.

2. Viruses multiply inside living cells.
Although viruses are not alive, they can **replicate** by taking over the cell machinery of a **host cell** and forcing it to make more viruses. In the diagram below, a bacteriophage, a virus that infects bacteria, attaches to a bacterium.

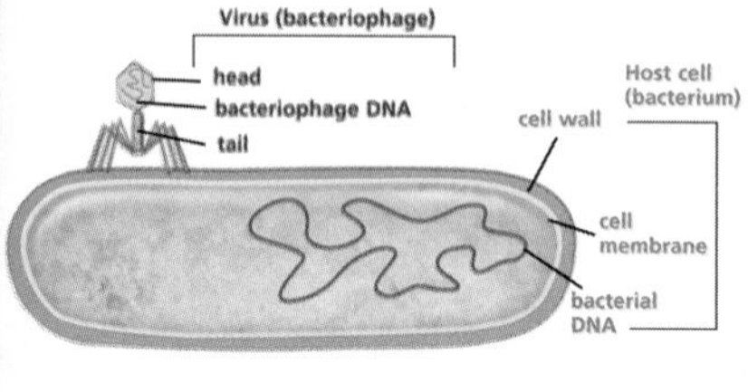

The viral DNA is injected into the cell, where it uses the host machinery to make more viral DNA and capsids. The viral parts are assembled into new viruses and released when the host cell bursts open.

3. Viruses may harm host cells.
Viruses use a host cell's material, energy, and cell processes to make more viruses. This reproductive process usually kills host cells and thus causes disease that may kill the organism. Scientists are currently trying to find ways to use viruses in positive ways, such as in gene therapy.

SECTION

1.4 Protists are a diverse group of organisms. pp. 30–35

1. Most protists are single-celled.
Protists include all organisms whose cells have nuclei and that are not plants, animals, or fungi. Most protists are single-celled. Some protists are producers and provide oxygen, some are parasites that cause disease, and some are decomposers.

2. Protists obtain their energy in three ways.
Protists can be classified by the way they get energy.

- **Algae** capture sunlight and convert it to chemical energy. They have chlorophyll and photosynthesize, releasing oxygen gas into the air. Thus, algae are producers. Seaweed, euglenas, diatoms, volvox, and the *Chlamydomonas* pictured below are algae.

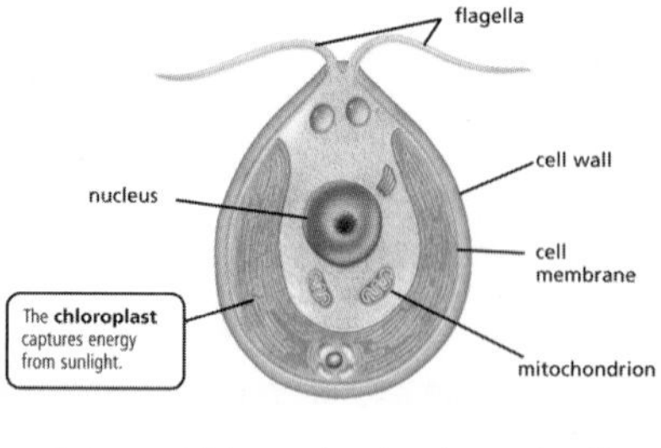

- **Protozoa** get their energy by eating other organisms. They are all single-celled, and most move about with cilia or flagella to search for food. Some protozoa are parasites and cause disease. Paramecia and the organism that causes malaria are protozoa.
- Other protists absorb materials that contain stored energy. Cellular slime molds, plasmodial slime molds, and water molds are members of this group.

Common Misconceptions

MISCONCEPTION DATABASE
CLASSZONE.COM Background on student misconceptions

CATCHING COLDS Many people think that you catch a cold by becoming chilled, going out without a coat, or going out in the rain without boots. The common cold is caused by viruses, which must infect body cells to produce the symptoms of a cold.

TE This misconception is addressed on p. 26.

Previewing Chapter Resources

- Section-by-section listing of all print and technology resources
- Suggested pacing
- Correlations to National Science Education Standards

Previewing Chapter Resources

KEY TO ICONS: CD/CD-ROM · Teacher Edition · INTERNET · Pupil Edition · UNIT RESOURCE BOOK

CHAPTER 1
Single-Celled Organisms and Viruses

INTEGRATED TECHNOLOGY		READING AND REINFORCEMENT	ASSESSMENT
CLASSZONE.COM • eEdition Plus • EasyPlanner Plus • Misconception Database • Content Review • Test Practice • Visualizations • Resource Centers • Internet Activity: Microscopic Life and You • Math Tutorial CD-ROM • eEdition • EasyPlanner • Power Presentations • Content Review • Lab Generator • Test Generator AUDIO CDS • Audio Readings • Audio Readings in Spanish SCILINKS.ORG	• Where Can You Find Microscopic Life? • How Quickly Do Bacteria Multiply? • Internet Activity: Microscopic Life and You UNIT RESOURCE BOOK • Family Letter, p. ix • Spanish Family Letter, p. x • Unit Projects, pp. 5–10 Lab Generator CD-ROM Generate customized labs.	• Description Wheel, B20–21 • Main Idea Web, C38–39 • Daily Vocabulary Scaffolding, H1–8 UNIT RESOURCE BOOK • Vocabulary Practice, pp. 56–57 • Decoding Support, p. 58 • Summarizing the Chapter, pp. 79–80 Audio Readings CD Listen to Pupil Edition. Audio Readings in Spanish CD Listen to Pupil Edition in Spanish.	• Chapter Re • Standardize UNIT ASSESS • Diagnostic • Chapter Tes • Alternative Spanish Chap Test Generat Generate cus Lab Generat Rubrics for La
SECTION 1.1 Single-celled organisms have all the characteristics • VISUALIZATION, Binary Fission • MATH TUTORIAL UNIT TRANSPARENCY BOOK	• EXPLORE Organisms, p. 9 • INVESTIGATE Microorganisms, p. 13 • Math in Science, p. 15 K nisms, p. 20	UNIT RESOURCE BOOK • Reading Study Guide, A & B, pp. 13–16 • Spanish Reading Study Guide, pp. 17–18 • Challenge and Extension, p. 19 • Reinforcing Key Concepts, p. 21 • Challenge Reading, pp. 54–55	Ongoing Asse Section 1.1 R UNIT ASSESS Section 1.1 Q
	N, Bacteria, K N, Bacteria,	UNIT RESOURCE BOOK • Reading Study Guide, A & B, pp. 24–27 • Spanish Reading Study Guide, pp. 28–29 • Challenge and Extension, p. 30 • Reinforcing Key Concepts, p. 31	Ongoing Asse Section 1.2 R UNIT ASSESS Section 1.2 Q
	4 p. 25 9 K 1	UNIT RESOURCE BOOK • Reading Study Guide, A & B, pp. 34–37 • Spanish Reading Study Guide, pp. 38–39 • Challenge and Extension, p. 40 • Reinforcing Key Concepts, p. 42	Ongoing Asse Section 1.3 R UNIT ASSESS Section 1.3 Q
	31 K 52 TION, Growth of 70–78	UNIT RESOURCE BOOK • Reading Study Guide, A & B, pp. 45–48 • Spanish Reading Study Guide, pp. 49–50 • Challenge and Extension, p. 51 • Reinforcing Key Concepts, p. 53	Ongoing Asse Section 1.4 R UNIT ASSESS Section 1.4 Q

e-celled orga
p. 16–23
smallest living thin
life.
structures needed for
NA, and cytoplasm can

Previewing Labs

Lab Generator CD-ROM
Edit these Pupil Edition labs and generate alternative labs.

EXPLORE the BIG idea	
Where Can You Find Microscopic Life? p. 7 Students look for microscopic life, using a hand lens or microscope.	**TIME** 20 minutes **MATERIALS** hand lens or microscope
How Quickly Do Bacteria Multiply? p. 7 Students use beans to model binary fission and exponential growth.	**TIME** 10 minutes **MATERIALS** funnel, 2-L bottle, duct tape, beans, clock or stopwatch
Internet Activity: Microscopic Life and You, p. 7 Students investigate the microscopic life that lives on, in, and around people.	**TIME** 20 minutes **MATERIALS** computer with Internet access
SECTION 1.1 **EXPLORE Organisms,** p. 9 Students list living things found in a classroom.	**TIME** 10 minutes **MATERIALS** paper, pencil
INVESTIGATE Microorganisms, p. 13 Students observe the responses of hydra and daphnia to their environment.	**TIME** 30 minutes **MATERIALS** microscope, slide, hydra culture, cover slip, daphnia culture, warm water
SECTION 1.2 **CHAPTER INVESTIGATION** Bacteria, pp. 22–23 Students sample bacteria that live on their hands and in the environment to determine how widespread bacteria are.	**TIME** 40 minutes **MATERIALS** 3 covered petri dishes with sterile nutrient agar, marker, tape, 3 sterile cotton swabs, small object (such as a coin or eraser), hand lens (for Challenge: 1 additional covered petri dish with agar, sterile swabs)
SECTION 1.3 **EXPLORE Viruses,** p. 24 Students model the way viruses were discovered, by passing a mixture through a filter.	**TIME** 10 minutes **MATERIALS** small kitchen sieve, paper plate, table salt, sesame seeds, small container, hand lens
INVESTIGATE Viruses, p. 25 Students use a model to analyze how infections spread.	**TIME** 30 minutes **MATERIALS** class set of paper cups, water, baking soda, phenolphthalein (solution A), and 0.1 M dilute hydrochloric acid (solution B), 2 droppers, measuring spoons
SECTION 1.4 **INVESTIGATE Protists,** p. 31 Students observe the movement and behavior of protists in pond water.	**TIME** 40 minutes **MATERIALS** pond water, microscope, hand lens, dropper, slide, cover slip

Additional INVESTIGATION, Growth of Algae, A, B, & C, pp. 70–78; Teacher Instructions, pp. 360–361

Chapter 1: Single-Celled Organisms and Viruses 5D C

Previewing Labs

- Brief descriptions of all chapter labs and activities
- Time and materials required for each activity

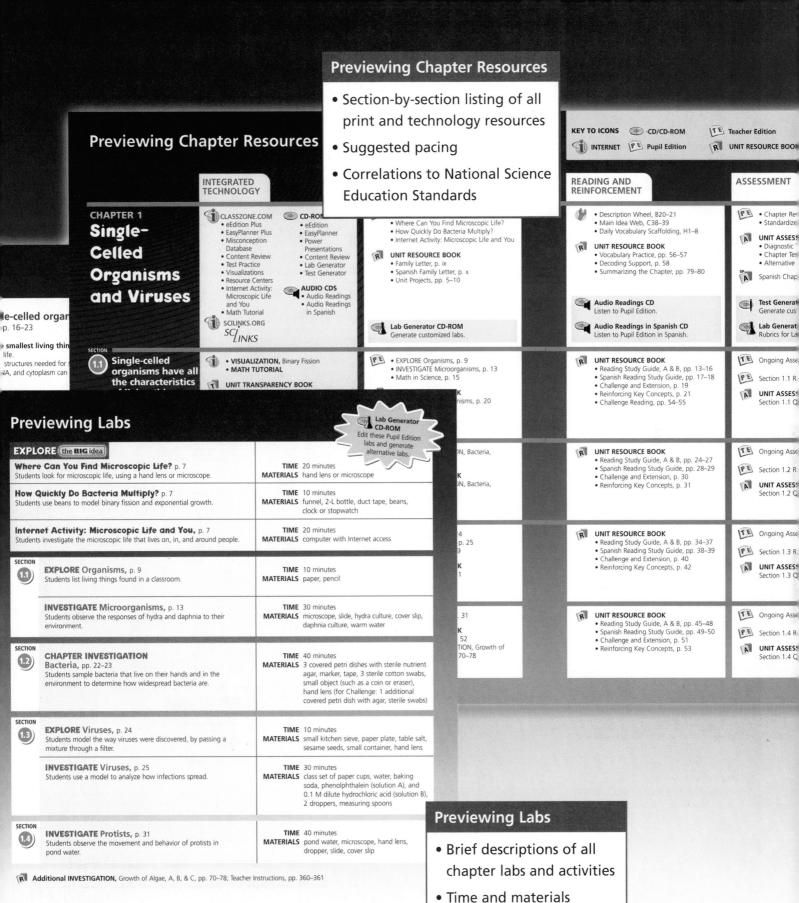

Planning the Lesson

Point-of-use support for each lesson provides a wealth of teaching options.

1. Prepare

- Concept and vocabulary review
- Note-taking and vocabulary strategies

2. Focus

- Set Learning Goals
- 3-Minute Warm-up

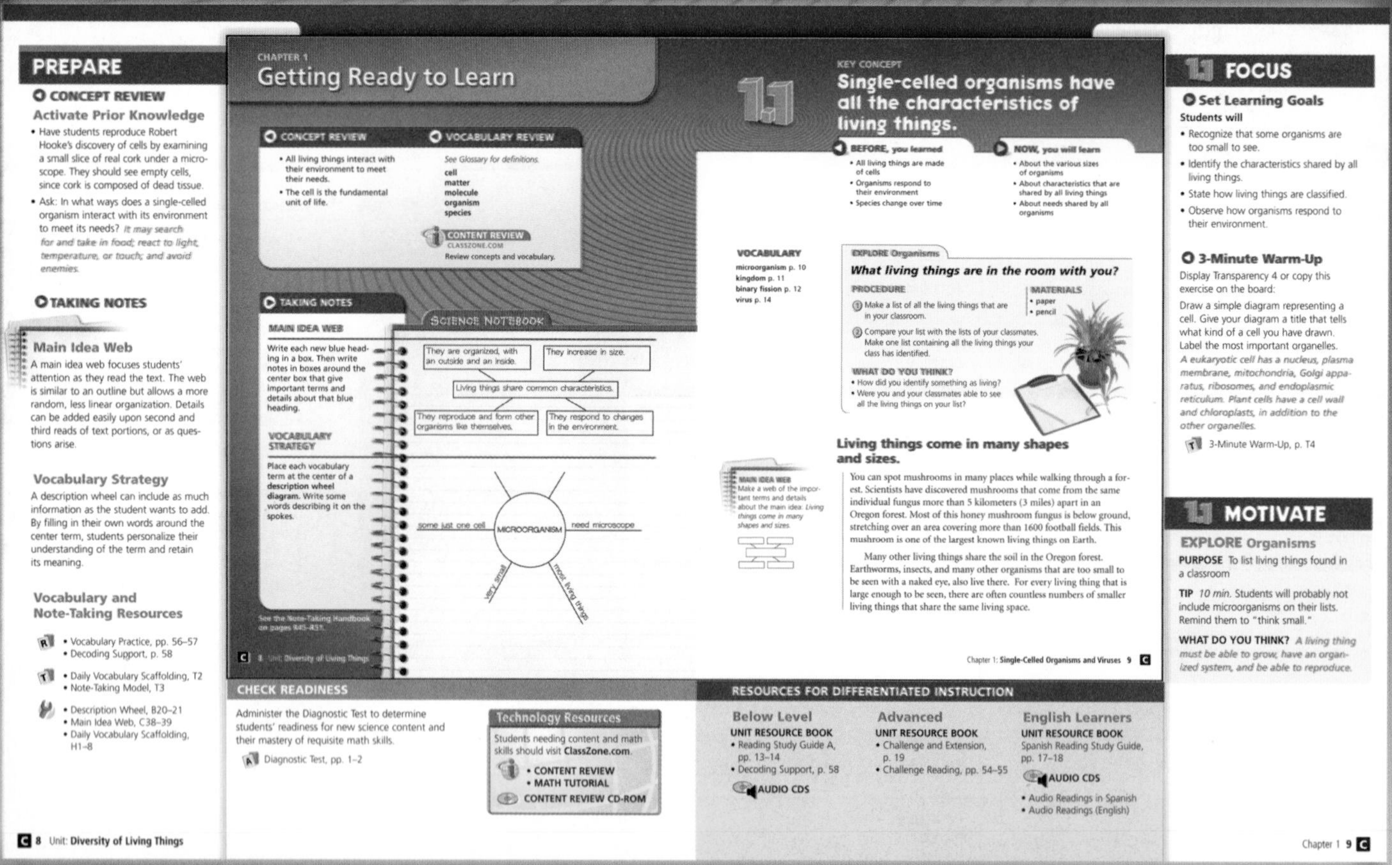

3. Motivate

- Engaging entry into the section
- Explore activity or Think About question

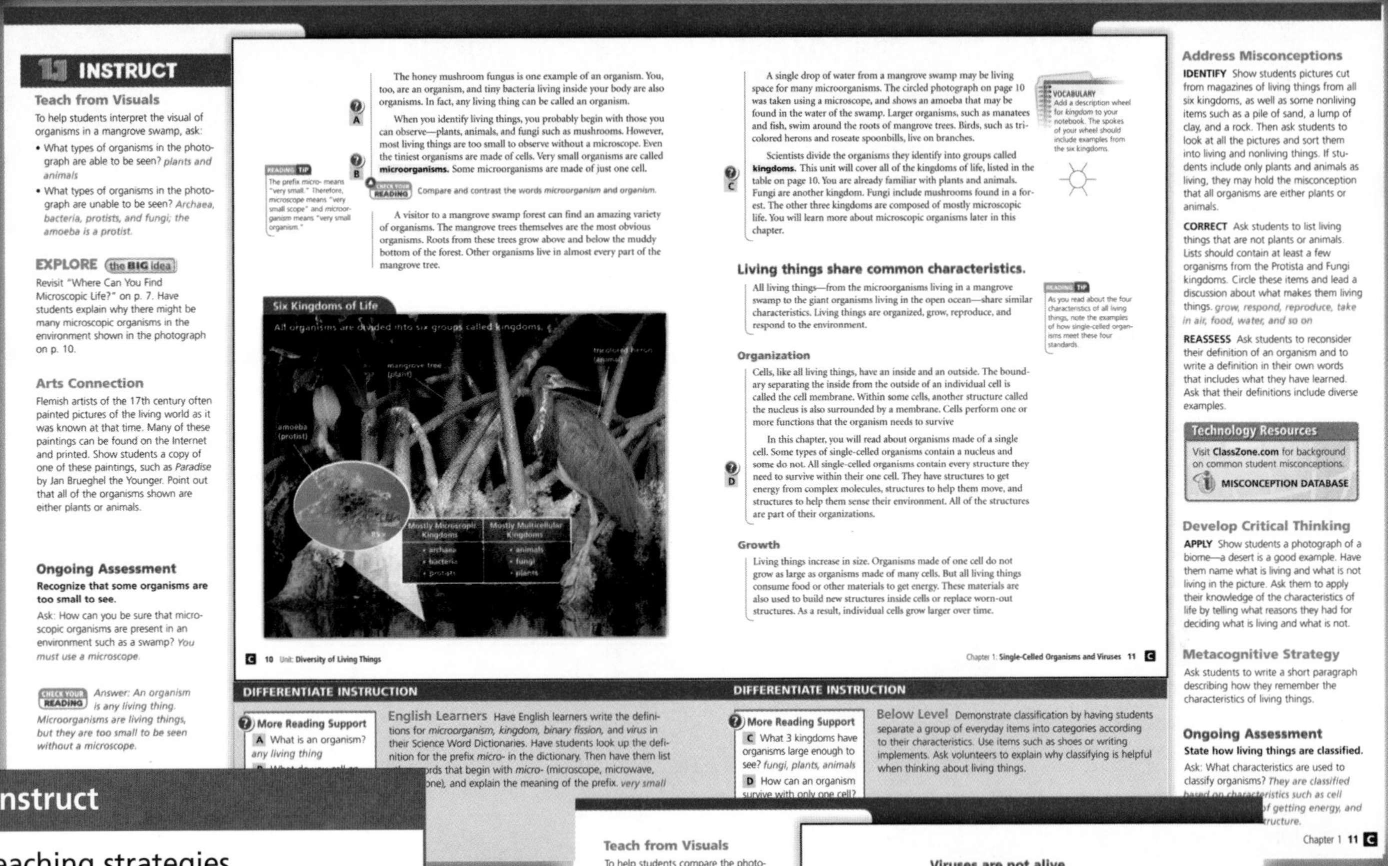

1.1 INSTRUCT

Teach from Visuals

To help students interpret the visual of organisms in a mangrove swamp, ask:

- What types of organisms in the photograph are able to be seen? *plants and animals*
- What types of organisms in the photograph are unable to be seen? *Archaea, bacteria, protists, and fungi; the amoeba is a protist.*

EXPLORE the BIG idea

Revisit "Where Can You Find Microscopic Life?" on p. 7. Have students explain why there might be many microscopic organisms in the environment shown in the photograph on p. 10.

Arts Connection

Flemish artists of the 17th century often painted pictures of the living world as it was known at that time. Many of these paintings can be found on the Internet and printed. Show students a copy of one of these paintings, such as *Paradise* by Jan Brueghel the Younger. Point out that all of the organisms shown are either plants or animals.

Ongoing Assessment

Recognize that some organisms are too small to see.

Ask: How can you be sure that microscopic organisms are present in an environment such as a swamp? *You must use a microscope.*

CHECK YOUR READING *Answer: An organism is any living thing. Microorganisms are living things, but they are too small to be seen without a microscope.*

The honey mushroom fungus is one example of an organism. You, too, are an organism, and tiny bacteria living inside your body are also organisms. In fact, any living thing can be called an organism.

When you identify living things, you probably begin with those you can observe—plants, animals, and fungi such as mushrooms. However, most living things are too small to observe without a microscope. Even the tiniest organisms are made of cells. Very small organisms are called **microorganisms.** Some microorganisms are made of just one cell.

READING TIP: The prefix micro- means "very small." Therefore, microscope means "very small scope" and microorganism means "very small organism."

CHECK YOUR READING: Compare and contrast the words *microorganism* and *organism*.

A visitor to a mangrove swamp forest can find an amazing variety of organisms. The mangrove trees themselves are the most obvious organisms. Roots from these trees grow above and below the muddy bottom of the forest. Other organisms live in almost every part of the mangrove tree.

Six Kingdoms of Life

All organisms are divided into six groups called kingdoms.

A single drop of water from a mangrove swamp may be living space for many microorganisms. The circled photograph on page 10 was taken using a microscope, and shows an amoeba that may be found in the water of the swamp. Larger organisms, such as manatees and fish, swim around the roots of mangrove trees. Birds, such as tricolored herons and roseate spoonbills, live on branches.

Scientists divide the organisms they identify into groups called **kingdoms.** This unit will cover all of the kingdoms of life, listed in the table on page 10. You are already familiar with plants and animals. Fungi are another kingdom. Fungi include mushrooms found in a forest. The other three kingdoms are composed of mostly microscopic life. You will learn more about microscopic organisms later in this chapter.

VOCABULARY: Add a description wheel for *kingdom* to your notebook. The spokes of your wheel should include examples from the six kingdoms.

Living things share common characteristics.

All living things—from the microorganisms living in a mangrove swamp to the giant organisms living in the open ocean—share similar characteristics. Living things are organized, grow, reproduce, and respond to the environment.

READING TIP: As you read about the four characteristics of all living things, note the examples of how single-celled organisms meet these four standards.

Organization

Cells, like all living things, have an inside and an outside. The boundary separating the inside from the outside of an individual cell is called the cell membrane. Within some cells, another structure called the nucleus is also surrounded by a membrane. Cells perform one or more functions that the organism needs to survive.

In this chapter, you will read about organisms made of a single cell. Some types of single-celled organisms contain a nucleus and some do not. All single-celled organisms contain every structure they need to survive within their one cell. They have structures to get energy from complex molecules, structures to help them move, and structures to help them sense their environment. All of the structures are part of their organizations.

Growth

Living things increase in size. Organisms made of one cell do not grow as large as organisms made of many cells. But all living things consume food or other materials to get energy. These materials are also used to build new structures inside cells or replace worn-out structures. As a result, individual cells grow larger over time.

Address Misconceptions

IDENTIFY Show students pictures cut from magazines of living things from all six kingdoms, as well as some nonliving items such as a pile of sand, a lump of clay, and a rock. Then ask students to look at all the pictures and sort them into living and nonliving things. If students include only plants and animals as living, they may hold the misconception that all organisms are either plants or animals.

CORRECT Ask students to list living things that are not plants or animals. Lists should contain at least a few organisms from the Protista and Fungi kingdoms. Circle these items and lead a discussion about what makes them living things. *grow, respond, reproduce, take in air, food, water, and so on*

REASSESS Ask students to reconsider their definition of an organism and to write a definition in their own words that includes what they have learned. Ask that their definitions include diverse examples.

Technology Resources

Visit ClassZone.com for background on common student misconceptions.

MISCONCEPTION DATABASE

Develop Critical Thinking

APPLY Show students a photograph of a biome—a desert is a good example. Have them name what is living and what is not living in the picture. Ask them to apply their knowledge of the characteristics of life by telling what reasons they had for deciding what is living and what is not.

Metacognitive Strategy

Ask students to write a short paragraph describing how they remember the characteristics of living things.

Ongoing Assessment

State how living things are classified.

Ask: What characteristics are used to classify organisms? *They are classified based on characteristics such as cell ... of getting energy, and ... structure.*

DIFFERENTIATE INSTRUCTION

More Reading Support

A What is an organism? *any living thing*

English Learners Have English learners write the definitions for *microorganism, kingdom, binary fission,* and *virus* in their Science Word Dictionaries. Have students look up the definition for the prefix *micro-* in the dictionary. Then have them list ... words that begin with *micro-* (microscope, microwave, ... phone), and explain the meaning of the prefix. *very small*

DIFFERENTIATE INSTRUCTION

More Reading Support

C What 3 kingdoms have organisms large enough to see? *fungi, plants, animals*

D How can an organism survive with only one cell?

Below Level Demonstrate classification by having students separate a group of everyday items into categories according to their characteristics. Use items such as shoes or writing implements. Ask volunteers to explain why classifying is helpful when thinking about living things.

4. Instruct

- Teaching strategies
- Reading support
- Ongoing assessment
- Addressing misconceptions
- Differentiated instruction activities and tips

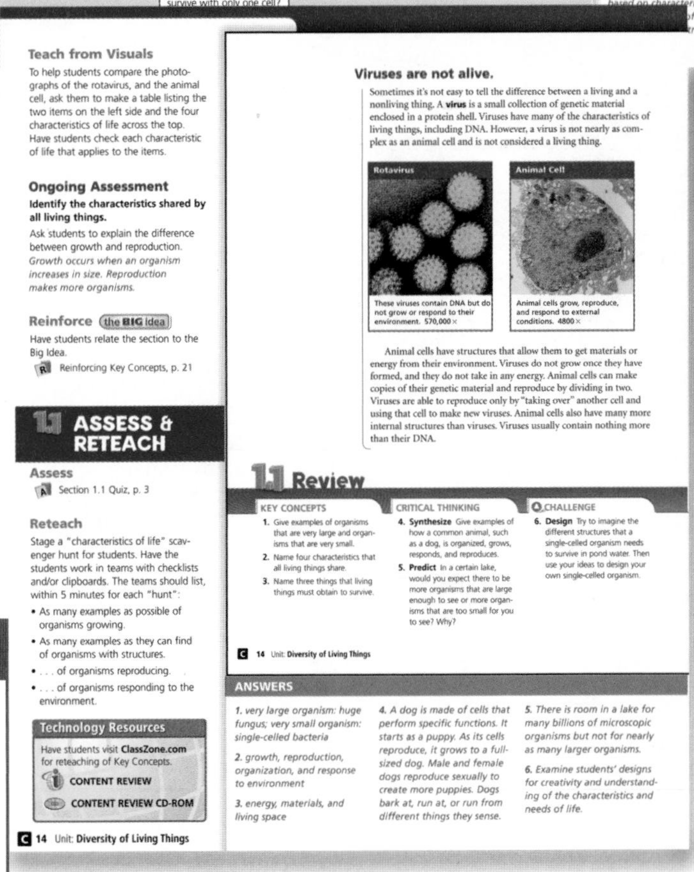

Teach from Visuals

To help students compare the photographs of the rotavirus, and the animal cell, ask them to make a table listing the two items on the left side and the four characteristics of life across the top. Have students check each characteristic of life that applies to the items.

Ongoing Assessment

Identify the characteristics shared by all living things.

Ask students to explain the difference between growth and reproduction. *Growth occurs when an organism increases in size. Reproduction makes more organisms.*

Reinforce the BIG idea

Have students relate the section to the Big Idea.

Reinforcing Key Concepts, p. 21

1.1 ASSESS & RETEACH

Assess

Section 1.1 Quiz, p. 3

Reteach

Stage a "characteristics of life" scavenger hunt for students. Have the students work in teams with checklists and/or clipboards. The teams should list, within 5 minutes for each "hunt":

- As many examples as possible of organisms growing.
- As many examples as they can find of organisms with structures.
- . . . of organisms reproducing.
- . . . of organisms responding to the environment.

Technology Resources

Have students visit ClassZone.com for reteaching of Key Concepts.

CONTENT REVIEW

CONTENT REVIEW CD-ROM

Viruses are not alive.

Sometimes it's not easy to tell the difference between a living and a nonliving thing. A **virus** is a small collection of genetic material enclosed in a protein shell. Viruses have many of the characteristics of living things, including DNA. However, a virus is not nearly as complex as an animal cell and is not considered a living thing.

These viruses contain DNA but do not grow or respond to their environment. 570,000×

Animal cells grow, reproduce, and respond to external conditions. 4800×

Animal cells have structures that allow them to get materials or energy from their environment. Viruses do not grow once they have formed, and they do not take in any energy. Animal cells can make copies of their genetic material and reproduce by dividing in two. Viruses are able to reproduce only by "taking over" another cell and using that cell to make new viruses. Animal cells also have many more internal structures than viruses. Viruses usually contain nothing more than their DNA.

1.1 Review

KEY CONCEPTS

1. Give examples of organisms that are very large and organisms that are very small.
2. Name four characteristics that all living things share.
3. Name three things that living things must obtain to survive.

CRITICAL THINKING

4. **Synthesize** Give examples of how a common animal, such as a dog, is organized, grows, responds, and reproduces.
5. **Predict** In a certain lake, would you expect there to be more organisms that are large enough to see or more organisms that are too small for you to see? Why?

CHALLENGE

6. **Design** Try to imagine the different structures that a single-celled organism needs to survive in pond water. Then use your ideas to design your own single-celled organism.

ANSWERS

1. *very large organism: huge fungus; very small organism: single-celled bacteria*
2. *growth, reproduction, organization, and response to environment*
3. *energy, materials, and living space*
4. *A dog is made of cells that perform specific functions. It starts as a puppy. As its cells reproduce, it grows to a full-sized dog. Male and female dogs reproduce sexually to create more puppies. Dogs bark at, run at, or run from different things they sense.*
5. *There is room in a lake for many billions of microscopic organisms but not for nearly as many larger organisms.*
6. *Examine students' designs for creativity and understanding of the characteristics and needs of life.*

5. Assess & Reteach

- Answers to Section Review
- Reteaching activity
- Resources for review and assessment

Lab Materials List

The following charts list the consumables, nonconsumables, and equipment needed for all activities. Quantities are per group of four students. Lab aprons, goggles, water, books, paper, pens, pencils, and calculators are assumed to be available for all activities.

Materials kits are available. For more information, please call McDougal Littell at 1-800-323-5435.

Consumables

Description	Quantity per Group	Explore *page*	Investigate *page*	Chapter Investigation *page*
apple	1	107		
artificial sweetener	3 tsp			72
bag, 1/2 gallon, zip-top	2		184	
baking soda	16 oz		25	
bottle, 1/2 liter plastic, clear	1		94	
carrot	1/4		146	
celery	1/4	51		
clothespin	8			72
cotton swab, sterile	3			22
cover slip	2		13, 31	
culture, daphnia	1 mL		13	
culture, hydra	1 mL		13	
cup, clear plastic	5	24, 51, 66		
cup, paper	25		25, 52	104
earthworm	6–11	128		134
egg, hardboiled	1		170	
feather, assorted	3–5	173		
filter paper, fine, 4"	1			134
flour	1 1/2 cup			72
flower	1		111	
hydrochloric acid, 0.1 M dilute	50 mL		25	
ice	2–4 cups		184	
iodine	5 mL	51		
kelp granules	2 tbs		94	
knife, plastic	3	107	124, 170	
label, adhesive	10			104
leaf	1	85		
marker, colored	1			72
marker, permanent black	1			22, 104
mealworm	10		146	
millet seed	2 cups			180

Description	Quantity per Group	Explore page	Investigate page	Chapter Investigation page
mushroom, fresh	1	66		
nutrient agar, sterile	100 mL			22
oat bran	2 cups		146	
owl pellet	1		60	
paper, white, 8.5" x 11"	1	66		
paper clip	40		44	
paper towel	10	107	102	104
pasta, dry	2 cups			180
pea pod	1	107		
pear	1/8	51		
phenolphthalein	50 mL		25	
pill bug	5	142		
pine cone, dry and open	1		102	
plant, bean seedling	3		52	
plant, live moss	1	92		
plate, paper	1	24		
potato, raw	2	51	124, 146	
salt, table	3/4 cup	24		72
sand	1 lb			134
seed, vegetable, 5–10 varieties	3 each			104
sesame seed	100 mL	24		
shoebox, cardboard with lid	1	142		
slide	2		13, 31	
soil, potting	10 lbs	128	52	104, 134
spoon, plastic	1		124	
straw, clear drinking	8			72
string	18"		94	
sugar	3 tsp			72
tape, masking	1 roll		124	22, 72
vegetable shortening	1 lb		184	
water, distilled	850 mL	128		134
water, pond	1 mL		31	
yeast, quick-rise	1 tsp			72

Nonconsumables

Description	Quantity per Group	Explore page	Investigate page	Chapter Investigation page
aquarium, small	1	157		134
baking sheet	1			72
balance, triple beam	1		52	
beaker, 250 mL	1			134
beaker, 500 mL	1		102	

Description	Quantity per Group	Explore *page*	Investigate *page*	Chapter Investigation *page*
bowl, extra-large plastic	1		184	
box, small cardboard	2		44	
button, white	1		94	
container, clear plastic	4–6	142		134, 180
eyedropper	1–2	51	25, 31	180
flashlight with batteries	1			134
hand lens	1	24, 66, 85, 92, 142	31, 111	22
jar, 26 oz glass	1			180
jar, 26 oz glass with air holes in lid	1		146	
knife, sharp	1	66		
measuring cup, 1/4 cup	1			72
measuring spoon, 1/4 teaspoon	1			72
measuring spoon, tablespoon	1		94	
measuring spoon, teaspoon	1		25	72
meter stick	1	164		
microscope	1		13, 31	
mollusk shell	1–2		138	
needle tool	1		60	
petri dish	4		146	22
pliers	1–2			180
rubber band	50			180
ruler, metric	1		94	72
sand dollar	1–2		138	
sea star	1–2		138	
sieve	1	24		
spoon, slotted	1–2			180
spray bottle, 250 mL	1	128		134
stopwatch	1			134, 180
test tube	4–6			180
test tube rack	1			180
tray, dissection	1	128	60, 146	
tweezers	1		60, 146	180

Unit Resource Book Datasheet

Description		Explore *page*	Investigate *page*	Chapter Investigation *page*
Bone Identification Key			60	

Safety Equipment

Description		Explore *page*	Investigate *page*	Chapter Investigation *page*
gloves		51, 128	60, 124, 146	104, 134

McDougal Littell Science

Diversity of Living Things

LIFE SCIENCE

A ▶ Cells and Heredity
B ▶ Life Over Time
C ▶ Diversity of Living Things
D ▶ Ecology
E ▶ Human Biology

EARTH SCIENCE

A ▶ Earth's Surface
B ▶ The Changing Earth
C ▶ Earth's Waters
D ▶ Earth's Atmosphere
E ▶ Space Science

PHYSICAL SCIENCE

A ▶ Matter and Energy
B ▶ Chemical Interactions
C ▶ Motion and Forces
D ▶ Waves, Sound, and Light
E ▶ Electricity and Magnetism

Copyright © 2005 by McDougal Littell, a division of Houghton Mifflin Company.

No part of this work may be reproduced or transmitted in any form or by any means, electronic or mechanical, including photocopy and recording, or by any information storage or retrieval system without the prior written permission of McDougal Littell unless such copying is expressly permitted by federal copyright law. Address inquiries to Supervisor, Rights and Permissions, McDougal Littell, P.O. Box 1667, Evanston, IL 60204.

ISBN: 0-618-33434-3 1 2 3 4 5 6 7 8 VJM 08 07 06 05 04

Internet Web Site: http://www.mcdougallittell.com

ii

Science Consultants

Chief Science Consultant

James Trefil, Ph.D. is the Clarence J. Robinson Professor of Physics at George Mason University. He is the author or co-author of more than 25 books, including *Science Matters* and *The Nature of Science.* Dr. Trefil is a member of the American Association for the Advancement of Science's Committee on the Public Understanding of Science and Technology. He is also a fellow of the World Economic Forum and a frequent contributor to *Smithsonian* magazine.

Rita Ann Calvo, Ph.D. is Senior Lecturer in Molecular Biology and Genetics at Cornell University, where for 12 years she also directed the Cornell Institute for Biology Teachers. Dr. Calvo is the 1999 recipient of the College and University Teaching Award from the National Association of Biology Teachers.

Kenneth Cutler, M.S. is the Education Coordinator for the Julius L. Chambers Biomedical Biotechnology Research Institute at North Carolina Central University. A former middle school and high school science teacher, he received a 1999 Presidential Award for Excellence in Science Teaching.

Instructional Design Consultants

Douglas Carnine, Ph.D. is Professor of Education and Director of the National Center for Improving the Tools of Educators at the University of Oregon. He is the author of seven books and over 100 other scholarly publications, primarily in the areas of instructional design and effective instructional strategies and tools for diverse learners. Dr. Carnine also serves as a member of the National Institute for Literacy Advisory Board.

Linda Carnine, Ph.D. consults with school districts on curriculum development and effective instruction for students struggling academically. A former teacher and school administrator, Dr. Carnine also co-authored a popular remedial reading program.

Sam Miller, Ph.D. is a middle school science teacher and the Teacher Development Liaison for the Eugene, Oregon, Public Schools. He is the author of curricula for teaching science, mathematics, computer skills, and language arts.

Donald Steely, Ph.D. serves as principal investigator at the Oregon Center for Applied Science (ORCAS) on federal grants for science and language arts programs. His background also includes teaching and authoring of print and multimedia programs in science, mathematics, history, and spelling.

Vicky Vachon, Ph.D. consults with school districts throughout the United States and Canada on improving overall academic achievement with a focus on literacy. She is also co-author of a widely used program for remedial readers.

iii

Content Reviewers

John Beaver, Ph.D.
Ecology
Professor, Director of Science Education Center
College of Education and Human Services
Western Illinois University
Macomb, IL

Donald J. DeCoste, Ph.D.
Matter and Energy, Chemical Interactions
Chemistry Instructor
University of Illinois
Urbana-Champaign, IL

Dorothy Ann Fallows, Ph.D., MSc
Diversity of Living Things, Microbiology
Partners in Health
Boston, MA

Michael Foote, Ph.D.
The Changing Earth, Life Over Time
Associate Professor
Department of the Geophysical Sciences
The University of Chicago
Chicago, IL

Lucy Fortson, Ph.D.
Space Science
Director of Astronomy
Adler Planetarium and Astronomy Museum
Chicago, IL

Elizabeth Godrick, Ph.D.
Human Biology
Professor, CAS Biology
Boston University
Boston, MA

Isabelle Sacramento Grilo, M.S.
The Changing Earth
Lecturer, Department of the Geological Sciences
Montana State University
Bozeman, MT

David Harbster, MSc
Diversity of Living Things
Professor of Biology
Paradise Valley Community College
Phoenix, AZ

Richard D. Norris, Ph.D.
Earth's Waters
Professor of Paleobiology
Scripps Institution of Oceanography
University of California, San Diego
La Jolla, CA

Donald B. Peck, M.S.
Motion and Forces; Waves, Sound, and Light; Electricity and Magnetism
Director of the Center for Science Education (retired)
Fairleigh Dickinson University
Madison, NJ

Javier Penalosa, Ph.D.
Diversity of Living Things, Plants
Associate Professor, Biology Department
Buffalo State College
Buffalo, NY

Raymond T. Pierrehumbert, Ph.D.
Earth's Atmosphere
Professor in Geophysical Sciences (Atmospheric Science)
The University of Chicago
Chicago, IL

Brian J. Skinner, Ph.D.
Earth's Surface
Eugene Higgins Professor of Geology and Geophysics
Yale University
New Haven, CT

Nancy E. Spaulding, M.S.
Earth's Surface, The Changing Earth, Earth's Waters
Earth Science Teacher (retired)
Elmira Free Academy
Elmira, NY

Steven S. Zumdahl, Ph.D.
Matter and Energy, Chemical Interactions
Professor Emeritus of Chemistry
University of Illinois
Urbana-Champaign, IL

Susan L. Zumdahl, M.S.
Matter and Energy, Chemical Interactions
Chemistry Education Specialist
University of Illinois
Urbana-Champaign, IL

Safety Consultant

Juliana Texley, Ph.D.
Former K–12 Science Teacher and School Superintendent
Boca Raton, FL

English Language Advisor

Judy Lewis, M.A.
Director, State and Federal Programs for reading proficiency and high risk populations
Rancho Cordova, CA

iv

Teacher Panel Members

Carol Arbour
Tallmadge Middle School,
Tallmadge, OH

Patty Belcher
Goodrich Middle School,
Akron, OH

Gwen Broestl
Luis Munoz Marin Middle School,
Cleveland, OH

Al Brofman
Tehipite Middle School,
Fresno, CA

John Cockrell
Clinton Middle School,
Columbus, OH

Jenifer Cox
Sylvan Middle School,
Citrus Heights, CA

Linda Culpepper
Martin Middle School,
Charlotte, NC

Kathleen Ann DeMatteo
Margate Middle School,
Margate, FL

Melvin Figueroa
New River Middle School,
Ft. Lauderdale, FL

Doretha Grier
Kannapolis Middle School,
Kannapolis, NC

Robert Hood
Alexander Hamilton Middle School,
Cleveland, OH

Scott Hudson
Coverdale Elementary School,
Cincinnati, OH

Loretta Langdon
Princeton Middle School,
Princeton, NC

Carlyn Little
Glades Middle School,
Miami, FL

Ann Marie Lynn
Amelia Earhart Middle School,
Riverside, CA

James Minogue
Lowe's Grove Middle School,
Durham, NC

Joann Myers
Buchanan Middle School,
Tampa, FL

Barbara Newell
Charles Evans Hughes Middle School,
Long Beach, CA

Anita Parker
Kannapolis Middle School,
Kannapolis, NC

Greg Pirolo
Golden Valley Middle School,
San Bernardino, CA

Laura Pottmyer
Apex Middle School,
Apex, NC

Lynn Prichard
Booker T. Washington Middle Magnet
School, Tampa, FL

Jacque Quick
Walter Williams High School,
Burlington, NC

Robert Glenn Reynolds
Hillman Middle School,
Youngstown, OH

Theresa Short
Abbott Middle School,
Fayetteville, NC

Rita Slivka
Alexander Hamilton Middle School,
Cleveland, OH

Marie Sofsak
B F Stanton Middle School,
Alliance, OH

Nancy Stubbs
Sweetwater Union Unified School District,
Chula Vista, CA

Sharon Stull
Quail Hollow Middle School,
Charlotte, NC

Donna Taylor
Okeeheelee Middle School,
West Palm Beach, FL

Sandi Thompson
Harding Middle School,
Lakewood, OH

Lori Walker
Audubon Middle School & Magnet Center,
Los Angeles, CA

Teacher Lab Evaluators

Jill Brimm-Byrne
Albany Park Academy,
Chicago, IL

Gwen Broestl
Luis Munoz Marin Middle School,
Cleveland, OH

Al Brofman
Tehipite Middle School,
Fresno, CA

Michael A. Burstein
The Rashi School,
Newton, MA

Trudi Coutts
Madison Middle School,
Naperville, IL

Stacy Covert
Lufkin Road Middle School,
Apex, NC

Jenifer Cox
Sylvan Middle School,
Citrus Heights, CA

Larry Cwik
Madison Middle School,
Naperville, IL

Jennifer Donatelli
Kennedy Junior High School,
Lisle, IL

Paige Fullhart
Highland Middle School,
Libertyville, IL

Sue Hood
Glen Crest Middle School,
Glen Ellyn, IL

Ann Min
Beardsley Middle School,
Crystal Lake, IL

Aileen Mueller
Kennedy Junior High School,
Lisle, IL

Nancy Nega
Churchville Middle School,
Elmhurst, IL

Oscar Newman
Sumner Math and Science Academy,
Chicago, IL

Marina Penalver
Moore Middle School,
Portland, ME

Lynn Prichard
Booker T. Washington Middle Magnet
School, Tampa, FL

Jacque Quick
Walter Williams High School,
Burlington, NC

Seth Robey
Gwendolyn Brooks Middle School,
Oak Park, IL

Kevin Steele
Grissom Middle School,
Tinley Park, IL

v

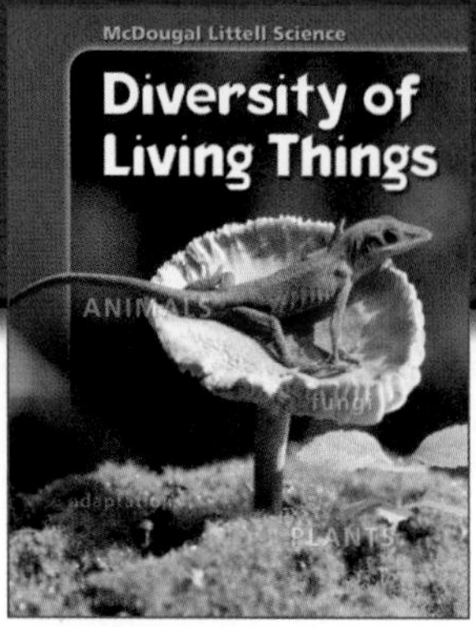

Diversity of Living Things

eEdition

Unit Features

SCIENTIFIC AMERICAN

1 Single-Celled Organisms and Viruses 6

the BIG idea

Bacteria and protists have the characteristics of living things, while viruses are not alive.

2 Introduction to Multicellular Organisms 40

the BIG idea

Multicellular organisms live in and get energy from a variety of environments.

How does an organism get energy and materials from its environment?

vii

Features

Visual Highlights

Internet Resources @ ClassZone.com

INVESTIGATIONS AND ACTIVITIES

EXPLORE THE BIG IDEA

Chapter Opening Inquiries

CHAPTER INVESTIGATION

Full-Period Labs

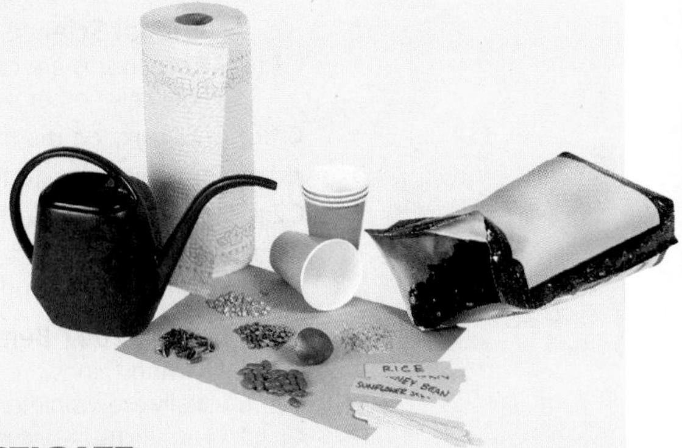

EXPLORE

Introductory Inquiry Activities

INVESTIGATE

Skill Labs

Standards and Benchmarks

Each chapter in **Diversity of Life** covers some of the learning goals that are described in the *National Science Education Standards* (NSES) and the Project 2061 *Benchmarks for Science Literacy.* Selected content and skill standards are shown below in shortened form. The following National Science Education Standards are covered on pages xii-xvii, in Frontiers in Science, and in Timelines in Science, as well as in chapter features and laboratory investigations: Understandings About Scientific Inquiry (A.9), Science and Technology in Society (F.5), Understandings About Science and Technology (E.6), Science as a Human Endeavor (G.1), Nature of Science (G.2), and History of Science (G.3)

Content Standards

1 Single-Celled Organisms and Viruses

	National Science Education Standards
C.1.b	All organisms are composed of cells—the fundamental unit of life. Most organisms are single cells; other organisms are multicellular.
C.1.c	Cells carry on the many functions needed to sustain life.
C.1.f	Some diseases are the result of damage by infection by other organisms.
C.2.a	Reproduction is a characteristic of all living systems.
C.3.a	All organisms must be able to obtain and use resources, grow, reproduce, and maintain stable internal conditions.

	Project 2061 Benchmarks
5.C.1	All living things are composed of cells, from just one to many millions, whose details usually are visible only through a microscope.
5.C.3	Many of the basic functions of organisms—such as extracting energy from food and getting rid of waste—are carried out within cells.
6.E.3	Viruses, bacteria, fungi, and parasites may infect the human body and interfere with normal body functions.

2 Introduction to Multicellular Organisms

	National Science Education Standards
C.1.a	Important levels of organization for structure and function include cells, organs, tissues, organ systems, whole organisms, and ecosystems.
C.3.c	Behavior is one kind of response an organism makes to an internal or external stimulus.
F.1.e	Food provides energy and nutrients for growth and development.

	Project 2061 Benchmarks
5.A.1	One of the most general distinctions among organisms is between plants, which use sunlight to make their own food, and animals, which consume energy-rich foods.
5.B.2	In sexual reproduction, a single specialized cell from a female merges with a specialized cell from a male.
5.C.2	Various organs and tissues serve the needs of cells for food, air, and waste removal.
5.D.2	Organisms may interact with one another as producer/consumer, predator/prey, or one organism may scavenge or decompose another.
5.E.3	Energy can change from one form to another in living things. Animals get energy from oxidizing their food, releasing some of its energy as heat.
6.C.1	Organs and organ systems are composed of cells and help provide cells with basic needs.
6.C.2	For the body to use food for energy and building materials, the food must first be digested into molecules that are absorbed and transported to cells.

3 Plants

National Science Education Standards

C.2.b Plants can reproduce sexually—the egg and sperm are produced in the flowers of flowering plants.

C.4.c Energy in the form of sunlight is transferred by producers into chemical energy through photosynthesis.

Project 2061 Benchmarks

4.C.6 The composition of soil is greatly influenced by plant roots and debris, bacteria, fungi, worms, insects, rodents, and other organisms.

5.A.5 All organisms are part of global food webs. One web includes land plants, the animals that feed on them, and so forth.

5.E.1 Plants use the energy in light to make sugars out of carbon dioxide and water. This food can be used immediately for fuel or materials or it may be stored for later use.

4 Invertebrate Animals

National Science Education Standards

C.1.d Specialized cells perform specialized functions in multicellular organisms.

C.5.a Although different species might look dissimilar, the unity among organisms becomes apparent from an analysis of internal structures.

Project 2061 Benchmarks

5.A.2 Animals have a great variety of body plans and internal structures that contribute to their being able to find food and reproduce.

5.A.3 Similarities among organisms are found in internal anatomical features.

5 Vertebrate Animals

National Science Education Standards

C.3.b Regulation of an organism's internal environment keeps conditions within the range required to survive.

C.5.b Biological adaptations include changes in structures, behaviors, or physiology that enhance survival in a particular environment.

Project 2061 Benchmarks

5.F.2 Changes in environmental conditions affect survival of organisms and entire species.

6.B.3 Patterns of human development are similar to those of other vertebrates.

6.C.3 Lungs take in oxygen for combustion of food, they eliminate the carbon dioxide produced.

Skill Standards

National Science Education Standards

A.1 Identify questions that can be answered through scientific methods.

A.2 Design and conduct a scientific investigation.

A.3 Use appropriate tools and techniques to gather and analyze data.

A.4 Use evidence to describe, predict, explain, and model.

A.5 Think critically to find relationships between results and interpretations.

A.6 Give alternative explanations and predictions.

A.7 Communicate procedures, results, and conclusions.

A.8 Use mathematics in all aspects of scientific inquiry.

Project 2061 Benchmarks

9.A.3 Write numbers in different forms.

9.A.4 Use the operations addition and subtraction as inverses of each other—one undoing what the other does; likewise multiplication and division.

9.B.2 Use mathematics to describe change.

9.B.3 Use graphs to show relationships.

9.C.4 Use a graphic display of numbers to show patterns such as trends, varying rates of change, gaps, or clusters.

11.A.2 Think about things as systems.

11.A.3 See how systems are interconnected.

11.C.5 Understand how symmetry (or the lack of it) can determine properties of many objects, including organisms.

12.A.2 Investigate, using hypotheses.

12.A.3 See multiple ways to interpret results.

12.B.7 Determine units for answers.

12.D.2 Read and interpret tables and graphs.

12.D.3 Research books, periodicals, databases.

12.D.4 Understand charts, graphs and tables.

12.E.5 Criticize faulty reasoning.

Introducing Life Science

Scientists are curious. Since ancient times, they have been asking and answering questions about the world around them. Scientists are also very suspicious of the answers they get. They carefully collect evidence and test their answers many times before accepting an idea as correct.

In this book you will see how scientific knowledge keeps growing and changing as scientists ask new questions and rethink what was known before. The following sections will help get you started.

UNIFYING PRINCIPLES of Life Science

What Is Life Science?

Life science is the study of living things. As you study life science, you will observe and read about a variety of organisms, from huge redwood trees to the tiny bacteria that cause sore throats. Because Earth is home to such a great variety of living things, the study of life science is rich and exciting.

But life science doesn't simply include learning the names of millions of organisms. It includes big ideas that help us to understand how all these livings things interact with their environment. Life science is the study of characteristics and needs that all living things have in common. It's also a study of changes—both daily changes as well as changes that take place over millions of years. Probably most important, in studying life science, you will explore the many ways that all living things—including you—depend upon Earth and its resources.

The text and visuals in this book will invite you into the world of living things and provide you with the key concepts you'll need in your study. Activities offer a chance for you to investigate some aspects of life science on your own. The four unifying principles listed below provide a way for you to connect the information and ideas in this program.

- **All living things share common characteristics.**
- **All living things share common needs.**
- **Living things meet their needs through interactions with the environment.**
- **The types and numbers of living things change over time.**

the BIG idea

Each chapter begins with a big idea. Keep in mind that each big idea relates to one or more of the unifying principles.

UNIFYING PRINCIPLE

All living things share common characteristics.

Birds nest in a group of weeds as sunlight shines and a breeze blows. Which of these is alive? Warblers and weeds are living things, but sunlight and breezes are not. All living things share common characteristics that distinguish them from nonliving things.

What It Means

This unifying principle helps you explore one of the biggest questions in science, "What is life?" Let's take a look at four characteristics that distinguish living things from nonliving things : organization, growth, reproduction, and response.

Organization

If you stand a short distance from a reed warbler's nest, you can observe the largest level of organization in a living thing—the **organism** itself. Each bird is an organism. If you look at a leaf under a microscope, you can observe the smallest level of organization capable of performing all the activities of life, a **cell.** All living things are made of cells.

Growth

Most living things grow and develop. Growth often involves not only an increase in size, but also an increase in complexity, such as a tadpole growing into a frog. If all goes well, the small warblers in the picture will grow to the size of their parent.

Reproduction

Most living things produce offspring like themselves. Those offspring are also able to reproduce. That means that reed warblers produce reed warblers which in turn reproduce more reed warblers.

Response

You've probably noticed that your body adjusts to changes in your surroundings. If you are exploring outside on a hot day, you may notice that you sweat. On a cold day, you may shiver. Sweating and shivering are examples of response.

Why It's Important

People of all ages experience the urge to explore and understand the living world. Understanding the characteristics of living things is a good way to start this exploration of life. In addition, knowing about the characteristics of living things helps you identify:

- similarities and differences among various organisms
- key questions to ask about any organism you study

xv

UNIFYING PRINCIPLE

All living things share common needs.

What do you need to stay alive? What does an animal like a fish or a coral need to stay alive? All living things have common needs.

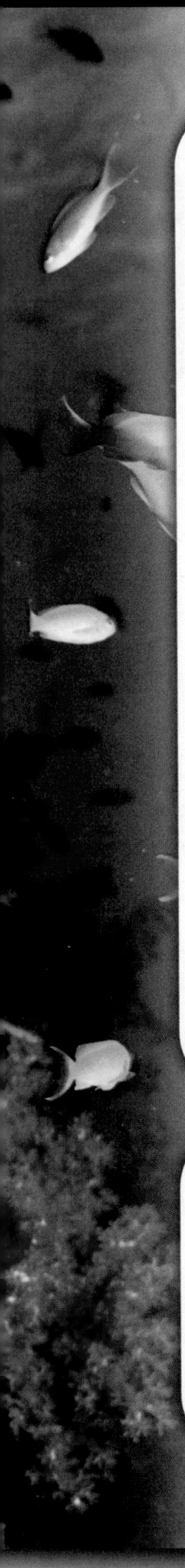

What It Means

Inside every living thing, chemical reactions constantly change materials into new materials. For these reactions to occur, an organism needs energy, water and other materials, and living space.

Energy

You use energy all the time. Movement, growth, and sleep all require energy, which you get from food. Plants use the energy of sunlight to make their own food. All animals get their energy by eating either plants or other animals that eat plants.

Water and Other Materials

Water is the main ingredient in the cells of all living things. The chemical reactions inside cells take place in water, and water plays a part in moving materials around within organisms.

Other materials are also essential for life. For example, plants must have carbon dioxide from the air to make their own food. Plants and animals both use oxygen to release the energy stored in their food. You and other animals that live on land get oxygen when you breathe in air. The fish swimming around the coral reef in the picture have gills, which allow them to get oxygen that is dissolved in the water.

Living Space

You can think of living space as a home—a space that protects you from external conditions and a place where you can get materials such as water and air. The ocean provides living space for the coral that makes up this coral reef. The coral itself provides living space for many other organisms.

Why It's Important

Understanding the needs of living things helps people make wise decisions about resources. This knowledge can also help you think carefully about

- the different ways in which various organisms meet their needs for energy and materials
- the effects of adding chemicals to the water and air around us
- the reasons why some types of plants or animals may disappear from an area

xvii

UNIFYING PRINCIPLE

Living things meet their needs through interactions with the environment.

A moose chomps on the leaves of a plant. This ordinary event involves many interactions among living and nonliving things within the forest.

What It Means

To understand this unifying principle, take a closer look at the words *environment* and *interactions.*

Environment

The **environment** is everything that surrounds a living thing. An environment is made up of both living and nonliving factors. For example, the environment in this forest includes rainfall, rocks, and soil as well as the moose, the evergreen trees, and the birch trees. In fact, the soil in these forests is called "moose and spruce" soil because it contains materials provided by the animals and evergreens in the area.

Interaction

All living things in an environment meet their needs through interactions. An **interaction** occurs when two or more things act in ways that affect one another. For example, trees and other forest plants can meet their need for energy and materials through interactions with materials in soil and air, and light from the Sun. New plants get living space as birds, wind, and other factors carry seeds from one location to another.

Animals like this moose meet their need for food through interactions with other living things. The moose gets food by eating leaves off trees and other plants. In turn, the moose becomes food for wolves.

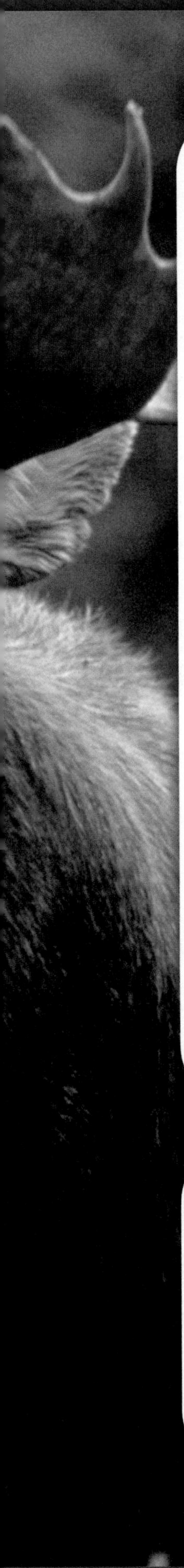

Why It's Important

Learning about living things and their environment helps scientists and decision makers address issues such as:

- developing land for human use without damaging the environment
- predicting how a change in the moose population would affect the soil in the forest
- determining the ways in which animals harm or benefit the trees in a forest

UNIFYING PRINCIPLE

The types and numbers of living things change over time.

The story of life on Earth is a story of changes. Some changes take place over millions of years. At one time, animals similar to modern fish swam in the area where this lizard now runs.

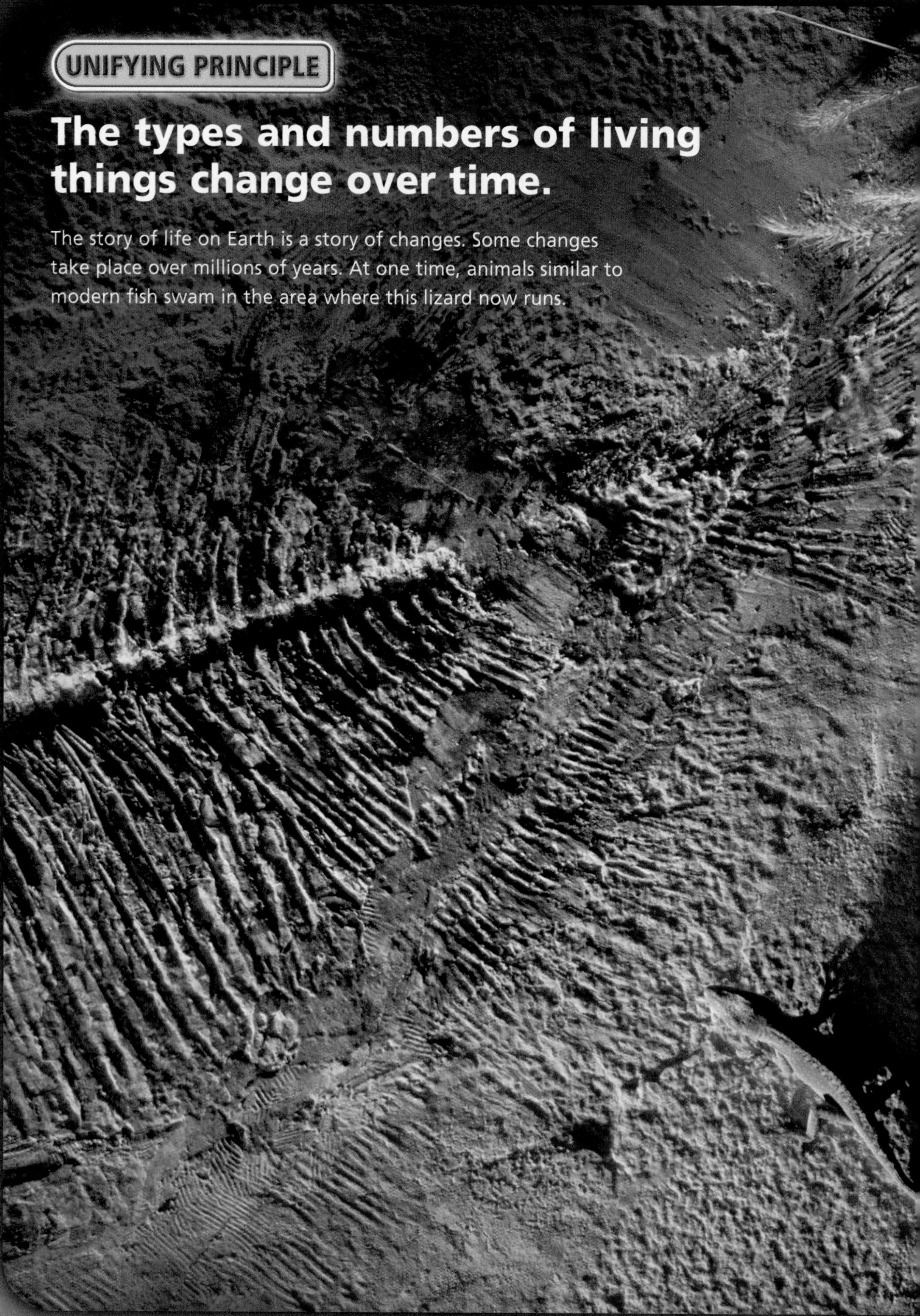

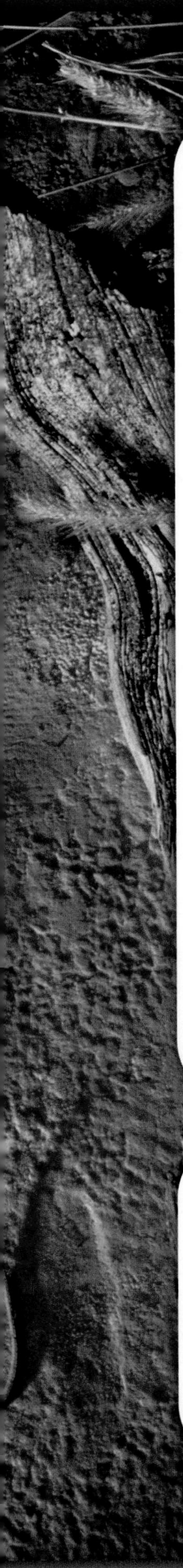

What It Means

To understand how living things change over time, let's look closely at the terms *diversity* and *adaptation.*

Diversity

You are surrounded by an astonishing variety of living things. This variety is called **biodiversity.** Today, scientists have described and named 1.4 million species. There are even more species that haven't been named. Scientists use the term *species* to describe a group of closely related living things. Members of a **species** are so similar that they can reproduce offspring that are able to reproduce. Lizards, such as the one you see in the photograph, are so diverse that they make up many different species.

Over the millions of years that life has existed on Earth, new species have originated and others have disappeared. The disappearance of a species is called **extinction.** Fossils, like the one in the photograph, provide evidence that some types of organisms that lived millions of years ago became extinct.

Adaptation

Scientists use the term **adaptation** to mean a characteristic of a species that allows members of that species to survive in a particular environment. Adaptations are related to needs. A salamander's legs are an adaptation that allows it to move on land.

Over time, species either develop adaptations to changing environments or they become extinct. The history of living things on Earth is related to the history of the changing Earth. The presence of a fish-like fossil indicates that the area shown in this photograph was once covered by water.

Why It's Important

By learning how living things change over time, you will gain a better understanding of the life that surrounds you and how it survives. Discovering more about the history of life helps scientists to

- identify patterns of relationships among various species
- predict how changes in the environment may affect species in the future

xxi

The Nature of Science

You may think of science as a body of knowledge or a collection of facts. More important, however, science is an active process that involves certain ways of looking at the world.

Scientific Habits of Mind

Scientists are curious. They are always asking questions. A scientist who observes that the number of plants in a forest preserve has decreased might ask questions such as, "Are more animals eating the plants?" or "Has the way the land is used affected the numbers of plants?" Scientists around the world investigate these and other important questions.

Scientists are observant. They are always looking closely at the world around them. A scientist who studies plants often sees details such as the height of a plant, its flowers, and how many plants live in a particular area.

Scientists are creative. They draw on what they know to form a possible explanation for a pattern, event, or behavior that they have observed. Then scientists create a plan for testing their ideas.

Scientists are skeptical. Scientists don't accept an explanation or answer unless it is based on evidence and logical reasoning. They continually question their own conclusions as well as conclusions suggested by other scientists. Scientists trust only evidence that is confirmed by other people or methods.

A white-tailed deer feeds on many plants including the trillium shown here.

By measuring the growth of this tree, a scientist can study interactions in the ecosystem.

Science Processes at Work

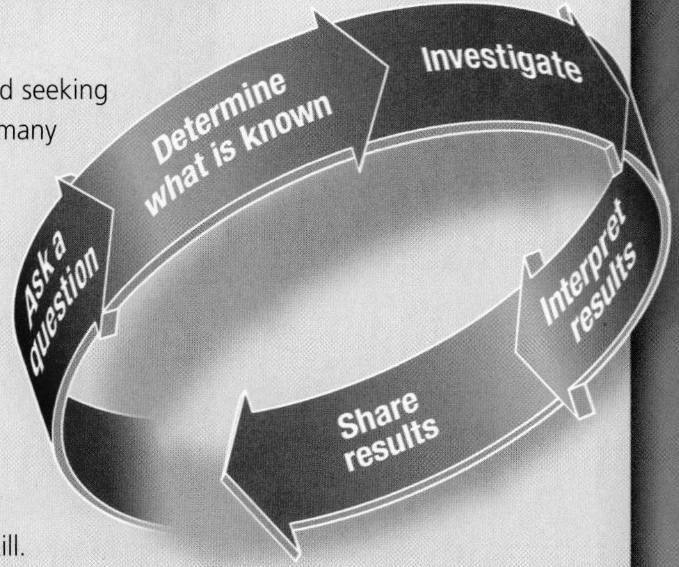

You can think of science as a continuous cycle of asking and seeking answers to questions about the world. Although there are many processes that scientists use, scientists typically do each of the following:

- Ask a question
- Determine what is known
- Investigate
- Interpret results
- Share results

Ask a Question

It may surprise you that asking questions is an important skill. A scientific investigation may start when a scientist asks a question. Perhaps scientists observe an event or process that they don't understand, or perhaps answering one question leads to another.

Determine What Is Known

When beginning an inquiry, scientists find out what is already known about a question. They study results from other scientific investigations, read journals, and talk with other scientists. A biologist who is trying to understand how the change in the number of deer in an area affects plants will study reports of censuses taken for both plants and animals.

Investigate

Investigating is the process of collecting evidence. Two important ways of collecting evidence are observing and experimenting.

Observing is the act of noting and recording an event, characteristic, behavior, or anything else detected with an instrument or with the senses. For example, a scientist notices that plants in one part of the forest are not thriving. She sees broken plants and compares the height of the plants in one area with those in another.

An **experiment** is an organized procedure during which all factors but the one being studied are controlled. For example, the scientist thinks the reason that some plants in the forest are not thriving may be because deer are eating the flowers off the plants. An experiment she might try is to mark two similar parts of an area where the plants grow and then to build a fence around one part so the deer can't get to the plants there. The fence must be constructed so the same amount of light, air, and water reach the plants. The only factor that changes is contact between plants and the deer.

Close observation of the Colorado Potatobeetle led scientists to answers that help farmers control this insect pest.

Forming hypotheses and making predictions are two other skills involved in scientific investigations. A **hypothesis** is a tentative explanation for an observation or a scientific problem that can be tested by further investigation. For example, since at least 1900, Colorado Potatobeetles were able to resist chemical insecticides. It was hypothesized that bacteria living in the beetles' environment were killing many beetles, because otherwise the beetles would be found in larger numbers. A **predicton** is an expectation of what will be observed or what will happen, and can be used to test a hypothesis. It was predicted that certain bacteria would kill Colorado Potatobeetles. This prediction was confirmed when bacteria called *Bt* was discovered to kill Colorado Potatobeetles and other insect pests.

Interpret Results

As scientists investigate, they analyze their evidence, or data, and begin to draw conclusions. **Analyzing data** involves looking at the evidence gathered through observations or experiments and trying to identify any patterns that might exist in the data. Often scientists need to make additional observations or perform more experiments before they are sure of their conclusions. Many times scientists make new predictions or revise their hypotheses.

Computers help scientists analyze the sequence of base pairs in the DNA molecule.

Share Results

An important part of scientific investigation is sharing results of experiments. Scientists read and publish in journals and attend conferences to communicate with other scientists around the world. Sharing data and procedures gives them a way to test one another's results. They also share results with the public through newspapers, television, and other media.

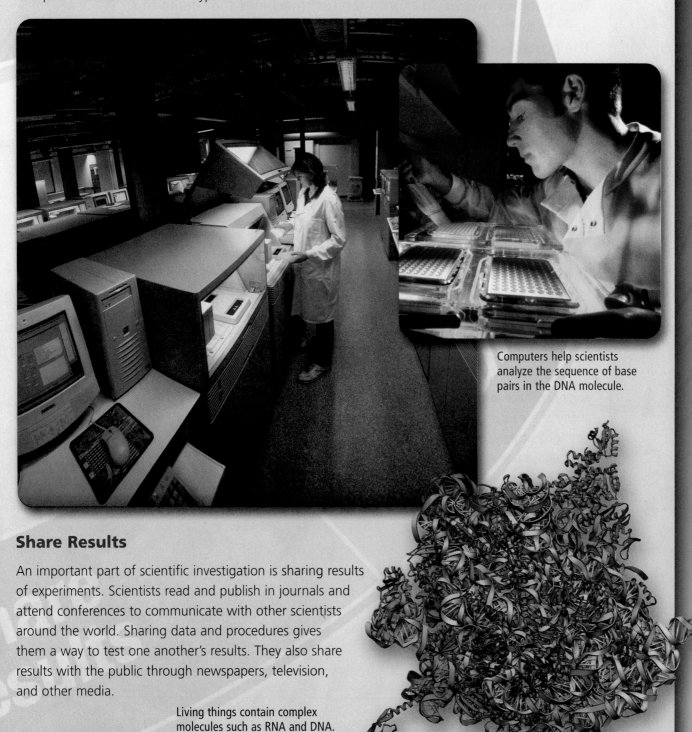

Living things contain complex molecules such as RNA and DNA. To study them scientists often use models like the one shown here.

The Nature of Technology

Imagine what life would be like without cars, computers, and cell phones. Imagine having no refrigerator or radio. It's difficult to think of a world without these items we call technology. Technology, however, is more than just machines that make our daily activities easier. Like science, technology is also a process. The process of technology uses scientific knowledge to design solutions to real-world problems.

Science and Technology

Science and technology go hand in hand. Each depends upon the other. Even designing a device as simple as a toaster requires knowledge of how heat flows and which materials are the best conductors of heat. Scientists also use a number of devices to help them collect data. Microscopes, telescopes, spectrographs, and computers are just a few of the tools that help scientists learn more about the world. The more information these tools provide, the more devices can be developed to aid scientific research and to improve modern lives.

The Process of Technological Design

Heart disease is among the leading causes of death today. Doctors have successfully replaced damaged hearts with hearts from donors. Medical engineers have developed pacemakers that improve the ability of a damaged heart to pump blood. But none of these solutions is perfect. Although it is very complex, the heart is really a pump for blood, thus, using technology to build a better replace ment pump should be possible. The process of technological design involves many choices. In the case of an artificial heart, choices about how and what to develop involve cost, safety, and patient preference. What kind of technology will result in the best quality of life for the patient?

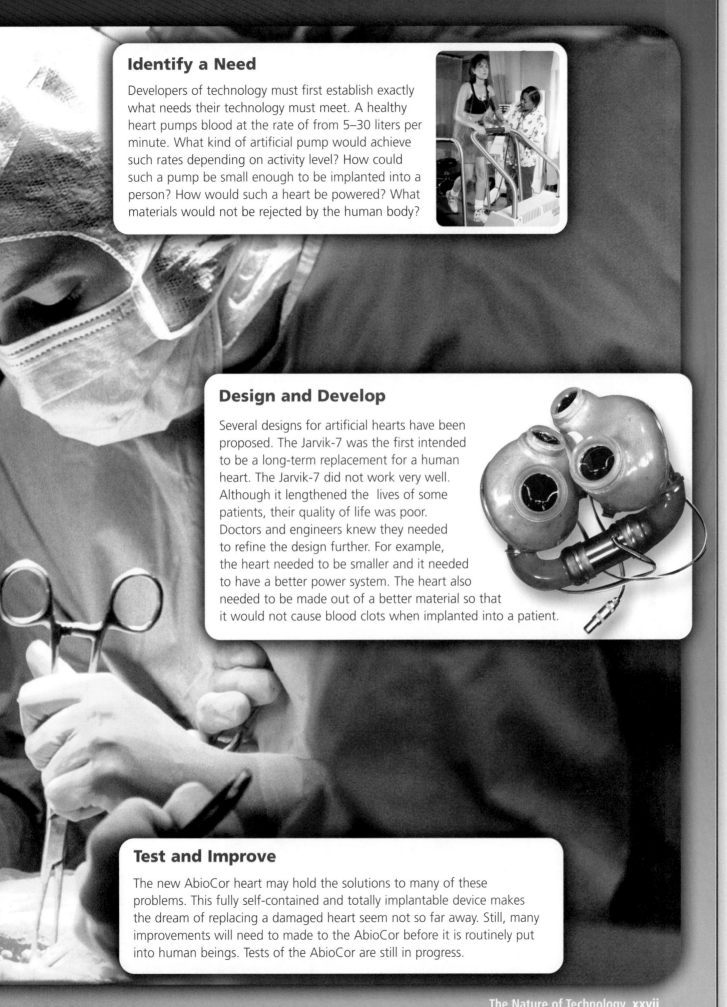

Identify a Need

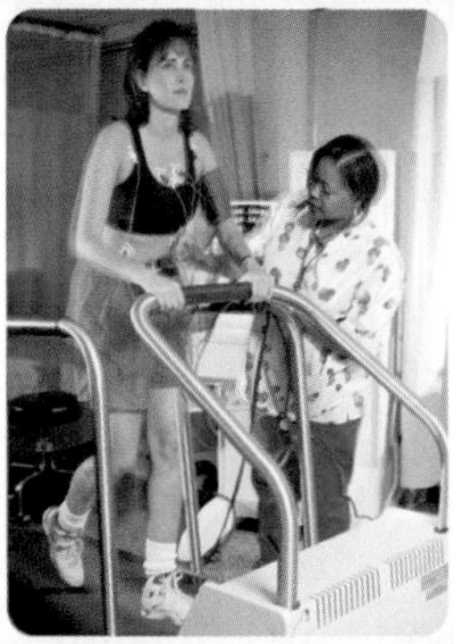

Developers of technology must first establish exactly what needs their technology must meet. A healthy heart pumps blood at the rate of from 5–30 liters per minute. What kind of artificial pump would achieve such rates depending on activity level? How could such a pump be small enough to be implanted into a person? How would such a heart be powered? What materials would not be rejected by the human body?

Design and Develop

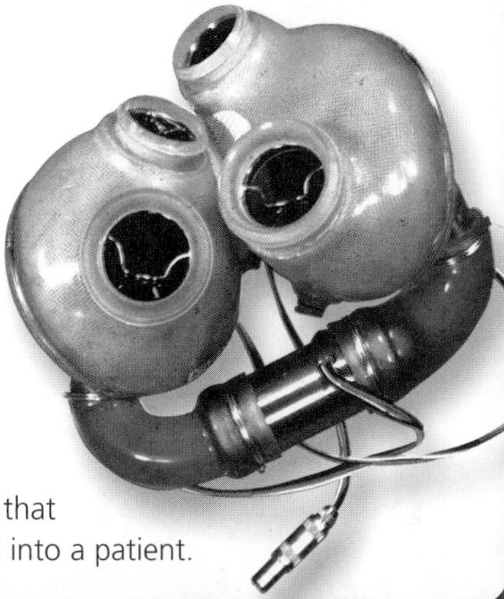

Several designs for artificial hearts have been proposed. The Jarvik-7 was the first intended to be a long-term replacement for a human heart. The Jarvik-7 did not work very well. Although it lengthened the lives of some patients, their quality of life was poor. Doctors and engineers knew they needed to refine the design further. For example, the heart needed to be smaller and it needed to have a better power system. The heart also needed to be made out of a better material so that it would not cause blood clots when implanted into a patient.

Test and Improve

The new AbioCor heart may hold the solutions to many of these problems. This fully self-contained and totally implantable device makes the dream of replacing a damaged heart seem not so far away. Still, many improvements will need to made to the AbioCor before it is routinely put into human beings. Tests of the AbioCor are still in progress.

xxvii

Using McDougal Littell Science

Reading Text and Visuals

This book is organized to help you learn. Use these boxed pointers as a path to help you learn and remember the **Big Ideas** and **Key Concepts**.

Take notes.

Use the strategies on the **Getting Ready to Learn** page.

Read the Big Idea.

As you read **Key Concepts** for the chapter, relate them to **the Big Idea**.

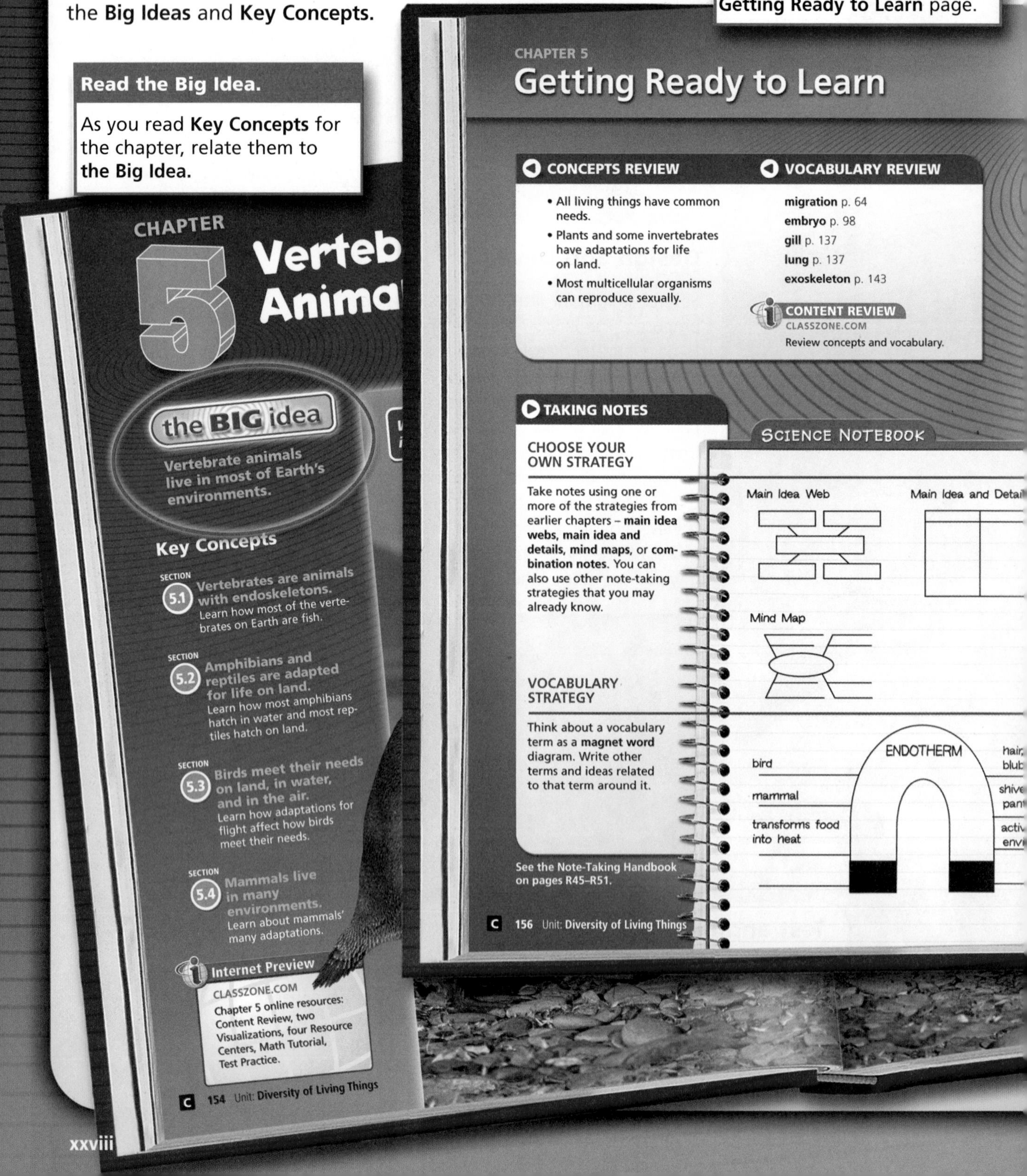

CHAPTER
5
Verteb
Anima
the BIG idea
Vertebrate animals live in most of Earth's environments.
Key Concepts
SECTION 5.1 Vertebrates are animals with endoskeletons. Learn how most of the vertebrates on Earth are fish.
SECTION 5.2 Amphibians and reptiles are adapted for life on land. Learn how most amphibians hatch in water and most reptiles hatch on land.
SECTION 5.3 Birds meet their needs on land, in water, and in the air. Learn how adaptations for flight affect how birds meet their needs.
SECTION 5.4 Mammals live in many environments. Learn about mammals' many adaptations.
Internet Preview
CLASSZONE.COM
Chapter 5 online resources: Content Review, two Visualizations, four Resource Centers, Math Tutorial, Test Practice.
C 154 Unit: Diversity of Living Things
xxviii
CHAPTER 5
Getting Ready to Learn
CONCEPTS REVIEW
• All living things have common needs.
• Plants and some invertebrates have adaptations for life on land.
• Most multicellular organisms can reproduce sexually.
VOCABULARY REVIEW
migration p. 64
embryo p. 98
gill p. 137
lung p. 137
exoskeleton p. 143
CONTENT REVIEW
CLASSZONE.COM
Review concepts and vocabulary.
TAKING NOTES
CHOOSE YOUR OWN STRATEGY
Take notes using one or more of the strategies from earlier chapters – main idea webs, main idea and details, mind maps, or combination notes. You can also use other note-taking strategies that you may already know.
VOCABULARY STRATEGY
Think about a vocabulary term as a magnet word diagram. Write other terms and ideas related to that term around it.
See the Note-Taking Handbook on pages R45–R51.
C 156 Unit: Diversity of Living Things
SCIENCE NOTEBOOK
Main Idea Web
Main Idea and Detail
Mind Map
ENDOTHERM
bird
mammal
transforms food into heat

Read each heading.

See how it fits in the outline of the chapter.

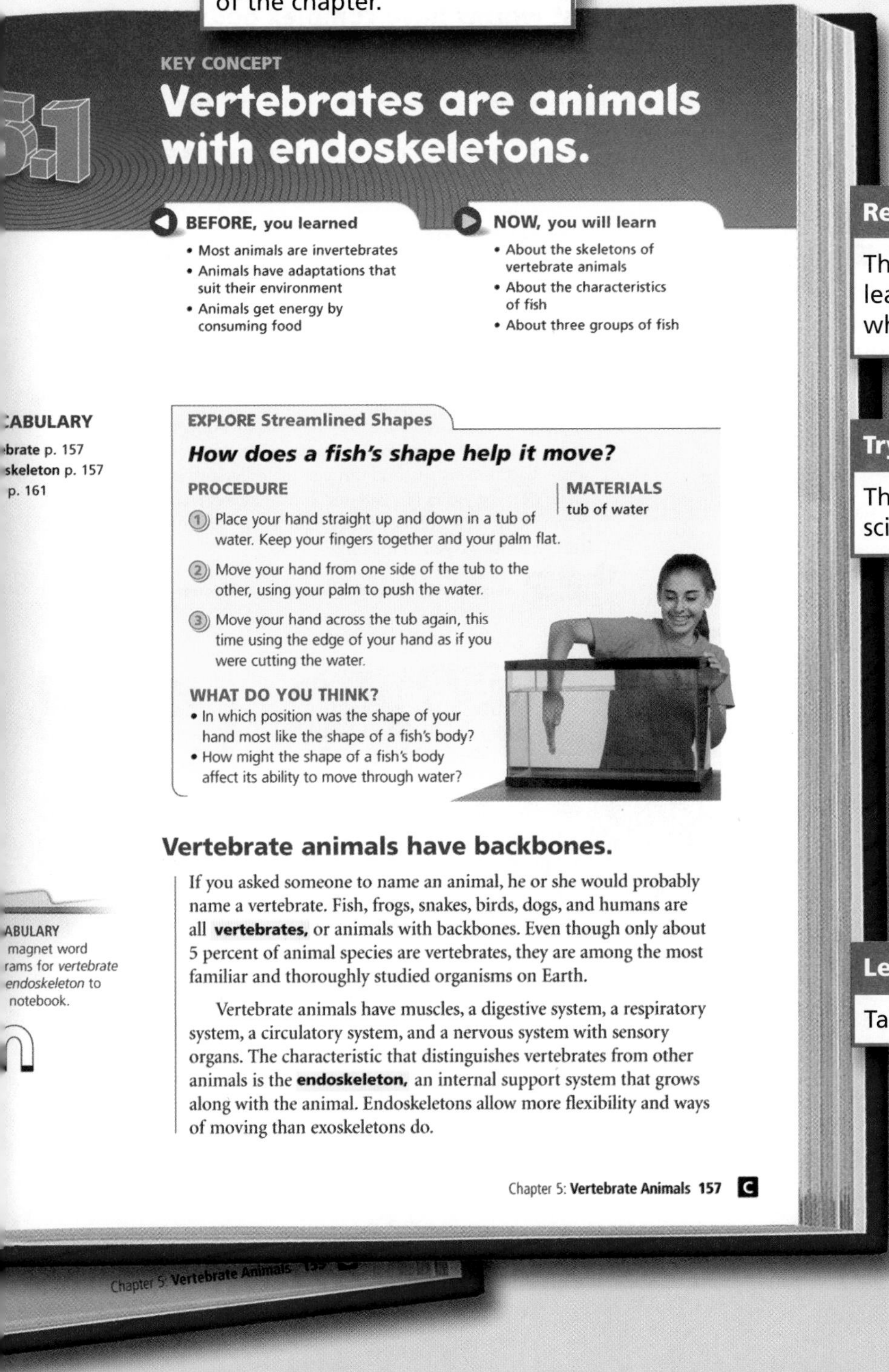

KEY CONCEPT

5.1 Vertebrates are animals with endoskeletons.

BEFORE, you learned
- Most animals are invertebrates
- Animals have adaptations that suit their environment
- Animals get energy by consuming food

NOW, you will learn
- About the skeletons of vertebrate animals
- About the characteristics of fish
- About three groups of fish

:ABULARY
·brate p. 157
skeleton p. 157
p. 161

EXPLORE Streamlined Shapes

How does a fish's shape help it move?

PROCEDURE

1. Place your hand straight up and down in a tub of water. Keep your fingers together and your palm flat.
2. Move your hand from one side of the tub to the other, using your palm to push the water.
3. Move your hand across the tub again, this time using the edge of your hand as if you were cutting the water.

MATERIALS
tub of water

WHAT DO YOU THINK?
- In which position was the shape of your hand most like the shape of a fish's body?
- How might the shape of a fish's body affect its ability to move through water?

Vertebrate animals have backbones.

If you asked someone to name an animal, he or she would probably name a vertebrate. Fish, frogs, snakes, birds, dogs, and humans are all **vertebrates,** or animals with backbones. Even though only about 5 percent of animal species are vertebrates, they are among the most familiar and thoroughly studied organisms on Earth.

Vertebrate animals have muscles, a digestive system, a respiratory system, a circulatory system, and a nervous system with sensory organs. The characteristic that distinguishes vertebrates from other animals is the **endoskeleton,** an internal support system that grows along with the animal. Endoskeletons allow more flexibility and ways of moving than exoskeletons do.

ABULARY
magnet word
rams for *vertebrate*
endoskeleton to
notebook.

Chapter 5: **Vertebrate Animals 157** C

Remember what you know.

Think about concepts you learned earlier and preview what you'll learn now.

Try the activities.

They will introduce you to science concepts.

Learn the vocabulary.

Take notes on each term.

Reading Text and Visuals

Most worms have complex body systems.

RESOURCE CENTER
CLASSZONE.COM
Learn more about the many types of worms.

Some worms have simple bodies. Others have well-developed body systems. Worms have a tube-shaped body, with bilateral symmetry. In many worms, food enters at one end and is processed as it moves through a digestive tract. Worms take in oxygen, dissolved in water, through their skin. Because of this, worms must live moist environments. Many live in water.

Segmented Worms

Segmented worms have bodies that are divided into individual compartments, or segments. These worms are referred to as annelids (AN-uh-lihdz), which means "ringed animals." One annelid you might be familiar with is the earthworm. As the diagram below shows, an earthworm's segments can be seen on the outside of its body.

An earthworm has organs that are organized into body systems. The digestive system of an earthworm includes organs for digestion and food storage. It connects to the excretory system, which removes waste. Earthworms pass soil through their digestive system. They digest decayed pieces of plant and animal matter from the soil and excrete what's left over. A worm's feeding and burrowing activity adds nutrients and oxygen to the soil.

CHECK YOUR READING Name two body systems found in earthworms.

Read one paragraph at a time.

Look for a topic sentence that explains the main idea of the paragraph. Figure out how the details relate to that idea. One paragraph might have several important ideas; you may have to reread to understand.

Answer the questions.

Check Your Reading questions will help you remember what you read.

Inside an Earthworm

An earthworm has organs and body systems.

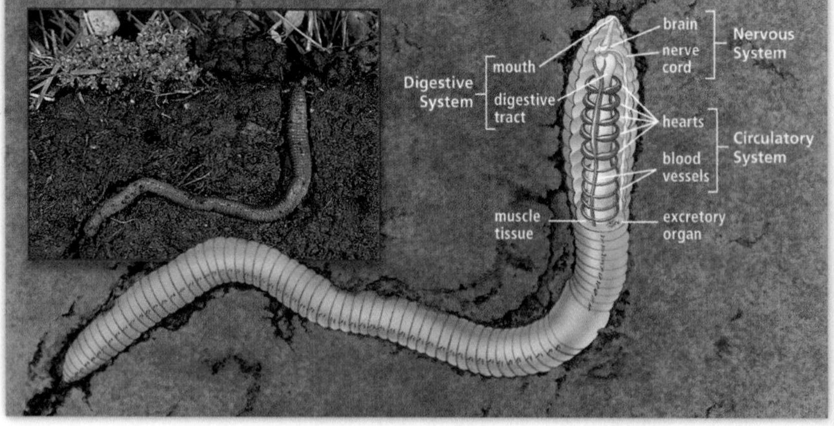

Study the visuals.

- Read the title.
- Read all labels and captions.
- Figure out what the picture is showing. Notice the information in the captions.

132 Unit: Diversity of Living Things

Doing Labs

To understand science, you have to see it in action. Doing labs helps you understand how things really work.

① **Read the entire lab first.**

② **Form a hypothesis.**

③ **Follow the procedure.**

④ **Record the data.**

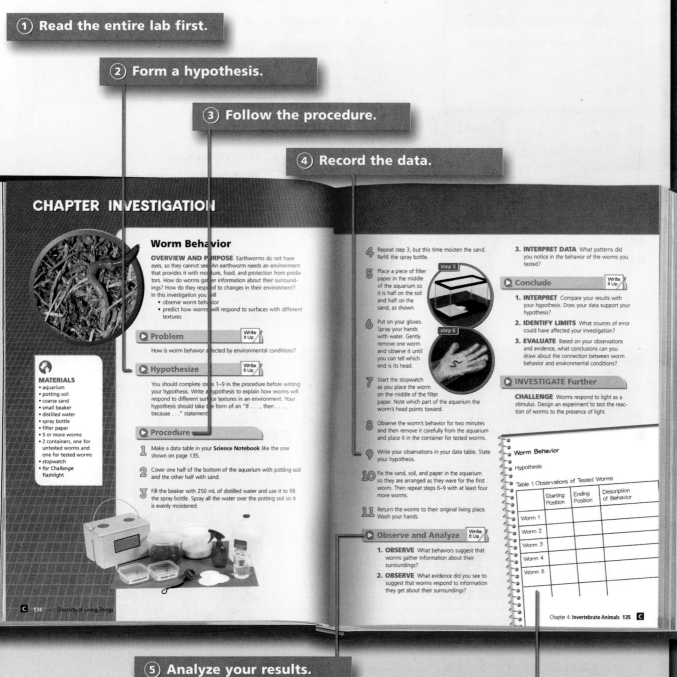

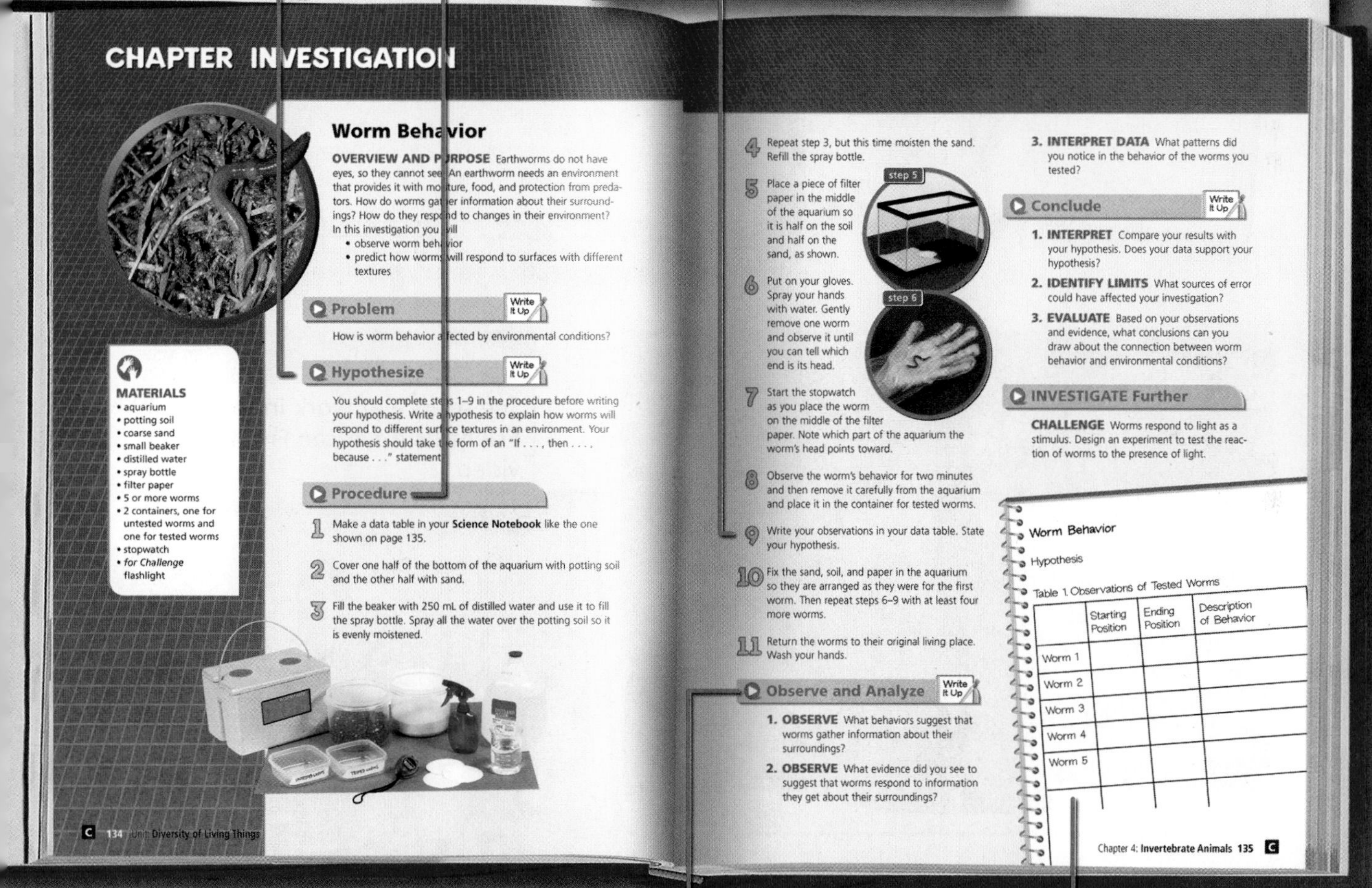

CHAPTER INVESTIGATION

Worm Behavior

OVERVIEW AND PURPOSE Earthworms do not have eyes, so they cannot see. An earthworm needs an environment that provides it with moisture, food, and protection from predators. How do worms gather information about their surroundings? How do they respond to changes in their environment? In this investigation you will
- observe worm behavior
- predict how worms will respond to surfaces with different textures

Problem Write It Up

How is worm behavior affected by environmental conditions?

Hypothesize Write It Up

You should complete steps 1–9 in the procedure before writing your hypothesis. Write a hypothesis to explain how worms will respond to different surface textures in an environment. Your hypothesis should take the form of an "If . . . , then . . . , because . . ." statement.

MATERIALS
- aquarium
- potting soil
- coarse sand
- small beaker
- distilled water
- spray bottle
- filter paper
- 5 or more worms
- 2 containers, one for untested worms and one for tested worms
- stopwatch
- *for Challenge* flashlight

Procedure

1. Make a data table in your **Science Notebook** like the one shown on page 135.
2. Cover one half of the bottom of the aquarium with potting soil and the other half with sand.
3. Fill the beaker with 250 mL of distilled water and use it to fill the spray bottle. Spray all the water over the potting soil so it is evenly moistened.
4. Repeat step 3, but this time moisten the sand. Refill the spray bottle.
5. Place a piece of filter paper in the middle of the aquarium so it is half on the soil and half on the sand, as shown.
6. Put on your gloves. Spray your hands with water. Gently remove one worm and observe it until you can tell which end is its head.
7. Start the stopwatch as you place the worm on the middle of the filter paper. Note which part of the aquarium the worm's head points toward.
8. Observe the worm's behavior for two minutes and then remove it carefully from the aquarium and place it in the container for tested worms.
9. Write your observations in your data table. State your hypothesis.
10. Fix the sand, soil, and paper in the aquarium so they are arranged as they were for the first worm. Then repeat steps 6–9 with at least four more worms.
11. Return the worms to their original living place. Wash your hands.

step 5

step 6

Observe and Analyze Write It Up

1. **OBSERVE** What behaviors suggest that worms gather information about their surroundings?
2. **OBSERVE** What evidence did you see to suggest that worms respond to information they get about their surroundings?
3. **INTERPRET DATA** What patterns did you notice in the behavior of the worms you tested?

Conclude Write It Up

1. **INTERPRET** Compare your results with your hypothesis. Does your data support your hypothesis?
2. **IDENTIFY LIMITS** What sources of error could have affected your investigation?
3. **EVALUATE** Based on your observations and evidence, what conclusions can you draw about the connection between worm behavior and environmental conditions?

INVESTIGATE Further

CHALLENGE Worms respond to light as a stimulus. Design an experiment to test the reaction of worms to the presence of light.

Worm Behavior

Hypothesis

Table 1. Observations of Tested Worms

	Starting Position	Ending Position	Description of Behavior
Worm 1			
Worm 2			
Worm 3			
Worm 4			
Worm 5			

C 134 Unit: Diversity of Living Things

Chapter 4: Invertebrate Animals 135 C

⑤ **Analyze your results.**

⑥ **Write your lab report.**

Using Technology

The Internet is a great source of information about up-to-date science. The ClassZone Web site and NSTA SciLinks have exciting sites for you to explore. Video clips and simulations can make science come alive.

Look for red banners.

Go to **ClassZone.com** to see simulations, visualizations, resource centers, and content review.

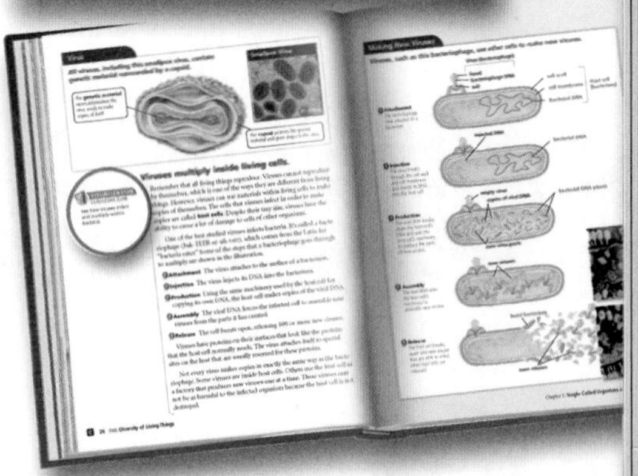

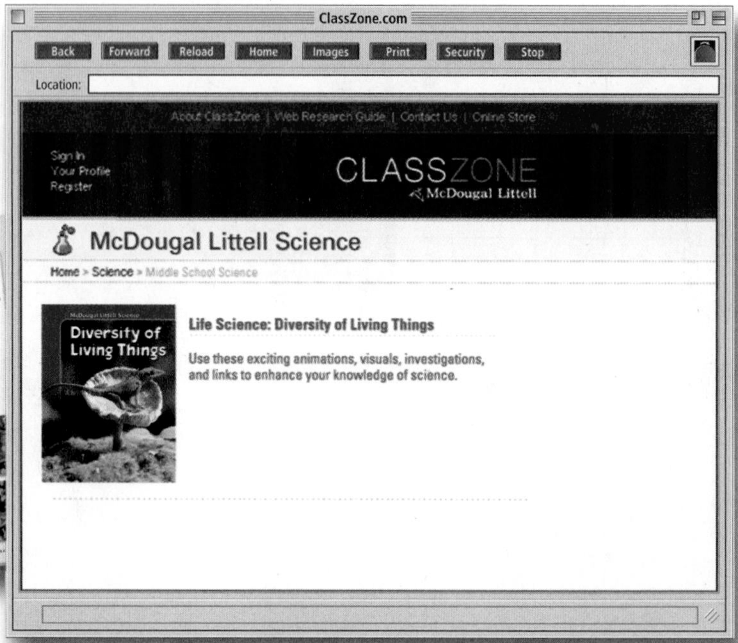

Watch the video.

See science at work in the **Scientific American Frontiers** video.

Look up SciLinks.

Go to **scilinks.org** to explore the topic.

NSTA scilinks.org SciLinks

Animal Behavior **Code: MDL040**

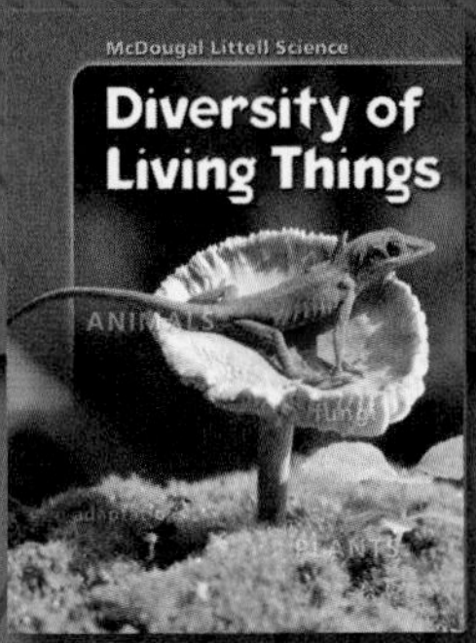

Diversity of Living Things
Contents Overview

Unit Features

VIDEO SUMMARY

SCIENTIFIC AMERICAN FRONTIERS

"Frozen Alive," a segment of the *Scientific American Frontiers* series that aired on PBS, explores animal species that can survive freezing temperatures. Scientists hope that understanding the adaptations of these animals might lead to new technologies. It might be possible to freeze donated human organs and tissues. If the organs could be frozen safely without destroying their functions, they could be preserved until needed.

The two animals featured in the video are the flounder and the wood frog. When the days shorten, a flounder's body begins to produce a protein that acts like antifreeze in the bloodstream. This keeps the fish's blood from freezing at low temperatures. Similarly, when temperatures drop, wood frogs dig into the ground, and fluids in their bodies freeze solid. Extra glucose in the frog's liver lowers the freezing point, preserves the organs, and later promotes simultaneous thawing throughout the body. By studying these natural responses to extreme conditions, scientists hope to save more human lives.

National Science Education Standards

A.1–8 Abilities Necessary to Do Scientific Inquiry

A.9.a–b, A.9.d–g Understandings About Scientific Inquiry

F.5.e Science and Technology in Society

G.1.a–b Science as a Human Endeavor

G.2.a, G.2.c Nature of Science

FRONTIERS in Science

Chilling Changes

How do organisms survive in this chilly Arctic landscape? Scientists are studying how organisms have adapted to the extreme cold of Earth's northern climates.

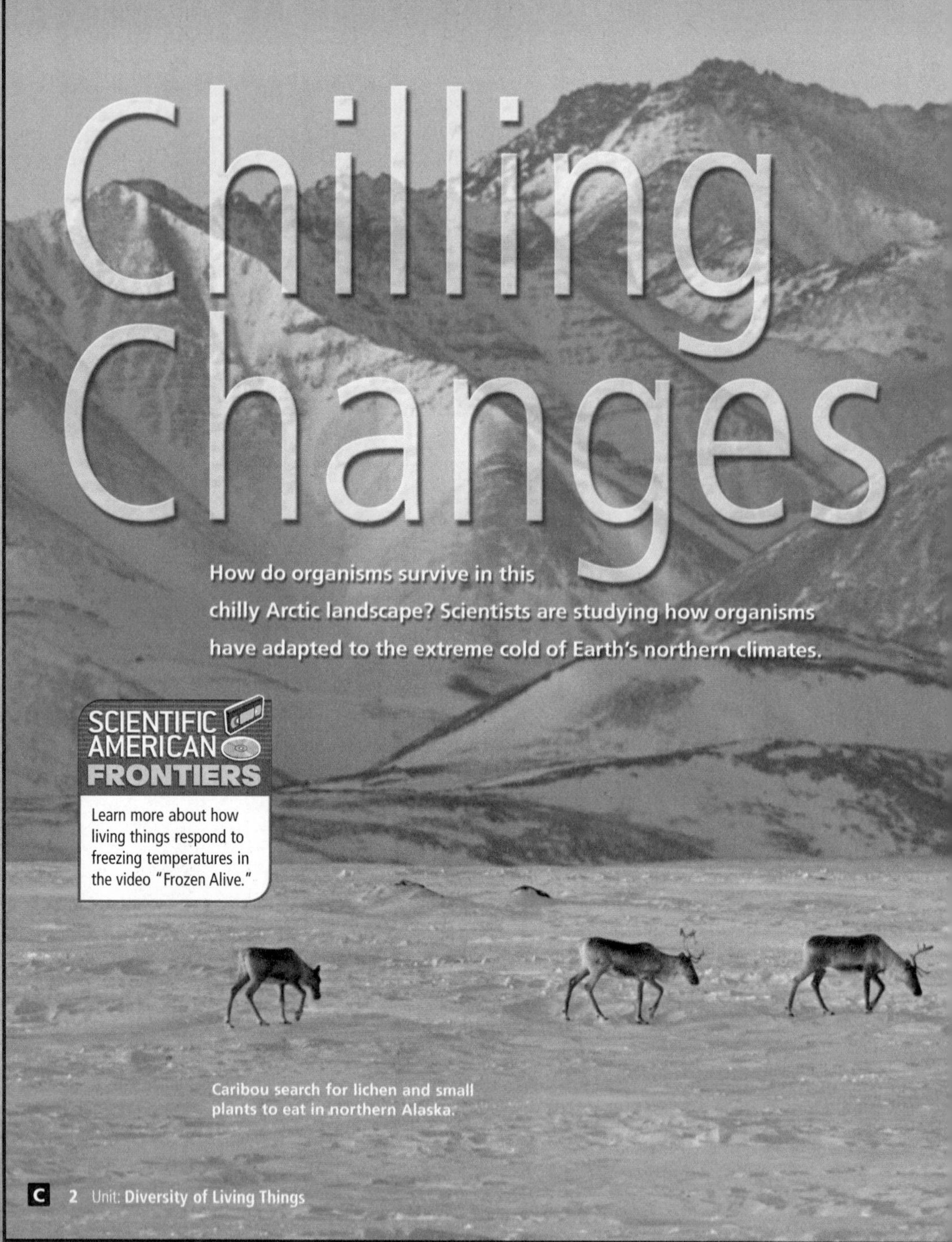

SCIENTIFIC AMERICAN FRONTIERS

Learn more about how living things respond to freezing temperatures in the video "Frozen Alive."

Caribou search for lichen and small plants to eat in northern Alaska.

C 2 Unit: Diversity of Living Things

ADDITIONAL RESOURCES

Technology Resources

Scientific American Frontiers Video: *Frozen Alive:* 11-minute video segment that introduces the unit.

ClassZone.com
CAREER LINK, Zoologist

Guide student viewing and comprehension of the video:

R Video Teaching Guide, pp. 1–2; Video Viewing Guide, p. 3; Video Wrap-Up, p. 4

Scientific American Frontiers Video Guide, pp. 7–10

Unit projects procedures and rubrics:

R Unit Projects, pp. 5–10

Lichen find materials and living space on rocks in frozen environments.

When Life Chills Out

When faced with the approach of winter, some animals hurry up, and some slow down. You may have seen animals getting a move on, migrating to warmer places. Other animals stay alive at home by hibernating, or entering a sleep-like state. Hibernation conserves energy until spring by slowing down many body processes. But what happens when an organism's body temperature dips below freezing? (?) A

Ice formation inside an organism's body is one of the most serious, and potentially damaging, threats to winter survival. If ice forms inside cells, it can damage the cell's operation and tear holes in the cell membrane, almost certainly leading to death. (?) B

An arctic ground squirrel peers above snow-covered ground in Alaska.

Some animals and plants take cold weather survival to an extreme. Some frogs and insects can allow their bodies to freeze solid. Some fish have fluids in their bodies that can *cool* below the freezing point of water and still flow. How do they do it?

Scientists studying survivors of extreme cold are not just learning about the ways organisms respond in the wild. Research on cold weather survival helps scientists understand problems faced by human society. Growing crops year-round and using bacteria to clean up pollution in a frozen environment are two examples. There are two main responses that allow living things to live in a subfreezing world.

DIFFERENTIATE INSTRUCTION

(?) More Reading Support

A What do you call passing the winter in a sleeplike state? *hibernation*

B What happens if ice forms inside a cell? *The cell dies.*

Advanced Lichens are composite organisms of fungi in symbiosis with algae or cyanobacteria or both. Have students research lichens, focusing on how the symbiosis enables lichens to grow in extreme climates. Students could present their findings on a poster.

FOCUS

Set Learning Goals

Students will

- Analyze the harmful effects of very cold temperatures.
- Examine ways that animals and plants survive in extreme cold.
- Explore how various organisms respond to changes in their environment.

Point out that extremely cold environments are geographic frontiers as well as scientific ones. In order to study them, scientists have to first develop equipment and techniques for survival. Ask: Where are the extremely cold climates on Earth? *the Arctic, Antarctica, winter in temperate forests, ocean depths, mountaintops*

INSTRUCT

Scientific Process

Emphasize that, to determine how organisms survive in extreme cold, scientists must make very close, thorough observations. Also, observations of wildlife must sometimes take place for an extended time, requiring patience and accurate monitoring and recording of detail.

Technology Design

Learning how an animal's body changes in response to heat requires technologies that allow scientists to observe body processes without harming the animal or changing the course of the process that is being observed. In response to this challenge, scientists make use of noninvasive medical imaging tools such as CAT scans and MRIs.

Scientific Process

When Ken Storey began studying wood frogs, he theorized that they thaw in the way objects do, warming from the outside in. Once he was able to scan frogs that were thawing, however, he had to reject that hypothesis. His observation showed that the internal organs, such as the heart and liver, are the first to thaw.

Integrate the Sciences

For a long time, scientists thought that life could survive only on Earth. Now, because researchers have found that microorganisms can live under extreme conditions, scientists have revised their knowledge about life and the conditions that can support it. In early 2004 *Spirit*, a NASA rover, looked for signs of water on Mars. Water is one need of all living things.

Determining What Is Known

Ask: How could scientists use what they know about human reactions to cold to learn about other types of organisms? *Knowing how cold affects humans gives scientists a basis for comparing those responses to the responses of animals and other living things.*

Arctic woolly caterpillars may take many years to grow to full size.

When the caterpillar reaches full size, it forms a cocoon and changes into a moth.

Avoid Ice, or Tolerate It

Organisms can either avoid the forming of ice, or tolerate it. Some animals, like a fish called a flounder, produce antifreeze proteins in their bodies. The proteins keep ice from forming in the blood and other fluids. The fluids thicken, but still flow. They never turn to ice crystals. Scientists call this response "freeze-avoidance."

Others, like some frogs, many types of insects, and some trees do allow ice to form inside their bodies. In these organisms, ice forms, but only in the spaces outside the cells. "Freeze-tolerant" organisms pump fluid out of their cells. Specialized proteins located outside of the cells encourage ice to form there, rather than inside the cells.

Perhaps the champion of freeze tolerance is a small Arctic insect called the Arctic woolly caterpillar. It is only active for a few weeks out of the year, during the brief Arctic summer. It can take this caterpillar up to fifteen years to mature into a moth!

Temperatures during Arctic winters often fall as low as −70°C, and, because the ground is frozen year-round, the caterpillars cannot burrow. Instead, their bodies freeze solid in winter, shutting down nearly all of their cellular activity and

SCIENTIFIC AMERICAN FRONTIERS

View the "Frozen Alive" segment of your Scientific American Frontiers video to see how scientists studying fish and frogs have begun to solve the mysteries of life surviving a freeze.

IN THIS SCENE FROM THE VIDEO How do the frog's systems start up again, after being shut down during freezing. This scan shows activity in the liver.

WHAT BRINGS ON THE THAW? Ken Storey collects wood frogs. The frogs have an automatic response to cold temperatures, which allows Ken and his team to keep them in a freezer without killing them. Storey and his team know the "what" of freeze tolerance. They want to learn the "how." How does the frog stay alive during and after shutdown? Storey expects that the thawing process works in the usual way, that the frogs will begin to warm from the outside in. Instead he finds that internal organs begin the thaw. The liver is the first organ to activate. It produces glucose or sugar, which then gets other cells, those in the heart, to begin their work.

DIFFERENTIATE INSTRUCTION

More Reading Support

C What is one animal that avoids freezing? *Sample answer: flounder*

D What is one animal that tolerates freezing? *Sample answer: Arctic woolly caterpillar*

Below Level Have students construct a chart to organize information about cold-weather adaptations. The chart might include different animals and plants, the conditions in which they live, and the behavioral and physical adaptations that suit them to a cold environment. Point out that what seems like a "difficult environment" to us may be the best possible environment for an organism such as the Arctic wooly caterpillar. Discuss why this so.

expending almost no energy. Scientists have known for centuries that some insects can freeze over the winter, but only recently have they begun to understand exactly how insects' bodies change to allow them to freeze and thaw.

Scientist Ken Storey has discovered a set of genes that turn on, like a switch, in response to freezing temperatures. These genes are called master control genes, because they have the ability to turn other genes in the body on or off. In the case of freeze-tolerant animals, master control genes first turn off nearly all the genes in the body. Arctic caterpillars have many thousands of genes, but during the winter most of those genes are turned off. Keeping genes turned off is an adaptation that helps the caterpillar conserve energy throughout the winter.

Is It Really Extreme?

Organisms have adapted to a wide variety of environments. The Arctic woolly caterpillar is suited to life near the Arctic Circle. Some organisms must be able to survive in the exact opposite—the extreme heat of deserts. Others thrive in extremely salty conditions. Some plants and bacteria can even live in places with dangerously high levels of poisonous chemicals.

An environment doesn't have to be extreme to present challenges to the organisms that live there. Animals in a forest or meadow still need to find food and shelter, avoid predators, and respond to changes. The organisms living in any environment face a unique set of challenges.

UNANSWERED Questions

As scientists learn more about how animals survive in extremely cold climates, they also uncover additional questions:

- Do animals respond to temperature or a lack of light?
- How do cellular processes that halt during the winter get started up again?
- Can scientists grow freeze resistant crops?

UNIT PROJECTS

As you study this unit, work alone or in a group on one of the projects listed below. Use the bulleted steps to guide your project.

Museum Exhibit

Plan a museum exhibit showing different ways in which animals respond to extreme environments.

- Research three to five types of organisms and find out what kind of environmental changes these organisms face.
- Design, make visuals, and write text to accompany your exhibit.

Grow a Fast Plant

Observe the entire life cycle of a plant in less than six weeks. Record your observations in a journal.

- Gather information and any materials necessary to grow and care for fast plants. Obtain fast plant seeds and plant them.
- Observe the plant for a few minutes every day. Make notes in your journal, and write weekly summaries.

Local Field Study

Report on organisms that live in your local area.

Identify three organisms you see on a regular basis and learn more about how they survive throughout the year.

Pick one plant or animal and observe it for a few minutes once a week in the morning and in the late afternoon. How does it change over time?

CAREER CENTER
CLASSZONE.COM
Learn more about careers in biology.

DIFFERENTIATE INSTRUCTION

More Reading Support

E What types of environments present challenges for living things? *all environments*

Differentiate Unit Projects Projects are appropriate for varying abilities. Allow students to choose the ones that interest them most and let them vary their product. Encourage below level students to give visual or oral presentations or to record audio presentations about their topic.

Below Level Encourage students to try "Grow a Fast Plant."

Advanced Challenge students to complete "Local Field Study."

UNANSWERED Questions

Have students read the questions and think of some of their own. Remind them that scientists always end up with more questions—that inquiry is the driving force of science.

- With the class, generate on the board a list of new questions.
- Students can add to the list after they watch the Scientific American Frontiers Video.
- Students can use the list as a springboard for choosing their Unit Projects.

UNIT PROJECTS

Encourage students to pick the project that most appeals to them. Point out that each is long-term and will take several weeks to complete. You might group or pair students to work on projects and in some cases guide student choice. Some of the projects have student choice built into them. Each project has two worksheet pages, including a rubric. Use the pages to guide students through criteria, process, and schedule.

R Unit Projects, pp. 5–10

REVISIT concepts introduced in this article:

Chapter 1
- Single-celled organisms have all the characteristics of living things, pp. 9–15
- Bacteria and Archaea are found in many environments, pp. 18–19

Chapter 2
- Multicellular organisms meet their needs in different ways, pp. 43–49
- Animals interact with the environment; animals respond to seasonal changes, pp. 62–64

Chapter 4
- Insects undergo metamorphosis, p. 146

Chapter 5
- Mammals are endotherms, pp. 183–184

CHAPTER

1 Single-Celled Organisms and Viruses

Life Science

UNIFYING PRINCIPLES

PRINCIPLE 1
All living things share common characteristics.

PRINCIPLE 2
All living things share common needs.

PRINCIPLE 3
Living things meet their needs through interactions with the environment.

PRINCIPLE 4
The types and numbers of living things change over time.

Unit: Diversity of Living Things

BIG IDEAS

CHAPTER 1
Single-Celled Organisms and Viruses
Bacteria and protists have the characteristics of living things, while viruses are not alive.

CHAPTER 2
Introduction to Multicellular Organisms
Multicellular organisms live in and get energy from a variety of environments.

CHAPTER 3
Plants
Plants are a diverse group of organisms that live in many land environments.

CHAPTER 4
Invertebrate Animals
Invertebrate animals have a variety of body plans and adaptations.

CHAPTER 5
Vertebrate Animals
Vertebrate animals live in most of Earth's environments.

CHAPTER 1 KEY CONCEPTS

SECTION 1.1

Single-celled organisms have all the characteristics of living things.

1. Living things come in many shapes and sizes.
2. Living things share common characteristics.
3. Living things need energy, materials, and living space.
4. Viruses are not alive.

SECTION 1.2

Bacteria are single-celled organisms without nuclei.

1. Bacteria and archaea are the smallest living things.
2. Archaea and bacteria are found in many environments.
3. Bacteria may help or harm other organisms.

SECTION 1.3

Viruses are not alive but affect living things.

1. Viruses share some characteristics with living things.
2. Viruses multiply inside living cells.
3. Viruses may harm host cells.

SECTION 1.4

Protists are a diverse group of organisms.

1. Most protists are single-celled.
2. Protists obtain their energy in three ways.

The Big Idea Flow Chart is available on p. T1 in the **UNIT TRANSPARENCY BOOK.**

Previewing Content

SECTION

1.1 Single-celled organisms have all the characteristics of living things.
pp. 9–15

1. Living things come in many shapes and sizes.
Living things range in size from the enormous honey mushroom fungus that is more than 5 kilometers wide to microorganisms that can be seen only with a microscope. This diversity of life is divided into six **kingdoms.**
- Archaea
- Bacteria
- Protists
- Animals
- Fungi
- Plants

2. Living things share common characteristics.
Certain characteristics are shared by all living things.
- Living things are organized into cells. Organisms that have only one cell have all the structures they need to survive.
- All living things grow. They consume food, build structures, and repair or replace worn-out structures.
- All living things reproduce, making more organisms like themselves. Single-celled organisms reproduce by **binary fission.**
- All living things respond to changes in their environment.

3. Living things need energy, materials, and living space.
Food supplies an organism with the energy it needs to move, grow, and develop. Organisms also need water and other materials.
- Water is an ingredient in many cellular reactions, it provides structure, and it makes the watery environment that cells need.
- Most organisms require oxygen from the air.

4. Viruses are not alive.
A **virus** contains genetic material enclosed in a protein shell. However, viruses are not as complex as living cells. For example, viruses do not grow or respond to the environment.

SECTION

1.2 Bacteria are single-celled organisms without nuclei. pp. 16–23

1. Bacteria and archaea are the smallest living things.
Bacteria are the simplest kind of life.
- A bacterial cell contains all the structures needed for survival. Its cell wall, cell membrane, DNA, and cytoplasm can all be seen below.

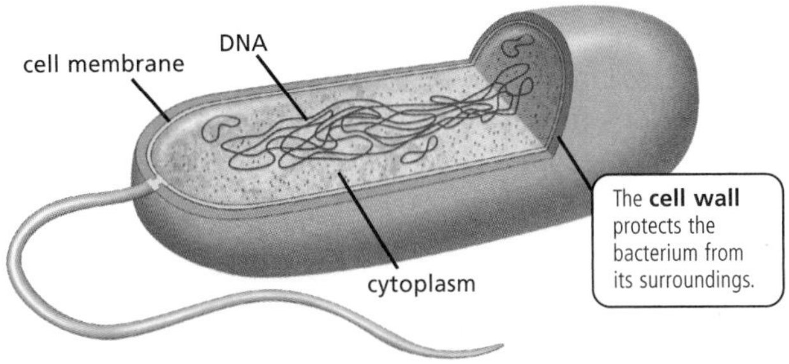

DNA
cell membrane
cytoplasm
The cell wall protects the bacterium from its surroundings.

- Bacteria are often classified by their shape. They may be spirals, rods, or spheres.

2. Archaea and bacteria are found in many environments.
Archaea and bacteria are both single-celled with no nuclei, but are sufficiently different chemically to be in separate kingdoms. Archaea can survive in a number of extreme environments.
- Methanogens produce methane and die if exposed to O_2.
- Halophiles must live in very salty water.
- Thermophiles live in either extreme heat or cold.

Bacteria are found almost everywhere.
- Some, called **producers,** use sunlight as an energy source.
- Others, **decomposers,** get energy from dead organic matter.
- **Parasites** get energy by living on or inside living organisms.

3. Bacteria may help or harm other organisms.
Helpful bacteria break down organic matter in soil and sewage. They also fix nitrogen, making it available to plants. They often are an essential part of animal digestive systems. Harmful bacteria cause disease in plants and animals by invading body parts or by poisoning the host organism with chemicals they release or with their cells.

Common Misconceptions

MISCONCEPTION DATABASE
CLASSZONE.COM Background on student misconceptions

ORGANISMS Students commonly think that all organisms are either plants or animals. In fact, there are four other kingdoms of living things.

TE This misconception is addressed on p. 11.

BACTERIA Many students think that bacteria are not living things. They are, in fact, alive, and have all the characteristics of living things.

TE This misconception is addressed on p. 17.

HELPFUL BACTERIA Students may hold the misconception that all bacteria cause disease and are bad. In reality, many bacteria are helpful. Some help produce food, some produce oxygen, some decompose waste material, and some fix nitrogen.

TE This misconception is addressed on p. 20.

Previewing Content

SECTION

1.3 Viruses are not alive but affect living things. pp. 24–29

1. **Viruses share some characteristics with living things.**
   Viruses are tiny infectious particles that consist of genetic material surrounded by a protein coat called a **capsid.** Viruses are not alive and do not grow or respond to the environment.

2. **Viruses multiply inside living cells.**
   Although viruses are not alive, they can **replicate** by taking over the cell machinery of a **host cell** and forcing it to make more viruses. In the diagram below, a bacteriophage, a virus that infects bacteria, attaches to a bacterium.

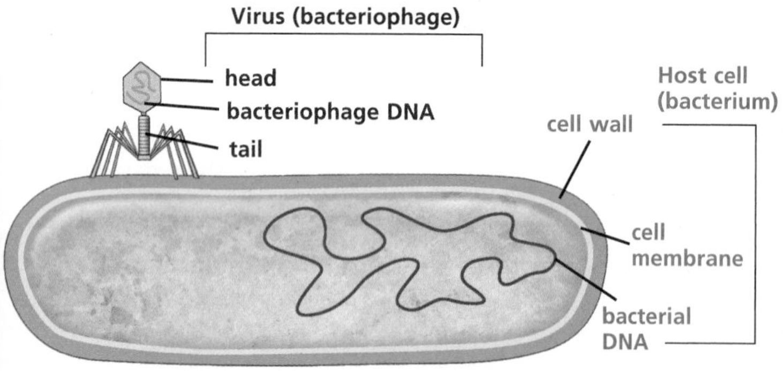

   The viral DNA is injected into the cell, where it uses the host machinery to make more viral DNA and capsids. The viral parts are assembled into new viruses and released when the host cell bursts open.

3. **Viruses may harm host cells.**
   Viruses use a host cell's material, energy, and cell processes to make more viruses. This reproductive process usually kills host cells and thus causes disease that may kill the organism. Scientists are currently trying to find ways to use viruses in positive ways, such as in gene therapy.

SECTION

1.4 Protists are a diverse group of organisms. pp. 30–35

1. **Most protists are single-celled.**
   Protists include all organisms whose cells have nuclei and that are not plants, animals, or fungi. Most protists are single-celled. Some protists are producers and provide oxygen, some are parasites that cause disease, and some are decomposers.

2. **Protists obtain their energy in three ways.**
   Protists can be classified by the way they get energy.
   - **Algae** capture sunlight and convert it to chemical energy. They have chlorophyll and photosynthesize, releasing oxygen gas into the air. Thus, algae are producers. Seaweed, euglenas, diatoms, volvox, and the *Chlamydomonas* pictured below are algae.

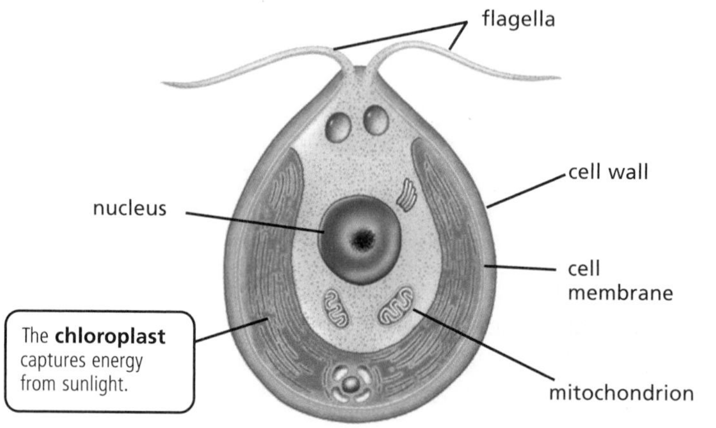

   - **Protozoa** get their energy by eating other organisms. They are all single-celled, and most move about with cilia or flagella to search for food. Some protozoa are parasites and cause disease. Paramecia and the organism that causes malaria are protozoa.
   - Other protists absorb materials that contain stored energy. Cellular slime molds, plasmodial slime molds, and water molds are members of this group.

Common Misconceptions

MISCONCEPTION DATABASE
CLASSZONE.COM Background on student misconceptions

CATCHING COLDS Many people think that you catch a cold by becoming chilled, going out without a coat, or going out in the rain without boots. The common cold is caused by viruses, which must infect body cells to produce the symptoms of a cold.

TE This misconception is addressed on p. 26.

Previewing Labs

Lab Generator CD-ROM
Edit these Pupil Edition labs and generate alternative labs.

EXPLORE the BIG idea

Where Can You Find Microscopic Life? p. 7 Students look for microscopic life, using a hand lens or microscope.	**TIME** 20 minutes **MATERIALS** hand lens or microscope
How Quickly Do Bacteria Multiply? p. 7 Students use beans to model binary fission and exponential growth.	**TIME** 10 minutes **MATERIALS** funnel, 2-L bottle, duct tape, beans, clock or stopwatch
Internet Activity: Microscopic Life and You, p. 7 Students investigate the microscopic life that lives on, in, and around people.	**TIME** 20 minutes **MATERIALS** computer with Internet access

SECTION 1.1

EXPLORE Organisms, p. 9 Students list living things found in a classroom.	**TIME** 10 minutes **MATERIALS** paper, pencil
INVESTIGATE Microorganisms, p. 13 Students observe the responses of hydra and daphnia to their environment.	**TIME** 30 minutes **MATERIALS** microscope, slide, hydra culture, cover slip, daphnia culture, warm water

SECTION 1.2

CHAPTER INVESTIGATION **Bacteria,** pp. 22–23 Students sample bacteria that live on their hands and in the environment to determine how widespread bacteria are.	**TIME** 40 minutes **MATERIALS** 3 covered petri dishes with sterile nutrient agar, marker, tape, 3 sterile cotton swabs, small object (such as a coin or eraser), hand lens (for Challenge: 1 additional covered petri dish with agar, sterile swabs)

SECTION 1.3

EXPLORE Viruses, p. 24 Students model the way viruses were discovered, by passing a mixture through a filter.	**TIME** 10 minutes **MATERIALS** small kitchen sieve, paper plate, table salt, sesame seeds, small container, hand lens
INVESTIGATE Viruses, p. 25 Students use a model to analyze how infections spread.	**TIME** 30 minutes **MATERIALS** class set of paper cups, water, baking soda, phenolphthalein (solution A), and 0.1 M dilute hydrochloric acid (solution B), 2 droppers, measuring spoons

SECTION 1.4

INVESTIGATE Protists, p. 31 Students observe the movement and behavior of protists in pond water.	**TIME** 40 minutes **MATERIALS** pond water, microscope, hand lens, dropper, slide, cover slip

R **Additional INVESTIGATION,** Growth of Algae, A, B, & C, pp. 70–78; Teacher Instructions, pp. 360–361

Previewing Chapter Resources

CHAPTER 1
Single-Celled Organisms and Viruses

INTEGRATED TECHNOLOGY

CLASSZONE.COM
- eEdition Plus
- EasyPlanner Plus
- Misconception Database
- Content Review
- Test Practice
- Visualizations
- Resource Centers
- Internet Activity: Microscopic Life and You
- Math Tutorial

SCILINKS.ORG

SCILINKS

CD-ROMS
- eEdition
- EasyPlanner
- Power Presentations
- Content Review
- Lab Generator
- Test Generator

AUDIO CDS
- Audio Readings
- Audio Readings in Spanish

LABS AND ACTIVITIES

EXPLORE the Big Idea, p. 7
- Where Can You Find Microscopic Life?
- How Quickly Do Bacteria Multiply?
- Internet Activity: Microscopic Life and You

UNIT RESOURCE BOOK
- Family Letter, p. ix
- Spanish Family Letter, p. x
- Unit Projects, pp. 5–10

Lab Generator CD-ROM
Generate customized labs.

SECTION

1.1 Single-celled organisms have all the characteristics of living things.
pp. 9–15

Time: 2 periods (1 block)

Lesson Plan, pp. 11–12

Integrated Technology
- **VISUALIZATION,** Binary Fission
- **MATH TUTORIAL**

UNIT TRANSPARENCY BOOK
- Big Idea Flow Chart, p. T1
- Daily Vocabulary Scaffolding, p. T2
- Note-Taking Model, p. T3
- 3-Minute Warm-Up, p. T4

Labs and Activities
- EXPLORE Organisms, p. 9
- INVESTIGATE Microorganisms, p. 13
- Math in Science, p. 15

UNIT RESOURCE BOOK
- Datasheet, Microorganisms, p. 20
- Math Support, p. 59
- Math Practice, p. 60

SECTION

1.2 Bacteria are single-celled organisms without nuclei.
pp. 16–23

Time: 3 periods (1.5 blocks)

Lesson Plan, pp. 22–23

Integrated Technology

RESOURCE CENTER, Bacteria

UNIT TRANSPARENCY BOOK
- Daily Vocabulary Scaffolding, p. T2
- 3-Minute Warm-Up, p. T4

Labs and Activities

CHAPTER INVESTIGATION, Bacteria, pp. 22–23

UNIT RESOURCE BOOK
CHAPTER INVESTIGATION, Bacteria, A, B, & C, pp. 61–69

SECTION

1.3 Viruses are not alive but affect living things.
pp. 24–29

Time: 2 periods (1 block)

Lesson Plan, pp. 32–33

Integrated Technology
- **RESOURCE CENTER,** Viruses
- **VISUALIZATION,** Virus Replication

UNIT TRANSPARENCY BOOK
- Daily Vocabulary Scaffolding, p. T2
- 3-Minute Warm-Up, p. T5
- "Making New Viruses" Visual, p. T6

Labs and Activities
- EXPLORE Viruses, p. 24
- INVESTIGATE Viruses, p. 25
- Extreme Science, p. 29

UNIT RESOURCE BOOK
Datasheet, Viruses, p. 41

SECTION

1.4 Protists are a diverse group of organisms.
pp. 30–35

Time: 3 periods (1.5 blocks)

Lesson Plan, pp. 43–44

Integrated Technology

UNIT TRANSPARENCY BOOK
- Big Idea Flow Chart, p. T1
- Daily Vocabulary Scaffolding, p. T2
- 3-Minute Warm-Up, p. T5
- Chapter Outline, pp. T7–T8

Labs and Activities

INVESTIGATE Protists, p. 31

UNIT RESOURCE BOOK
- Datasheet, Protists, p. 52
- Additional INVESTIGATION, Growth of Algae, A, B, & C, pp. 70–78

KEY TO ICONS
CD/CD-ROM · Teacher Edition · UNIT TRANSPARENCY BOOK · SPANISH ASSESSMENT BOOK
INTERNET · Pupil Edition · UNIT RESOURCE BOOK · UNIT ASSESSMENT BOOK · SCIENCE TOOLKIT

READING AND REINFORCEMENT	ASSESSMENT	STANDARDS
• Description Wheel, B20–21 • Main Idea Web, C38–39 • Daily Vocabulary Scaffolding, H1–8 **UNIT RESOURCE BOOK** • Vocabulary Practice, pp. 56–57 • Decoding Support, p. 58 • Summarizing the Chapter, pp. 79–80 **Audio Readings CD** Listen to Pupil Edition. **Audio Readings in Spanish CD** Listen to Pupil Edition in Spanish.	• Chapter Review, pp. 37–38 • Standardized Test Practice, p. 39 **UNIT ASSESSMENT BOOK** • Diagnostic Test, pp. 1–2 • Chapter Test, A, B, & C, pp. 7–18 • Alternative Assessment, pp. 19–20 Spanish Chapter Test, pp. 41–44 **Test Generator CD-ROM** Generate customized tests. **Lab Generator CD-ROM** Rubrics for Labs	National Standards A.2–8, A.9.a–c, A.9.e–f, C.1.a–b, G.1.b See p. 6 for the standards.
UNIT RESOURCE BOOK • Reading Study Guide, A & B, pp. 13–16 • Spanish Reading Study Guide, pp. 17–18 • Challenge and Extension, p. 19 • Reinforcing Key Concepts, p. 21 • Challenge Reading, pp. 54–55	Ongoing Assessment, pp. 10–12, 14 Section 1.1 Review, p. 14 **UNIT ASSESSMENT BOOK** Section 1.1 Quiz, p. 3	National Standards A.2–8, A.9.a–c, A.9.e–f, C.1.a–b, G.1.b
UNIT RESOURCE BOOK • Reading Study Guide, A & B, pp. 24–27 • Spanish Reading Study Guide, pp. 28–29 • Challenge and Extension, p. 30 • Reinforcing Key Concepts, p. 31	Ongoing Assessment, pp. 16–19, 21 Section 1.2 Review, p. 21 **UNIT ASSESSMENT BOOK** Section 1.2 Quiz, p. 4	National Standards A.2–7, A.9.a–b, A.9.e–f, C.1.a–b, G.1.b
UNIT RESOURCE BOOK • Reading Study Guide, A & B, pp. 34–37 • Spanish Reading Study Guide, pp. 38–39 • Challenge and Extension, p. 40 • Reinforcing Key Concepts, p. 42	Ongoing Assessment, pp. 25–27 Section 1.3 Review, p. 28 **UNIT ASSESSMENT BOOK** Section 1.3 Quiz, p. 5	National Standards A.2–7, A.9.a–b, A.9.e–f, C.1.a–b, G.1.a–b
UNIT RESOURCE BOOK • Reading Study Guide, A & B, pp. 45–48 • Spanish Reading Study Guide, pp. 49–50 • Challenge and Extension, p. 51 • Reinforcing Key Concepts, p. 53	Ongoing Assessment, pp. 31–32, 34–35 Section 1.4 Review, p. 35 **UNIT ASSESSMENT BOOK** Section 1.4 Quiz, p. 6	National Standards A.2–7, A.9.a–b, A.9.e–f, C.1.a–b, G.1.a–b

Previewing Resources for Differentiated Instruction

CHAPTER INVESTIGATION

Leveled resources present the same concepts for different abilities.

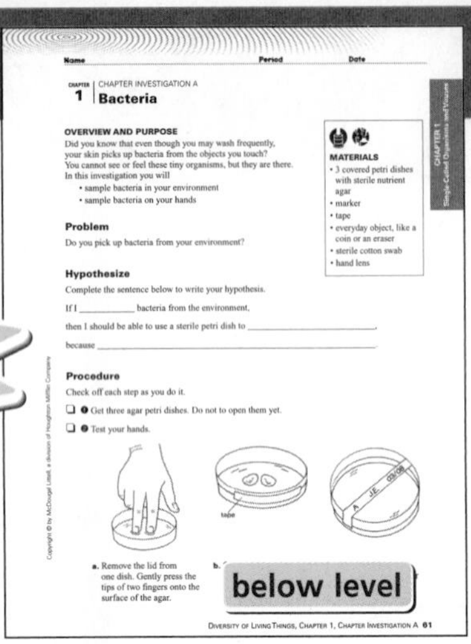

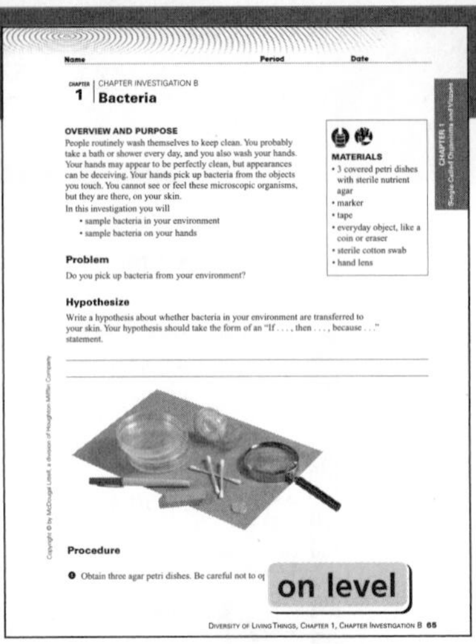

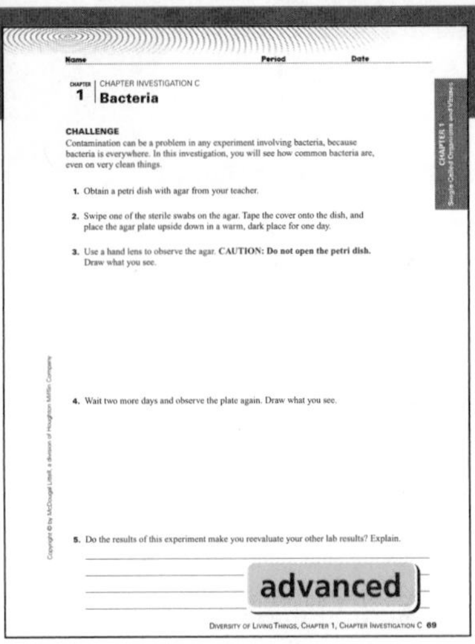

UNIT RESOURCE BOOK, pp. 61–64 | pp. 65–68 | pp. 65–69

READING STUDY GUIDE

Reading Study Guide is also in Spanish.

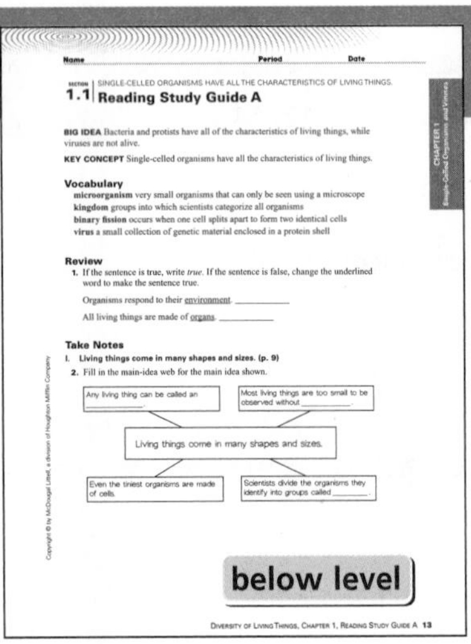

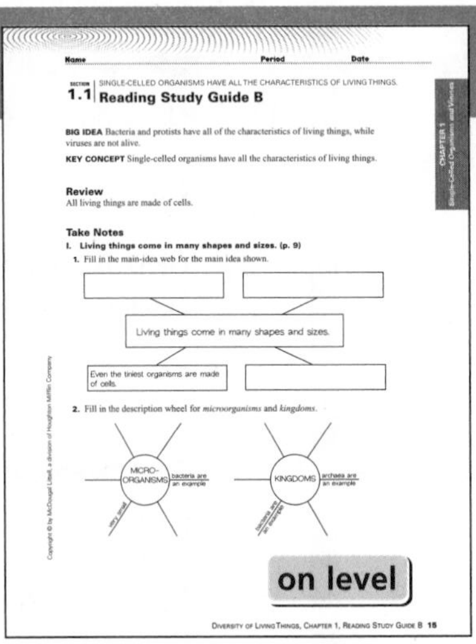

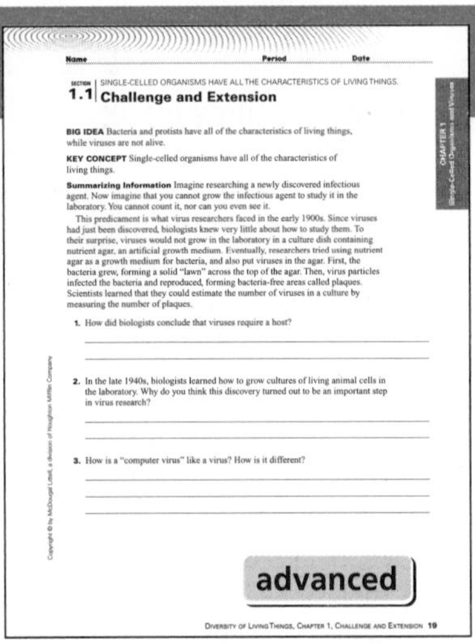

UNIT RESOURCE BOOK, pp. 13–14 | pp. 15–16 | p. 19

CHAPTER TEST

Chapter Test is also in Spanish.

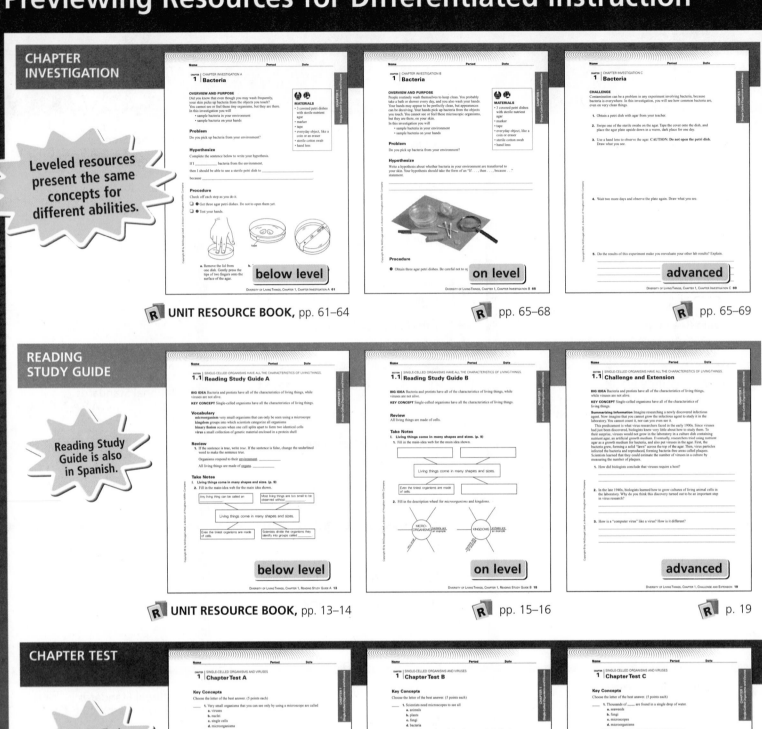

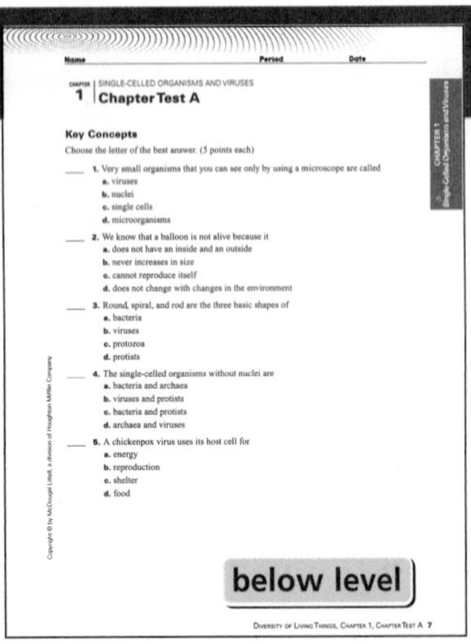

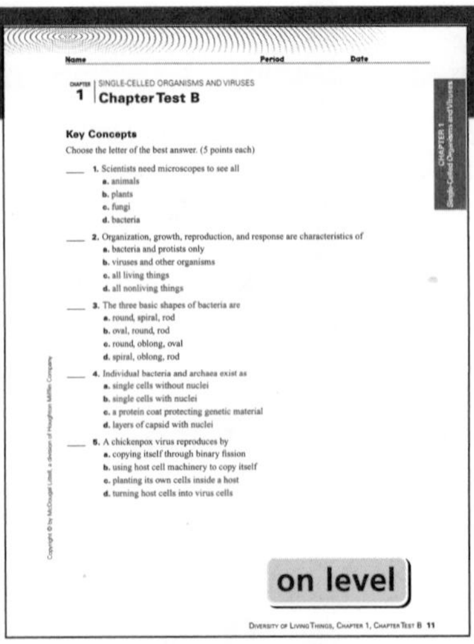

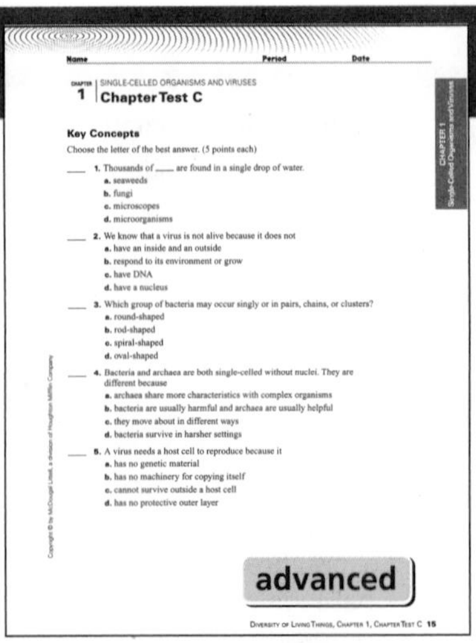

UNIT ASSESSMENT BOOK, pp. 7–10 | pp. 11–14 | pp. 15–18

TECHNOLOGY

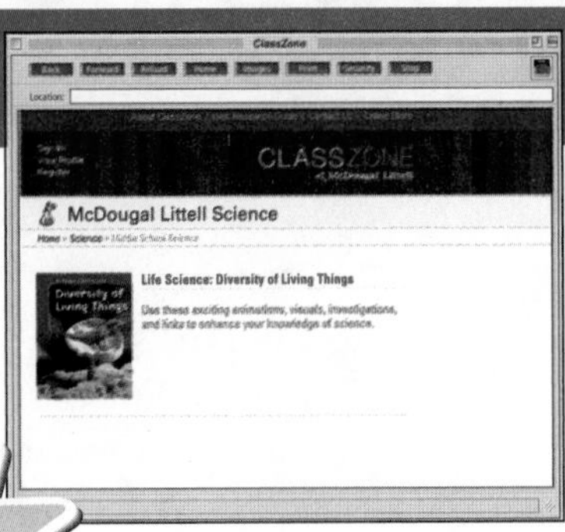
CLASSZONE
McDougal Littell Science
Life Science: Diversity of Living Things
Use these exciting animations, visuals, investigations, and links to enhance your knowledge of science.

There are two Visualizations for this chapter.

CLASSZONE.COM

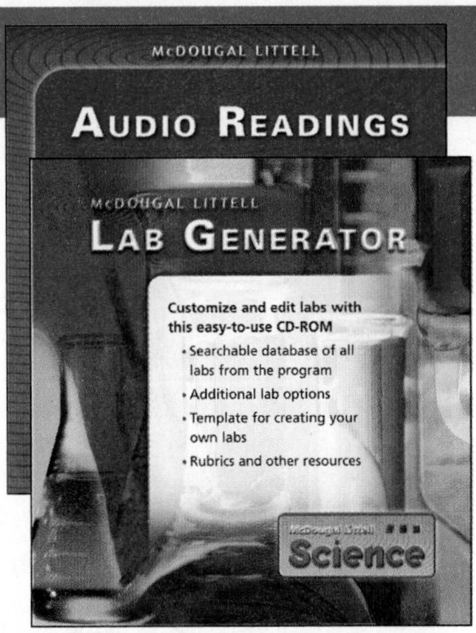
McDOUGAL LITTELL
AUDIO READINGS
McDOUGAL LITTELL
LAB GENERATOR
Customize and edit labs with this easy-to-use CD-ROM
• Searchable database of all labs from the program
• Additional lab options
• Template for creating your own labs
• Rubrics and other resources
McDougal Littell
Science

CD/CD-ROMS

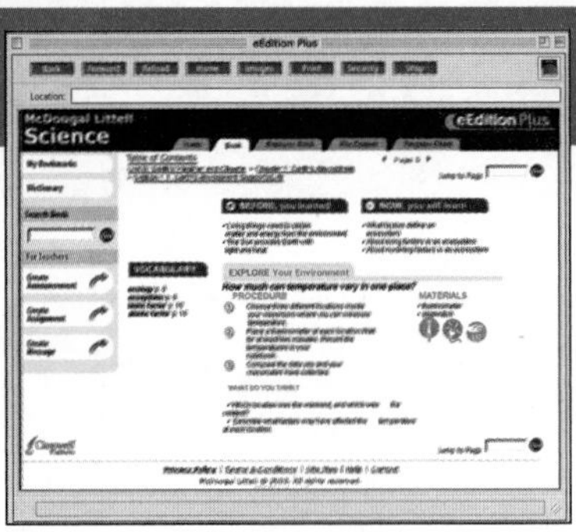
eEdition Plus
McDougal Littell
Science

CLASSZONE.COM

VISUAL CONTENT

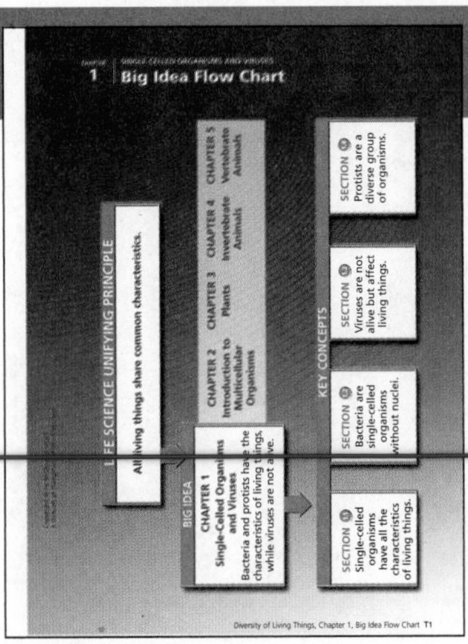
1 Big Idea Flow Chart
LIFE SCIENCE UNIFYING PRINCIPLE
All living things share common characteristics.
BIG IDEA
CHAPTER 1 Single-Celled Organisms and Viruses
Bacteria and protists have the characteristics of living things, while viruses are not alive.
CHAPTER 2 Introduction to Multicellular Organisms
CHAPTER 3 Plants
CHAPTER 4 Invertebrate Animals
CHAPTER 5 Vertebrate Animals
KEY CONCEPTS
SECTION 1.1 Single-celled organisms have all the characteristics of living things.
SECTION Bacteria are single-celled organisms without nuclei.
SECTION Viruses are not alive but affect living things.
SECTION Protists are a diverse group of organisms.
Diversity of Living Things, Chapter 1, Big Idea Flow Chart T1

UNIT TRANSPARENCY BOOK, p. T1

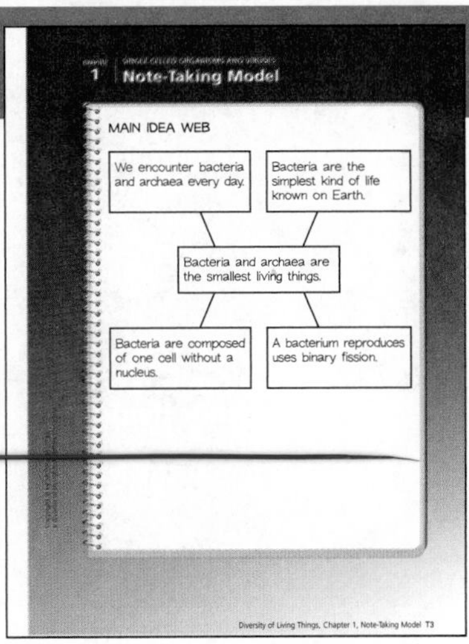
1 Note-Taking Model
MAIN IDEA WEB
We encounter bacteria and archaea every day.
Bacteria are the simplest kind of life known on Earth.
Bacteria and archaea are the smallest living things.
Bacteria are composed of one cell without a nucleus.
A bacterium reproduces uses binary fission.
Diversity of Living Things, Chapter 1, Note-Taking Model T3

p. T3

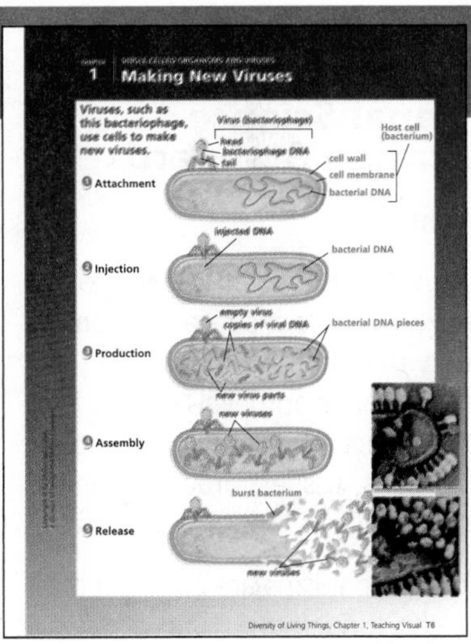
1 Making New Viruses
Viruses, such as this bacteriophage, use cells to make new viruses.
Virus (bacteriophage)
Host cell (bacterium)
cell wall
cell membrane
bacterial DNA
Attachment
Injection
injected DNA
bacterial DNA
Production
bacterial DNA pieces
Assembly
new viruses
Release
burst bacterium
new viruses
Diversity of Living Things, Chapter 1, Teaching Visual T6

p. T6

MORE SUPPORT

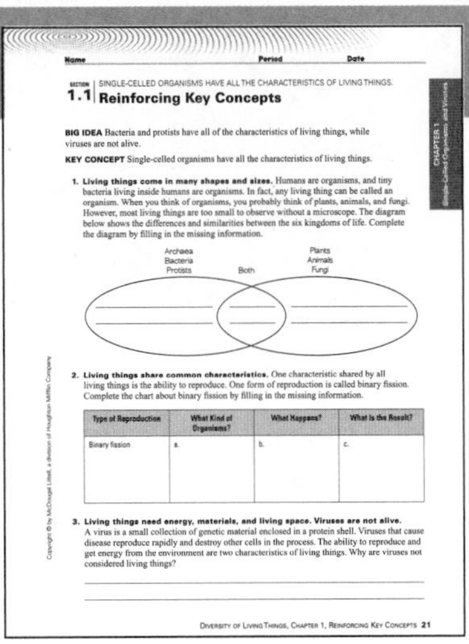
1.1 Reinforcing Key Concepts
Diversity of Living Things, Chapter 1, Reinforcing Key Concepts 21

Reinforcing Key Concepts for each section

UNIT RESOURCE BOOK, p. 21

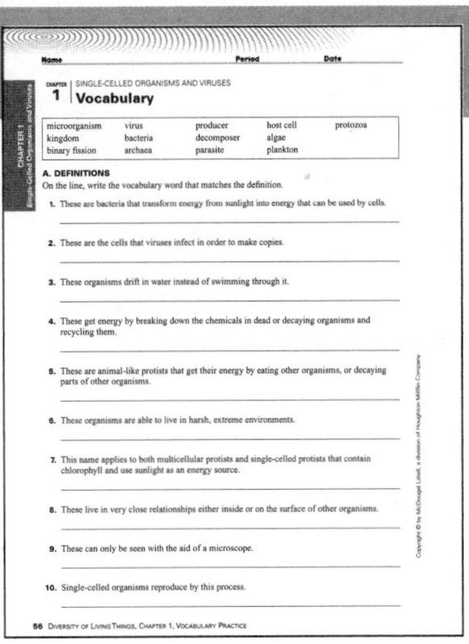
1 Vocabulary
A. DEFINITIONS
56 Diversity of Living Things, Chapter 1, Vocabulary Practice

pp. 56–57

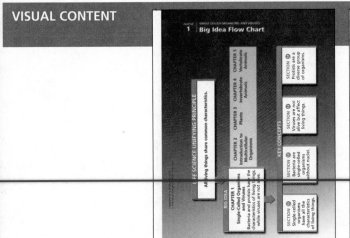

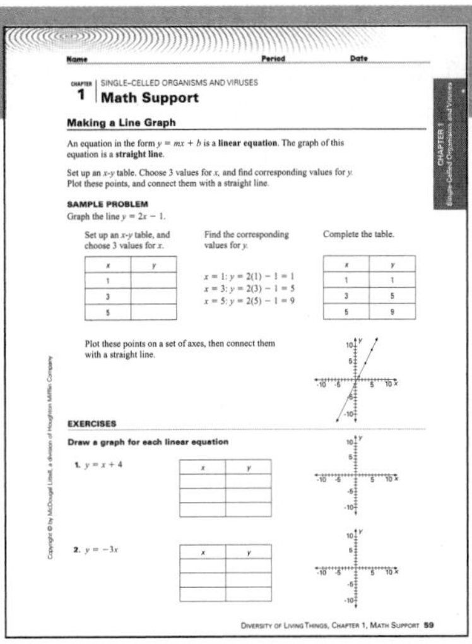
1 Math Support
Making a Line Graph
SAMPLE PROBLEM
EXERCISES
Diversity of Living Things, Chapter 1, Math Support 59

p. 59

CHAPTER 1

Single-Celled Organisms and Viruses

INTRODUCE

the BIG idea

Have students look at the microscopic structures in the photograph and discuss how the question in the box links to the Big Idea:

- Do the structures in the photograph look like they are alive? Why or why not?
- How do you think you could determine if the structures are alive?

National Science Education Standards

Content

C.1.a Living systems at all levels of organization demonstrate the complementary nature of structure and function. Important levels of organization for structure and function include cells, organs, tissues, organ systems, whole organisms, and ecosystems.

C.1.b All organisms are composed of cells—the fundamental unit of life. Most organisms are single cells; other organisms, including humans, are multicellular.

Process

A.2–8 Design and conduct an investigation; use tools to gather and interpret data; use evidence to describe, predict, explain, model; think critically to make relationships between evidence and explanation; recognize different explanations and predictions; communicate scientific procedures and explanations; use mathematics.

A.9.a–c, A.9.e–f Understand scientific inquiry by using different investigations, methods, mathematics, and explanations based on logic, evidence, and skepticism.

G.1.b Science requires different abilities.

CHAPTER 1

Single-Celled Organisms and Viruses

the BIG idea

Bacteria and protists have the characteristics of living things, while viruses are not alive.

Key Concepts

SECTION 1.1 **Single-celled organisms have all the characteristics of living things.** Learn about characteristics shared by all living things.

SECTION 1.2 **Bacteria are single-celled organisms without nuclei.** Learn about the characteristics of bacteria and archaea.

SECTION 1.3 **Viruses are not alive but affect other living things.** Learn about virus structure and how they affect other cells.

SECTION 1.4 **Protists are a diverse group of organisms.** Learn about protists and how they affect the environment.

Internet Preview

CLASSZONE.COM

Chapter 1 online resources: Content Review, two Visualizations, three Resource Centers, Math Tutorial, Test Practice.

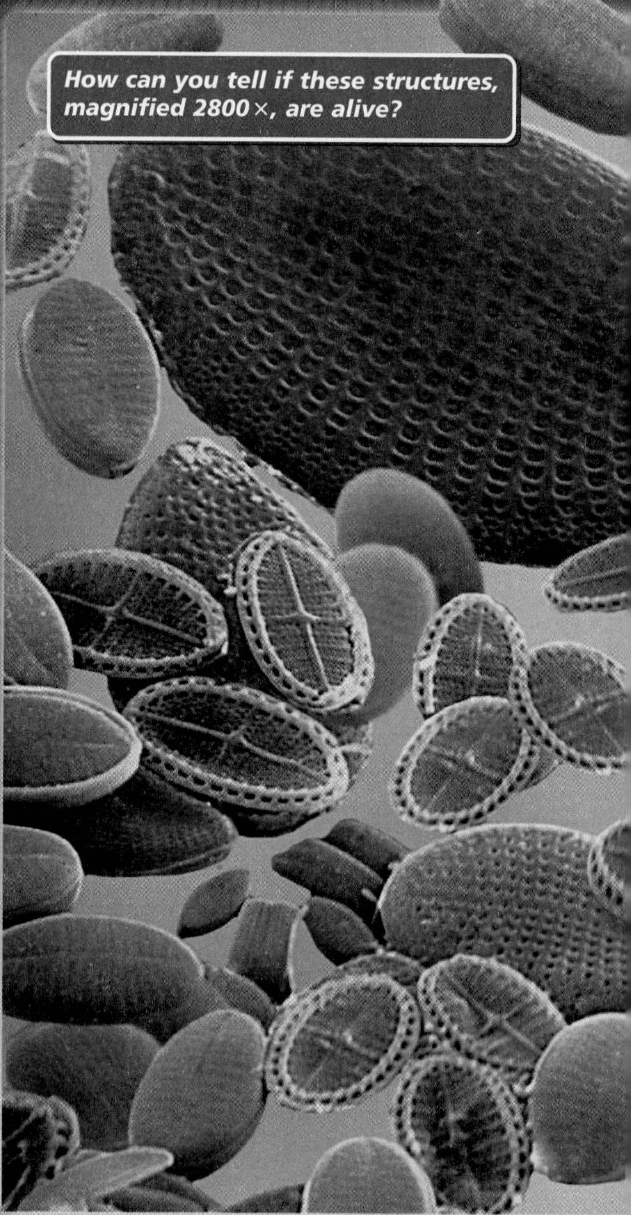

How can you tell if these structures, magnified 2800×, are alive?

INTERNET PREVIEW

CLASSZONE.COM For student use with the following pages:

Review and Practice
- Content Review, pp. 8, 36
- Math Tutorial: Making a Line Graph, p. 15
- Test Practice, p. 39

Activities and Resources
- Internet Activity: Microscopic Life and You, p. 7
- Visualizations: Binary Fission, p. 12; Virus Replication, p. 26
- Resource Centers: Bacteria, p. 18; Viruses, p. 25

NSTA SCILINKS scilinks.org

Kingdom Protista
Code: MDL039

C 6 Unit: Diversity of Living Things

EXPLORE the BIG idea

Where Can You Find Microscopic Life?

Make a list of places where you might find living things that are too small to be seen by your unaided eye. Then use a hand lens, magnifying glass, or microscope, to investigate some of the places on your list.

Observe and Think What do microscopic organisms look like. Why are there so many places where you can find microscopic life?

How Quickly Do Bacteria Multiply?

Tape a funnel to the top of a two-liter bottle. Place one bean in the funnel. After one minute, drop two more beans into the funnel. Continue adding beans to the bottle every minute, adding twice as many beans as you did before. When it is time to add 64 beans, use 1/8 of a cup, and then continue to double the amounts.

Observe and Think How long did it take to fill the bottle?

Internet Activity: Microscopic Life and You

Go to **ClassZone.com** to learn about the single-celled organisms.

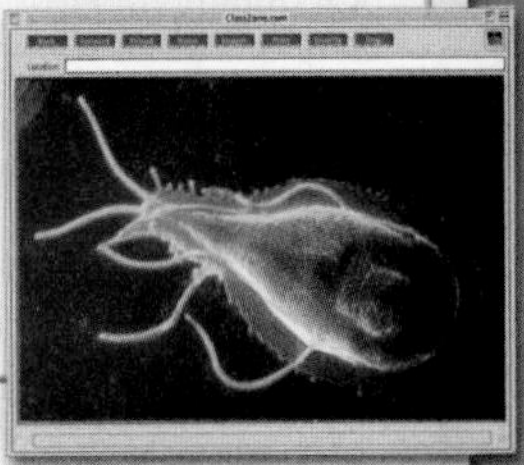

Observe and Think What types of organism live in the human body?

NSTA scilinks.org SCILINKS
Kingdom Protista **Code: MDL039**

EXPLORE the BIG idea

These inquiry-based activities are appropriate for use at home or as a supplement to classroom instruction.

Where Can You Find Microscopic Life?

PURPOSE To look at microscopic life using a hand lens or microscope.

TIP *20 min.* Have groups of students brainstorm to make their lists.

Answer: Microscopic organisms have diverse sizes and shape, although most have only one cell. They can survive in a large range of environments.

REVISIT after p. 10.

How Quickly Do Bacteria Multiply?

PURPOSE To model how quickly bacteria reproduce.

TIP *10 min.* Students should add large quantities of beans slowly, so the funnel does not clog.

Answer: The bottle should fill up in the 12th minute, when 4 cups of beans are added.

REVISIT after p. 12.

Internet Activity: Microscopic Life and You

PURPOSE To investigate the microscopic life that lives on, in, and around people.

TIP *20 min.* Have students list the places in the body where the organisms they read about are found.

Answer: Helpful bacteria live in many parts of the body. Harmful bacteria, fungi spores, and small insects can harm a human host.

REVISIT after p. 19.

TEACHING WITH TECHNOLOGY

PC Microscopes or Microscopes with Cameras If computer microscopes are available, use them for the investigations on pp. 13 and 31 to view microscopic organisms on a computer screen. Save screen-grabs on the PC capturing magnified images of the organisms. If a microscope with attached camera is available, take photographs of these microorganisms for later identification.

PREPARE

CONCEPT REVIEW

Activate Prior Knowledge

- Have students reproduce Robert Hooke's discovery of cells by examining a small slice of real cork under a microscope. They should see empty cells, since cork is composed of dead tissue.
- Ask: In what ways does a single-celled organism interact with its environment to meet its needs? *It may search for and take in food; react to light, temperature, or touch; and avoid enemies.*

TAKING NOTES

Main Idea Web

A main idea web focuses students' attention as they read the text. The web is similar to an outline but allows a more random, less linear organization. Details can be added easily upon second and third reads of text portions, or as questions arise.

Vocabulary Strategy

A description wheel can include as much information as the student wants to add. By filling in their own words around the center term, students personalize their understanding of the term and retain its meaning.

Vocabulary and Note-Taking Resources

R
- Vocabulary Practice, pp. 56–57
- Decoding Support, p. 58

T
- Daily Vocabulary Scaffolding, T2
- Note-Taking Model, T3

- Description Wheel, B20–21
- Main Idea Web, C38–39
- Daily Vocabulary Scaffolding, H1–8

CHAPTER 1

Getting Ready to Learn

CONCEPT REVIEW

- All living things interact with their environment to meet their needs.
- The cell is the fundamental unit of life.

VOCABULARY REVIEW

See Glossary for definitions.

cell
matter
molecule
organism
species

CONTENT REVIEW
CLASSZONE.COM
Review concepts and vocabulary.

TAKING NOTES

SCIENCE NOTEBOOK

MAIN IDEA WEB

Write each new blue heading in a box. Then write notes in boxes around the center box that give important terms and details about that blue heading.

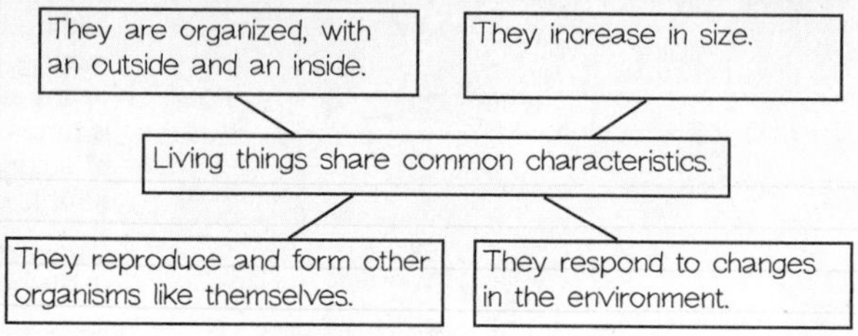

They are organized, with an outside and an inside.
They increase in size.
Living things share common characteristics.
They reproduce and form other organisms like themselves.
They respond to changes in the environment.

VOCABULARY STRATEGY

Place each vocabulary term at the center of a **description wheel diagram.** Write some words describing it on the spokes.

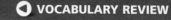

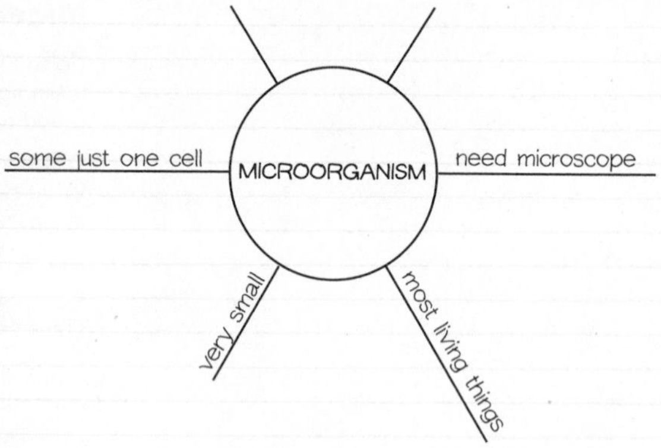

See the Note-Taking Handbook on pages R45–R51.

C 8 Unit: Diversity of Living Things

CHECK READINESS

Administer the Diagnostic Test to determine students' readiness for new science content and their mastery of requisite math skills.

A Diagnostic Test, pp. 1–2

Technology Resources

Students needing content and math skills should visit **ClassZone.com**.

- **CONTENT REVIEW**
- **MATH TUTORIAL**

CONTENT REVIEW CD-ROM

KEY CONCEPT

Single-celled organisms have all the characteristics of living things.

BEFORE, you learned

- All living things are made of cells
- Organisms respond to their environment
- Species change over time

NOW, you will learn

- About the various sizes of organisms
- About characteristics that are shared by all living things
- About needs shared by all organisms

VOCABULARY

microorganism p. 10
kingdom p. 11
binary fission p. 12
virus p. 14

EXPLORE Organisms

What living things are in the room with you?

PROCEDURE

1. Make a list of all the living things that are in your classroom.
2. Compare your list with the lists of your classmates. Make one list containing all the living things your class has identified.

MATERIALS

- paper
- pencil

WHAT DO YOU THINK?

- How did you identify something as living?
- Were you and your classmates able to see all the living things on your list?

Living things come in many shapes and sizes.

MAIN IDEA WEB
Make a web of the important terms and details about the main idea: *Living things come in many shapes and sizes.*

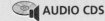

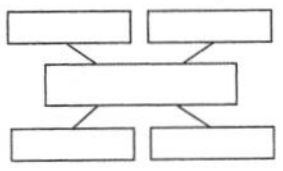

You can spot mushrooms in many places while walking through a forest. Scientists have discovered mushrooms that come from the same individual fungus more than 5 kilometers (3 miles) apart in an Oregon forest. Most of this honey mushroom fungus is below ground, stretching over an area covering more than 1600 football fields. This mushroom is one of the largest known living things on Earth.

Many other living things share the soil in the Oregon forest. Earthworms, insects, and many other organisms that are too small to be seen with a naked eye, also live there. For every living thing that is large enough to be seen, there are often countless numbers of smaller living things that share the same living space.

1.1 FOCUS

Set Learning Goals

Students will

- Recognize that some organisms are too small to see.
- Identify the characteristics shared by all living things.
- State how living things are classified.
- Observe how organisms respond to their environment.

3-Minute Warm-Up

Display Transparency 4 or copy this exercise on the board:

Draw a simple diagram representing a cell. Give your diagram a title that tells what kind of a cell you have drawn. Label the most important organelles. *A eukaryotic cell has a nucleus, plasma membrane, mitochondria, Golgi apparatus, ribosomes, and endoplasmic reticulum. Plant cells have a cell wall and chloroplasts, in addition to the other organelles.*

T 3-Minute Warm-Up, p. T4

1.1 MOTIVATE

EXPLORE Organisms

PURPOSE To list living things found in a classroom

TIP *10 min.* Students will probably not include microorganisms on their lists. Remind them to "think small."

WHAT DO YOU THINK? *A living thing must be able to grow, have an organized system, and be able to reproduce.*

RESOURCES FOR DIFFERENTIATED INSTRUCTION

Below Level

UNIT RESOURCE BOOK

- Reading Study Guide A, pp. 13–14
- Decoding Support, p. 58

AUDIO CDS

Advanced

UNIT RESOURCE BOOK

- Challenge and Extension, p. 19
- Challenge Reading, pp. 54–55

English Learners

UNIT RESOURCE BOOK

Spanish Reading Study Guide, pp. 17–18

AUDIO CDS

- Audio Readings in Spanish
- Audio Readings (English)

1.1 INSTRUCT

Teach from Visuals

To help students interpret the visual of organisms in a mangrove swamp, ask:

- What types of organisms in the photograph are able to be seen? *plants and animals*
- What types of organisms in the photograph are unable to be seen? *Archaea, bacteria, protists, and fungi; the amoeba is a protist.*

EXPLORE the BIG idea

Revisit "Where Can You Find Microscopic Life?" on p. 7. Have students explain why there might be many microscopic organisms in the environment shown in the photograph on p. 10.

Arts Connection

Flemish artists of the 17th century often painted pictures of the living world as it was known at that time. Many of these paintings can be found on the Internet and printed. Show students a copy of one of these paintings, such as *Paradise* by Jan Brueghel the Younger. Point out that all of the organisms shown are either plants or animals.

Ongoing Assessment

Recognize that some organisms are too small to see.

Ask: How can you be sure that microscopic organisms are present in an environment such as a swamp? *You must use a microscope.*

CHECK YOUR READING *Answer: An organism is any living thing. Microorganisms are living things, but they are too small to be seen without a microscope.*

The honey mushroom fungus is one example of an organism. You, too, are an organism, and tiny bacteria living inside your body are also organisms. In fact, any living thing can be called an organism. **?A**

When you identify living things, you probably begin with those you can observe—plants, animals, and fungi such as mushrooms. However, most living things are too small to observe without a microscope. Even the tiniest organisms are made of cells. Very small organisms are called **microorganisms.** Some microorganisms are made of just one cell. **?B**

READING TIP The prefix *micro-* means "very small." Therefore, *microscope* means "very small scope" and *microorganism* means "very small organism."

CHECK YOUR READING Compare and contrast the words *microorganism* and *organism.*

A visitor to a mangrove swamp forest can find an amazing variety of organisms. The mangrove trees themselves are the most obvious organisms. Roots from these trees grow above and below the muddy bottom of the forest. Other organisms live in almost every part of the mangrove tree.

Six Kingdoms of Life

All organisms are divided into six groups called kingdoms.

Mostly Microscopic Kingdoms	Mostly Multicellular Kingdoms
• archaea	• animals
• bacteria	• fungi
• protists	• plants

DIFFERENTIATE INSTRUCTION

More Reading Support

A What is an organism? *any living thing*

B What do you call an organism that cannot be seen with the naked eye? *a microorganism*

English Learners Have English learners write the definitions for *microorganism, kingdom, binary fission,* and *virus* in their Science Word Dictionaries. Have students look up the definition for the prefix *micro-* in the dictionary. Then have them list other words that begin with *micro-* (microscope, microwave, microphone), and explain the meaning of the prefix. *very small*

A single drop of water from a mangrove swamp may be living space for many microorganisms. The circled photograph on page 10 was taken using a microscope, and shows an amoeba that may be found in the water of the swamp. Larger organisms, such as manatees and fish, swim around the roots of mangrove trees. Birds, such as tri-colored herons and roseate spoonbills, live on branches.

Scientists divide the organisms they identify into groups called **kingdoms.** This unit will cover all of the kingdoms of life, listed in the table on page 10. You are already familiar with plants and animals. Fungi are another kingdom. Fungi include mushrooms found in a forest. The other three kingdoms are composed of mostly microscopic life. You will learn more about microscopic organisms later in this chapter.

VOCABULARY
Add a description wheel for *kingdom* to your notebook. The spokes of your wheel should include examples from the six kingdoms.

Living things share common characteristics.

All living things—from the microorganisms living in a mangrove swamp to the giant organisms living in the open ocean—share similar characteristics. Living things are organized, grow, reproduce, and respond to the environment.

READING TIP
As you read about the four characteristics of all living things, note the examples of how single-celled organisms meet these four standards.

Organization

Cells, like all living things, have an inside and an outside. The boundary separating the inside from the outside of an individual cell is called the cell membrane. Within some cells, another structure called the nucleus is also surrounded by a membrane. Cells perform one or more functions that the organism needs to survive

In this chapter, you will read about organisms made of a single cell. Some types of single-celled organisms contain a nucleus and some do not. All single-celled organisms contain every structure they need to survive within their one cell. They have structures to get energy from complex molecules, structures to help them move, and structures to help them sense their environment. All of the structures are part of their organizations.

Growth

Living things increase in size. Organisms made of one cell do not grow as large as organisms made of many cells. But all living things consume food or other materials to get energy. These materials are also used to build new structures inside cells or replace worn-out structures. As a result, individual cells grow larger over time.

DIFFERENTIATE INSTRUCTION

More Reading Support

C What 3 kingdoms have organisms large enough to see? *fungi, plants, animals*

D How can an organism survive with only one cell? *It has all the structures needed in the cell.*

Below Level Demonstrate classification by having students separate a group of everyday items into categories according to their characteristics. Use items such as shoes or writing implements. Ask volunteers to explain why classifying is helpful when thinking about living things.

Address Misconceptions

IDENTIFY Show students pictures cut from magazines of living things from all six kingdoms, as well as some nonliving items such as a pile of sand, a lump of clay, and a rock. Then ask students to look at all the pictures and sort them into living and nonliving things. If students include only plants and animals as living, they may hold the misconception that all organisms are either plants or animals.

CORRECT Ask students to list living things that are not plants or animals. Lists should contain at least a few organisms from the Protista and Fungi kingdoms. Circle these items and lead a discussion about what makes them living things. *grow, respond, reproduce, take in air, food, water, and so on*

REASSESS Ask students to reconsider their definition of an organism and to write a definition in their own words that includes what they have learned. Ask that their definitions include diverse examples.

Technology Resources

Visit **ClassZone.com** for background on common student misconceptions.

MISCONCEPTION DATABASE

Develop Critical Thinking

APPLY Show students a photograph of a biome—a desert is a good example. Have them name what is living and what is not living in the picture. Ask them to apply their knowledge of the characteristics of life by telling what reasons they had for deciding what is living and what is not.

Metacognitive Strategy

Ask students to write a short paragraph describing how they remember the characteristics of living things.

Ongoing Assessment

State how living things are classified.

Ask: What characteristics are used to classify organisms? *They are classified based on characteristics such as cell structure, means of getting energy, and physical or body structure.*

Teach from Visuals

To help students interpret the photograph of binary fission, ask:

- What must happen before a single-celled organism can divide? ***Its DNA must first be copied.***
- What would happen if a single-celled organism divided before it had two copies of its DNA? ***Only one of the two resulting cells would contain DNA. The other cell could not survive.***

Teach Difficult Concepts

The concept of binary fission is difficult because students are used to thinking of linear relationships. Making a graph of an exponential increase might help illustrate how an unchecked population could grow. You might also use manipulatives, as in the Teacher Demo, below.

Teacher Demo

Using beans, demonstrate how bacteria divide by binary fission. Start by placing one bean on the table. Call the bean the "parent" bean. Have a student place two beans next to the parent bean. Toothpicks can be used to connect the parent bean to the offspring beans. Have another student provide the next generation by placing four beans on the table. Have other students add further generations.

EXPLORE (the BIG idea)

Revisit "How Quickly Do Bacteria Multiply?" on p. 7. Have students describe how fast the beans "reproduced," and why this is a good model of binary fission.

Ongoing Assessment

CHECK YOUR READING *Answer: Single-celled organisms have all the structures they need in one cell. They grow by increasing in size. They reproduce by binary fission, in which the organism divides into two organisms.*

Binary Fission

These bacteria make exact copies of themselves through the process of binary fission.

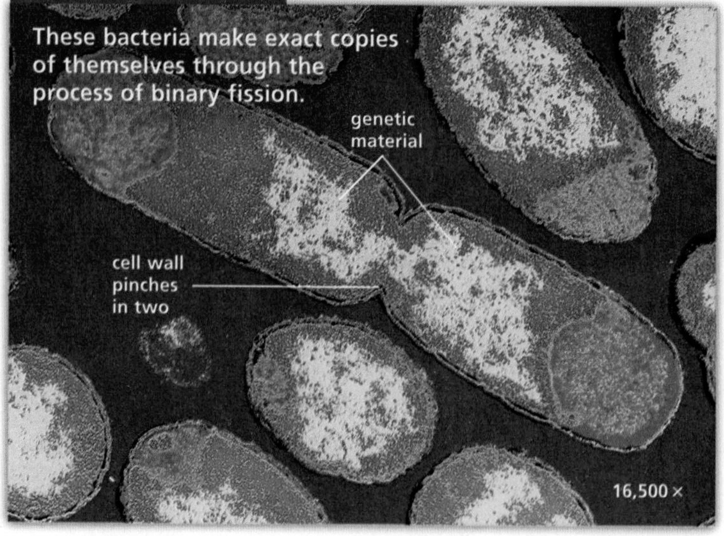

Reproduction

Living things reproduce, forming other organisms like themselves. Every organism contains genetic material, which is a code contained in a special molecule called DNA. The code contains every characteristic of the individual organism. In order to reproduce, an organism must make a copy of this material, which is passed on to its offspring.

VISUALIZATION CLASSZONE.COM
Observe the process of binary fission.

Single-celled organisms reproduce by a process called **binary fission.** In binary fission, material from one cell is broken apart into two cells. The genetic material of the original cell doubles so that each daughter cell has an exact copy of the DNA of the original cell. You might say that single-celled organisms multiply by dividing. One cell divides into 2 cells, 2 cells divide into 4, 4 into 8, 16, 32, 64, and so on. In some cells, binary fission can repeat in as little as 20 minutes.

CHECK YOUR READING **Describe how a single-celled organism is organized, grows, and reproduces.**

Response

Organisms respond to changes in the environment. Even microscopic organisms respond to conditions such as light, temperature, and touch. The ability to respond allows organisms to find food, avoid being eaten, or perform other tasks necessary to survive.

C 12 Unit: **Diversity of Living Things**

DIFFERENTIATE INSTRUCTION

More Reading Support

E What molecule contains the cell's genetic material? *DNA*

F What process do single-celled organisms use to reproduce? *binary fission*

English Learners The activity on p. 13 uses the imperative. The word "you" in the directions is understood, or implied. For example, "Add a drop of the daphnia culture," really means "*You* add a drop of the daphnia culture." Make sure English learners understand how to read and follow these directions. You may wish to model the procedure for them, or have an advanced student do so.

INVESTIGATE Microorganisms

How do these organisms respond to their environment?

SKILL FOCUS
Observing

MATERIALS
- microscope
- slide
- hydra culture
- daphnia culture
- water

TIME
30 minutes

PROCEDURE

1. Place a drop of the hydra culture on a microscope slide. Using the microscope, find a hydra under medium power and sketch what you see.
2. Add a drop of warm water to the culture on the slide. How does the hydra respond? Record your observations.
3. Add a drop of the daphnia culture. Record your observations.

WHAT DO YOU THINK?

- Which observations, if any, indicate that hydras respond to their environment?
- Daphnia are organisms. What is the relationship between hydra and daphnia?

CHALLENGE What other experiments could you do to observe the responses of hydra or daphnia to their environment?

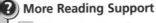

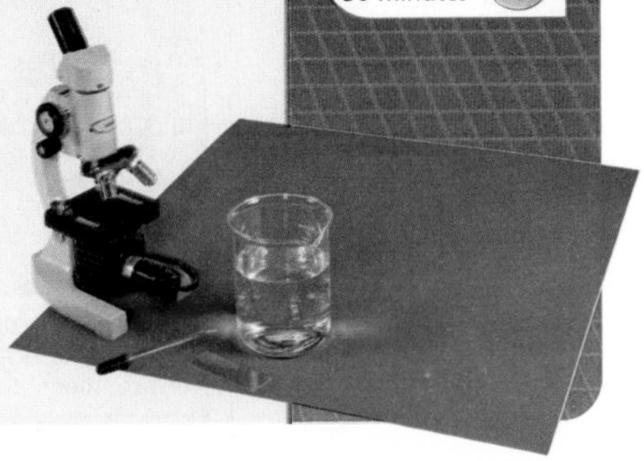

Living things need energy, materials, and living space.

Have you ever wondered why you need to eat food, breathe air, and drink water? All living things need energy and materials. For most organisms, water and air are the materials necessary for life.

Food supplies you with energy. You—like all living things—need energy to move, grow, and develop. All animals have systems for breaking down food into usable forms of energy and materials. Plants have structures that enable them to transform sunlight into usable energy. Some microorganisms transform sunlight, while others need to use other organisms as sources of energy.

Most of the activities of living things take place in water. Water is also an ingredient for many of the reactions that take place in cells. In addition, water helps support an organism's body. If you add water to the soil around a wilted plant, you will probably see the plant straighten up as water moves into its cells.

Materials in the air include gases such as carbon dioxide and oxygen. Many of the processes that capture and release energy involve these gases. Some organisms—such as those found around hydrothermal vents—use other chemicals to capture and release energy.

DIFFERENTIATE INSTRUCTION

More Reading Support

G Where do living things get energy? *food and sunlight*

H Why is water so important to living things? *Most activities take place in water.*

Advanced Have students learn more about the processes of cellular respiration and photosynthesis. Ask them to draw the cycle of energy in the environment, including cellular respiration and photosynthesis.

R Challenge and Extension, p. 19

Have students who are interested in the lives of single-celled organisms read the following article:

R Challenge Reading, pp. 54–55

INVESTIGATE Microorganisms

PURPOSE To observe the responses of hydra and daphnia to their environment

TIPS *30 min.* Warm water—between 25°C and 30°C—will give the best response. Students can best view the activity by using cover slips on the slides. Suggest that students record observations by drawing field of view circles and sketching within them. Assist students with disabilities by setting up pre-focused stations. Note that about 50 percent of hydra may show no response.

WHAT DO YOU THINK? *Hydras should move away from the warm water. They should capture daphnia with their tentacles and stuff them into their mouths. Both behaviors are responses to the environment. The daphnia may also try to get away from the hydra. The daphnia are food for the hydra.*

CHALLENGE *Answers may include testing the organisms' responses to light, darkness, various chemicals, and cold.*

R Datasheet, Microorganisms, p. 20

Technology Resources

Customize this student lab as needed or look for an alternative. Print rubrics to assess student lab reports.

Lab Generator CD-ROM

Teaching with Technology

If a PC microscope is available, show the organisms on a computer screen. If a microscope fitted with a camera is available, photograph hydra and daphnia in action.

Integrate the Sciences

All organisms that use oxygen get energy through a complex series of chemical reactions known as cellular respiration. In this process, glucose breaks down, releasing the energy stored in its chemical bonds. The energy is then packaged in small quantities that the cell can use.

Teach from Visuals

To help students compare the photographs of the rotavirus, and the animal cell, ask them to make a table listing the two items on the left side and the four characteristics of life across the top. Have students check each characteristic of life that applies to the items.

Ongoing Assessment

Identify the characteristics shared by all living things.

Ask students to explain the difference between growth and reproduction. *Growth occurs when an organism increases in size. Reproduction makes more organisms.*

Reinforce (the BIG idea)

Have students relate the section to the Big Idea.

R Reinforcing Key Concepts, p. 21

1.1 ASSESS & RETEACH

Assess

A Section 1.1 Quiz, p. 3

Reteach

Stage a "characteristics of life" scavenger hunt for students. Have the students work in teams with checklists and/or clipboards. The teams should list, within 5 minutes for each "hunt":

- As many examples as possible of organisms growing.
- As many examples as they can find of organisms with structures.
- . . . of organisms reproducing.
- . . . of organisms responding to the environment.

Technology Resources

Have students visit **ClassZone.com** for reteaching of Key Concepts.

CONTENT REVIEW

CONTENT REVIEW CD-ROM

Viruses are not alive.

Sometimes it's not easy to tell the difference between a living and a nonliving thing. A **virus** is a small collection of genetic material enclosed in a protein shell. Viruses have many of the characteristics of living things, including DNA. However, a virus is not nearly as complex as an animal cell and is not considered a living thing.

Rotavirus

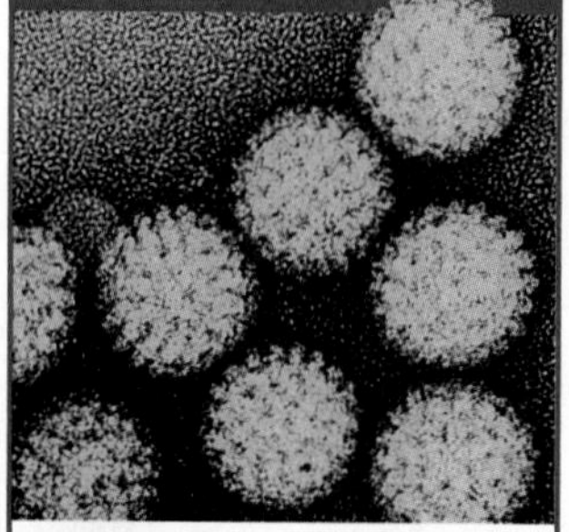

These viruses contain DNA but do not grow or respond to their environment. 570,000×

Animal Cell

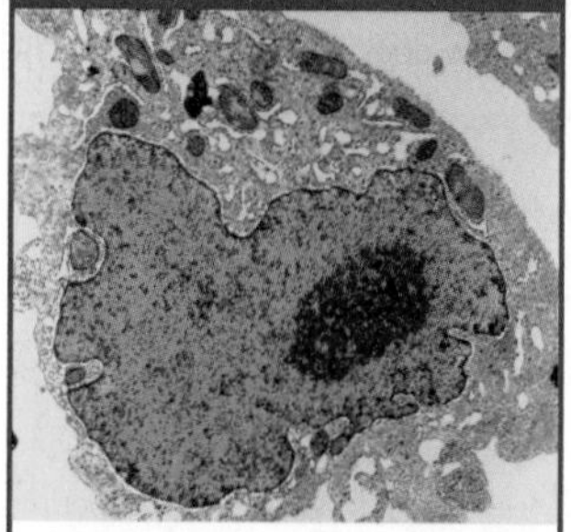

Animal cells grow, reproduce, and respond to external conditions. 4800×

Animal cells have structures that allow them to get materials or energy from their environment. Viruses do not grow once they have formed, and they do not take in any energy. Animal cells can make copies of their genetic material and reproduce by dividing in two. Viruses are able to reproduce only by "taking over" another cell and using that cell to make new viruses. Animal cells also have many more internal structures than viruses. Viruses usually contain nothing more than their DNA.

1.1 Review

KEY CONCEPTS

1. Give examples of organisms that are very large and organisms that are very small.
2. Name four characteristics that all living things share.
3. Name three things that living things must obtain to survive.

CRITICAL THINKING

4. **Synthesize** Give examples of how a common animal, such as a dog, is organized, grows, responds, and reproduces.
5. **Predict** In a certain lake, would you expect there to be more organisms that are large enough to see or more organisms that are too small for you to see? Why?

CHALLENGE

6. **Design** Try to imagine the different structures that a single-celled organism needs to survive in pond water. Then use your ideas to design your own single-celled organism.

ANSWERS

1. very large organism: huge fungus; very small organism: single-celled bacteria

2. growth, reproduction, organization, and response to environment

3. energy, materials, and living space

4. A dog is made of cells that perform specific functions. It starts as a puppy. As its cells reproduce, it grows to a full-sized dog. Male and female dogs reproduce sexually to create more puppies. Dogs bark at, run at, or run from different things they sense.

5. There is room in a lake for many billions of microscopic organisms but not for nearly as many larger organisms.

6. Examine students' designs for creativity and understanding of the characteristics and needs of life.

MATH in SCIENCE

SKILL: MAKING A LINE GRAPH

Graphing Growth

MATH TUTORIAL
CLASSZONE.COM
Click on Math Tutorial for more help with making a line graph.

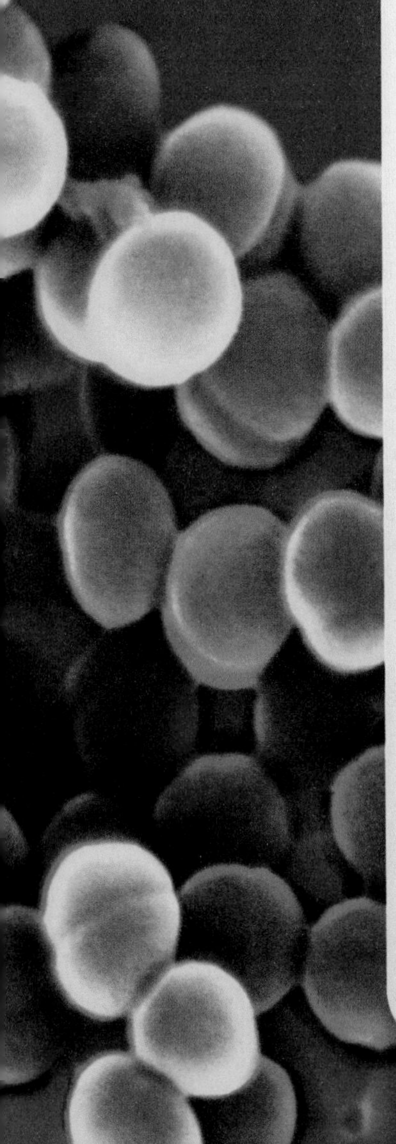

If you hold marbles in your hand and drop them into a bowl, each drop into the marble collection adds the same amount. If you plot this growth on a line graph, you will have a straight line.

By contrast, a bacteria colony's growth expands as it grows. All the bacteria divide in two. Every time all the bacteria divide, the colony doubles in size.

EXAMPLE

Compare the two types of growth on a graph.

Graph 1. Suppose the marble collection begins with one marble, and after every minute, one marble is added.

The graph shows: $x + 1 = y$
The slope of the line stays the same at each interval.

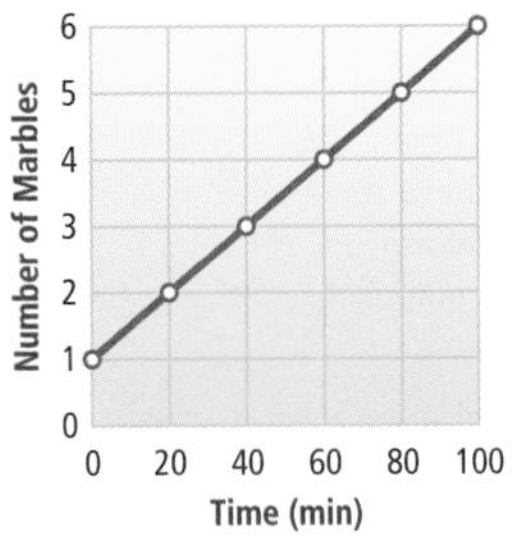

Graph 2. Suppose a bacteria colony begins with one bacteria, and every minute, all the bacteria divide, forming two.

The graph shows: $2x = y$
The slope of the line gets steeper at each interval.

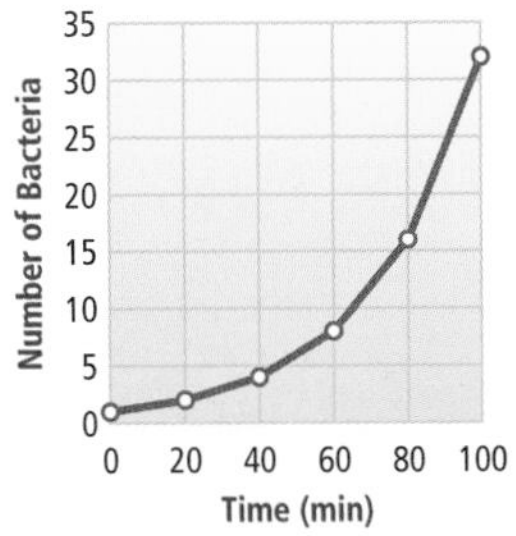

Examine the graphs and answer the questions.

1. How many marbles are in the collection after 3 sec? How many bacteria are in the colony after 3 sec?
2. After 7 sec, what number of marbles would show in graph 1? Name the coordinates for this point. What number of bacteria would be shown in graph 2? Name the coordinates.
3. Copy the two graphs on graph paper. Extend each graph to 10 sec. Plot the growth according to the pattern or formula given.

CHALLENGE Suppose the bacteria have a lifespan of 10 sec. How many bacteria will be in the colony after 20 sec?

15 C

ANSWERS

1. 4 marbles; 8 bacteria

2. 8 marbles; (7, 8); 128 bacteria; (7, 128)

3. Graph 1 will continue to be a straight line. Graph 2 will be a steep curved line.

CHALLENGE *about 1,000,000 (2^{20} = 1,048,576). Since each bacteria divides ten times before it dies, the lifespan limitation does not significantly affect the size of the colony.*

MATH IN SCIENCE
Math Skills Practice for Science

Set Learning Goal

To compare line graphs of linear and exponential growth

Present the Science

When a diluted sample of bacteria is spread on a nutrient agar surface, the individual bacterial cells begin to divide. In 24 to 48 hours, they form bacterial colonies that can be seen with the naked eye. Each colony contains many generations of bacteria that arose from a single cell.

Develop Graphing Skills

- Remind students that the independent variable is plotted along the *x*-axis and the dependent variable is plotted on the *y*-axis.
- In both graphs, time is the independent variable. Number of bacteria and number of marbles are the dependent variables.
- Discuss the meaning of *x* and *y* in the equations.

DIFFERENTIATION TIP Review the concept of variables with students who need extra help.

Close

Ask: How are the two growth patterns similar at the beginning of growth? After growth has occurred for several minutes? *Graph 2 starts increasing similar to graph 1, but numbers of bacteria soon exceed the number of marbles.*

R
- Math Support, p. 59
- Math Practice, p. 60

Technology Resources

Students can visit **ClassZone.com** for practice making a line graph.

MATH TUTORIAL

1.2 FOCUS

Set Learning Goals

Students will

- Give reasons that bacteria are the simplest living things.
- Identify two groups of single-celled organisms.
- Recognize that bacteria may help or harm other organisms.

3-Minute Warm-Up

Display Transparency 4 or copy this exercise on the board:

Decide if these statements are true. If not true, correct them.

1. If an object cannot grow, it is not alive. *true*
2. Viruses are alive because they can reproduce. *Viruses are not alive because they reproduce only within host cells.*
3. Living things are divided into three kingdoms. *Living things are divided into six kingdoms.*

3-Minute Warm-Up, p. T4

1.2 MOTIVATE

THINK ABOUT

PURPOSE To examine the idea that bacteria live everywhere

DEMONSTRATE Show students the well-known electron micrograph of bacteria on the point of a pin, which can be found in many biology texts. See if students can identify what it shows. Ask why a pinprick can become infected.

Ongoing Assessment

Give reasons that bacteria are the simplest living things.

Ask: Why are bacteria thought to be simple? *They are one cell and have no nucleus.*

1.2

KEY CONCEPT

Bacteria are single-celled organisms without nuclei.

BEFORE, you learned
- Organisms come in all shapes and sizes
- All living things share common characteristics
- Living things may be divided into six kingdoms

NOW, you will learn
- About the simplest living things
- About bacteria and archaea
- That bacteria may help or harm other organisms

VOCABULARY

bacteria p. 16
archaea p. 18
producer p. 19
decomposer p. 19
parasite p. 19

THINK ABOUT

Where are bacteria?

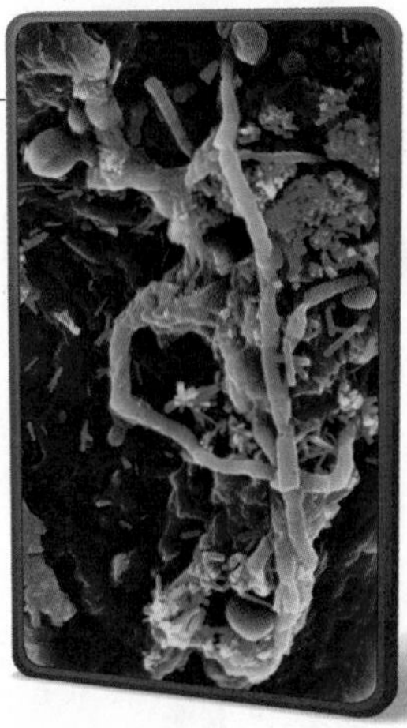

Bacteria are the simplest form of life. But that doesn't mean they're not important or numerous. As you look about the room you're sitting in, try to think of places where you might find bacteria. In fact, bacteria are on the walls, in the air, on the floor, and on your skin. It's hard to think of a place where you wouldn't find bacteria. The photograph shows a magnification of bacteria living on a sponge. The bacteria are magnified 580 ×. There are hundreds of millions of bacteria on your skin right now. And there are trillions of bacteria that live inside your intestines and help you digest food.

Bacteria and archaea are the smallest living things.

MAIN IDEA WEB
Make a web of the important terms and details about the main idea: *Bacteria and archaea are the smallest living things.* Be sure to include how bacteria are classified.

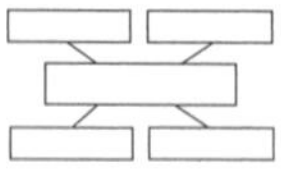

The names of the organisms belonging in the kingdoms Archaea and Bacteria are probably unfamiliar. Yet you actually encounter these organisms every day. Bacteria are everywhere: on your skin, in the ground, in puddles and ponds, in the soil, and in the sea. About 300 species of bacteria are living in your mouth right now.

Bacteria are the simplest kind of life known on Earth. All bacteria are composed of just one cell without a nucleus. Their genetic material is contained in a single loop within the cell. A bacterium reproduces using binary fission.

RESOURCES FOR DIFFERENTIATED INSTRUCTION

Below Level

UNIT RESOURCE BOOK
- Reading Study Guide A, pp. 24–25
- Decoding Support, p. 58

AUDIO CDS

Advanced

UNIT RESOURCE BOOK
Challenge and Extension, p. 30

English Learners

UNIT RESOURCE BOOK
Spanish Reading Study Guide, pp. 28–29

AUDIO CDS
- Audio Readings in Spanish
- Audio Readings (English)

Bacteria

All bacteria are single cells without nuclei.

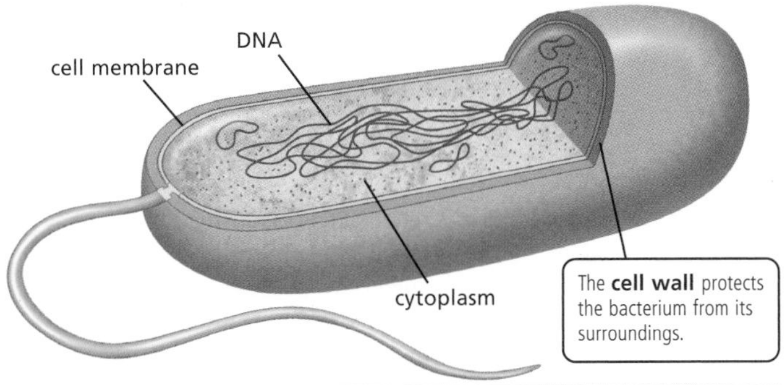
DNA
cell membrane
cytoplasm
The **cell wall** protects the bacterium from its surroundings.

Bacterial cells are different from the cells of other organisms. A bacterial cell is about 1/10 to 1/20 the size of a typical cell from organisms such as animals, plants, fungi, or protists. These four groups include organisms made up of cells with true nuclei. The nucleus is a structure that is enclosed by a membrane and that holds the genetic material.

READING TIP The plural of *bacterium* is *bacteria,* and the plural of *nucleus* is *nuclei.*

Despite their small size, bacteria are simple only when compared with more complex cells. Bacteria are much more complex than viruses, because they have many internal structures that viruses do not have. For example, one important feature of most bacteria is a covering called a cell wall, which surrounds and protects the soft cell membrane like a rain jacket. Bacterial cells contain many large molecules and structures that are not found in viruses.

Spiral Bacteria

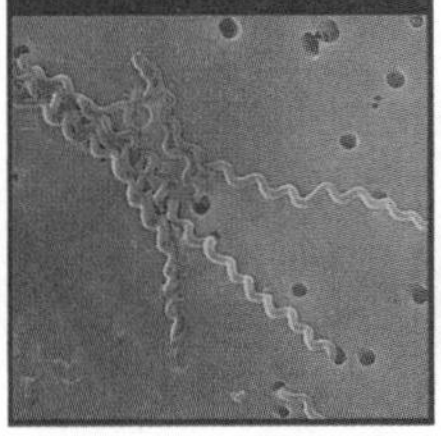

Rod Bacteria

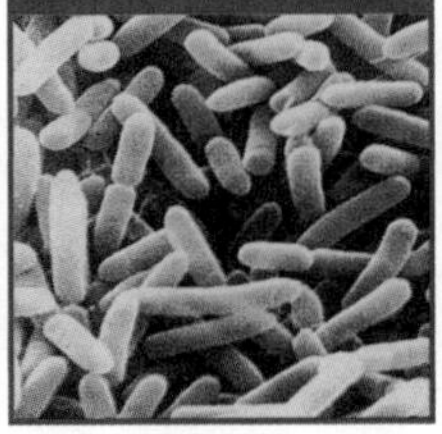

Round Bacteria

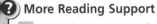

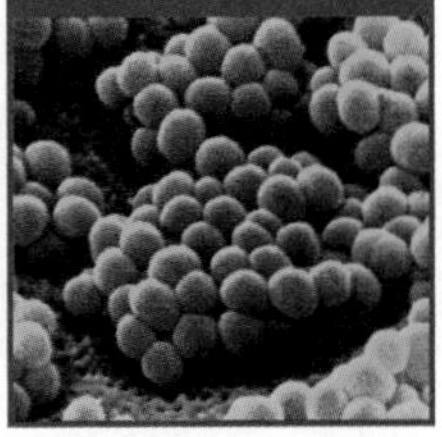

Scientists often classify bacteria by their external shapes.

- Spiral-shaped bacteria occur in single strands.
- Rod-shaped bacteria may occur singly or in chains.
- Round-shaped bacteria may occur singly or in pairs, chains, or clusters.

CHECK YOUR READING Name two features that all bacteria share.

1.2 INSTRUCT

Teach from Visuals

To help students interpret the diagram of bacterial structure, ask:

- Where is the bacterium's DNA located? *in the cytoplasm*
- Why do you think a bacterial cell needs both a cell wall and a cell membrane? *The cell wall protects the cell from injury. The cell membrane regulates what materials move in and out of the cell.*

Address Misconceptions

IDENTIFY Ask: What do bacteria do? Students may answer that bacteria cause illness or help other organisms. If students do not mention growing, breathing, eating, reproducing, or other functions characteristic of life, they may think that bacteria are not alive.

CORRECT Have students list the characteristics of living things and determine how bacteria fulfill each of them. Have reference books available for student use.

REASSESS Ask: How do bacteria get food? *Some bacteria make their own food. Others get nutrients from their environment.*

Technology Resources

Visit **ClassZone.com** for background on common student misconceptions.

MISCONCEPTION DATABASE

Ongoing Assessment

Recognize that bacteria come in three basic shapes.

Ask: What three shapes can bacteria have? *spirals, rods, and spheres*

CHECK YOUR READING *Answer: Bacteria are single-celled and have no nuclei.*

DIFFERENTIATE INSTRUCTION

More Reading Support

A How does a bacterial cell compare in size with other cells? *Bacterial cells are much smaller.*

B Are bacteria more or less complex than viruses? *more*

English Learners Place the words *producer, decomposer,* and *parasite* on the classroom's Science Word Wall with brief definitions of each. Have students draw and label their own models of bacteria as shown above. They can choose to draw the spiral, rod, or round bacteria shapes. They should label the *cell membrane, DNA, cytoplasm,* and *cell wall.*

Teach from Visuals

To help students interpret the photographs of archaea, ask: What do the environments of the organisms in the three photographs have in common? *They all have extreme conditions that most organisms cannot tolerate.*

Integrate the Sciences

In general, archaea are more similar to the rest of the living world than to true bacteria. For this reason, it is thought that archaea are the ancestors of all living things except true bacteria. Most of the differences between true bacteria and archaea are chemical. Bacterial cell walls contain peptidoglycan; archaea cell walls do not. The lipids in bacterial cell membranes are all unbranched hydrocarbons; archaea cell membranes contain some branched hydrocarbons. True bacteria are affected by certain antibiotics that do not affect archaea.

Real World Example

Until recently, scientists thought archaea were found only in extreme environments. In the past 10 years, however, many members of the kingdom Archaea have been found living with other organisms in the more moderate habitats of the oceans. It is now thought that archaea may be as ubiquitous as true bacteria.

Ongoing Assessment

Identify two groups of single-celled organisms.

Ask: What two groups of organisms are single-celled and have no nuclei? *bacteria and archaea*

Archaea and bacteria are found in many environments.

RESOURCE CENTER CLASSZONE.COM

Find out more about the many different types of bacteria.

Two types of single-celled organisms do not have nuclei. Bacteria are the most common and can be found in nearly every environment. Archaea are similar in size to bacteria, but share more characteristics with the cells of complex organisms like plants and animals.

Archaea

Archaea (AHR-kee-uh) are single-celled organisms that can survive in the largest range of environments. These environments may be very hot, very cold, or contain so much of a substance such as salt that most living things would be poisoned. As a result, scientists often group archaea according to where they live.

Methanogens take their name from methane, the natural gas they produce. These archaea die if they are exposed to oxygen. They may live in the dense mud of swamps and marshes, and in the guts of animals such as cows and termites.

Halophiles live in very salty lakes and ponds. Some halophiles die if their water is not salty enough. When a salty pond dries up, so do the halophiles. They can survive drying and begin dividing again when water returns to the pond.

Thermophiles are archaea that thrive in extreme heat or cold. They may live in hot environments such as hot springs, near hot vents deep under the sea, or buried many meters deep in the ice.

READING TIP

The word halophile is formed using the root word *halo-* which means "salt," and the suffix *–phile* which means "love." Therefore, a *halophile* is a "salt lover."

Archaea

Archaea are organisms that can live in extreme environments.

Methanogens

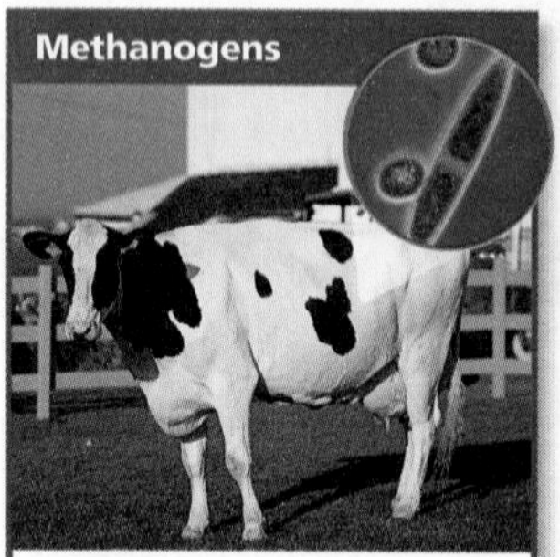

Methanogens maybe found in a cows' stomach where they help with digestion.

Halophiles

Halophiles can be found in extremely salty bodies of water such as the Dead Sea.

Thermophiles

Thermophiles may be found in hot geysers such as this one in Yellowstone National Park.

DIFFERENTIATE INSTRUCTION

More Reading Support

C Where do halophiles live? *in very salty lakes and ponds*

D Where do thermophiles live? *in extreme heat or cold*

English Learners This section has a variety of introductory clauses and phrases. For example, "When a salty pond dries up, so do the halophiles." Have students find the comma that separates the introductory clause or phrase from the rest of the sentence. Then, help them locate the subject of the sentence.

C 18 Unit: Diversity of Living Things

Bacteria

Most single-celled organisms without a nuclei are classified as bacteria. Bacteria are found in almost every environment and perform a variety of tasks. Some bacteria contain chlorophyll. Using sunlight for energy, these bacteria are an important food source in oceans. These bacteria also release oxygen gas, which animals need to breathe.

CHECK YOUR READING What are some common traits of bacteria and archaea?

Bacteria without chlorophyll perform different tasks. Some bacteria break down parts of dead plants and animals to help recycle matter. Some bacteria release chemicals into the environment, providing a food source for other organisms. Scientists often group bacteria by the roles they play in the environment. Three of the most common roles are listed below.

Bacteria that transform energy from sunlight into energy that can be used by cells are called **producers.** These bacteria are a food source for organisms that cannot make their own food.

Decomposers get energy by breaking down materials in dead or decaying organisms. Decomposers help other organisms reuse materials found in decaying matter.

Parasites live in very close relationships either inside or on the surface of other organisms. Parasites harm their host organisms or host cells. Other bacteria live in close relationships with host organisms but are helpful to their hosts, or do not affect them.

VOCABULARY
Be sure to add description wheels for *producer, decomposer,* and *parasite* to your notebook.

Bacteria

Three roles bacteria play in the environment are shown below.

Producers

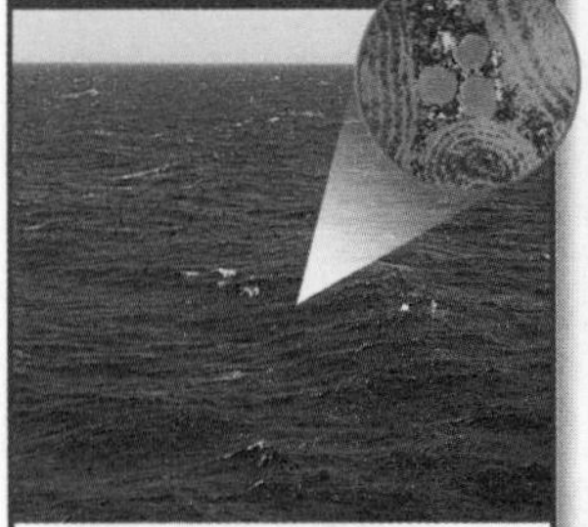

Cyanobacteria in Earth's oceans provide oxygen for animals to breathe.

Decomposers

Some bacteria break down dead wood into materials used by other organisms.

Parasites

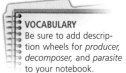

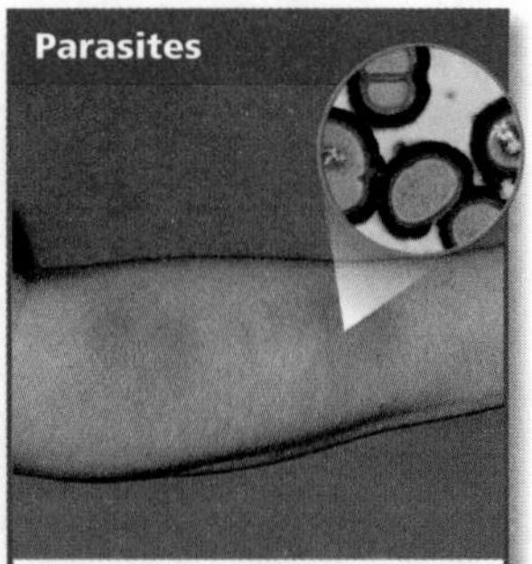

Staphylococcus bacteria cause infections such as these boils in humans.

Teach from Visuals

To help students interpret the photographs showing the roles that bacteria play, ask:

- How would our environment be different if cyanobacteria did not exist? *There probably would not be enough oxygen in the atmosphere for animals to breathe.*
- How would our environment be different if bacteria did not decompose materials? *Plants, insects, and fungus might not grow for lack of food and space.*
- How would our lives be different if no bacteria were parasites? *There would be no diseases caused by bacteria. Diseases caused by viruses would still exist, however.*

EXPLORE the BIG idea

Revisit "Internet Activity: Microscopic Life and You" on p. 7. Have students describe what they have learned about microscopic life.

Teacher Demo

Set up a microscope so students can observe *Lactobacillus,* the bacteria that is used to make yogurt. Use distilled water to thin some yogurt containing live bacteria. Put a drop of diluted yogurt on a microscope slide. Add a cover slip. Have students observe the bacteria under the highest power of the microscope.

Ongoing Assessment

CHECK YOUR READING *Answer: Archaea and bacteria are both single-celled and have no nuclei.*

DIFFERENTIATE INSTRUCTION

More Reading Support

E How do producers get energy? *from sunlight*

F Where do bacterial parasites live? *inside or on other organisms*

Advanced Have students research the organisms usually found in a compost pile. What conditions are there, and what is the source of the bacteria? Students can set up a small compost pile in a plastic or glass container.

R Challenge and Extension, p. 30

Below Level Have students use a Venn diagram to compare archaea and bacteria.

Teach from Visuals

To help students interpret the photograph of helpful bacteria, ask:

- Where does the nitrogen gas that bacteria fix come from? *from the air*
- Why do plants need bacteria to fix nitrogen? *Plants cannot use nitrogen gas to make proteins. The nitrogen must be converted into a form that the plants can use.*

Address Misconceptions

IDENTIFY Ask students what a "germ" is and whether all bacteria are germs. If students respond that germs cause disease and that all bacteria are germs, they probably hold the misconception that all bacteria are harmful.

CORRECT Have students list the ways bacteria are helpful to other organisms or to the environment. *They release oxygen; fix nitrogen; break down dead organisms into chemicals that living things can use; clean up sewage, water, and oil spills; and aid in digestion.* Have students explain how each of the listed activities is helpful to other organisms. For example, bacteria that release oxygen help animals to breathe. Use a two-column chart such as the one below:

Bacteria's Action	How Helpful
O_2 release	animals breathe
fix nitrogen	nutrient for plants

REASSESS Ask: Why do you think people have this misconception that all bacteria are bad? *People think they are more affected by disease organisms than by helpful bacteria.*

Technology Resources

Visit **ClassZone.com** for background on common student misconceptions.

MISCONCEPTION DATABASE

Real World Example

Bacteria help produce sauerkraut, pickles, vinegar, yogurt, cheese, sour cream, and other foods. Our diets would be limited without the help of bacteria.

Bacteria may help or harm other organisms.

Some bacteria, such as producers and decomposers, are helpful to other organisms. But other bacteria can be harmful. These bacteria can causes diseases in animals and plants.

Helpful Bacteria

One shovelful of ordinary soil contains trillions of bacteria, and every fallen leaf or dead animal is covered with bacteria. These bacteria break down the matter in dead bodies and waste materials. Broken-down materials may become available for other organisms to build their bodies.

Cities use bacteria to break down sewage. Bacteria in sewage-treatment plants live on the material dissolved in liquid sewage. After the bacteria have finished, the water is clean enough to sterilize. Then water can be released into rivers or oceans. Other bacteria are used to clean up oil spills by decomposing oil suspended on the ocean's surface.

? G Bacteria can also change materials that do not come from living things and make them available for other organisms. For example, some bacteria can convert nitrogen gas to nitrogen compounds. This process, called nitrogen fixation, makes nitrogen available to plants in a form that is useful to them. Plants use this nitrogen in making proteins, which are an important part of every cell.

Helpful Bacteria

Bacteria inside the root nodules of soybean plants convert nitrogen into a form the plant can use.

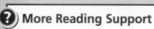

DIFFERENTIATE INSTRUCTION

More Reading Support

G What do you call the conversion of nitrogen gas into nitrogen compounds by bacteria? *nitrogen fixation*

Alternative Assessment Invite a restaurant owner or manager to visit your classroom. Before the visit, have students write questions about the precautions the restaurant takes to prevent the growth of bacteria in food. Have students discuss with the restaurant owner general practices to ensure food safety.

Like bacteria, certain types of archaea are helpful to other organisms. All animals that eat plants such as grass depend on archaea. Methanogens help break down the cellulose in cell walls. Termites and cows are two examples of animals that can digest cellulose because of the archaea in their stomachs.

CHECK YOUR READING **Name two helpful roles that bacteria can play in the environment.**

Harmful Bacteria

Not all bacteria are helpful to other organisms. Scientists first discovered that bacteria cause some diseases in the late 1800s. Much of the scientific research into harmful bacteria developed because bacteria cause disease in humans. Tuberculosis, cholera, and infant diarrhea are examples of disease caused by bacteria. Bacteria also may cause disease in many animals and in plants.

Bacteria can cause the symptoms of disease in three ways.

- They can invade parts of the body, multiplying in body tissues and dissolving cells.
- They can poison the body with chemicals they produce and release.
- They can poison the body with chemicals that are part of the bacteria itself.

Bacterial wilt causes disease in this pumpkin.

One way to fight bacterial disease is with vaccinations. Vaccines help individual organisms prepare to fight diseases they might encounter in the future. Humans, as well as cats and dogs, get vaccinations for bacterial diseases. A similar approach helps wildlife managers keep salmon safe from bacterial kidney disease.

1.2 Review

KEY CONCEPTS

1. Explain why bacteria are classified as living cells but viruses are not.
2. Name two main groups of bacteria.
3. Describe three ways that bacteria affect other organisms.

CRITICAL THINKING

4. **Visualize** Draw a diagram of a bacterium. Label the parts of the cell.
5. **Predict** Where in your neighborhood would you find most of the bacteria that cause decomposition?

CHALLENGE

6. **Analyze** Parasitic bacteria do not usually kill their hosts, at least not for a long period of time. Why is it better for parasites not to kill their hosts?

ANSWERS

1. Bacteria can grow, reproduce, and get nutrients from their environment. Viruses do not have structures that allow them to do these things.

2. bacteria and archaea

3. Producers provide oxygen and energy. Decomposers break down dead organisms. Parasites can cause disease in other organisms.

4. Diagrams should include the cell membrane, DNA, cytoplasm, and a cell wall. They should be spiral, round, or rod-shaped.

5. Sample answer: in soil and in a garbage can

6. If parasites kill their hosts, they cannot continue to grow and multiply in them and must find another host.

History of Science

Until the middle of the 19th century, many people died from bacterial infections after surgery or childbirth. Ignaz Philipp Semmelweis, a physician working in Vienna, demonstrated that the infection rate decreased when physicians washed their hands before treating patients. Later, Joseph Lister and Louis Pasteur confirmed Semmelweis's observations and identified the bacteria that caused many of the infections, thus proving the germ theory of disease.

Ongoing Assessment

CHECK YOUR READING *Answer: decompose organic matter, fix nitrogen*

Reinforce the BIG idea

Have students relate the section to the Big Idea.

R Reinforcing Key Concepts, p. 31

1.2 ASSESS & RETEACH

Assess

A Section 1.2 Quiz, p. 4

Reteach

Provide various materials that students can use to make models of bacterial and archaea cells. Yarn or string can represent DNA and a flagellum, gelatin can represent cytoplasm, a cross-section of a ball or tube can represent the cell wall.

Technology Resources

Have students visit **ClassZone.com** for reteaching of Key Concepts.

CONTENT REVIEW

CONTENT REVIEW CD-ROM

CHAPTER INVESTIGATION

Focus

PURPOSE To take samples and determine quantities of bacteria that live on hands and in the environment

OVERVIEW Students will touch their fingers, a common object, and a swab of an area of their environment to petri dishes containing sterile nutrient agar. They will examine the petri dishes after they have incubated to see whether bacterial colonies have grown. Students will find the following:

- Their hands harbor bacteria.
- Common objects harbor bacteria.
- Bacteria are in all areas of the environment.

Lab Preparation

- Sterile nutrient agar petri dishes can be purchased from a biological supply house or can be made in your own lab following careful instructions.
- Sterile cotton swabs can be purchased in a drug store.
- Arrange to borrow an incubator if a dark, warm area is not available.
- Prior to the investigation, have students read through it. Have them prepare data tables. Or, distribute datasheets and rubrics.

UNIT RESOURCE BOOK, pp. 61–69

SCIENCE TOOLKIT, F14

Lab Management

- In steps 3–6, remind students to close lids immediately after adding fingerprints to or swabbing the agar.
- Remind students that each bacterial colony represents many generations of bacteria that arose from a single cell that was deposited on the agar.
- Note that any fuzzy growth in the petri dishes are molds (fungi), not bacteria.

SAFETY Remind students to wash their hands after handling the bacterial cultures.

CHAPTER INVESTIGATION

Bacteria

OVERVIEW AND PURPOSE People routinely wash themselves to keep clean. You probably take a bath or shower every day, and you also wash your hands. Your hands may appear to be perfectly clean, but appearances can be deceiving. Your hands pick up bacteria from the objects you touch. You cannot see or feel these microscopic organisms, but they are there, on your skin. In this activity you will

- sample bacteria in your environment
- sample bacteria on your hands

Problem (Write It Up)

Do you pick up bacteria from your environment?

Hypothesize (Write It Up)

Write a hypothesis about whether bacteria in your environment are transferred to your skin. Your hypothesis should take the form of an "If . . . , then . . . , because . . ." statement.

MATERIALS

- 3 covered petri dishes with sterile nutrient agar
- marker
- tape
- everyday object, like a coin or eraser
- sterile cotton swab
- hand lens

for Challenge:

- 1 covered petri dish with sterile nutrient agar
- sterile swab

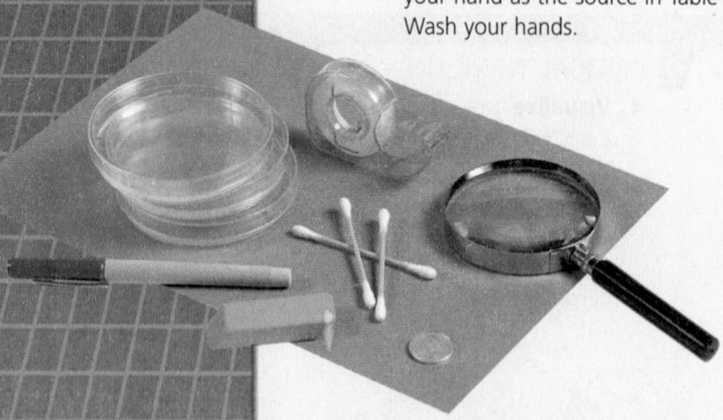

Procedure

1. Make a data table in your **Science Notebook** like the one shown on page 23.
2. Obtain three agar petri dishes. Be careful not to open the dishes.
3. Remove the lid from one dish and gently press two fingers onto the surface of the agar. Close the lid immediately. Tape the dish closed. Mark the tape with the letter A. Include your initials and the date. Mark your hand as the source in Table 1. Wash your hands.

step 3

INVESTIGATION RESOURCES

CHAPTER INVESTIGATION, Bacteria
- Level A, pp. 61–64
- Level B, pp. 65–68
- Level C, p. 69

Advanced students should complete Levels B & C.

Writing a Lab Report, D12–13

Technology Resources

Customize this student lab as needed or look for an alternative. Print rubrics to assess student lab reports.

Lab Generator CD-ROM

4 Choose a small object you handle every day, such as a coin or an eraser. Remove the lid from the second petri dish and swipe the object across the agar. You can instead use a sterile swab to rub on the object, and then swipe the swab across the agar. Close the lid immediately. Tape and mark the dish B, as in step 3. Include the source in Table 1.

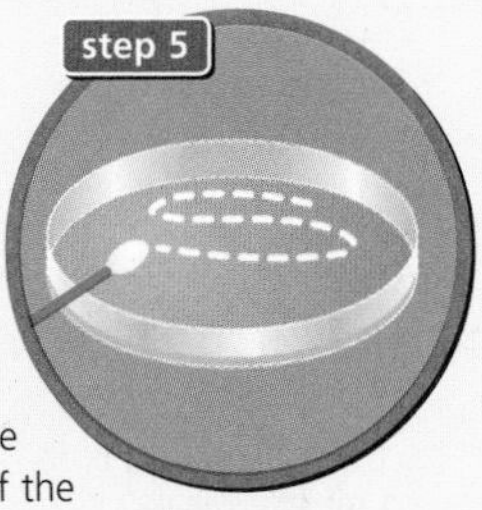

5 Choose an area of the classroom you have regular contact with, for example, the top of your desk or the classroom door. Use a clean swab to rub the area and then swipe the swab across the agar of the third petri dish. Tape and mark the dish as C, following the instructions in step 3. Dispose of the swab according to your teacher's instructions.

6 Place the agar plates upside down in a dark, warm place for two to three days. **CAUTION: Do not open the dishes. Wash your hands when you have finished.**

Observe and Analyze

Write It Up

1. **OBSERVE** Observe the dishes with the hand lens. You may want to pull the tape aside, but do not remove the covers. Include a description of the bacteria in Table 1. Are the bacteria in one dish different from the others?
2. **OBSERVE** Observe the amounts of bacterial growth in each dish and record your observations in Table 1. Which dish has the most bacterial growth? the least growth?
3. Return the petri dishes to your teacher for disposal. **CAUTION: Do not open the dishes. Wash your hands thoroughly with warm water and soap when you have finished.**

Conclude

Write It Up

1. **INFER** Why is it necessary for the agar to be sterile before you begin the experiment?
2. **INFER** What function does the agar serve?
3. **INTERPRET** Compare your results with your hypothesis. Do your observations support your hypothesis?
4. **IDENTIFY LIMITS** What limits are there in making a connection between the bacteria in dish A and those in dishes B and C?
5. **EVALUATE** Why is it important to keep the petri dishes covered?
6. **APPLY** Why is it important to use separate petri dishes for each sample?

INVESTIGATE Further

CHALLENGE Contamination can be a problem in any experiment involving bacteria, because bacteria are everywhere. Obtain a petri dish from your teacher. Swipe a sterile swab on the agar and place the agar plate upside down in a dark, warm place for two to three days. Do the results of this test make you reevaluate your other lab results?

Bacteria

Table 1. Observations of Bacteria

Petri Dish	Source	Description of Bacteria	Amount of Bacteria
A	hand		
B			
C			

Observe and Analyze

Write It Up

1. *The colonies in different dishes may look alike, although they may be different bacteria. Most likely they will be different.*
2. *Results will depend on which surface has been cleaned most recently. The least growth will likely be in dish A or dish C. The greatest might be in any of A, B, or C.*

Conclude

Write It Up

1. *You need to be sure that the bacteria growing on the agar came from your sample and not from the agar itself.*
2. *It offers food and living space to the bacteria.*
3. *Answers should give any evidence that supported the hypothesis, or evidence that did not support it.*
4. *You cannot say definitely that an object or location is the source of the bacteria on your hands because bacteria are everywhere.*
5. *to prevent bacteria in the air from falling on the agar*
6. *to prevent cross-contamination*

INVESTIGATE Further

CHALLENGE This petri dish serves as a control. If bacterial colonies grow on this agar dish, you can infer that these bacteria came from the air or from the swab. If there is no growth on this dish, you can have confidence that most or all of the bacteria in dishes A–C came from the sample.

Post-Lab Discussion

- Ask: Why was it important to place the petri dishes in a warm place? *If the temperature was too low, bacteria might not grow.*
- Ask: Why do you think it was necessary to store the petri dishes upside down? *Water usually condenses on the inside of the petri dishes. This water might drip on the agar surface and spread the bacteria to other areas of the agar.*
- Ask: If bacteria are everywhere, how can you avoid getting sick? *Not all bacteria cause disease; wash your hands, keep hands away from face.*

1.3 FOCUS

Set Learning Goals

Students will

- Recognize that viruses consist of only genes and a protein coat.
- Describe how viruses harm host cells.
- Summarize how viruses use a cell's machinery to reproduce.
- Use a model to analyze how viral infections spread.

3-Minute Warm-Up

Display Transparency 5 or copy this exercise on the board:

Match the definition to the correct term.

Definitions

1. a very small living thing *d*
2. living thing that lives in or on another organism *c*
3. nonliving infectious particle *a*

Terms

a. virus
b. binary fission
c. parasite
d. microorganism
e. host cell

3-Minute Warm-Up, p. T5

1.3 MOTIVATE

EXPLORE Viruses

PURPOSE To model the way viruses were discovered

TIP *10 min.* The sieves must have a fine enough mesh to prevent sesame seeds from falling through. Make sure that there are no tear holes in the mesh.

WHAT DO YOU THINK? *Particle size; make the mesh finer or add a second type of filter such as a piece of filter paper.*

KEY CONCEPT

1.3 Viruses are not alive but affect all living things.

BEFORE, you learned

- Most organisms are made of a single cell
- Living things share common characteristics
- Viruses are not living things

NOW, you will learn

- About the structure of viruses
- How viruses use a cell's machinery to reproduce
- How viruses affect host cells

VOCABULARY

host cell p. 26

EXPLORE Viruses

How were viruses discovered?

PROCEDURE

1. Fill a small container with mixed sesame seeds and salt.
2. Holding the sieve over the paper plate, pour the mixture into the sieve.
3. Gently shake the sieve until nothing more falls through.
4. Using a hand lens, examine the material that fell through the sieve and the material that stayed in the sieve.

MATERIALS

- small container
- sesame seeds
- table salt
- small kitchen sieve
- paper plate
- hand lens

WHAT DO YOU THINK?

- What is the most important difference between the particles that got through the sieve and the particles that remained behind?
- How could you change your sieve to make it not let through both kinds of particles?

Viruses share some characteristics with living things.

MAIN IDEA WEB
Remember to make a web of the important terms and details about the main idea: *Viruses share some characteristics with living things.*

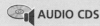

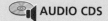

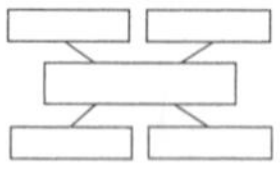

In the late 1800s, scientists such as Louis Pasteur showed that some small organisms can spoil food and cause disease. Once the cause was found, scientists looked for ways to prevent spoilage and disease. One method of prevention they found was removing these harmful organisms from liquids.

Bacteria may be removed from liquids by pouring the liquid through a filter, like a coffee filter or a sieve. To remove bacteria, a filter must have holes smaller than one millionth of a meter in diameter.

RESOURCES FOR DIFFERENTIATED INSTRUCTION

Below Level

UNIT RESOURCE BOOK

- Reading Study Guide A, pp. 34–35
- Decoding Support, p. 58

AUDIO CDS

Advanced

UNIT RESOURCE BOOK

Challenge and Extension, p. 40

English Learners

UNIT RESOURCE BOOK

Spanish Reading Study Guide, pp. 38–39

AUDIO CDS

- Audio Readings in Spanish
- Audio Readings (English)

INVESTIGATE Viruses

How do infections spread?

SKILL FOCUS
Analyzing

MATERIALS
- paper cup
- sample liquid
- solution A
- solution B

TIME
30 minutes

PROCEDURE

1. Get a cup of sample liquid from your teacher. Pour half the liquid from your cup into the cup of a classmate, then pour the same amount back into the original cup. Your cup should then contain a mixture of the liquids from both cups.
2. Repeat step 1 with at least two other classmates.
3. Drop one drop of solution A into your paper cup. If it changes color, you are "infected." If you were "infected," add drops of solution B until your liquid turns clear again. Count how many drops it takes to "cure" you.

WHAT DO YOU THINK?

- If you were "infected," can you figure out who "infected" you?
- If you were not "infected," is it possible for anyone who poured liquid into your cup to be "infected"?

CHALLENGE Only one person in your class started out with an "infection." Try to figure out who it was.

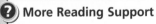

RESOURCE CENTER CLASSZONE.COM
Learn more about viruses.

When a filter had removed all of the harmful organisms from a liquid, the liquid no longer caused any illnesses. This method worked when there was only bacteria in the liquid. Sometimes filtering did not prevent disease. Something much smaller than bacteria was in the liquid. Scientists called these disease-causing particles viruses, from the Latin word for "slimy liquid" or "poison."

CHECK YOUR READING How does the size of viruses compare with the size of bacteria?

A Scientists have learned much about viruses, and can even make images of them with specialized microscopes. Viruses consist of genetic material contained inside a protective protein coat called a capsid. The protein coat may be a simple tube, such as the coat of an ebola virus, or have many layers, such as the smallpox virus shown on page 26.

Viruses may come in many shapes and sizes, but all viruses consist of a capsid and genetic material. The ability of viruses to make copies of their genetic material is one way that viruses are similar to living things. Also the protein coat is similar to a cell's outer membrane. But viruses do not grow, and viruses do not respond to changes in their environment. Therefore, viruses are not living organisms.

DIFFERENTIATE INSTRUCTION

More Reading Support

A How can viruses be seen? *with specialized microscopes*

English Learners In "Investigate Viruses" on p. 25, the repeated use of quotation marks around the word *infected* may be confusing to English learners. Explain that because the activity really is a model, no one is being infected. The quotes indicate that the infection is happening in thought only, or in an imagined situation.

1.3 INSTRUCT

INVESTIGATE Viruses

PURPOSE Use a model to analyze how viral infections spread

TIPS *30 min.* Give each student a cup of water, and one student a cup filled with a mixture of baking soda and water. (1000 mg per 100 mL). You may need 2 or 3 drops of solution *A* (phenolphthalein) to get a pink color for "infected" students. Tap water may also turn pink, so use distilled water if possible. If students repeat step 1 two more times, about eight people will be infected.

WHAT DO YOU THINK? *You could narrow down the possibilities but you couldn't figure out for sure who infected you. It is not possible for anyone who poured liquid into your cup to be infected if you are not.*

CHALLENGE *Students who were uninfected at the end can rule out all the people they shared liquid with. Students who were infected at the end will have trouble ruling out anyone they shared liquid with.*

R Datasheet, Viruses, p. 41

Technology Resources

Customize this student lab as needed or look for an alternative. Print rubrics to assess student lab reports.

Lab Generator CD-ROM

Metacognitive Strategy

Have students write a short paragraph explaining any difficulties or particular interest they found while doing the Challenge analysis of "Investigate Viruses." Ask: How does this type of analysis apply to the real world? *Sample answer: Epidemiologists use surveys about recent contact to track down the origins of outbreaks.*

Ongoing Assessment

CHECK YOUR READING *Answer: Viruses are much smaller than bacteria.*

Teach from Visuals

To help students interpret the visual of the structure of a virus, ask:

- Where is the virus's genetic material? *inside the virus*
- What part of the virus attaches to the host cell? *the capsid*

Address Misconceptions

IDENTIFY Ask: Why might it be a bad idea to go without a coat in cold or rainy weather? If students say it causes a cold or flu, they may hold misconceptions about how viruses infect living things.

CORRECT Have students review the process of virus infection (steps 1–5 on p. 26). Discuss whether the infectious process can be affected by cold or rainy weather.

REASSESS Have students write a short paragraph explaining why going without a coat in cold or rainy weather cannot give you a cold.

Technology Resources

Visit **ClassZone.com** for background on common student misconceptions.

MISCONCEPTION DATABASE

Teach Difficult Concepts

To help students understand how virus genetic material can take over the machinery of the host cell, ask:

- How can a long molecule of host DNA be destroyed? *It can be chopped up so no intact genes are left.*
- How can virus genes take over the cell? *They can copy themselves and code only for virus proteins.*

Ongoing Assessment

Recognize that viruses consist of only genes and a protein coat.

Ask: What is the function of the protein coat of a virus? *It protects the genes inside the virus.*

Describe how viruses harm host cells.

Ask: What happens to the host cell after the virus reproduces? *It usually bursts open and dies.*

Virus

All viruses, including this smallpox virus, contain genetic material surrounded by a capsid.

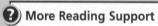

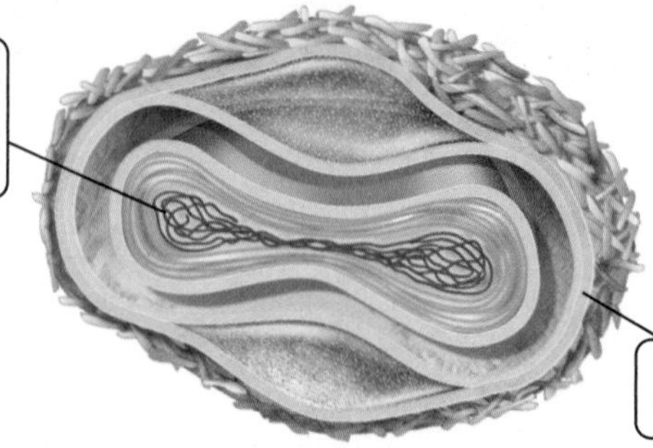
The **genetic material** stores information the virus needs to make copies of itself.
The **capsid** protects the genetic material and gives shape to the virus.

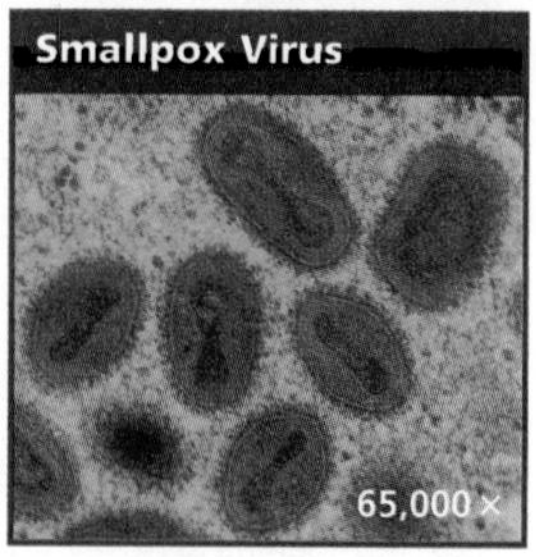
Smallpox Virus
65,000×

Viruses multiply inside living cells.

VISUALIZATION CLASSZONE.COM

See how viruses infect and multiply within bacteria.

Remember that all living things reproduce. Viruses cannot reproduce by themselves, which is one of the ways they are different from living things. However, viruses can use materials within living cells to make copies of themselves. The cells that viruses infect in order to make copies are called **host cells.** Despite their tiny size, viruses have the ability to cause a lot of damage to cells of other organisms. (B)

One of the best studied viruses infects bacteria. It's called a bacteriophage (bak-TEER-ee-uh-FAYJ), which comes from the Latin for "bacteria eater." Some of the steps that a bacteriophage goes through to multiply are shown in the illustration.

1. **Attachment** The virus attaches to the surface of a bacterium.
2. **Injection** The virus injects its DNA into the bacterium. (C)
3. **Production** Using the same machinery used by the host cell for copying its own DNA, the host cell makes copies of the viral DNA.
4. **Assembly** The viral DNA forces the infected cell to assemble new viruses from the parts it has created.
5. **Release** The cell bursts open, releasing 100 or more new viruses.

Viruses have proteins on their surfaces that look like the proteins that the host cell normally needs. The virus attaches itself to special sites on the host that are usually reserved for these proteins.

Not every virus makes copies in exactly the same way as the bacteriophage. Some viruses are inside host cells. Others use the host cell as a factory that produces new viruses one at a time. These viruses may not be as harmful to the infected organism because the host cell is not destroyed.

DIFFERENTIATE INSTRUCTION

More Reading Support

B What are host cells? *cells that are infected by viruses*

C How does virus DNA get into the host cell? *It is injected.*

Below Level Have students make a poster showing the steps of virus reproduction. They can use the poster for a class presentation, describing each of the five steps.

Making New Viruses

Viruses, such as this bacteriophage, use other cells to make new viruses.

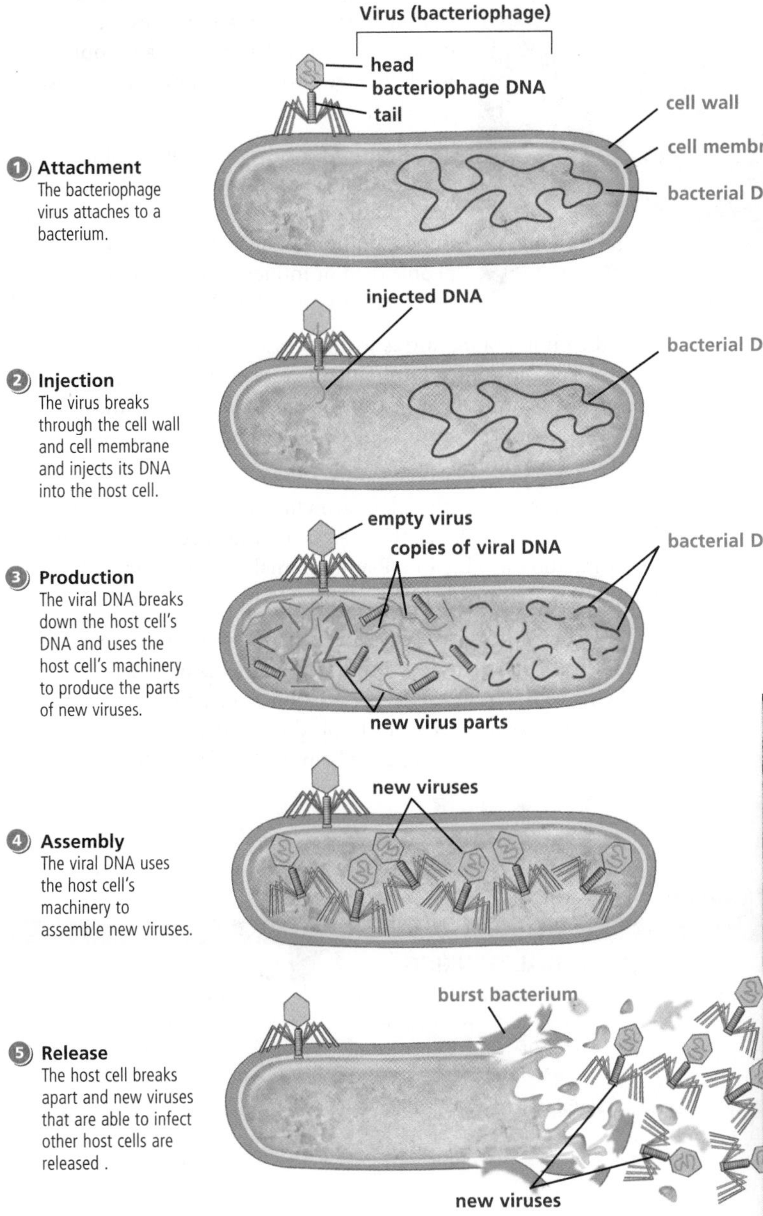

Virus (bacteriophage)
head
bacteriophage DNA
tail
cell wall
cell membrane
bacterial DNA
Host cell (bacterium)
1 Attachment
The bacteriophage virus attaches to a bacterium.
injected DNA
bacterial DNA
2 Injection
The virus breaks through the cell wall and cell membrane and injects its DNA into the host cell.
empty virus
copies of viral DNA
bacterial DNA pieces
3 Production
The viral DNA breaks down the host cell's DNA and uses the host cell's machinery to produce the parts of new viruses.
new virus parts
new viruses
4 Assembly
The viral DNA uses the host cell's machinery to assemble new viruses.
burst bacterium
5 Release
The host cell breaks apart and new viruses that are able to infect other host cells are released .
new viruses

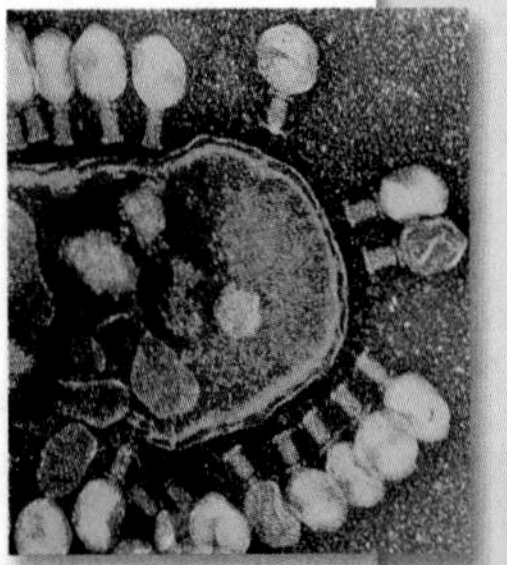

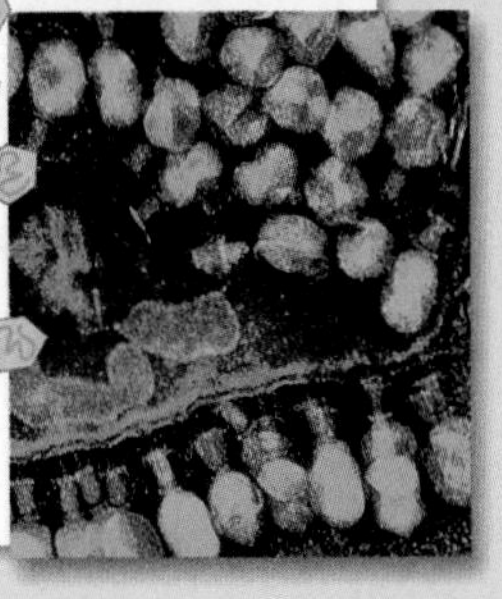

DIFFERENTIATE INSTRUCTION

Below Level Students can make a model of a bacteriophage using a large bolt, two nuts, and three pieces of wire. Screw the nuts onto the bolt so they touch the bolt head. Wrap the wires around the bottom of the bolt. Ask: What do the bolt head and nuts represent? *the head of the bacteriophage* The wires? *the tail fibers of the bacteriophage*

Advanced Have students find out how scientists study viruses. Ask them to assess whether it's possible to study them in the classroom.

R Challenge and Extension, p. 40

Teach from Visuals

To help students interpret the diagram of virus replication, ask:

- How does the bacteriophage in the diagram attach to the host cell? *Its tail attaches to special sites on the host.*
- What parts of the virus does the host cell's machinery make? *tails and DNA*
- How are new viruses assembled? *New virus DNA is inserted inside new capsids.*
- What happens to the host cell when the new viruses are released? *It bursts and dies.*

T The visual "Making New Viruses" is available as T6 in the Unit Transparency Book.

Teacher Demo

To demonstrate how easily viruses can spread from person to person, place some glitter on various classroom surfaces. Throughout the day, have students observe where the glitter has traveled. They will find glitter on their hands, clothing, and anything else they touch.

Develop Critical Thinking

COMPARE Have students compare computer viruses and the viruses they have been studying in this lesson. Discuss whether *virus* is an appropriate name for a computer virus. Ask: How does the method of "infection" of computer viruses compare with that of the viruses you are studying? *They can attach themselves to programs and destroy the program after reproducing themselves.*

Ongoing Assessment

Summarize how viruses use a cell's machinery to reproduce.

Ask: How does the virus disable the host cell's DNA? *The virus breaks up the DNA into pieces so that it can no longer function.*

History of Science

In 1967, the World Health Organization (WHO) began an international effort to eradicate smallpox. In 1980, WHO certified that smallpox had been eradicated worldwide. It has been estimated that if the eradication effort had not been made, 350 million additional victims and 40 million deaths would have occurred worldwide from 1980 to 2000.

Reinforce the BIG idea

Have students relate the section to the Big Idea.

R Reinforcing Key Concepts, p. 42

1.3 ASSESS & RETEACH

Assess

A Section 1.3 Quiz, p. 5

Reteach

Have students make a flipbook showing virus replication. They should make a title page and five labeled pages showing each step: attachment, injection, production, assembly, release. To draw the steps, they can refer to the diagram on p. 27. You might wish to have students create four to six "in between" sketches between each of the five "extreme" or "key" frames in their flipbooks.

Technology Resources

Have students visit **ClassZone.com** for reteaching of Key Concepts.

CONTENT REVIEW

CONTENT REVIEW CD-ROM

Rows of hospital beds are filled with Massachusetts influenza patients in 1918.

Viruses may harm host cells.

A host cell does not often benefit from providing living space for a virus. The virus uses the cell's material, energy, and processes. In many cases, after a virus has made many copies of itself, the new viruses burst out of the host cell and destroy it.

Harmful viruses cause huge problems. Viruses that cause diseases such as polio, smallpox, diphtheria, or AIDS have had a major impact on human history. About 25 million people died of influenza in an outbreak that occurred just after World War I.

In the photograph, nurses work to ease the symptoms of infected patients. The most infectious patients were enclosed in tents. Others were made as comfortable as possible on beds outside. Since viruses such as influenza can spread quickly, the camp was isolated from the rest of the community.

Plant viruses can stunt plant growth and kill plants. When plant viruses invade crop plants, they can cause much economic damage, decreasing food production. Plants, animals, bacteria, and all other living things are capable of being infected by viruses.

Today, scientists are discovering ways to use viruses in a positive way. Scientists use viruses to insert certain pieces of genetic material into living cells. For example, the portion of genetic material that allows some marine organisms to produce a chemical that glows can be inserted into tissue samples to help with their identification.

1.3 Review

KEY CONCEPTS

1. What are the two parts that every virus has?
2. Why are viruses not considered to be living things?
3. Explain how copies of viruses are produced.

CRITICAL THINKING

4. **Compare and Contrast** What features do viruses and cells have in common? How are they different?
5. **Explain** Summarize the steps by which a bacteriophage makes new viruses.

CHALLENGE

6. **Synthesize** What characteristics of viruses can make them so dangerous to humans and other living organisms?

ANSWERS

1. genetic material and a capsid

2. They cannot grow, respond, or reproduce outside a host cell.

3. Viruses invade host cells and use them as a factory to make new viruses.

4. Both contain genetic material. A viral capsid is like a cell wall. Cells have structures that grow and respond.

5. The bacteriophage attaches to the host cell and injects DNA into the cell. The genetic material takes over the cell's machinery to make new virus parts, which are assembled into new viruses. The new viruses escape when the host cell bursts.

6. Viruses easily spread from one living thing to another.

EXTREME SCIENCE

TULIPOMANIA

The Virus and the Tulip

The people of Holland around the 1620s were trading fortunes and farmland for one beautiful flower. Tulips had arrived from Turkey by way of Vienna, and interest in the flower spread through Holland like a fever. In a frenzy called "tulipomania," collectors paid as much as 5400 guilders, the price of a city house, for just one tulip bulb—but not any ordinary tulip.

Broken Flower Bulbs

Tulip traders searched for tulips with patterns, stripes, or feathery petals. These plants were called broken bulbs. Within a field of colored tulips, suddenly an odd or patterned flower grew. Once a color break showed up, it stayed with that flower's line into each new generation, until the line died off. And die off it did. The patterned petals were caused by a virus inside the plant, and the virus caused the flowers to weaken. The blooms got smaller in each generation.

The Mystery Source

While the trade in tulips rose, growers tried many tricks to produce the crazy patterns. Still, the broken bulbs grew rarely and randomly, or so it seemed. Viruses weren't discovered until 300 years later. Scientists then figured out that a virus had caused the broken bulbs.

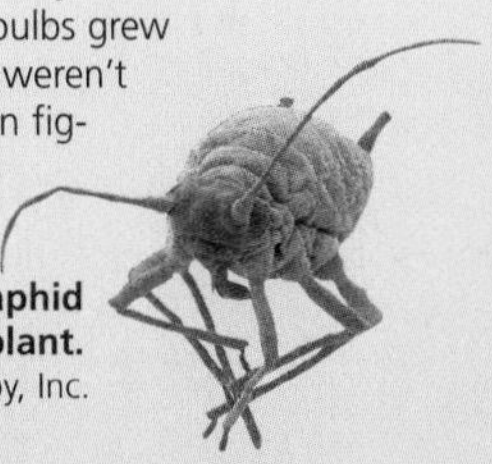

A small leaf-eating insect called an aphid had carried the virus from plant to plant.
© Dennis Kunkel/Dennis Kunkel Microscopy, Inc.

Back Down to the Ground

Like any goldrush, tulipomania crashed. People lost fortunes and fought over claims. In 1637, the government stopped all trading. Today, tulips with striped and feathery patterns grow around the world, but the patterns observed are not caused by viruses. Still, every now and then, a strange looking bulb appears. Instead of prizing it, growers remove it. They don't want a field of virus-infested, weakened plants.

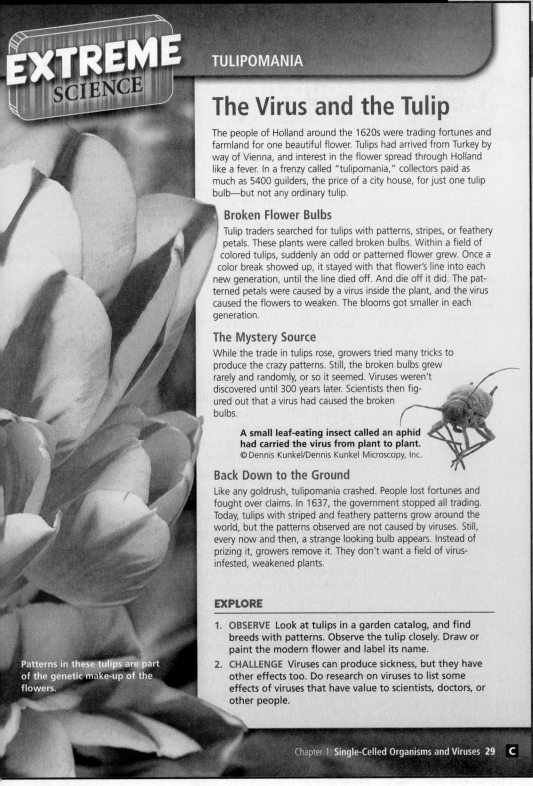

Patterns in these tulips are part of the genetic make-up of the flowers.

EXPLORE

1. **OBSERVE** Look at tulips in a garden catalog, and find breeds with patterns. Observe the tulip closely. Draw or paint the modern flower and label its name.
2. **CHALLENGE** Viruses can produce sickness, but they have other effects too. Do research on viruses to list some effects of viruses that have value to scientists, doctors, or other people.

EXPLORE

1. ***OBSERVE*** *Check students' drawings to be sure they are detailed and labeled.*
2. ***CHALLENGE*** *Viruses are currently being used in gene therapy and other forms of DNA technology. Some harmless viruses are used to make vaccines against serious diseases.*

EXTREME SCIENCE

Fun and Motivating Science

Set Learning Goal

To read independently to learn about one impact on history viruses have had

Present the Science

Tulips first appeared in western and central Asia, primarily in Armenia and Persia. From there, the tulip spread into areas along the Black Sea, the Mediterranean, and eastward into China. The Turks were growing tulips over a thousand years ago.

Discussion Questions

- Show students a map of the Netherlands. Ask them why they think that country became such a large commercial grower of tulips. *central location, port location for trading*
- Ask students how they think aphids carried the virus from plant to plant. *Aphids punch holes in plants with their mouthparts so they can suck the plant sap. When they puncture the leaves of an infected plant, they pick up virus particles on their mouthparts. When they bite another plant, they inject viruses into the plant.*

DIFFERENTIATION TIP This "Extreme Science" independent activity can be done in cooperative groups. Have groups of four students act as "Reader," "Recorder," "Leader," and "Reporter" to read the feature together, answer the questions, and present answers to the class.

Close

Ask: Do you think tulip collectors in the 17th century were wise to spend so much money on a rare tulip? Why or why not?

1.4 FOCUS

Set Learning Goals

Students will

- Recognize that most protists are single-celled.
- Recognize that all protists have cells with nuclei.
- Summarize how protists get energy in three different ways.
- Observe microscopic organisms in pond water.

3-Minute Warm-Up

Display Transparency 5 or copy this exercise on the board:

A student examines a drop of pond water under a microscope and notices a single-celled organism he has never seen before. The organism does not have a nucleus, but it has a cell wall and cell membrane. It swims around in the water. Can you conclude that this organism is a bacterium or a protist? Draw and label a diagram of this organism, based on your conclusion.
It is a bacterium.

3-Minute Warm-Up, p. T5

1.4 MOTIVATE

THINK ABOUT

PURPOSE To think about how limestone originated from ancient microorganisms

DISCUSS Have students suggest ways that limestone could have formed from protist shells. *The shells could have been greatly compacted under the weight of water and of other shells and debris on the ocean floor.*

KEY CONCEPT

1.4 Protists are a diverse group of organisms.

BEFORE, you learned

- Organisms are grouped into six kingdoms
- Bacteria are single-celled organisms without a nucleus
- Viruses are not living things

NOW, you will learn

- About characteristics of protists
- About the cell structure of protists
- How protists get their energy

VOCABULARY

algae p. 31
plankton p. 33
protozoa p. 34

THINK ABOUT

Where can protists be found?

Protists include the most complex single-celled organisms found on Earth. Fifty million years ago, a spiral-shelled protist called a nummulitid was common in some oceans. Even though its shell was the size of a coin, the organism inside was microscopic and single celled. When the organism died, the shells accumulated on the ocean floor. Over millions of years, the shells were changed into the rock called limestone. This limestone was used to build the great pyramids of Egypt. Some of the most monumental structures on Earth would not exist without organisms made of just a single cell.

Most protists are single-celled.

When Anton van Leeuwenhoek began using one of the world's first microscopes, he looked at pond water, among other things. He described, in his words, many "very little animalcules." Some of the organisms he saw probably were animals, microscopic but multicellular animals. However, many of the organisms Leeuwenhoek saw moving through the pond water had only a single cell. Today, more than 300 years later, scientists call these single-celled organisms protists.

Protists include all organisms with cells having nuclei and not belonging to the animal, plant, or fungi kingdoms. In other words, protists may be considered a collection of leftover organisms. As a result, protists are the most diverse of all the kingdoms.

C 30 Unit: **Diversity of Living Things**

RESOURCES FOR DIFFERENTIATED INSTRUCTION

Below Level

UNIT RESOURCE BOOK

- Reading Study Guide A, pp. 45–46
- Decoding Support, p. 58

AUDIO CDS

Additional INVESTIGATION, Growth of Algae, A, B, & C, pp. 70–78; Teacher Instructions, pp. 360–361

Advanced

UNIT RESOURCE BOOK
Challenge and Extension, p. 51

English Learners

UNIT RESOURCE BOOK
Spanish Reading Study Guide, pp. 49–50

AUDIO CDS

- Audio Readings in Spanish
- Audio Readings (English)

INVESTIGATE Protists

What lives in pond water?

PROCEDURE

1. Using a dropper, place one small drop of pond water in the center of a slide. Try to include some of the material from the bottom of the container.
2. Gently place a cover slip on the drop of water.
3. Observe the slide with a hand lens first.
4. Starting with low power, observe the slide with a microscope. Be sure to follow microscope safety procedures as outlined by your teacher. Carefully focus up and down on the water. If you see moving organisms, try to follow them by gently moving the slide.

WHAT DO YOU THINK?

- Describe and draw what you could see with the hand lens.
- Describe and draw what you could see with the microscope.
- Compare your observations with those of other students.

CHALLENGE Choose one organism that moves and observe it for some time. Describe its behavior.

SKILL FOCUS
Observing

MATERIALS
- dropper
- pond water
- slide
- hand lenses
- microscope

TIME
40 minutes

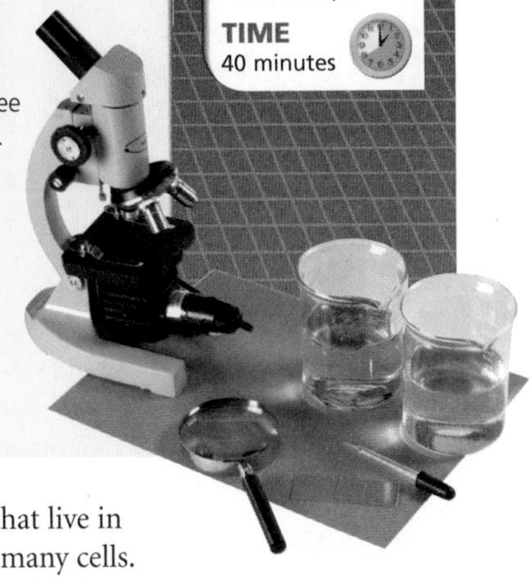

Most protists are single-celled, microscopic organisms that live in water. However, protists also include some organisms with many cells. These many-celled organisms have simpler structures than animals, plants, or fungi. They also have fewer types of cells in their bodies.

CHECK YOUR READING Why are protists considered the most diverse group of organisms?

The group of protists you're probably most familiar with is seaweeds. At first glance, seaweed looks like a plant. On closer inspection scientists see that it has a simpler structure. Some seaweeds called kelp can grow 100 meters long.

The name **algae** applies to both multicellular protists and single-celled protists that use sunlight as an energy source. Both seaweed and diatoms are types of algae. Slime molds are another type of multicellular protist.

VOCABULARY
Be sure to make a description wheel for *algae* and add to it as you read this section.

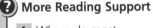

Given the many different types of organisms grouped together as protists, it is no surprise that protists play many roles in their environments. Algae are producers. They obtain energy from sunlight. Their cells provide food for many other organisms. These protists also produce oxygen, which is beneficial to many other organisms. Both of these roles are similar to those played by plants. Other protists act as parasites and can cause disease in many organisms, including humans.

DIFFERENTIATE INSTRUCTION

More Reading Support

A Where do most protists live? *in water*

B What are algae? *protists that use sunlight for energy*

English Learners The words *should, would,* and *could* are not universal in all languages but are common in academic English. Help students understand sentences that contain these. For example, "Some of the most monumental structures on Earth would not exist without organisms made of just a single cell."

1.4 INSTRUCT

INVESTIGATE Protists

PURPOSE To observe the movement and behavior of protists in pond water

TIPS *40 min.* Students can add a drop of methyl cellulose to their wet mounts to slow down the movement of rapidly swimming protists. Try modeling use of the microscope before students begin. Field of view circles may be used to aid viewing.

WHAT DO YOU THINK? *Students may see multicellular algae; many ciliates and flagellates moving rapidly, some green, others transparent; and amoebae moving slowly by extending. Students will see images more clearly and in greater detail with the microscope. Students should note new details after comparing observations.*

CHALLENGE *Students could see a paramecium, which moves rapidly. When it bumps into anything, it stops, backs away, and changes direction.*

R Datasheet, Protists, p. 52

Technology Resources

Customize this student lab as needed or look for an alternative. Print rubrics to assess student lab reports.

Lab Generator CD-ROM

Teaching with Technology

If a PC microscope is available, students can observe the organisms on a computer screen. If a microscope fitted with a camera is available, photograph the protists.

Ongoing Assessment

Recognize that most protists are single-celled.

Ask: Are all protists single-celled? Explain. *No; some are multicellular.*

CHECK YOUR READING *Answer: Protists include all organisms that don't fit into the other kingdoms.*

Teach from Visuals

To help students interpret the photographs of protists, ask:

- Which of the three protists in the photographs can move itself? *the euglena*
- How do you know? *It has a flagellum.*
- What is unusual about the shape of diatoms? *They are quite varied.*

Social Studies Connection

Seaweed and products derived from seaweed are important to the food processing industry and thus to the general economy. Large quantities of seaweed are cooked and eaten in Japan, China, Wales, and other countries. It is an excellent source of iron. A seaweed called nori is dried into sheets and used to make sushi. Red algae are a source of a gumlike substance called carrageenan, which is used to thicken marshmallows and ice cream and other dairy products.

Develop Critical Thinking

SYNTHESIZE Have students describe the different ways that protists move. After researching the topic, students can demonstrate how protists with flagella and cilia move and compare their movements with amoeboid movement, which is quite different. Have students make class presentations of their discoveries.

Ongoing Assessment

Recognize that all protists have cells with nuclei.

Ask: Is the structure of protist cells more similar to the cells of bacteria or plants and animals? Why? *plants and animals; all have nuclei*

CHECK YOUR READING *Answer: All three organisms capture the energy in sunlight and convert it to chemical energy. All three are green (contain chlorophyll).*

Diversity of Protists

Protists come in a variety of shapes and sizes.

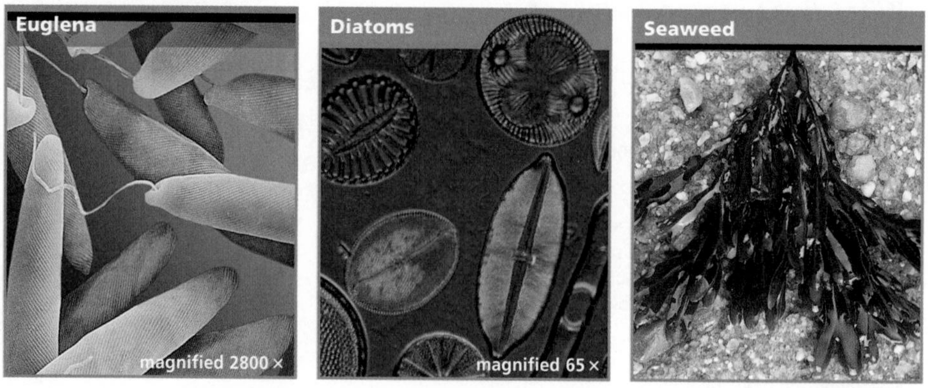

Protists live in any moist environment, including both freshwater and saltwater, and on the forest floor. Some protists move around in the water, some simply float in place, and some stick to surfaces. The photographs above show a small sample of the large variety of organisms that are called protists.

Seaweed is a multicellular protist that floats in the water and can be found washed up on beaches. Slime molds are organisms that attach to surfaces, absorbing nutrients from them. Diatoms are single-celled algae that float in water and are covered by hard shells. Euglena are single-celled organisms that can move like animals but also get energy from sunlight.

MAIN IDEA WEB
Remember to make a web of the important terms and details about the main idea: *Protists obtain their energy in three ways.* Include examples of each method of obtaining energy.

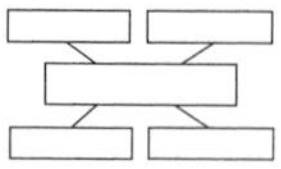

Protists obtain their energy in three ways.

D Protists can be classified by their way of getting energy. Some protists capture sunlight and convert it to usable energy. Another group of protists gets its energy from other organisms. A third group absorbs materials that contain stored energy.

Some protists, such as the euglena in the upper left photograph, can even switch from one mode of life to another. They swim rapidly through pond water like animals. If they receive enough sunlight, they look green and make their own food like plants. But if they are left in the dark long enough, they absorb nutrients from their environment like fungi.

CHECK YOUR READING Explain how the organisms in the photographs above get their energy.

DIFFERENTIATE INSTRUCTION

More Reading Support

C What is seaweed? *a multicellular protist that lives in the ocean*

D What is one way protists can be classified? *by the way they get energy*

Inclusion Have students make models of a euglena and a paramecium so that students can easily differentiate between flagella and cilia. A piece of string taped to a "euglena" made of clay can represent a flagellum. Shag carpeting, fake fur, or Velcro can represent cilia on a "paramecium."

Advanced Have students design an experiment to test a hypothesis about protists or pond water.

R Challenge and Extension, p. 51

Algae

Plantlike protists, called algae, get energy from sunlight. Like plants, they use the Sun's energy, water, and carbon dioxide from the air or water. Algae contain chlorophyll, a green pigment that is necessary to capture the Sun's energy. In the process of transforming energy from sunlight, algae release oxygen gas into the air. This important process, which is called photosynthesis, also takes place in plants and some bacteria. Organisms that perform photosynthesis also supply much of the food for other organisms.

Diatoms are examples of single-celled algae. Like all algae, a diatom contains a nucleus in which to store its genetic material. Diatoms also have chloroplasts, which are the energy-producing centers that contain chlorophyll.

Algae

Algae are plantlike protists. *Chlamydomonas* is an example of single-celled algae.

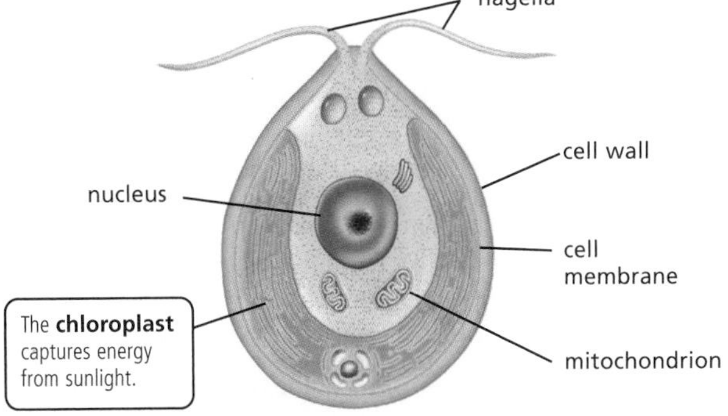

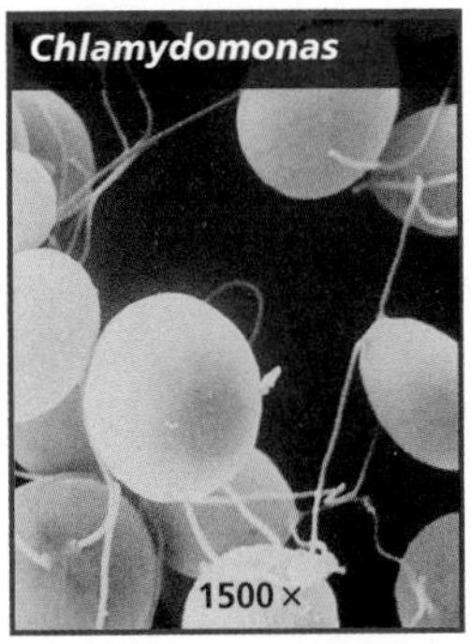

Another type of algae are microscopic colonies of nearly identical cells called volvox. These cells, arranged in a hollow ball, look like some single-celled algae. Sometimes cells break off from the hollow ball to form new colonies. The new colonies will eventually escape the parent colony. Colonial organisms such as volvox are the simplest kind of multicellular organisms. Seaweed is another example of multicellular algae.

All organisms that drift in water are called **plankton.** Plankton includes the young of many animals and some adult animals, as well as protists. Plankton that perform photosynthesis are called phytoplankton (plantlike plankton). Phytoplankton include algae and the cyanobacteria you learned about earlier. Phytoplankton live in all of the world's oceans and produce most of the oxygen animals breathe.

Teach from Visuals

To help students interpret the diagram of *Chlamydomonas* structure, ask:

- How does *Chlamydomonas* move? *using two flagella*
- Why is *Chlamydomonas* green? *Its chloroplast contains chlorophyll.*
- What is the function of the mitochondrion in the *Chlamydomonas* cell? *energy conversion*

Teacher Demo

Set up several microscopes around the room so students can observe a variety of protists. Show prepared slides of diatoms and wet mounts of volvox and amoebae.

Teach Difficult Concepts

Students may have trouble understanding that plankton is not a specific kind of organism. Tell students that *plankton* is a general term used to describe any organism that drifts in water currents. Many plankton organisms are single-celled bacteria and protists, but others are more complex multicellular plants and animals. Ask: Why is plankton so important to the environment? *It produces most of the oxygen that organisms breathe, and it is food for many marine animals.*

Arts Connection

Have students create an "underwater" drawing, painting, or model of pond life. An underwater effect can be made using oil pastels or crayons and washing over them with watercolor paint, or by simply using colored pencils and a watercolor wash. Their model or artwork should be based on sketching at an actual pond or from source materials, such as field guides. They should include at least one example each of algae and protozoa, as well as any animals or decaying matter found in the pond. *These works can be to scale or at an enlarged scale, but should specify the scale.*

DIFFERENTIATE INSTRUCTION

More Reading Support

E Where do algae get energy? *from sunlight*

F What are plankton? *tiny organisms that drift in water*

Additional Investigation To reinforce Section 1.4 learning goals, use the following full-period investigation:

R **Additional INVESTIGATION,** Growth of Algae, A, B, & C, pp. 70–78, 360–361
(Advanced students should complete Levels B and C.)

Below Level Give students a diagram of an alga like the one on this page. Have students label the parts of the organism.

Teach from Visuals

To help students interpret the diagram of paramecium structure, ask:

- How do you think food enters the oral groove? *Cilia sweep the food particles into the groove.*
- What clue in the photograph leads you to suspect that paramecia can move very rapidly? *The paramecium has thousands of cilia.*

History of Science

Anton van Leeuwenhoek was able to observe "little animalcules" with a microscope that he made himself. Although it had only a single lens, compared to the two or more lenses found in modern compound microscopes, the lens was very finely made, and van Leeuwenhoek could see many details. He spent many hours examining the life swimming in a drop of pond water. He kept a careful record of his observations, which still make interesting reading today.

Ongoing Assessment

CHECK YOUR READING *Answer: Protozoa eat other organisms and get energy from the nutrients in the organisms they eat. Algae get energy from sunlight.*

Protozoa

Protozoa, such as this *Paramecium*, are animallike protists.

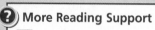

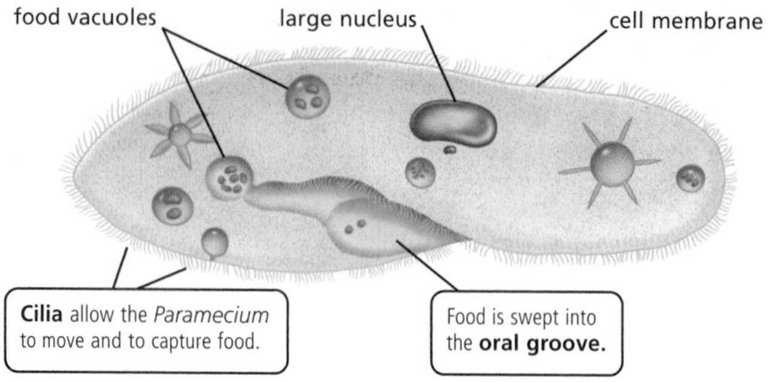
food vacuoles
large nucleus
cell membrane
Cilia allow the *Paramecium* to move and to capture food.
Food is swept into the **oral groove.**

Paramecium
40 ×

Protozoa

? G Protists that eat other organisms, or decaying parts of other organisms, are animal-like protists, or **protozoa.** They include many forms, all single-celled. Protozoa cannot use sunlight as a source of energy and they must move around to obtain the energy they need to survive. Certain chemicals in protozoa can recognize when a particle of food is nearby. The food particle is usually another organism or a part of one. The protozoan ingests the food and breaks it down to obtain energy.

Some animal-like protists swim rapidly, sweeping bacteria or other protists into a groove that looks like a mouth. One example, called a paramecium, is shown above. A paramecium moves about using thousands of short wavy strands called cilia.

Another group of protozoa swim with one or more long whiplike structures called flagella. A third group of protozoa has very flexible cells. Organisms such as the amoeba oozes along surfaces. When it encounters prey, the amoeba spreads out and wraps around its food.

REMINDER: A parasite is an organism that lives inside or on another organism and causes the organism harm.

? H A number of protists live as parasites, some of which cause disease in animals and humans. One of the world's most significant human diseases, malaria, is caused by a protist. A single species of mosquito carries the parasite from human to human. When the mosquito bites an infected human, it sucks up some of the parasite along with the blood. When that same mosquito bites another human, it passes on some of the parasite. In the human blood stream, the parasite goes through a complex life cycle, eventually destroying red blood cells.

CHECK YOUR READING How do protozoa and algae differ in the way they obtain energy?

DIFFERENTIATE INSTRUCTION

? More Reading Support

G What are protozoa? *animal-like protists that eat other organisms*

H What causes the disease malaria? *a protist that is carried by mosquitoes*

Advanced Have students design an experiment that will demonstrate that euglenas can switch the way they obtain energy from capturing sunlight to absorbing nutrients from their food.

Other Protists

Protists that absorb food from their environment can be called funguslike protists. These protists take in materials from the soil or from other organisms and break materials down in order to obtain energy. They are called decomposers.

The term *mold* refers to many organisms that produce a fuzzy-looking growth. Most of the molds you might be familiar with, like bread mold, are fungi. But three groups of protists are also called molds. These molds have structures that are too simple to be called fungi, and they are single celled for a portion of their lives. One example of a funguslike protist is water mold, which forms a fuzzy growth on food. This food may be decaying animal or plant tissue or living organisms. Water molds live mainly in fresh water.

Slime molds live on decaying plants on the forest floor. One kind of slime mold consists of microscopic single cells that ooze around, eating bacteria. When their food is scarce, however, many of the cells group together to produce a multicellular colony. The colony eventually produces a reproductive structure to release spores. Wind can carry spores about, and they sprout where they land.

A walk in a moist forest might give you a chance to see a third kind of mold. This organism looks like a fine net, like lace, several centimeters across, on rotting logs. These slime molds are not multicellular, but instead one giant cell with many nuclei. They are the plasmodial slime molds.

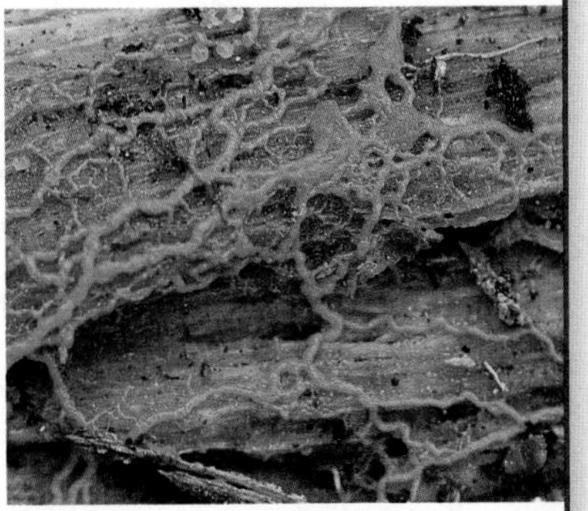

Plasmodial slime mold may grow on decaying wood after a period of rainy weather.

CHECK YOUR READING How do funguslike protists get energy?

1.4 Review

KEY CONCEPTS

1. What are two characteristics of protists?
2. What feature do all protists have in common?
3. What are the three different types of protists?

CRITICAL THINKING

4. **Provide Examples** Some protists are dependent on other organisms and cannot live independently. Other protists are independent and have organisms that depend on them. Give an example of each.

CHALLENGE

5. **Hypothesize** Scientists are considering reclassifying protists into many kingdoms. How might they decide how many kingdoms to use and how to place organisms in these kingdoms?

ANSWERS

1. Most protists are single-celled; all protist cells have nuclei.

2. All protist cells have nuclei.

3. animal-like protists, plant-like protists, and funguslike protists

4. A parasite such as the malaria-causing protist is dependent on its host. Free-swimming algae are independent, but provide oxygen for other organisms.

5. Students are likely to choose a classification scheme with three parts, based on energy sources, but other classifications are reasonable.

Real World Example

Many slime molds have very descriptive common names. Scrambled egg slime, toothpaste slime, wolf's milk slime, dog vomit slime, and troll butter are all slime molds that grow on damp rotted wood or garden mulch.

Ongoing Assessment

Summarize how protists get energy in three different ways.

Ask: In what three ways do protists get energy? *They capture energy from sunlight, get energy from other organisms, or absorb materials that contain stored energy.*

CHECK YOUR READING *Answer: Funguslike protists are decomposers that break down dead organisms and other organic matter to get energy.*

Reinforce the BIG idea

Have students relate the section to the Big Idea.

R Reinforcing Key Concepts, p. 53

1.4 ASSESS & RETEACH

Assess

A Section 1.4 Quiz, p. 6

Reteach

Have students make a chart summarizing the different groups of protists, if and how they move, how they get energy, and whether they are single-celled or multicellular.

Technology Resources

Have students visit **ClassZone.com** for reteaching of Key Concepts.

CONTENT REVIEW

CONTENT REVIEW CD-ROM

BACK TO

the BIG idea

Ask: What tests could scientists do to determine whether a microscopic object discovered on a plant is alive? *They could observe the object to see if it grows or reproduces. They could change the light and temperature conditions to see if it responds. They could examine it under a microscope to see if it has cellular organization.*

KEY CONCEPTS SUMMARY

SECTION 1.1
Ask: How are organisms grouped? *All living things are divided into six kingdoms according to their different characteristics.*

Ask: Are all organisms totally different from each other? *All living things share certain characteristics such as growth, organization, and reproduction.*

SECTION 1.2
Ask: How would you classify the bacteria shown in the photograph? *round bacteria that live in clusters*

Ask: What are three tasks bacteria perform in the environment? *producer, decomposer, parasite*

SECTION 1.3
Ask: What materials in the host cell does this virus use to make copies of itself? *host-cell machinery and DNA*

SECTION 1.4
Ask: In what way do algae, protozoans, and slime molds get energy? *Algae get energy from sunlight, protozoans get energy by eating other organisms, and slime molds get energy from the decomposed food they absorb.*

Review Concepts

- Big Idea Flow Chart, p. T1
- Chapter Outline, pp. T7–T8

1 Chapter Review

the BIG idea

Bacteria and protists have the characteristics of living things, while viruses are not alive.

CONTENT REVIEW
CLASSZONE.COM

KEY CONCEPTS SUMMARY

1.1 Single-celled organisms have all the characteristics of living things.

Scientists divide organisms into six **kingdoms.** All living things, including **microorganisms,** are organized, grow, reproduce, and respond to the environment.

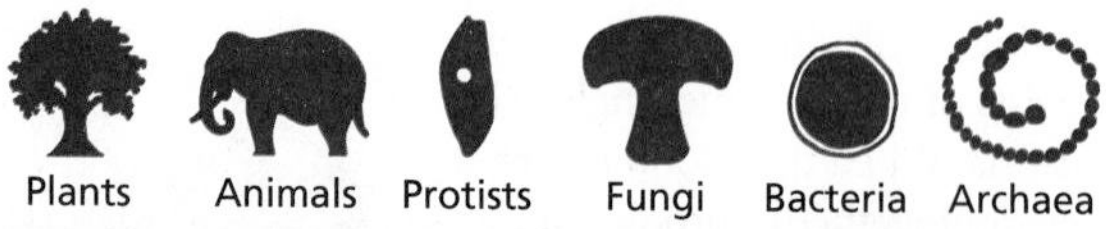

VOCABULARY
microorganism p. 10
kingdom p. 11
binary fission p. 12
virus p. 14

1.2 Bacteria are single-celled organisms without nuclei.

- Bacteria and archaea are the smallest living things.
- Archaea and bacteria are found in many environments.
- Bacteria may help or harm other organisms.

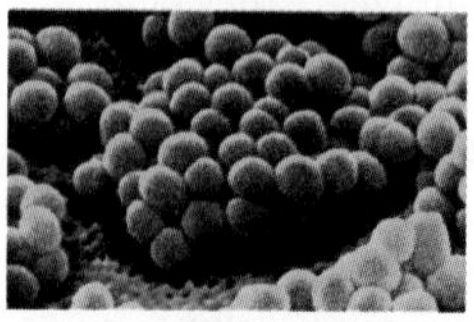

VOCABULARY
bacteria p. 16
archaea p. 17
producer p. 19
decomposer p. 19
parasite p. 19

1.3 Viruses are not alive but affect other living things.

A virus consists of genetic material enclosed in a protein coat. Viruses cannot reproduce on their own, but they use materials within living cells to make copies of themselves.

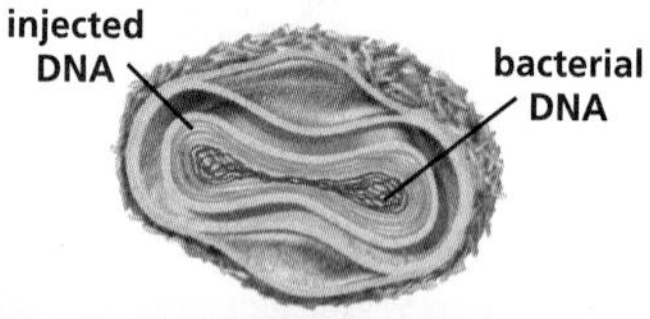

VOCABULARY
host cell p. 26

1.4 Protists are a diverse group of organisms.

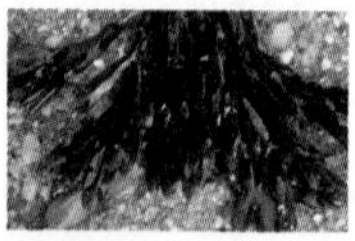

Plantlike **algae** get energy from sunlight.

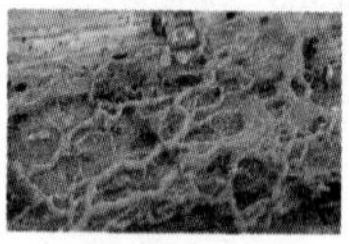

Funguslike protists are decomposers.

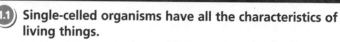

Protozoa, animal-like protists, eat other organisms.

VOCABULARY
algae p. 31
plankton p. 33
protozoa p. 34

C 36 Unit: Diversity of Living Things

Technology Resources

Have students visit **ClassZone.com** or use the CD-ROM for a cumulative review of concepts.

CONTENT REVIEW

CONTENT REVIEW CD-ROM

Engage students in a whole-class interactive review of Key Concepts. Edit content as you wish.

POWER PRESENTATIONS

Reviewing Vocabulary

Draw a triangle for each of the terms listed below. Define the term, use it in a sentence, and draw a picture to help you remember the term. An example is completed for you.

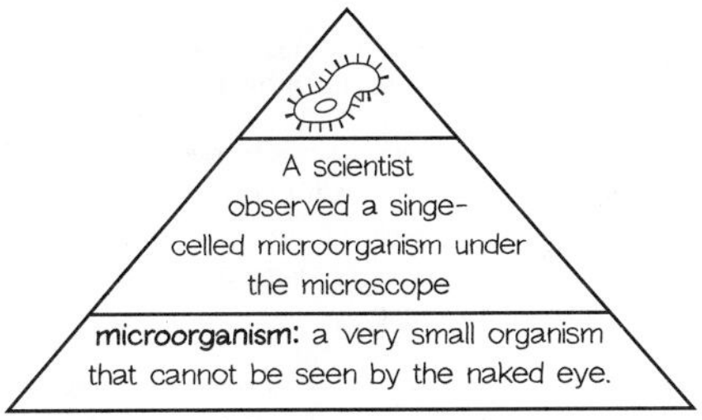
A scientist
observed a singe-
celled microorganism under
the microscope
microorganism: a very small organism
that cannot be seen by the naked eye.

1. binary fission

2. producer

3. virus

4. host cell

Describe how the vocabulary terms in the following pairs of words are related to each other. Explain the relationship in a one- or two-sentence answer. Underline each vocabulary term in your answers.

5. archaea, bacteria

6. microorganism, organism

7. decomposers, parasite

8. protists, algae

Reviewing Key Concepts

Multiple Choice *Choose the letter of the best answer.*

9. Which group is *not* a microscopic kingdom?
- **a.** fungi
- **b.** bacteria
- **c.** archaea
- **d.** protists

10. What happens in binary fission?
- **a.** DNA is combined into one cell.
- **b.** The daughter cells differ from the parent cell.
- **c.** Material from one cell is broken into two cells.
- **d.** One cell divides into four exact cells.

11. Which is a characteristic of a virus?
- **a.** obtains energy from sunlight
- **b.** responds to light and temperature
- **c.** doesn't contain DNA
- **d.** reproduces only within other cells

12. Which is the simplest type of organism on Earth?
- **a.** protists
- **b.** bacteria
- **c.** viruses
- **d.** parasites

13. Which statement about bacteria is *not* true?
- **a.** Bacteria reproduce using binary fission.
- **b.** Bacteria do not have a nucleus.
- **c.** Bacteria do not contain genetic material.
- **d.** Bacteria are either rod-, cone-, or spiral-shaped.

14. Archaea that can survive only in extreme temperatures are the
- **a.** methanogens
- **b.** halophiles
- **c.** thermophiles
- **d.** bacteria

15. A weakened viral or bacterial disease that is injected into the body is
- **a.** a filter
- **b.** a diatom
- **c.** a bacteriophage
- **d.** a vaccine

16. Which group of protists absorbs food from their environment?
- **a.** diatoms
- **b.** molds
- **c.** protozoa
- **d.** plankton

ASSESSMENT RESOURCES

UNIT ASSESSMENT BOOK
- Chapter Test A, pp. 7–10
- Chapter Test B, pp. 11–14
- Chapter Test C, pp. 15–18
- Alternative Assessment, pp. 19–20

SPANISH ASSESSMENT BOOK
Spanish Chapter Test, pp. 41–44

Technology Resources

Edit test items and answer choices.

Test Generator CD-ROM

Visit **ClassZone.com** to extend test practice.

Test Practice

Reviewing Vocabulary

1. Binary fission is a method of reproduction in single-celled organisms. Bacteria reproduce by a process called binary fission.

2. A producer is a type of bacteria that transforms energy from sunlight into energy that is used by the cell. Producers in Earth's oceans provide oxygen for animals to breathe.

3. A virus is a particle that contains genetic material surrounded by a capsid. Viruses cause diseases such as smallpox and polio.

4. Host cells are cells infected by a virus. Bacteria are host cells for bacteriophages.

5. Archaea and bacteria are both single-celled organisms without nuclei. Archaea are more similar to plant and animal cells and can survive in extreme environments.

6. A microorganism is a type of organism that is very small and can be seen only with a microscope.

7. Decomposers get energy by breaking down dead or decaying material. Parasites get energy from a living cell.

8. Protists include all organisms with nuclei that are not plants, animals, or fungi. Algae are protists that use sunlight as an energy source.

Reviewing Key Concepts

9. a
10. c
11. d
12. b
13. c
14. c
15. d
16. b

17. a

18. cellular organization, grow, reproduce, respond to their environment

19. Some bacteria cause disease.

20. protists and tiny animals that drift in the water

Thinking Critically

21. It would have organization, ability to grow and reproduce, respond to its environment.

22. Binary fission; cell duplicates its own DNA into two sides. Cell wall pinches in, forming two new cells.

23. Archaea may survive in extreme conditions. Their cells are more similar to other cells than bacteria.

24. They get energy by breaking down dead or decaying materials and recycling these materials.

25. 2^{10} bacteria. The bacteria double in number each hour.

26. less growth from washed finger than from unwashed finger

27. Viruses do not have the cell structures of bacteria, do not obtain energy from their environment, and cannot reproduce on their own.

28. provide living cells as hosts for virus reproduction

29. Algae act as producers and provide food for other organisms, including people. These release oxygen into the air for other organisms to use. Harmful protists such as some protozoa cause disease.

the BIG idea

30. Test structures to see if they meet the characteristics of life: organization, growth, response, reproduction. They are diatoms, single-celled algae protected by hard plates.

UNIT PROJECTS

Give students the appropriate Unit Project worksheets. Both directions and rubrics can be used as a guide.

R Unit Projects, pp. 5–10

---

17. Which obtains energy by feeding on other organisms?

a. amoeba **b.** algae **c.** phytoplankton **d.** mushroom

Short Answer *Write a short answer to each question.*

18. Briefly describe the characteristics that all living things share.

19. How are bacteria harmful to humans?

20. What are plankton?

Thinking Critically

21. APPLY Imagine you are a scientist on location in a rain forest in Brazil. You discover what you think might be a living organism. How would you be able to tell if the discovery is a living thing?

22. COMMUNICATE What process is shown in this photograph? Describe the sequence of events in the process shown.

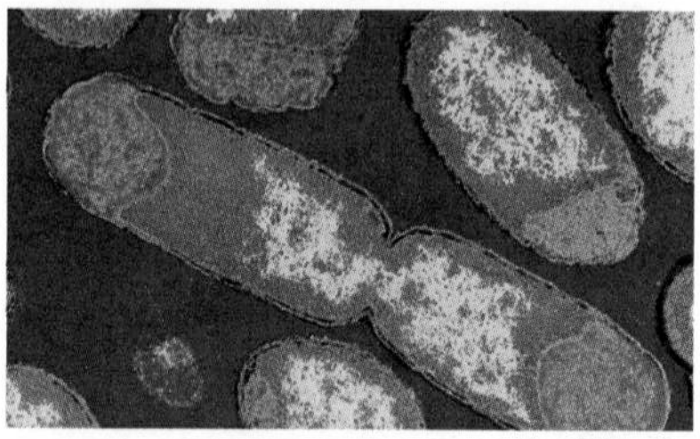

23. CLASSIFY Why are archaea classified in a separate kingdom from bacteria?

24. ANALYZE Why are some bacteria considered "nature's recyclers"? Explain the role that these bacteria play in the environment.

25. CALCULATE A bacterium reproduces every hour. Assuming the bacteria continue to reproduce at that rate, how many bacteria will there be after 10 hours? Explain how you know.

26. HYPOTHESIZE A student conducts an experiment to determine the effectiveness of washing hands on bacteria growth. He rubs an unwashed finger across an agar plate, then washes his hands and rubs the same finger across a second plate. What hypothesis might the student make for this experiment? Explain.

27. COMPARE AND CONTRAST Describe three ways that viruses differ from bacteria.

28. ANALYZE A scientist has grown cultures of bacteria on agar plates for study. Now the scientist wants to grow a culture of viruses in a laboratory for study. How might this be possible? Give an example.

29. PROVIDE EXAMPLES How are protists both helpful and harmful to humans? Give examples in your answer.

the BIG idea

31. INFER Look again at the picture on pages 6–7. Now that you have finished the chapter, how would you change or add details to your answer to the question on the photograph?

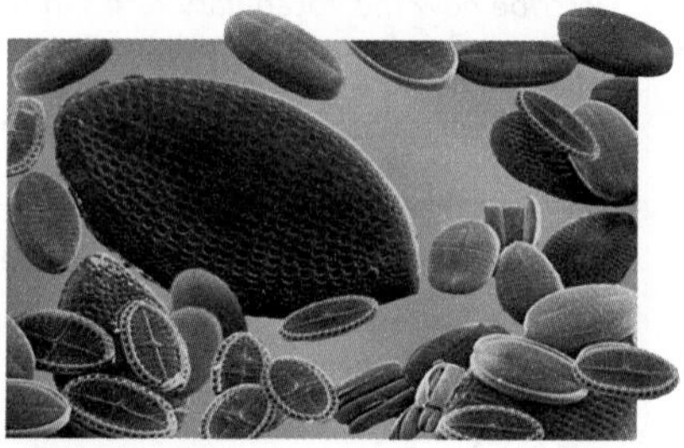

UNIT PROJECTS

If you are doing a unit project, make a folder for your project. Include in your folder a list of resources you will need, the date on which the project is due, and a schedule to track your progress. Begin gathering data.

MONITOR AND RETEACH

If students have trouble applying the concepts in items 22 and 25, review the process of binary fission. Have students draw a series of diagrams showing binary fission. The first diagram should show the cell's DNA being duplicated. The second diagram should show the cell's contents separating. The third diagram should show the cell wall pinching in to form two cells.

Students may benefit from summarizing one or more sections of the chapter.

R Summarizing the Chapter, pp. 79–80

Standardized Test Practice

For practice on your state test, go to . . . TEST PRACTICE CLASSZONE.COM

Analyzing Data

The graph below shows growth rates of bacteria at different temperatures.

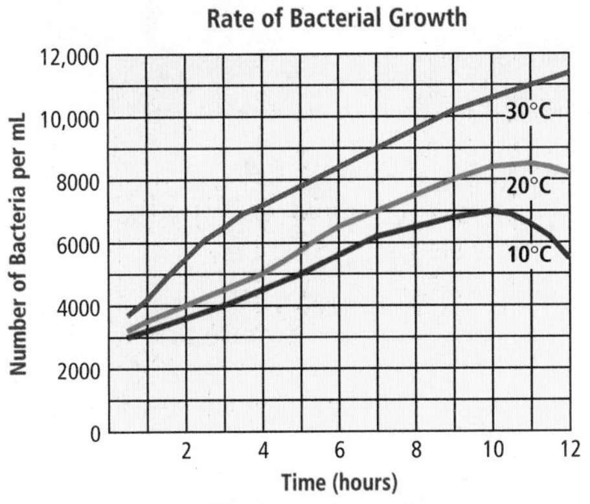

Choose the letter of the best answer.

1. At which temperature did growth of bacteria occur at the greatest rate?
- **a.** 0°C
- **b.** 10°C
- **c.** 20°C
- **d.** 30°C

2. Which statement is true about the growth rate of bacteria at 10°C?
- **a.** Bacteria grew rapidly at first, then declined after 6 hours.
- **b.** Bacteria growth increased at a steady rate.
- **c.** Bacteria grew slowly, then declined rapidly after 8 hours.
- **d.** Bacteria showed neither an increase nor decrease in growth rate.

3. What is the concentration of bacteria at a temperature of 20°C after 4 hours?
- **a.** about 5000 per mL
- **b.** about 6000 per mL
- **c.** about 7000 per mL
- **d.** about 8000 per mL

4. During which hour was the concentration of bacteria at 20°C the greatest?
- **a.** hour 2
- **b.** hour 4
- **c.** hour 6
- **d.** hour 8

5. Which conclusion can be drawn from the data in the graph?
- **a.** The rate of bacterial growth is the greatest at the highest temperature.
- **b.** The rate of bacterial growth is the least at the highest temperature.
- **c.** The rate of bacterial growth is the greatest at the lowest temperature.
- **d.** The rate of bacterial growth does not change depending on temperature.

6. How much did the rate of bacterial growth increase between 2 hours and 8 hours at 30°C?
- **a.** 2000 per mL
- **b.** 4000 per mL
- **c.** 6000 per mL
- **d.** 8000 per mL

Extended Response

7. A scientist wants to test the effect of temperature on the same bacteria shown in the graph at higher temperatures. The scientist tests the growth rate of bacteria at 50°C, 75°C, and 100°C. Based on the information in the graph and your knowledge of bacteria, what results might the scientist get? Explain your reasoning.

8. Antibiotics are drugs that are used to inhibit the growth of bacterial infections. A scientist wants to test the ability of three different antibiotics to control the growth of a certain type of bacteria. The scientist has isolated the bacteria in test tubes. Each antibiotic is prepared in a tablet form. Design an experiment that will test the effectiveness of the antibiotic tablets on the bacteria. Your experiment should include a hypothesis, a list of materials, a procedure, and a method of recording data.

METACOGNITIVE ACTIVITY

Have students answer the following questions in their **Science Notebook:**

1. What misconceptions about bacteria and viruses did you have? How have your ideas changed?
2. Which topics in this chapter would you like to study further? Why?
3. Have any of the concepts in this chapter made you want to change your behavior or everyday activities? What changes would you make?

Analyzing Data

1. d *3. a* *5. a*

2. c *4. d* *6. b*

Extended Response

7. RUBRIC

4 points for a response that correctly answers question and demonstrates an understanding of growth requirements of bacteria.

Sample answer: The bacterial population would most likely increase at 50°C since the trend in the graph indicates a better rate of growth at higher temperatures. However, many bacteria might not be able to survive at extremely high temperatures, such as 100°C, since that is the temperature of boiling water. Some archaea might be able to survive at such extreme temperatures.

3 points correctly answers question and demonstrates partial understanding of growth requirements of bacteria
2 points correctly answers question but demonstrates little understanding of growth requirements of bacteria
1 point incorrectly answers question and demonstrates little understanding of growth requirements of bacteria

8. RUBRIC

4 points for a response that demonstrates complete understanding of problem and includes an appropriate hypothesis, materials list, procedure, and method of recording data.

Sample answer: hypothesis: one of antibiotic samples will be more effective in controlling bacterial growth; materials: three petri dishes, agar, bacteria, antibiotic tablets; procedure: prepare petri dishes with agar, place bacteria in each dish, introduce a different antibiotic into each dish, observe rate of bacterial growth; data: record in a table observations about bacterial growth in each dish on succeeding days

3 points demonstrates partial understanding of problem and includes all requirements of the task as listed above
2 points demonstrates little understanding of problem. Some requirements of the task are missing.
1 point demonstrates minimal understanding of problem. Most requirements of the task are missing.

CHAPTER

2 Introduction to Multicellular Organisms

Life Science UNIFYING PRINCIPLES

PRINCIPLE 1
All living things share common characteristics.

PRINCIPLE 2
All living things share common needs.

PRINCIPLE 3
Living things meet their needs through interactions with the environment.

PRINCIPLE 4
The types and numbers of living things change over time.

Unit: Diversity of Living Things BIG IDEAS

CHAPTER 1
Single-Celled Organisms and Viruses
Bacteria and protists have all of the characteristics of living things, while viruses are not alive.

CHAPTER 2
Introduction to Multicellular Organisms
Multicellular organisms live in and get energy from a variety of environments.

CHAPTER 3
Plants
Plants are a diverse group of organisms that live in many land environments.

CHAPTER 4
Invertebrate Animals
Invertebrate animals have a variety of body plans and adaptations.

CHAPTER 5
Vertebrate Animals
Vertebrate animals live in most of Earth's environments.

CHAPTER 2 KEY CONCEPTS

SECTION 2.1

Multicellular organisms meet their needs in different ways.

1. Multicellular organisms have cells that are specialized.
2. Multicellular organisms are adapted to live in different environments.
3. Sexual reproduction leads to diversity.

SECTION 2.2

Plants are producers.

1. Plants capture energy from the Sun.
2. Plants are adapted to different environments.
3. Plants respond to their environment.
4. Plants respond to seasonal changes.

SECTION 2.3

Animals are consumers.

1. Animals obtain energy and materials from food.
2. Animals interact with the environment and with other organisms.
3. Animals respond to seasonal changes.

SECTION 2.4

Most fungi are decomposers.

1. Fungi absorb materials from the environment.
2. Fungi include mushrooms, molds, and yeasts.
3. Fungi can be helpful or harmful to other organisms.

The Big Idea Flow Chart is available on p. T9 in the **UNIT TRANSPARENCY BOOK.**

Previewing Content

SECTION

2.1 Multicellular organisms meet their needs in different ways. pp. 43–50

1. Multicellular organisms have cells that are specialized.
In multicellular organisms, different cells have different jobs.
- Cells that work together to carry out a job are organized into **tissue.**
- Tissues are organized into **organs.** Each organ has a particular function.
- Organs are part of different organ systems that meet specific needs of the organism.

2. Multicellular organisms are adapted to live in different environments.
All organisms have characteristics called adaptations that allow them to survive in their environment. An adaptation may relate to how an organism gets energy, it may relate to the organism's shape or structure, or it may be a type of behavior. Adaptations are the result of differences in genetic material.

3. Sexual reproduction leads to diversity.
In asexual reproduction, only one parent is involved and the offspring are identical to the parent. Most multicellular organisms reproduce sexually.
In **sexual reproduction,** the genetic material of two parents comes together to form an offspring.
- The first part of sexual reproduction is **meiosis.** Meiosis forms egg and sperm cells that contain half the amount of genetic material that body cells contain.
- The second part is **fertilization,** the union of a sperm and an egg, as seen in the diagram below.

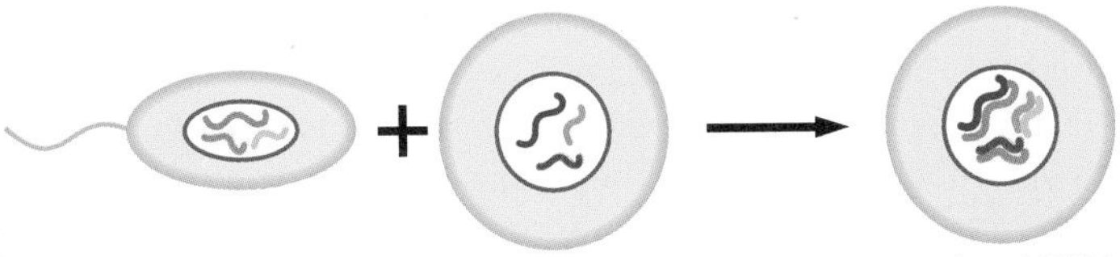

- The offspring differs genetically from both parents.

SECTION

2.2 Plants are producers. pp. 51–57

1. Plants capture energy from the Sun.
The Sun provides the energy almost all organisms need to live. Plants capture the Sun's energy and convert it to chemical energy during **photosynthesis.**
- During photosynthesis, carbon dioxide and water convert to sugars and oxygen.
- Plant cells store the sugars as granules of starch.
- When the plant needs energy, it breaks down starch into sugars.
- Plants are called **autotrophs** because they don't need to consume food; instead, they produce food for the rest of the ecosystem.
- Plants can convert stored energy back into usable energy through **cellular respiration.**

2. Plants are adapted to different environments.
Plants have developed many adaptations that allow them to live in a variety of environments.
- Grasses have deep roots and produce seeds quickly so they can survive in areas with a wide range of temperatures and precipitation.
- Trees need more water than grasses. Conifers can grow in colder climates than deciduous trees.
- Reproductive adaptations, such as rapid life cycles, allow some plants to live in deserts or in other extreme climates.

3. Plants respond to their environment.
Plants respond to several **stimuli** in their environment.
- All plants respond to gravity. Their roots grow down, and their stems grow up.
- Many plants produce tendrils that wrap around an object that they touch.
- Plants respond to light by bending toward a light source. This response is due to the action of a hormone called auxin, which makes cells grow rapidly. The unequal growth of cells on opposite sides of a stem makes the stem bend.

4. Plants respond to seasonal changes.
Shorter periods of daylight cause some plants to become dormant in autumn. They temporarily stop growing, lose their leaves, and require less energy.

Common Misconceptions

PLANTS' ORGANS Students often think that only animals have organs and organ systems. In this chapter students learn that plant structures such as leaves and roots are organs.

TE This misconception is addressed on p. 45.

MISCONCEPTION DATABASE
CLASSZONE.COM Background on student misconceptions

PLANTS AS PRODUCERS Many students hold the misconception that carbon dioxide, water, and soil are food that plants eat. In fact, plants are producers and make carbon-containing compounds from carbon dioxide and water, using the energy from sunlight.

TE This misconception is addressed on p. 53.

Previewing Content

SECTION

2.3 Animals are consumers. pp. 58–65

1. Animals obtain energy and materials from food.
Animals are **heterotrophs**—they feed on other organisms. Animals consume the energy found in other organisms.
- Animals digest their food. Digestion breaks down large chunks of food into small compounds, including sugars, that can be taken into cells.
- Cellular respiration breaks the bonds in sugars and releases the energy stored in the bonds. Cellular respiration requires oxygen. Animals have specialized structures for taking oxygen into their bodies.

2. Animals interact with the environment and with other organisms.
Animals have many adaptations that allow them to interact with their environment. Animals respond to stimuli such as temperature changes, hunger and thirst, and other animals. These responses are called **behavior.** Behavior can be inherited or learned.
Some adaptive behaviors are interactions between animals of different species. **Predators** are animals that hunt other animals for food. **Prey** are the animals that are hunted. Other adaptive behaviors work for the benefit of both animals.

3. Animals respond to seasonal changes.
Some animals, such as the monarch butterfly, respond to seasonal changes by **migrating**. They spend the winter months in warmer climates, then return to their northern breeding grounds in the spring. Some animals, such as bears and frogs, hibernate in the cold months by slowing down their body systems to conserve energy.

SECTION

2.4 Most fungi are decomposers. pp. 66–73

1. Fungi absorb materials from the environment.
Fungi bodies are composed of threadlike **hyphae.** A mass of hyphae is a mycelium.
- The cells in the hyphae release digestive enzymes into the environment. Organic material is digested, and the cells absorb the nutrients that they need.
- Many fungi are decomposers. In the soil, simple compounds left over from fungal digestion are taken up by plant roots and recycled through the biosphere.

Fungi reproduce by forming **spores** that are released into the air and spread by wind. Spores can survive for many years. They begin to grow only when water and food are available.

2. Fungi include mushrooms, molds, and yeasts.
Fungi commonly have one of three forms.
- Mushrooms are the reproductive bodies of a fungus that grows on trees or soil. They are filled with hyphae. Spores are produced in the cap of a mushroom.
- Molds form a fuzzy growth on food and other substrates. This growth is a mass of hyphae. Some hyphae produce spores. Many molds cause diseases, such as Dutch elm disease. Other molds are of use to humans because they produce antibiotics such as penicillin.
- Yeasts are single-celled fungi. Some yeasts cause disease. Others are very useful in making food, such as bread.

3. Fungi can be helpful or harmful to other organisms.
Hyphae can grow into any material made by another organism. For this reason, fungi can be either helpful or harmful.
- Fungi produce harmful chemicals called toxins and can cause diseases in food crops and trees.
- Fungi form beneficial relationships with plants by surrounding plant roots with hyphae. The fungus provides digested nutrients for the plant, and the plant provides food for the fungus.
- A **lichen** is a relationship between single-celled algae and fungi. As seen in the diagram, the algae are held in place by a mass of hyphae. Lichens can live where neither organism alone can live.

Common Misconceptions

MISCONCEPTION DATABASE
CLASSZONE.COM Background on student misconceptions

CELLULAR RESPIRATION Students may hold the misconception that only animals undergo cellular respiration. Actually, almost all organisms undergo cellular respiration and, thus, require oxygen.

TE This misconception is addressed on p. 61.

DECOMPOSERS Students often hold misconceptions about the role of decomposers in the environment. They may not consider decomposers as part of feeding relationships, or may think of decomposers as the "bottom" of a chain or hierarchy. Decomposers are responsible for recycling nutrients in the biosphere.

TE This misconception is addressed on p. 70.

Previewing Labs

Lab Generator CD-ROM
Edit these Pupil Edition labs and generate alternative labs.

EXPLORE the BIG idea

Where Does It Come From? p. 41 Students examine everyday objects to identify the sources of materials.	**TIME** 10 minutes **MATERIALS** 3 common objects, 3 index cards
How Can Multicellular Organisms Reproduce? p. 41 Students explore asexual reproduction by sprouting potato halves.	**TIME** 10 minutes (planting), 2 weeks (wait time), 5 minutes (observation) **MATERIALS** old potato with eyes, knife, 2 pots, soil, water
Internet Activity: Bee Dance, p. 41 Students explore how bees communicate.	**TIME** 20 minutes **MATERIALS** computer with Internet access
SECTION 2.1 **INVESTIGATE Specialization,** p. 44 Students model cell specialization and explore its advantages and disadvantages.	**TIME** 10 minutes **MATERIALS** 2 boxes, 40 paper clips, 4 pieces of paper, 2 pencils, watch with second hand
SECTION 2.2 **EXPLORE Stored Energy,** p. 51 Students test potatoes, celery, and pears with iodine, to see if they contain starch.	**TIME** 10 minutes **MATERIALS** pieces of potato, celery, and pears; 3 plastic cups; iodine solution; dropper
INVESTIGATE What Plants Need to Grow, p. 52 Students design an experiment that will test the hypothesis that plants grow by absorbing material from soil.	**TIME** 30 minutes **MATERIALS** potting soil, pots or paper cups, bean seedlings or beans, triple-beam balance, water
SECTION 2.3 **INVESTIGATE Owl Pellets,** p. 60 Students infer what an owl eats and how well it digests its food by examining an owl pellet.	**TIME** 30 minutes **MATERIALS** owl pellet, toothpick or dissecting needle, tweezers, tray, (for Challenge: bone identification key—available when ordering owl pellets)
SECTION 2.4 **EXPLORE Mushrooms,** p. 66 Students make spore prints to examine spores in mushroom caps.	**TIME** 10 minutes on 2 different days **MATERIALS** fresh store-bought mushrooms, sharp knife, clear plastic cup, white paper, hand lens
CHAPTER INVESTIGATION **Effect of Yeast on Bread Dough,** pp. 72–73 Students grow yeast on different substances to draw conclusions about how sugar, salt, and sweetener affect the growth of yeast.	**TIME** 40 minutes **MATERIALS** 4 sheets of notebook paper; baking sheet; 1 cup flour; measuring cups, 1 cup and 1/4 cup; measuring spoons, 1 tsp and 1/4 tsp; 3 teaspoons of sugar; 3 teaspoons of sweetener; 3 teaspoons of salt; 1 teaspoon of quick-rise yeast; warm water; metric ruler; marker; masking tape; 8 clear plastic straws; 8 clothespins

Additional INVESTIGATION, Plant Growth and Light Conditions, A, B, & C, pp. 140–148; Teacher Instructions, pp. 360–361

Previewing Chapter Resources

	INTEGRATED TECHNOLOGY	LABS AND ACTIVITIES
CHAPTER 2 **Introduction to Multicellular Organisms**	**CLASSZONE.COM** • eEdition Plus • EasyPlanner Plus • Misconception Database • Content Review • Test Practice • Visualization • Resource Centers • Internet Activity: Bee Dance • Math Tutorial **SCILINKS.ORG** SCI LINKS **CD-ROMS** • eEdition • EasyPlanner • Power Presentations • Content Review • Lab Generator • Test Generator **AUDIO CDS** • Audio Readings • Audio Readings in Spanish	EXPLORE the Big Idea, p. 41 • Where Does It Come From? • How Can Multicellular Organisms Reproduce? • Internet Activity: Bee Dance **UNIT RESOURCE BOOK** Unit Projects, pp. 5–10 **Lab Generator CD-ROM** Generate customized labs.
SECTION 2.1 **Multicellular organisms meet their needs in different ways.** pp. 43–50 Time: 2 periods (1 block) Lesson Plan, pp. 81–82	**MATH TUTORIAL** **UNIT TRANSPARENCY BOOK** • Big Idea Flow Chart, p. T9 • Daily Vocabulary Scaffolding, p. T10 • Note-Taking Model, p. T11 • 3-Minute Warm-Up, p. T12 • "Adaptations in Different Environments" Visual, p. T14	• INVESTIGATE Specialization, p. 44 • Math in Science, p. 50 **UNIT RESOURCE BOOK** • Datasheet, Specialization, p. 90 • Math Support, p. 129 • Math Practice, p. 130
SECTION 2.2 **Plants are producers.** pp. 51–57 Time: 2 periods (1 block) Lesson Plan, pp. 92–93	**RESOURCE CENTER,** Plant Adaptations **UNIT TRANSPARENCY BOOK** • Daily Vocabulary Scaffolding, p. T10 • 3-Minute Warm-Up, p. T12	• EXPLORE Stored Energy, p. 51 • INVESTIGATE What Plants Need to Grow, p. 52 **UNIT RESOURCE BOOK** • Datasheet, What Plants Need to Grow, p. 101 • Additional INVESTIGATION, Plant Growth and Light Conditions, A, B, & C, pp. 140–148
SECTION 2.3 **Animals are consumers.** pp. 58–65 Time: 2 periods (1 block) Lesson Plan, pp. 103–104	• **RESOURCE CENTER,** Animal Adaptations • **VISUALIZATION,** How Bees Communicate **UNIT TRANSPARENCY BOOK** • Daily Vocabulary Scaffolding, p. T10 • 3-Minute Warm-Up, p. T13	• INVESTIGATE Owl Pellets, p. 60 • Science on the Job, p. 65 **UNIT RESOURCE BOOK** Datasheet, Owl Pellets, p. 112
SECTION 2.4 **Most fungi are decomposers.** pp. 66–73 Time: 4 periods (2 blocks) Lesson Plan, pp. 114–115	**RESOURCE CENTER,** Fungi **UNIT TRANSPARENCY BOOK** • Big Idea Flow Chart, p. T9 • Daily Vocabulary Scaffolding, p. T10 • 3-Minute Warm-Up, p. T13 • Chapter Outline, pp. T15–T16	• EXPLORE Mushrooms, p. 66 • CHAPTER INVESTIGATION, What Do Yeast Cells Use for Energy?, pp. 72–73 **UNIT RESOURCE BOOK** CHAPTER INVESTIGATION, What Do Yeast Cells Use for Energy?, A, B, & C, pp. 131–139

KEY TO ICONS: CD/CD-ROM · INTERNET · Pupil Edition · Teacher Edition · UNIT RESOURCE BOOK · UNIT TRANSPARENCY BOOK · UNIT ASSESSMENT BOOK · SPANISH ASSESSMENT BOOK · SCIENCE TOOLKIT

READING AND REINFORCEMENT	ASSESSMENT	STANDARDS
• Four Square, B22–23 • Main Idea and Detail Notes, C37 • Daily Vocabulary Scaffolding, H1–8 **UNIT RESOURCE BOOK** • Vocabulary Practice, pp. 126–127 • Decoding Support, p. 128 • Summarizing the Chapter, pp. 149–150 **Audio Readings CD** Listen to Pupil Edition. **Audio Readings in Spanish CD** Listen to Pupil Edition in Spanish.	• Chapter Review, pp. 75–76 • Standardized Test Practice, p. 77 **UNIT ASSESSMENT BOOK** • Diagnostic Test, pp. 21–22 • Chapter Test, A, B, & C, pp. 27–38 • Alternative Assessment, pp. 39–40 Spanish Chapter Test, pp. 45–48 **Test Generator CD-ROM** Generate customized tests. **Lab Generator CD-ROM** Rubrics for Labs	National Standards A.1–8, A.9.a–c, A.9.e–g, C.1.a, C.3.c, F.1.e, G.1.b See p. 40 for the standards.
UNIT RESOURCE BOOK • Reading Study Guide, A & B, pp. 83–86 • Spanish Reading Study Guide, pp. 87–88 • Challenge and Extension, p. 89 • Reinforcing Key Concepts, p. 90	Ongoing Assessment, pp. 43, 45–49 Section 2.1 Review, p. 49 **UNIT ASSESSMENT BOOK** Section 2.1 Quiz, p. 23	National Standards A.2–8, A.9.a–c, A.9.e–f, C.1.a, F.1.e, G.1.b
UNIT RESOURCE BOOK • Reading Study Guide, A & B, pp. 94–97 • Spanish Reading Study Guide, pp. 98–99 • Challenge and Extension, p. 100 • Reinforcing Key Concepts, p. 102	Ongoing Assessment, pp. 52–54, 56–57 Section 2.2 Review, p. 57 **UNIT ASSESSMENT BOOK** Section 2.2 Quiz, p. 24	National Standards A.2–7, A.9.a–b, A.9.e–f, F.1.e, G.1.b
UNIT RESOURCE BOOK • Reading Study Guide, A & B, pp. 105–108 • Spanish Reading Study Guide, pp. 109–110 • Challenge and Extension, p. 111 • Reinforcing Key Concepts, p. 113 • Challenge Reading, pp. 124–125	Ongoing Assessment, pp. 59–63 Section 2.3 Review, p. 64 **UNIT ASSESSMENT BOOK** Section 2.3 Quiz, p. 25	National Standards A.2–7, A.9.a–b, A.9.e–f, C.3.c, F.1.e, G.1.b
UNIT RESOURCE BOOK • Reading Study Guide, A & B, pp. 116–119 • Spanish Reading Study Guide, pp. 120–121 • Challenge and Extension, p. 122 • Reinforcing Key Concepts, p. 123	Ongoing Assessment, pp. 67–71 Section 2.4 Review, p. 71 **UNIT ASSESSMENT BOOK** Section 2.4 Quiz, p. 26	National Standards A.1–7, A.9.a–b, A.9.e–g, C.3.c, F.1.e, G.1.b

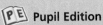

Previewing Resources for Differentiated Instruction

CHAPTER INVESTIGATION

Leveled resources present the same concepts for different abilities.

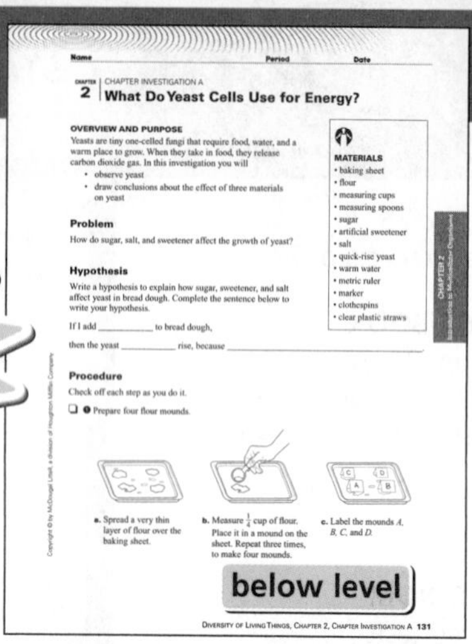
Name Period Date
CHAPTER 2 | CHAPTER INVESTIGATION A
What Do Yeast Cells Use for Energy?
OVERVIEW AND PURPOSE
Problem
How do sugar, salt, and sweetener affect the growth of yeast?
Hypothesis
Procedure
below level
DIVERSITY OF LIVING THINGS, CHAPTER 2, CHAPTER INVESTIGATION A 131

R **UNIT RESOURCE BOOK,** pp. 131–134

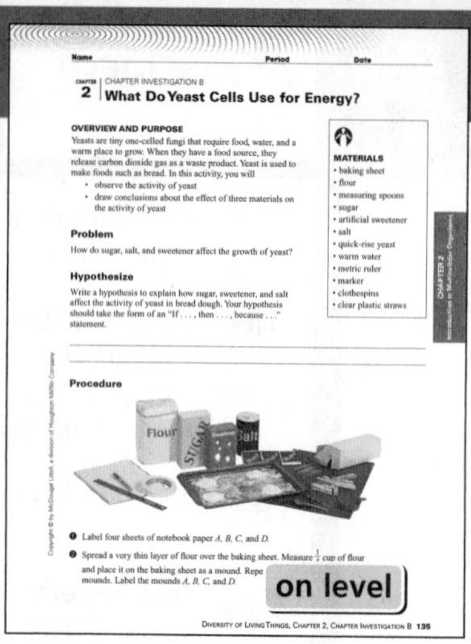
Name Period Date
CHAPTER 2 | CHAPTER INVESTIGATION B
What Do Yeast Cells Use for Energy?
OVERVIEW AND PURPOSE
Problem
How do sugar, salt, and sweetener affect the growth of yeast?
Hypothesize
Procedure
on level
DIVERSITY OF LIVING THINGS, CHAPTER 2, CHAPTER INVESTIGATION B 135

R pp. 135–138

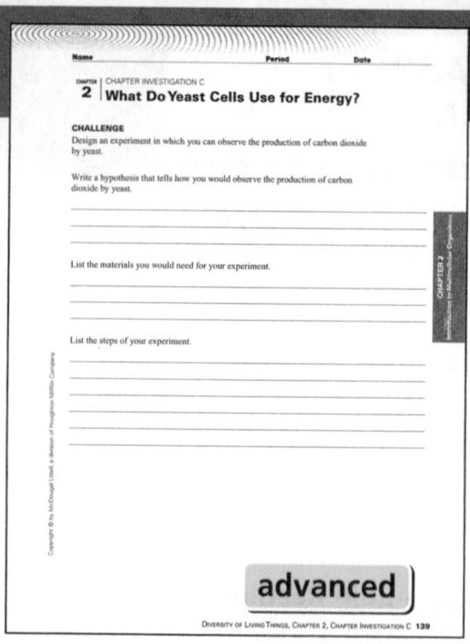
Name Period Date
CHAPTER 2 | CHAPTER INVESTIGATION C
What Do Yeast Cells Use for Energy?
CHALLENGE
Design an experiment in which you can observe the production of carbon dioxide by yeast.
Write a hypothesis that tells how you would observe the production of carbon dioxide by yeast.
List the materials you would need for your experiment.
List the steps of your experiment.
advanced
DIVERSITY OF LIVING THINGS, CHAPTER 2, CHAPTER INVESTIGATION C 139

R pp. 135–139

READING STUDY GUIDE

Reading Study Guide is also in Spanish.

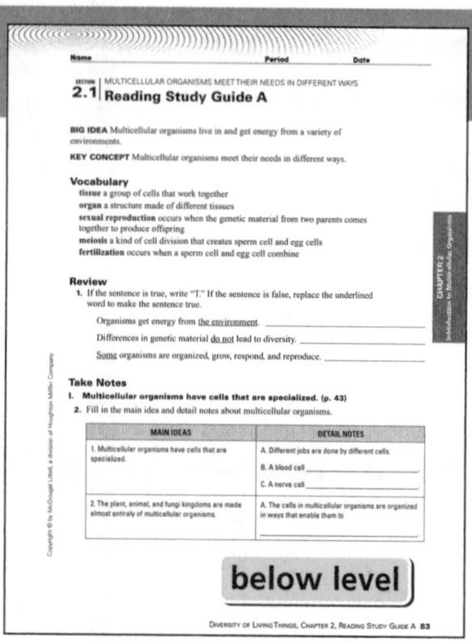
Name Period Date
SECTION 2.1 | MULTICELLULAR ORGANISMS MEET THEIR NEEDS IN DIFFERENT WAYS
Reading Study Guide A
Vocabulary
Review
Take Notes
below level
DIVERSITY OF LIVING THINGS, CHAPTER 2, READING STUDY GUIDE A 83

R **UNIT RESOURCE BOOK,** pp. 83–84

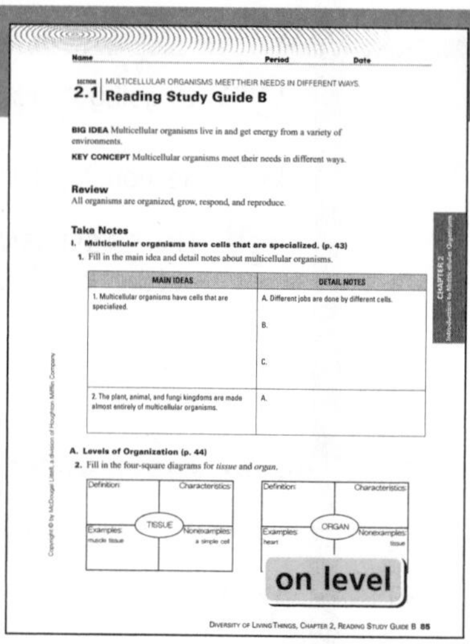
Name Period Date
SECTION 2.1 | MULTICELLULAR ORGANISMS MEET THEIR NEEDS IN DIFFERENT WAYS
Reading Study Guide B
Review
Take Notes
on level
DIVERSITY OF LIVING THINGS, CHAPTER 2, READING STUDY GUIDE B 85

R pp. 85–86

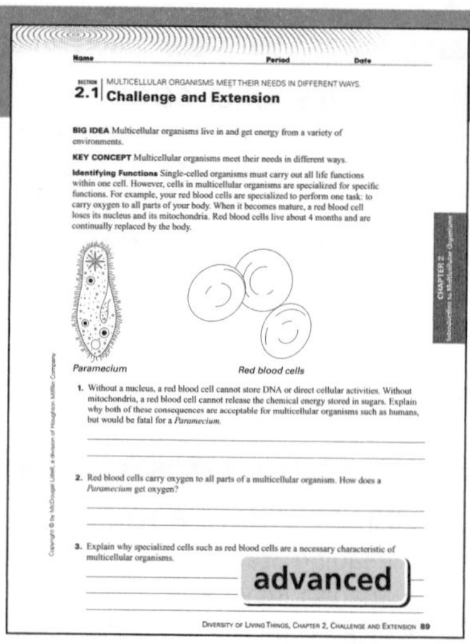
Name Period Date
SECTION 2.1 | MULTICELLULAR ORGANISMS MEET THEIR NEEDS IN DIFFERENT WAYS
Challenge and Extension
Paramecium
Red blood cells
advanced
DIVERSITY OF LIVING THINGS, CHAPTER 2, CHALLENGE AND EXTENSION 89

R p. 89

CHAPTER TEST

Chapter Test is also in Spanish.

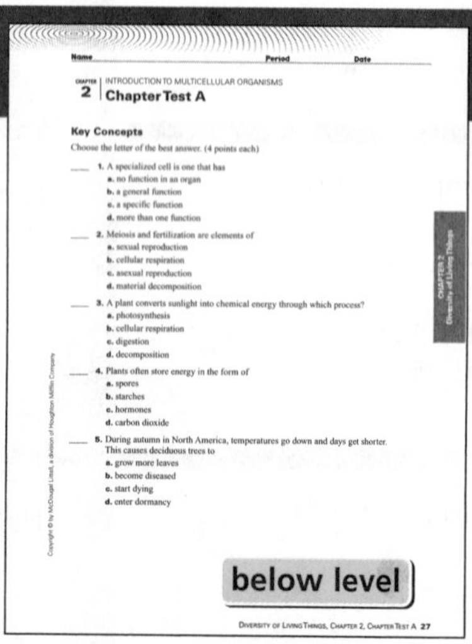
Name Period Date
CHAPTER 2 | INTRODUCTION TO MULTICELLULAR ORGANISMS
Chapter Test A
Key Concepts
Choose the letter of the best answer. (4 points each)
below level
DIVERSITY OF LIVING THINGS, CHAPTER 2, CHAPTER TEST A 27

A **UNIT ASSESSMENT BOOK,** pp. 27–30

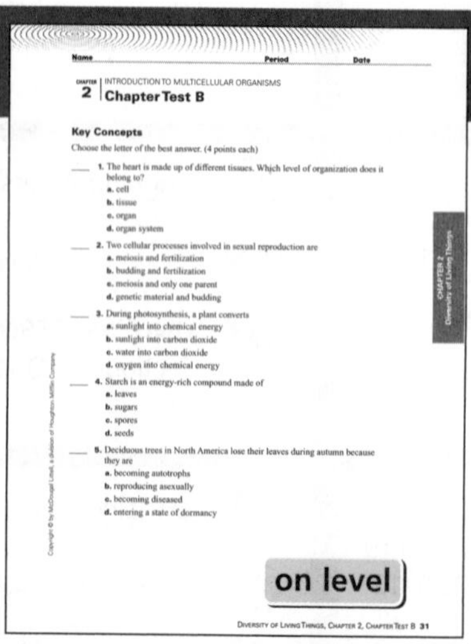
Name Period Date
CHAPTER 2 | INTRODUCTION TO MULTICELLULAR ORGANISMS
Chapter Test B
Key Concepts
Choose the letter of the best answer. (4 points each)
on level
DIVERSITY OF LIVING THINGS, CHAPTER 2, CHAPTER TEST B 31

A pp. 31–34

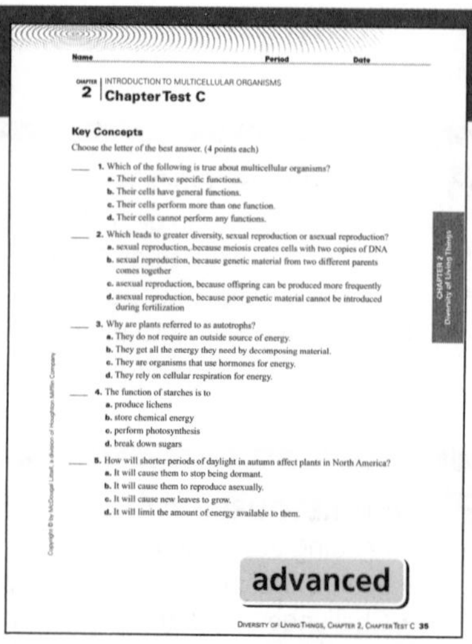
Name Period Date
CHAPTER 2 | INTRODUCTION TO MULTICELLULAR ORGANISMS
Chapter Test C
Key Concepts
Choose the letter of the best answer. (4 points each)
advanced
DIVERSITY OF LIVING THINGS, CHAPTER 2, CHAPTER TEST C 35

A pp. 35–38

TECHNOLOGY

There are three Resource Centers for this chapter.

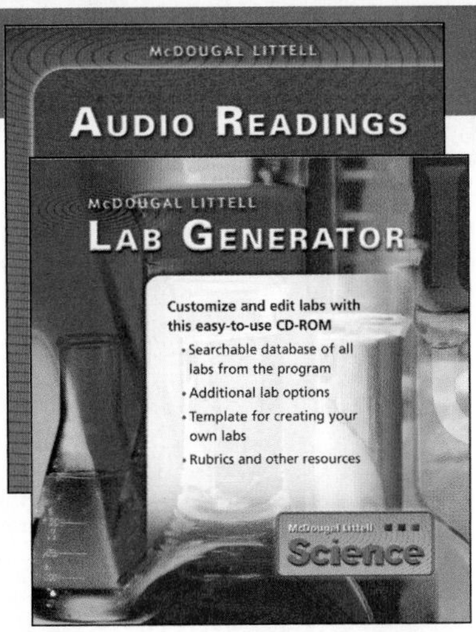

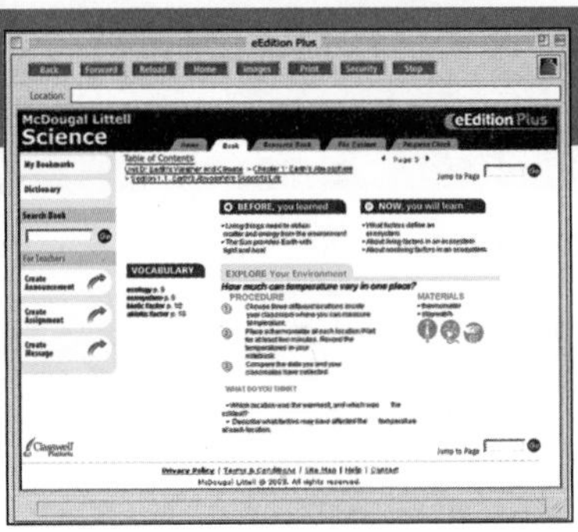

CLASSZONE.COM

CD/CD-ROMS

CLASSZONE.COM

VISUAL CONTENT

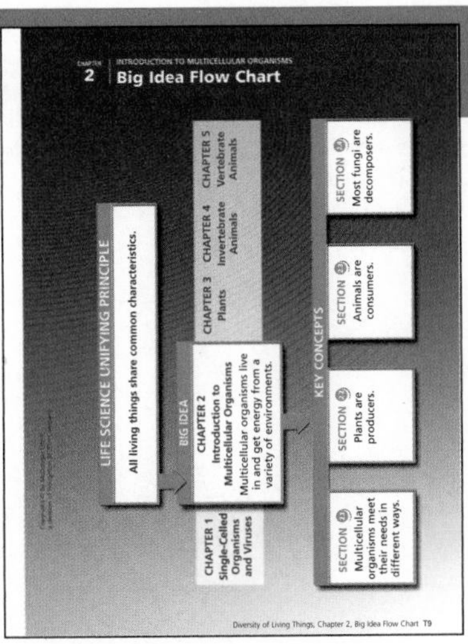

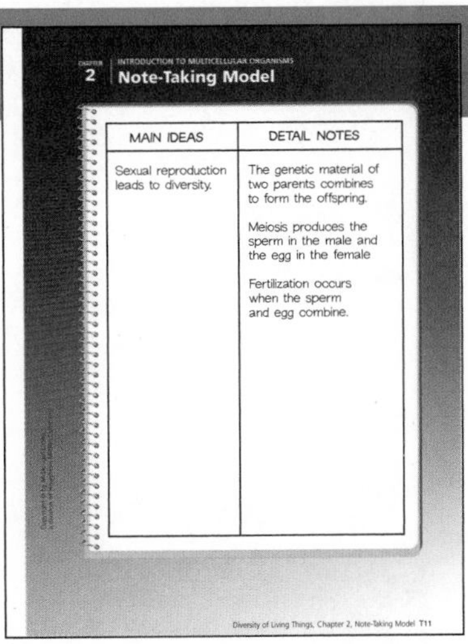

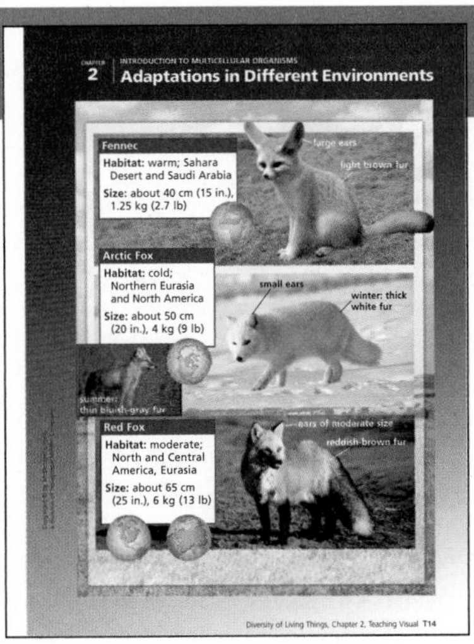

UNIT TRANSPARENCY BOOK, p. T9

p. T11

p. T14

MORE SUPPORT

Reinforcing Key Concepts for each section

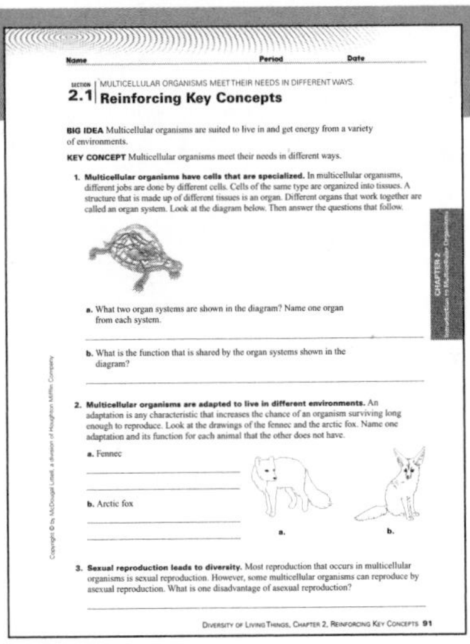
Name Period Date

SECTION 2.1 | MULTICELLULAR ORGANISMS MEET THEIR NEEDS IN DIFFERENT WAYS.

Reinforcing Key Concepts

BIG IDEA Multicellular organisms are suited to live in and get energy from a variety of environments.

KEY CONCEPT Multicellular organisms meet their needs in different ways.

1. **Multicellular organisms have cells that are specialized.** In multicellular organisms, different jobs are done by different cells. Cells of the same type are organized into tissues. A structure that is made up of different tissues is an organ. Different organs that work together are called an organ system. Look at the diagram below. Then answer the questions that follow.
   a. What two organ systems are shown in the diagram? Name one organ from each system.
   b. What is the function that is shared by the organ systems shown in the diagram?
2. **Multicellular organisms are adapted to live in different environments.** An adaptation is any characteristic that increases the chance of an organism surviving long enough to reproduce. Look at the drawings of the fennec and the arctic fox. Name one adaptation and its function for each animal that the other does not have.
   a. Fennec
   b. Arctic fox
3. **Sexual reproduction leads to diversity.** Most reproduction that occurs in multicellular organisms is sexual reproduction. However, some multicellular organisms can reproduce by asexual reproduction. What is one disadvantage of asexual reproduction?

Copyright © by McDougal Littell, a division of Houghton Mifflin Company

DIVERSITY OF LIVING THINGS, CHAPTER 2, REINFORCING KEY CONCEPTS 91

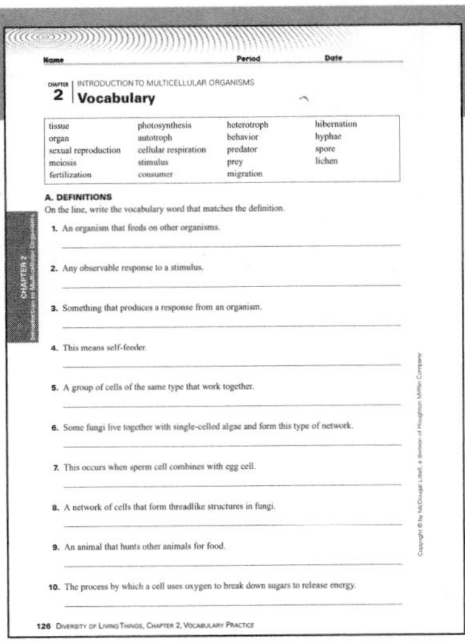
Name Period Date

CHAPTER 2 | INTRODUCTION TO MULTICELLULAR ORGANISMS

Vocabulary

tissue	photosynthesis	heterotroph	hibernation
organ	autotroph	behavior	hyphae
sexual reproduction	cellular respiration	predator	spore
meiosis	stimulus	prey	lichen
fertilization	consumer	migration	

A. DEFINITIONS

On the line, write the vocabulary word that matches the definition.

1. An organism that feeds on other organisms.
2. Any observable response to a stimulus.
3. Something that produces a response from an organism.
4. This means self-feeder.
5. A group of cells of the same type that work together.
6. Some fungi live together with single-celled algae and form this type of network.
7. This occurs when sperm cell combines with egg cell.
8. A network of cells that form threadlike structures in fungi.
9. An animal that hunts other animals for food.
10. The process by which a cell uses oxygen to break down sugars to release energy.

Copyright © by McDougal Littell, a division of Houghton Mifflin Company

126 DIVERSITY OF LIVING THINGS, CHAPTER 2, VOCABULARY PRACTICE

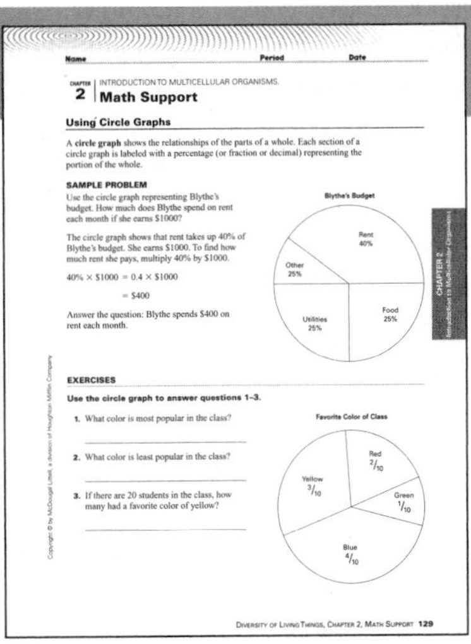
Name Period Date

CHAPTER 2 | INTRODUCTION TO MULTICELLULAR ORGANISMS

Math Support

Using Circle Graphs

A **circle graph** shows the relationships of the parts of a whole. Each section of a circle graph is labeled with a percentage (or fraction or decimal) representing the portion of the whole.

SAMPLE PROBLEM

Use the circle graph representing Blythe's budget. How much does Blythe spend on rent each month if she earns $1000?

The circle graph shows that rent takes up 40% of Blythe's budget. She earns $1000. To find how much rent she pays, multiply 40% by $1000.

$40\% \times \$1000 = 0.4 \times \$1000$

$= \$400$

Answer the question: Blythe spends $400 on rent each month.

Blythe's Budget: Rent 40%, Other 25%, Food 25%, Utilities 25%

EXERCISES

Use the circle graph to answer questions 1–3.

1. What color is most popular in the class?
2. What color is least popular in the class?
3. If there are 20 students in the class, how many had a favorite color of yellow?

Favorite Color of Class: Red $^2/_{10}$, Yellow $^3/_{10}$, Green $^1/_{10}$, Blue $^4/_{10}$

Copyright © by McDougal Littell, a division of Houghton Mifflin Company

DIVERSITY OF LIVING THINGS, CHAPTER 2, MATH SUPPORT 129

UNIT RESOURCE BOOK, p. 91

pp. 126–127

p. 129

CHAPTER 2

Introduction to Multicellular Organisms

INTRODUCE

the BIG idea

Have students look at the organisms in the photograph and discuss how the question in the box links to the Big Idea:

- Ask: How do you think the chipmunk gets energy?
- Ask: How do you think the mushroom gets energy?

National Science Education Standards

Content

C.1.a Important levels of organization for structure and function include cells, organs, tissues, organ systems, whole organisms, and ecosystems.

C.3.c Behavior is one response an organism can make to a stimulus. A behavioral response requires coordination and communication at many levels.

F.1.e Food provides energy and nutrients for growth and development. Nutrition requirements vary.

Process

A.1–8 Identify questions that can be answered through scientific investigations; design and conduct an investigation; use tools to gather and interpret data; use evidence to describe, predict, explain, and model; think critically to make relationships between evidence and explanation; recognize different explanations and predictions; communicate scientific procedures and explanations; use mathematics.

A.9.a–c, A.9.e–g Understand scientific inquiry by using different investigations, methods, mathematics, and explanations based on logic, evidence, and skepticism. Data often result in new investigations.

G.1.b Science requires different abilities.

CHAPTER 2

Introduction to Multicellular Organisms

the BIG idea

Multicellular organisms live in and get energy from a variety of environments.

How does an organism get energy and materials from its environment?

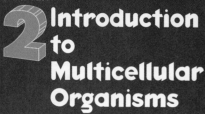

Key Concepts

SECTION 2.1 **Multicellular organisms meet their needs in different ways.**
Learn about specialized cells, tissues, and organs.

SECTION 2.2 **Plants are producers.**
Learn how plants get energy and respond to the environment.

SECTION 2.3 **Animals are consumers.**
Learn how animals get energy and how they interact with the environment.

SECTION 2.4 **Most fungi are decomposers.**
Learn about fungi and how they get energy.

Internet Preview

CLASSZONE.COM

Chapter 2 online resources: Content Review, Visualization, four Resource Centers, Math Tutorial, Test Practice

INTERNET PREVIEW

CLASSZONE.COM For student use with the following pages:

Review and Practice
- Content Review, pp. 42, 74
- Math Tutorial: Using Circle Graphs, p. 50
- Test Practice, p. 77

Activities and Resources
- Internet Activity: Bee Dance, p. 41
- Resource Centers: Plant and Animal Adaptations, pp. 54, 60; Fungi, p. 68
- Visualization, p. 56

NSTA scilinks.org SCiLINKS

Animal Behavior
Code: MDL040

EXPLORE the BIG idea

Where Does It Come From?

Think about the things you use every day. Just like any other organism, you depend on the environment to meet your needs. The food you eat comes from plants and animals. Also, much of what you use is made of materials processed from living matter.

Observe and Think Identify three nonfood items you come into contact with every day. Where does the material for these products come from?

How Can a Multicellular Organism Reproduce on Its Own?

Take an old potato and cut it in half, making sure that there are eyes on both halves. Plant each half in a pot of soil. Water the pots once a day. After two weeks, remove the potato halves from the pots and examine.

Observe and Think What happened to the potato halves?

Internet Activity: Bee Dance

Go to **ClassZone.com** to explore how bees communicate.

Observe and Think What type of information can a bee communicate to other bees in a hive?

NSTA scilinks.org SCILINKS
Animal Behavior **Code: MDL040**

TEACHING WITH TECHNOLOGY

Digital Camera Students can make a visual record of the materials they find in an owl pellet in the investigation on p. 60. If a tape recorder is available, they can record a narration to accompany the photographs.

CBL and Probeware Students can use a carbon dioxide probe stuck into a ball of dough to measure yeast fermentation activity in the Chapter Investigation on pp. 72–73. The probe replaces the straw apparatus.

EXPLORE the BIG idea

These inquiry-based activities are appropriate for use at home or as a supplement to classroom instruction.

Where Does It Come From?

PURPOSE To identify the sources of materials in everyday objects.

TIP *10 min.* Have groups of students brainstorm to name materials that come from plants. Suggest objects that are made in part from plant or animal matter.

Answers depend on objects chosen. Sample answers: cotton, wood, and paper from plants; silk and leather from animals

REVISIT after p. 57.

How Can Multicellular Organisms Reproduce?

PURPOSE To explore asexual reproduction. Students sprout potato halves in soil.

TIP *2 weeks* Position the potatoes in the soil with an eye at the top.

Answer: The eyes sprouted, and a new plant developed from each eye. This ability allows potatoes to reproduce asexually.

REVISIT after p. 49.

Internet Activity: Bee Dance

PURPOSE To explore how bees communicate with each other.

TIP *20 min.* Prompt students to discuss what kinds of information are important for a bee to know and communicate.

Answer: Bees communicate about the location of food.

REVISIT after p. 63.

PREPARE

CONCEPT REVIEW

Activate Prior Knowledge

Have students list the four characteristics of living things and draw a picture depicting each characteristic. *Drawings should portray organization, growth, response, and reproduction.*

TAKING NOTES

Main Idea and Detail Notes

Writing each new blue heading will emphasize the main ideas in each section. Students will probably fill the details column with the red headings. Blue and red headings form an outline that links the ideas of the section. Encourage students to use their charts as study tools.

Vocabulary Strategy

A four square diagram can include as much information as the student wants to add. By writing their own words around the center term, students personalize their understanding of the term and retain its meaning longer.

Vocabulary and Note-Taking Resources

- Vocabulary Practice, pp. 126–127
- Decoding Support, p. 128

- Daily Vocabulary Scaffolding, pp. T10
- Note-Taking Model, p. T11

- Four Square, B22–23
- Main Idea and Detail Notes, C37
- Daily Vocabulary Scaffolding, H1–8

CHAPTER 2

Getting Ready to Learn

CONCEPT REVIEW

- Living things are organized, grow, respond, and reproduce.
- Protists get energy in three different ways.
- Single-celled organisms reproduce when the cell divides.

VOCABULARY REVIEW

kingdom p. 11
producer p. 19
decomposer p. 19
See Glossary for definitions.

adaptation, DNA, genetic material

CONTENT REVIEW
CLASSZONE.COM
Review concepts and vocabulary.

TAKING NOTES

MAIN IDEA AND DETAILS

Make a two-column chart. Write the main ideas, such as those in the blue headings, in the column on the left. Write details about each of those main ideas in the column on the right.

VOCABULARY STRATEGY

Write each new vocabulary term in the center of a **four square** diagram. Write notes in the squares around each term. Include a definition, some characteristics, and some examples of the term. If possible, write some things that are not examples of the term.

See the Note-Taking Handbook on pages R45–R51.

SCIENCE NOTEBOOK

MAIN IDEAS	DETAILS
Plants respond to their environment.	1. Plants respond to different stimuli. 2. A stimulus is something that produces a response.

Definition	Characteristics
Group of same type of cells performing similar functions	Cells are similar. Different tissues do different jobs.
Examples	**Nonexamples**
skin tissue nerve tissue muscle tissue	a simple cell

TISSUE (center term)

CHECK READINESS

Administer the Diagnostic Test to determine students' readiness for new science content and their mastery of requisite math skills.

Diagnostic Test, pp. 21–22

Technology Resources

Students needing content and math skills should visit **ClassZone.com**.

- CONTENT REVIEW
- MATH TUTORIAL

CONTENT REVIEW CD-ROM

KEY CONCEPT

2.1 Multicellular organisms meet their needs in different ways.

BEFORE, you learned

- Organisms get energy and materials from the environment
- All organisms are organized, grow, respond, and reproduce
- Differences in genetic material lead to diversity

NOW, you will learn

- About the functions of cells in multicellular organisms
- How multicellular organisms are adapted to different environments
- About sexual reproduction

VOCABULARY

tissue p. 44
organ p. 44
sexual reproduction p. 48
meiosis p. 48
fertilization p. 48

THINK ABOUT

Why is teamwork important?

For any team to be successful, it is important for people to work well together. Within a team, each person has a different role. For example, the team in this restaurant includes people to greet diners and seat them, people to buy and cook the food, and people to take food orders and serve the food. By dividing different jobs among different people, a restaurant can serve more customers at the same time. What would happen in a large restaurant if the diners were seated, cooked for, and served by the same person?

Multicellular organisms have cells that are specialized.

MAIN IDEA AND DETAILS Make a chart and add notes about the main idea: *Multicellular organisms have cells that are specialized.*

In single-celled organisms, all the functions of life are performed by one cell. These functions include getting energy and materials, removing wastes, and responding to changes in the environment. In multicellular organisms, however, different jobs are done by different cells—the cells are specialized. A blood cell carries oxygen. A nerve cell sends or receives a signal. Just as the different jobs of running a restaurant are divided among different people, in multicellular organisms different functions are divided among different cells.

In this chapter, you will read about plants, animals, and fungi. These three kingdoms are made up almost entirely of multicellular organisms. The cells in multicellular organisms are organized in ways that enable them to survive and reproduce.

RESOURCES FOR DIFFERENTIATED INSTRUCTION

Below Level

UNIT RESOURCE BOOK
- Reading Study Guide A, pp. 83–84
- Decoding Support, p. 128

AUDIO CDS

Advanced

UNIT RESOURCE BOOK
Challenge and Extension, p. 89

English Learners

UNIT RESOURCE BOOK
Spanish Reading Study Guide, pp. 87–88

AUDIO CDS
- Audio Readings in Spanish
- Audio Readings (English)

2.1 FOCUS

Set Learning Goals

Students will

- Explain the functions of cells in multicellular organisms.
- Describe how multicellular organisms are adapted to different environments.
- Summarize sexual reproduction.
- Model specialization in organisms.

3-Minute Warm-Up

Display Transparency 12 or copy this exercise on the board:

Imagine that a bacterium begins to divide by binary fission. Each generation of daughter cells also divides, and soon there are millions of bacteria. How do you think the newest bacterium compares to the original cell? Explain. *The cells are identical. Binary fission produces daughter cells that are each identical to the parent cell.*

T 3-Minute Warm-Up, p. T12

2.1 MOTIVATE

THINK ABOUT

PURPOSE To consider how task specialization helps a team

DISCUSS Have students discuss what a large restaurant with one employee would be like. Ask: How could the employee improve the situation without hiring additional people? *Sample answer: improve efficiency, do lots of prep work, have customers get their own food and clean up their places*

Ongoing Assessment

Explain the functions of cells in multicellular organisms.

Ask: How are multicellular organisms like the staff of a restaurant? *Like the different staff members, each type of cell has a different job that it does, and the cells work together.*

2.1 INSTRUCT

INVESTIGATE Specialization

PURPOSE To model cell specialization and explore its advantages and disadvantages

TIP *10 min.* Each group of five students should designate one "single-celled organism," one "multicellular organism," and a timekeeper.

WHAT DO YOU THINK? *It takes less time to do all three tasks, and those members with special skills can do a special job quicker and better. Cells need to get materials (water, oxygen, food) quickly or they will die.*

CHALLENGE *The group could not survive unless all members were present and functioning.*

R Datasheet, Specialization, p. 90

Technology Resources

Customize this student lab as needed or look for an alternative. Print rubrics to assess student lab reports.

Lab Generator CD-ROM

History of Science

In 1838, almost 200 years after Robert Hooke first described cells, the German botanist Matthias Schleiden made careful studies of plant tissues and announced that cells are the fundamental unit of life in plants. A year later, the German physiologist Theodor Schwann made the same announcement about animals. These two discoveries are combined in the statement that all living things consist of cells. This generalization forms the basis of the cell theory, one of the foundations of biology.

INVESTIGATE Specialization

What are some advantages of specialization?

PROCEDURE

1. Form into two teams,each representing an organism. The single-celled team will be made up of just one person; the multicellular team will be made up of three. Each team obtains a box of materials from the teacher.
2. Each team must do the following tasks as quickly as possible: make a paper-clip chain, write the alphabet on both sides of one piece of paper, and make a paper airplane from the second piece of paper. The members of the three-person team must specialize, each person doing one task only.

WHAT DO YOU THINK?

- What are some advantages to having each person on the three-person team specialize in doing a different job?
- Why might time be a factor in the activities done by cells in a multicellular organism?

CHALLENGE Suppose the "life" of the multicellular team depended on the ability of one person to make a paper airplane. How would specialization be a disadvantage if that person were not at school?

SKILL FOCUS
Modeling

MATERIALS
- two boxes, each containing 20 paper clips, 2 pieces of paper, and 1 pencil

TIME
10 minutes

Levels of Organization

For any multicellular organism to survive, different cells must work together. The right type of cell must be in the right place to do the work that needs to be done.

Organization starts with the cell. Cells in multicellular organisms are specialized for a specific function. In animals, skin cells provide protection, nerve cells carry signals, and muscle cells produce movement. Cells of the same type are organized into **tissue,** a group of cells that work together. For example, what you think of as muscle is muscle tissue, made up of many muscle cells.

? A

VOCABULARY
Remember to add a four square for *tissue* and *organ* to your notebook.

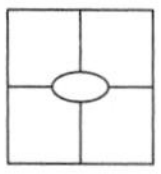

A structure that is made up of different tissues is called an **organ.** Organs have particular functions. The heart is an organ that functions as a pump. It has muscle tissue, which pumps the blood, and nerve tissue, which signals when to pump. Different organs that work together and have a common function are called an organ system. A heart and blood vessels are different organs that are both part of a circulatory system. These organs work together to deliver blood to all parts of a body. Together, cells, tissues, organs, and organ systems form an organism.

DIFFERENTIATE INSTRUCTION

More Reading Support

A What are tissues made of? *cells*

B What is an organ made of? *tissues*

English Learners English learners may have trouble with abstract or hypothetical questions. For example, in "Investigate Specialization," p. 44, "Suppose the 'life' of the multicellular team depended on the ability of one person to make a paper airplane." Help English learners by explaining that words such as *suppose, imagine,* and *assume* may be used to set up a hypothetical or "imagined" situation.

Organ Systems and the Organism

In almost all multicellular organisms, different organ systems take care of specific needs. Here are a few examples of organ systems found in many animals:

- nervous system enables a response to changing conditions
- muscular system produces movement and supplies heat
- respiratory system takes in oxygen and releases carbon dioxide
- circulatory system delivers oxygen and removes carbon dioxide
- digestive system breaks down food into a usable form

Organ systems allow multicellular organisms to obtain large amounts of energy, process large amounts of materials, respond to changes in the environment, and reproduce.

REMINDER
A system is a group of objects that interact, sharing energy and matter.

CHECK YOUR READING **How are the functions of organ systems related to the needs of an organism? Give an example.**

Different organ systems work together. For example, the respiratory system works with the circulatory system to deliver oxygen and remove carbon dioxide. When an animal such as a turtle breathes in, oxygen is brought into the lungs. Blood from the circulatory system picks up the oxygen, and the heart pumps the oxygen-rich blood out to the cells of the body. As oxygen is delivered, waste carbon dioxide is picked up. The blood is pumped back to the lungs. The carbon dioxide is released when the turtle breathes out. More oxygen is picked up when the turtle breathes in.

Organ Systems

Organ systems work together to meet the needs of an organism.

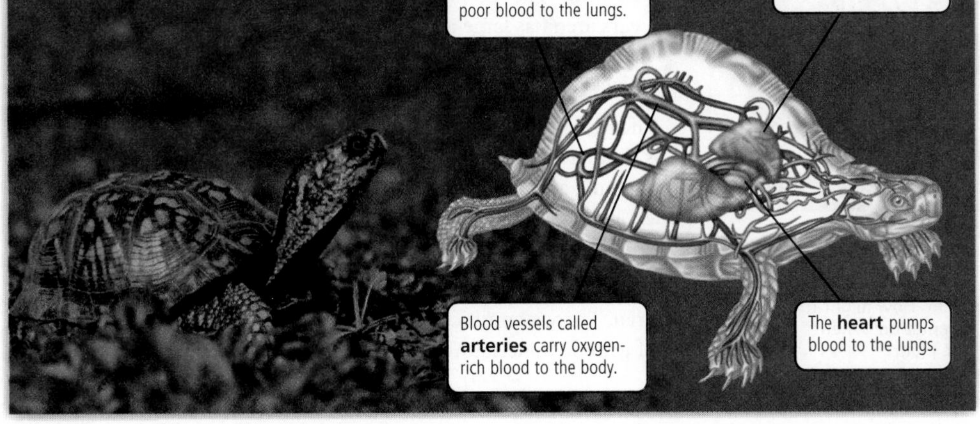

Blood vessels called **veins** return oxygen-poor blood to the lungs.
Each **lung** fills with air containing oxygen.
Blood vessels called **arteries** carry oxygen-rich blood to the body.
The **heart** pumps blood to the lungs.

DIFFERENTIATE INSTRUCTION

More Reading Support

C Why does a multicellular organism have several organ systems? *to take care of all its needs*

Below Level Have students draw a flow chart diagramming the path of oxygen into a turtle's body, through the blood, and into the cells. A second part of the flow chart can show the path of carbon dioxide away from the cells and out of the body.

Address Misconceptions

IDENTIFY Ask students what types of living things have organs. If students omit plants, or do not answer "multicellular organisms," they may hold the misconception that only animals have organs.

CORRECT Ask students to define an organ. ***a structure made of different tissues*** Display some leaves and have students examine them. Ask: Does a leaf fit this definition? What are the tissues in a leaf? ***yes; the layers in the blade, veins (vascular tissue)***

REASSESS Ask students to list other organs that a plant has. ***roots, flowers***

Technology Resources

Visit **ClassZone.com** for background on common student misconceptions.

MISCONCEPTION DATABASE

Teach from Visuals

To help students interpret the diagram of a turtle's circulatory system, ask:

- Why are some of the blood vessels red and others blue? *Red signifies blood that moves away from the heart and lungs. Blue indicates blood that returns to the heart.*
- What parts of the circulatory system are shown? *blood, arteries, veins, heart*
- What parts of the respiratory system are shown? *lungs*

Ongoing Assessment

CHECK YOUR READING *Answer: Each organ system takes care of a specific need. Together, the organ systems meet all the needs of the organism. For example, the organism's cells need oxygen, which is taken in by the respiratory system and delivered to the cells by the circulatory system.*

Teacher Demo

To demonstrate the advantages of color camouflage as an adaptation, use a hole punch to make a number of dots from different-colored construction paper. Spread out a square of colorful fabric and a square of plain white fabric on a desktop. Have students sprinkle the dots on each square and identify the background on which the dots are more visible. Ask students to apply this demo to how camouflage strategies work outdoors.

Ongoing Assessment

Describe how multicellular organisms are adapted to different environments.

Ask: Give an example of an adaptation an animal can have to cold climates. *Sample answer: thick fur*

CHECK YOUR READING *Answer: the way an organism gets energy or processes materials, how its body is shaped, and its behavior*

PHOTO CAPTION Answer: Flies are plant pollinators. The plant needs flies to reproduce.

Multicellular organisms are adapted to live in different environments.

READING TIP

Offspring is a word used to describe the new organisms produced by reproduction in any organism. Think of it as meaning "to spring off."

All organisms have characteristics that allow them to survive in their environment. An adaptation is any inherited characteristic that increases the chance of an organism's surviving and producing offspring that also reproduce. An adaptation may have to do with the way an organism gets its energy or processes materials. An adaptation may relate to the shape or structure of an organism's body. An adaptation can even be a form of behavior. D

CHECK YOUR READING The text above mentions different types of adaptations. Name three.

When most multicellular organisms reproduce, the offspring are not exact copies of the parents. There are differences. If a particular difference gives an organism an advantage over other members in its group, then that difference is referred to as an adaptation. Over time, the organism and its offspring do better and reproduce more.

You are probably familiar with the furry animal called a fox. Different species of fox have different adaptations that enable them to survive in different environments. Here are three examples:

- **Fennec** The fennec is a desert fox. Its large ears are an adaptation that helps the fox keep cool in the hot desert. As blood flows through the vessels in each ear, heat is released. Another adaptation is the color of its fur, which blends in with the desert sand.
- **Arctic fox** The Arctic fox lives in the cold north. Its small ears, legs, and nose are adaptations that reduce the loss of heat from its body. Its bluish-gray summer fur is replaced by a thick coat of white fur as winter approaches. Its winter coat keeps the fox warm and enables it to blend in with the snow.
- **Red fox** The red fox is found in grasslands and woodlands. Its ears aren't as large as those of the fennec or as small as those of the Arctic fox. Its body fur is reddish brown tipped in white and black, coloring that helps it blend into its environment.

The diversity of life on Earth is due to the wide range of adaptations that have occurred in different species. An elephant has a trunk for grasping and sensing. A female kangaroo carries its young in a pouch. The largest flower in the world, the rafflesia flower, is almost a meter wide, blooms for just a few days, and smells like rotting meat.

INFER The strong odor of the rafflesia flower attracts flies into the plant. How might this adaptation benefit the plant? E

Adaptations are the result of differences that can occur in genetic material. The way multicellular organisms reproduce allows for a mixing of genetic material. You will read about that next.

DIFFERENTIATE INSTRUCTION

More Reading Support

D What is an adaptation? *inherited characteristic that increases survival chances*

E How do adaptations arise? *differences in genes*

English Learners Help students with the numerous word forms they may encounter. For example, *adaptation* and *adapt.* Have students write a sentence using the noun *adaptation.* Then have them write another sentence using the verb *adapt.* Encourage them to verbally explain the difference between the two words.

Adaptations in Different Environments

READING VISUALS Foxes are hunters that feed on small animals. How might the coat color of each fox contribute to its survival?

DIFFERENTIATE INSTRUCTION

Advanced Have students design an experiment to demonstrate that coat color in foxes can contribute to their survival. The experiment should be feasible as a field study in the wild. Students should choose one species of fox and one environment when designing their experiment. They should consider any needs of the foxes that would be compromised by research, and avoid such compromise.

R Challenge and Extension, p. 89

Teach from Visuals

To help students interpret the visual of adaptations in different environments, ask:

- Why would having white fur in summer be a disadvantage for an arctic fox? *In summer there is no snow. The environment is a brownish color. White fur would make the fox more visible to both prey and predators.*
- What might trigger an arctic fox's body to grow white fur? *cold temperatures or shorter daylight hours*

T The "Adaptations in Different Environments" visual is available as T14 in the Unit Transparency Book.

Develop Critical Thinking

SYNTHESIZE Have students design an animal that lives in a hot, rainy tropical rainforest. Their designs should show adaptations that will keep the animal cool and dry, protect it against enemies that live in the trees, and enable it to capture insects for food. Adaptations can be structural or behavioral. Have students draw pictures of their animal or build a model.

Metacognitive Strategy

Ask students to explain which was a better tool for them: the visual and its captions on p. 47 or the text on p. 46. Which could help them review or study?

Ongoing Assessment

READING VISUALS *Answer: Camouflage makes them less likely to be seen by the animals hunting them and less likely to be seen by the animals they are hunting.*

Teach from Visuals

To help students interpret the diagrams of cells, explain that the colored squiggles represent chromosomes, a component of DNA. Normal cells, such as the cell on the bottom right, have pairs of matching chromosomes. Sperm cells and egg cells only receive half of the chromosomes during meiosis. Ask: What kind of cell division yields cells with two copies of DNA? *mitosis*

Teacher Demo

To help students understand the differences between sexual and asexual reproduction, try the following demo.

Root an African violet *leaf* in moist sand. When roots form, plant the new violet in soil. Show students the African violet and a flower. Point out the female part (pistil) and male parts (stamens) of the flower. Explain that fertilization takes place inside the female part and eventually produces seeds. Ask:

- Is the flower a site for sexual or asexual reproduction? *sexual*
- How about the African violet? *asexual*
- How many parents does the African violet have? *one*
- If the flower is pollinated and produces a seed, how many parents does the seed have? *two*

Ongoing Assessment

Summarize sexual reproduction.

Ask: Why is it necessary for meiosis to come before fertilization? *If meiosis did not cut the number of chromosomes in half before fertilization, the chromosome number would double in each offspring.*

Sexual reproduction leads to diversity.

One copy of DNA in cell after meiosis

F Most multicellular organisms reproduce sexually. In **sexual reproduction,** the genetic material of two parents comes together, and the resulting offspring have genetic material from both. Sexual reproduction leads to diversity because the DNA in the offspring is different from the DNA in the parents.

G Two different cellular processes are involved in sexual reproduction. The first is **meiosis** (my-OH-sihs), a special form of cell division that produces sperm cells in a male and egg cells in a female. Each sperm or egg cell contains only one copy of DNA, the genetic material. Most cells contain two copies of DNA.

The second process in sexual reproduction is **fertilization.** Fertilization occurs when the sperm cell from the male parent combines with the egg cell from the female parent. A fertilized egg is a single cell with DNA from both parents. Once the egg is fertilized, it divides. One cell becomes two, two cells become four, and so on. As the cells divide, they start to specialize, and different tissues and organs form.

Differences in genetic material and in the environment produce differences in offspring. Whether a tulip flower is red or yellow depends on the genetic material in its cells. How well the tulip grows depends on conditions in the environment as well as genetic materials.

DIFFERENTIATE INSTRUCTION

More Reading Support

F How many parents take part in sexual reproduction? *two*

G What two processes are involved in sexual reproduction? *meiosis and fertilization*

Inclusion Model the genetics of sexual reproduction for students with visual impairments. Cut out cells from sandpaper. Make chromosomes with different textures of yarn, twine, string, or rope. Lay the chromosomes on cell bases, in pairs and singly.

Sexual Reproduction The fertilized eggs of a salamander contain genetic material from two parents.

Asexual Reproduction The buds of a sea coral have the same genetic material as the parent.

Most reproduction that occurs in multicellular organisms is sexual reproduction. However, many multicellular organisms can reproduce by asexual reproduction. With asexual reproduction, a single parent produces offspring.

Budding is a form of asexual reproduction. In budding, a second organism grows off, or buds, from another. Organisms that reproduce asexually can reproduce more often. Asexual reproduction limits genetic diversity within a group because offspring have the same genetic material as the parent.

CHECK YOUR READING How do offspring produced by sexual reproduction compare with offspring produced by asexual reproduction?

With sexual reproduction, there is an opportunity for new combinations of characteristics to occur in the offspring. Perhaps these organisms process food more efficiently or reproduce more quickly. Or perhaps they have adaptations that allow them to survive a change in their environment. In the next three sections, you will read how plants, animals, and fungi have adapted to similar environments in very different ways.

2.1 Review

KEY CONCEPTS

1. How do specialized cells relate to the different levels of organization in a multicellular organism?
2. What is an adaptation? Give an example.
3. What two cellular processes are involved in sexual reproduction?

CRITICAL THINKING

4. **Compare and Contrast** How do the offspring produced by sexual reproduction compare to those produced by asexual reproduction?
5. **Predict** If fertilization occurred without meiosis, how many copies of DNA would be in the cells of the offspring?

CHALLENGE

6. **Synthesize** Do you consider the different levels of organization in a multicellular organism an adaptation? Explain your reasoning.

ANSWERS

1. Specialized cells make up different tissues.

2. An adaptation is a characteristic that helps an organism survive. Examples may include structures, behaviors, or processes.

3. meiosis and fertilization

4. Offspring of sexual reproduction have a combination of characteristics from two parents, increasing the species' diversity. Asexual reproduction produces identical offspring more often and with less energy.

5. four

6. The different levels of organization can be considered an adaptation. They help the organism survive by performing tasks more efficiently.

EXPLORE the BIG idea

Revisit "How Can Multicellular Organisms Reproduce?" on p. 41. Ask students what kind of reproduction their potatoes accomplished.

Develop Critical Thinking

INFER Ask students to give an example of when asexual reproduction might have advantages over sexual reproduction. *An organism would save energy and could reproduce more frequently.*

Ongoing Assessment

CHECK YOUR READING *Answer: An offspring produced by sexual reproduction has genetic material from two parents, so it has characteristics from each parent. An offspring produced asexually has identical genetic material and characteristics of the parent.*

Reinforce the BIG idea

Have students relate the section to the Big Idea.

R Reinforcing Key Concepts, p. 91

2.1 ASSESS & RETEACH

Assess

A Section 2.1 Quiz, p. 23

Reteach

Have students review pp. 44–45. Ask them to name an organ system in their body and list an organ, a tissue, and a type of cell in that system. *Sample answer: circulatory system, heart, muscle tissue, muscle cell*

Technology Resources

Have students visit **ClassZone.com** for reteaching of Key Concepts.

CONTENT REVIEW

CONTENT REVIEW CD-ROM

MATH IN SCIENCE
Math Skills Practice for Science

Set Learning Goal

To display part-to-whole relationships in a circle graph to present data about animal sightings

Present the Science

Counting the number of animals at a waterhole is a way to sample the total diversity of animals in the area.

Develop Graphing Skills

- Remind students that the bar in a fraction can be read as "out of" or "divided by."
- Review how to read a circle graph: Count up how many total parts there are (10), then determine how many parts are in the first shaded wedge (3). The shaded parts are numerators and the total number of parts is the denominator.

DIFFERENTIATION TIP Review fractions. Remind students that a fraction has two parts, a numerator (top number) and a denominator (bottom number).

Close

Ask: Why might a researcher want to count the number of animals at a waterhole each month for a year? *Sample answer: because animals migrate seasonally and may not be around each month of the year*

- Math Support, p. 129
- Math Practice, p. 130

Technology Resources

Students can visit **ClassZone.com** for practice using circle graphs.

MATH TUTORIAL

MATH in SCIENCE

SKILL: USING CIRCLE GRAPHS

MATH TUTORIAL
CLASSZONE.COM
Click on Math Tutorial for more help interpreting circle graphs.

Making Data Visual

A circle graph is a good way to see part-to-whole relationships. To use data presented in a circle graph, do the following:

Example

Suppose, at a waterhole in a game preserve, researchers observed ten animals throughout the day. What fraction of the sightings were giraffes?

(1) The circle graph shows the data for the sightings. The whole circle represents the total sightings, 10.

(2) 3 of the 10 equal parts are shaded for giraffes.

(3) Write "3 out of 10" as a fraction $\frac{3}{10}$.

ANSWER Giraffes = $\frac{3}{10}$ of the sightings.

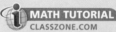

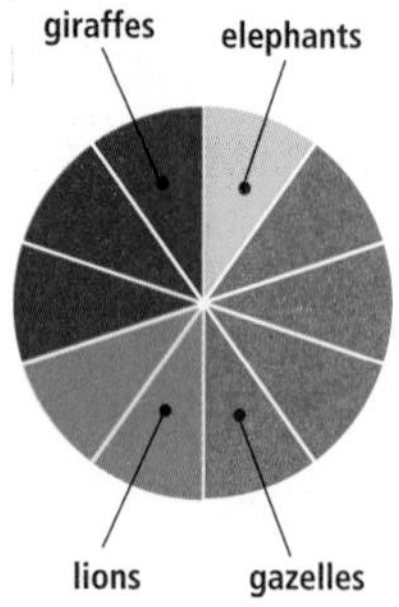

Answer the following questions.

1. What fraction of the sightings were lions?
2. What fraction of the sightings were gazelles?
3. Which animal did the researchers observe in greatest number?
4. Which animal did they observe in the least number? How many sightings occurred for that animal?
5. What fraction of the total does that animal represent?
6. If the researchers had seen one hundred animals, and the graph looked the same as it does, how many giraffe sightings would their graph represent?

CHALLENGE You also record the nighttime visitors: first, 2 young elephants; then a lioness with 2 thirsty cubs; then a giraffe, followed by a hyena and 3 gazelles. Calculate the fraction of the night's population that is represented by each type of animal. Use the fractions to draw a circle graph of the data on "Night Sightings at the Waterhole." Shade and label the graph with the types of animals.

ANSWERS

1. $\frac{2}{10}$ *or* $\frac{1}{5}$

2. $\frac{4}{10}$ *or* $\frac{2}{5}$

3. gazelles

4. elephants, 1

5. $\frac{1}{10}$

6. 30

CHALLENGE *elephants:* $\frac{2}{10}$*; lions:* $\frac{3}{10}$*; giraffe:* $\frac{1}{10}$*; hyena:* $\frac{1}{10}$*; gazelles:* $\frac{3}{10}$

KEY CONCEPT

2.2 Plants are producers.

BEFORE, you learned
- Multicellular organisms have tissues, organs, and systems
- Organisms have adaptations that can make them suited to their environment
- Sexual reproduction leads to genetic diversity

NOW, you will learn
- How plants obtain energy
- How plants store energy
- How plants respond to their environment

VOCABULARY

photosynthesis p. 52
autotroph p. 52
cellular respiration p. 53
stimulus p. 55

EXPLORE Stored Energy

In what form does a plant store energy?

PROCEDURE

1. Obtain pieces of potato, celery, and pear that have been placed in small plastic cups.
2. Place a few drops of the iodine solution onto the plant material in each cup. The iodine solution will turn dark blue in the presence of starch. It does not change color in the presence of sugar.

MATERIALS
- pieces of potato, celery, and pear
- 3 plastic cups
- iodine solution
- eye dropper

WHAT DO YOU THINK?
- Observing each sample, describe what happened to the color of the iodine solution after a few minutes.
- Starch and sugars are a source of energy for a plant. What do your observations suggest about how different plants store energy?

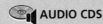

Plants capture energy from the Sun.

MAIN IDEA AND DETAILS Add the main idea *plants capture energy from the Sun* to your notebook and fill in details on photosynthesis and stored energy.

If you stand outside on a warm, sunny day, you may see and feel energy from the Sun. Without the Sun's energy, Earth would be a cold, dark planet. The Sun's heat and light provide the energy almost all organisms need to live.

However, energy from the Sun cannot drive cell processes directly. Light energy must be changed into chemical energy. Chemical energy is the form of energy all organisms use to carry out the functions of life. Plants are an important part of the energy story because plants capture energy from the Sun and convert it to chemical energy.

2.2 FOCUS

Set Learning Goals

Students will
- Describe how plants obtain energy.
- Describe how plants store energy.
- Identify how plants respond to their environment.
- Design an experiment to see if plants grow by absorbing material from soil.

3-Minute Warm-Up

Display Transparency 12 or copy this exercise on the board:

Match the definitions to the correct terms.

Definitions
1. sperm or eggs are created *c*
2. a sperm and an egg join *d*
3. a group of cells that work together *a*

Terms

a. tissue
b. organ
c. meiosis
d. fertilization
e. sexual reproduction

T 3-Minute Warm-Up, p. T12

2.2 MOTIVATE

EXPLORE Stored Energy

PURPOSE To determine whether potatoes, celery, and pears contain starch

TIPS *10 min.* Choose pears carefully. The starch content of a pear decreases as it ripens. Iodine Solution is available in drug stores in the appropriate concentration. Be sure students do not ingest it. Review guidelines for safe handling.

WHAT DO YOU THINK? *The potato turns blue black, and the celery does not change color. Ripe pears do not turn blue black, but unripe pears do. Tubers and unripe fruits store energy in starch. Celery stems do not store energy in starch.*

RESOURCES FOR DIFFERENTIATED INSTRUCTION

Below Level

UNIT RESOURCE BOOK
- Reading Study Guide A, pp. 94–95
- Decoding Support, p. 128

AUDIO CDS

R **Additional INVESTIGATION,** Plant Growth and Light Conditions, A, B, & C, pp. 140–148; Teacher Instructions, pp. 360–361

Advanced

UNIT RESOURCE BOOK
Challenge and Extension, p. 100

English Learners

UNIT RESOURCE BOOK
Spanish Reading Study Guide, pp. 98–99

AUDIO CDS
- Audio Readings in Spanish
- Audio Readings (English)

2.2 INSTRUCT

INVESTIGATE What Plants Need to Grow

PURPOSE To design an experiment to test whether plants grow by absorbing material from soil

TIPS *30 min.* Discuss what experimental variables are. Make sure students understand that they are not testing the hypothesis that plants need *only* soil to grow. To test this hypothesis, plants would have to be grown in a vacuum and could not be watered. Other types of plants may be used.

WHAT DO YOU THINK? *Sample answer: the masses of the plant and the soil it grows in*

CHALLENGE *Sample answer: The dependent variable—the mass of the soil—should be measured at the beginning and at the end of the experiment by removing the plant, wiping all the soil from the roots into the pot, and weighing the soil on the balance.*

R Datasheet, What Plants Need to Grow, p. 101

Technology Resources

Customize this student lab as needed or look for an alternative. Print rubrics to assess student lab reports.

Lab Generator CD-ROM

Ongoing Assessment

Describe how plants obtain energy.

Ask: What energy conversion takes place during photosynthesis? *The energy in sunlight changes to chemical energy.*

CHECK YOUR READING *Answer: Photosynthesis produces sugars and oxygen. The plant uses sugars for energy and as raw materials for growth.*

Producing Sugars

READING TIP: The roots for *photosynthesis* are *photo-*, which means "light," and *synthesis*, which means "to put together." Together they mean "put together by light."

Plants capture energy from sunlight and convert it to chemical energy through the process of **photosynthesis.** The plant takes in water and carbon dioxide from the environment and uses these simple materials to produce sugar, an energy-rich compound that contains carbon. Oxygen is also produced. Plants are referred to as producers because they produce energy-rich carbon compounds.

The cells, tissues, and organ systems in a plant work together to supply the materials needed for photosynthesis. Most photosynthesis takes place in the leaves. The leaves take in carbon dioxide from the air, and the stems support the leaves and hold them up toward the Sun. The roots of the plant anchor it in the soil and supply water. The sugars produced are used by the rest of the plant for energy and as materials for growth.

A CHECK YOUR READING What is the product of photosynthesis?

? B Another name for a plant is **autotroph** (AW-tuh-TRAHF). Autotroph means self-feeder. Plants do not require food from other organisms. Plants will grow if they have energy from the Sun, carbon dioxide from the air, and water and nutrients from the soil.

INVESTIGATE What Plants Need to Grow

Where does the material for plant growth come from?

DESIGN — YOUR OWN — EXPERIMENT

Until about 400 years ago, people thought that plants get everything they need from soil. Design an experiment to test this hypothesis: "If a plant grows by taking in material from soil, then the mass of the soil will decrease over time because soil material is taken into the plant."

PROCEDURE

1. Design an experiment, choosing from the materials listed.
2. Use the lab handbook, pages R28–32, to help you write your experimental procedure. Identify the variables and constants.

WHAT DO YOU THINK?

Measurement can be an important part of an experiment. What types of measurement do you use?

CHALLENGE An operational definition is a description of how you will measure the dependent variable. Give an operational definition for your experiment.

SKILL FOCUS
Designing experiments

MATERIALS
- potting soil
- pots or paper cups
- bean seedlings or beans
- triple beam balance
- water

TIME
30 minutes

DIFFERENTIATE INSTRUCTION

More Reading Support

A What are the raw materials of photosynthesis? *water and CO_2*

B What do you call an organism that does not consume food? *an autotroph*

Additional Investigation To reinforce Section 2.2 learning goals, use the following full-period investigation:

R **Additional INVESTIGATION,** Plant Growth and Light Conditions, A, B, & C, pp. 140–148, 360–361 (Advanced students should complete Levels B and C.)

English Learners Have students write the definitions for *autotroph, photosynthesis,* and *stimuli* in their Science Word Dictionaries. Pronounce these words for students.

Storing and Releasing Energy

Plants are not the only organisms that capture energy through photosynthesis. Algae and certain bacteria and protists also use photosynthesis. Plants are different from single-celled producers, however. Plants are multicellular organisms with parts of their bodies specialized for storing energy-rich material. Single-celled producers can store very little energy.

Only part of the energy captured by a plant is used as fuel for cellular processes. Some of the sugar produced is used as building material, enabling the plant to grow. The remaining sugar is stored. Often the sugars are stored as starches. Starch is an energy-rich compound made of many sugars. Starches can store a lot of chemical energy. When a plant needs energy, the starches are broken back down into sugars and energy is released. **Cellular respiration** is the process by which a cell uses oxygen to break down sugars to release the energy they hold.

Some plants, such as carrots and beets, store starch in their roots. Other plants, including celery and rhubarb, have stems adapted for storing starch. A potato is a swollen, underground stem called a tuber. Tubers have buds—the eyes of the potato—that can sprout into new plants. The starch stored in the tuber helps the new sprouts survive.

CHECK YOUR READING What is the original source of a plant's stored energy?

Plants are adapted to different environments.

Almost everywhere you look on land, you'll see plants. Leaves, stems, and roots are adaptations that enable plants, as producers, to live on land. Not all plants, however, look the same. Just as there are many different types of land environments, there are many different types of plants that have adapted to these environments.

Grasses are an example of plants that grow in several environments. Many grasses have deep roots, produce seeds quickly, and can grow in areas with a wide range of temperatures and different amounts of precipitation. Grasses can survive drought, fires, freezing temperatures, and grazing. As long as the roots of the plant survive, the grasses will grow again. Grasses are found in the Arctic tundra, as well as in temperate and tropical climates.

Teach from Visuals

To help students interpret the photograph of a potato plant, ask:

- Where are the plant's stems? *Some are aboveground, producing leaves. Some are underground, swollen structures that we call potatoes.*
- Where are the plant's roots? *underground*

Address Misconceptions

IDENTIFY Ask: How do plants get nutrients? If students' responses indicate that plants are consumers rather than producers, students may hold the misconception that carbon dioxide, water, and soil are food that plants eat.

CORRECT Ask students to define *food. stored energy* Have them explain why carbon dioxide, water, and soil are not food sources for plants. *They do not provide energy. The plants need to capture the sunlight to get their energy.*

REASSESS Ask: Why don't plants need food? *They get energy from sunlight and store it in the chemical bonds of sugars that they synthesize.*

Technology Resources

Visit **ClassZone.com** for background on common student misconceptions.

MISCONCEPTION DATABASE

Ongoing Assessment

Describe how plants store energy.

Ask: In what form do plants store the food they produce? *as sugar and starch*

CHECK YOUR READING *Answer: sunlight*

DIFFERENTIATE INSTRUCTION

More Reading Support

C What organisms other than plants use photosynthesis? *algae, some bacteria, and some protists*

D Where do plants store starch? *roots, stems, tubers*

Inclusion Bring in a small section of turf (lawn grasses) and a growing plant that has deep roots, such as most house plants. Let students who are visually impaired or who have cognitive disabilities examine the root systems of the plants and discover for themselves how shallow grass roots are compared with the roots of other plants.

Real World Example

Carnivorous plants usually grow in acid bogs that have little nitrogen. Although these plants make sugars by photosynthesis, they get the nitrogen and minerals they need by digesting insects. Pitcher plants attract insects with their bright colors. Once the insect enters the plant, it can only move down a long, water-filled tube. The insect drowns and is digested by enzymes that the plant secretes into the water. Sundews, which are found in temperate forests, secrete a sticky glue that traps insects on the leaves, where they are digested. Bladderworts are aquatic plants that trap insects in bladders, using a trapdoor that snaps shut like a spring. Enzymes digest the insect.

Teach from Visuals

To help students interpret the photographs of a Venus flytrap,

- Point out that the Venus flytrap has tiny hairs on the inner surfaces of the leaves. When an insect touches these hairs, the leaf snaps closed and traps the insect.
- Ask: Would you expect to find Venus flytraps growing in rich garden soil? Explain. *No; they grow in soil that lacks certain minerals. The digested insects supply these minerals.*

Ongoing Assessment

CHECK YOUR READING *Answer: reproductive adaptations such as the ability to produce seeds quickly, and protective adaptations such as thorns or odors*

PHOTO CAPTION Answer: Yes; it still produces sugars by photosynthesis; it is green because of chlorophyll. It gets its energy from sunlight absorbed by chlorophyll.

E Now compare trees to grasses. If the leaves and stems of a tree die away because of fire or drought, often the plant will not survive. Because of their size, trees require a large amount of water for photosynthesis. A coniferous (koh-NIHF-uhr-uhs) tree, like the pine, does well in colder climates. It has needle-shaped leaves that stay green throughout the year, feeding the plant continually. A deciduous (dih-SIHJ-oo-uhs) tree, like the maple, loses its leaves when temperatures turn cold. The maple needs a long growing season and plenty of water for new leaves to grow.

RESOURCE CENTER CLASSZONE.COM
Learn more about plant adaptations.

Plants have reproductive adaptations. It may surprise you to learn that flowering plants living on cold, snowy mountaintops have something in common with desert plants. When rain falls in the desert, wildflower seeds sprout very quickly. Within a few weeks, the plants grow, flower, and produce new seeds that will be ready to sprout with the next rainy season. The same thing happens in the mountains, where the snow may thaw for only a few weeks every summer. Seeds sprout, flowers grow, and new seeds are produced—all before the snow returns. You will read more about plant reproduction in Chapter 3.

F Some plants have adaptations that protect them. Plants in the mustard family give off odors that keep many plant-eating insects away. Other plants, such as poison ivy and poison oak, produce harmful chemicals. The nicotine in a tobacco plant is a poison that helps to keep the plant from being eaten.

CHECK YOUR READING Name two different types of adaptations plants have.

Some adaptations plants have relate to very specific needs. For example, the Venus flytrap is a plant that grows in areas where the soil lacks certain materials. The leaves of the Venus flytrap fold in the middle and have long teeth all around the edges. When an insect lands on an open leaf, the two sides of the leaf fold together. The teeth form a trap that prevents the captured insect from escaping. Fluids given off by the leaf digest the insect's body, providing materials the plant can't get from the soil.

ANALYZE An insect provides nutrients that this Venus flytrap cannot get from the soil. Is this plant still a producer? Ask yourself where the plant gets its energy.

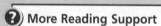

DIFFERENTIATE INSTRUCTION

More Reading Support

E Do trees or grasses need more water? *trees*

F Name a protective adaptation. *Sample answer: odors*

Below Level Have students research the many adaptations of cactus plants to their dry environment and prepare a poster in the style of the diagram for adaptation, p. 47.

Plant roots always grow downward and stems always grow upward. All plants respond to gravity as a stimulus.

The tendril of a climbing plant grows around a nearby object. The plant responds to touch as a stimulus.

Plants respond to their environment.

During a hot afternoon, parts of the flower known as the Mexican bird of paradise close. As the Sun goes down, the flower reopens. The plant is responding to a stimulus, in this case, sunlight. A **stimulus** is something that produces a response from an organism. Plants, like all organisms, respond to stimuli in their environment. This ability helps them to survive and grow.

VOCABULARY Make a four square for the term *stimulus* in your science notebook.

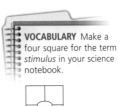

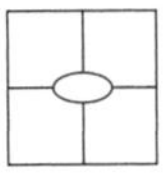

Gravity

Gravity is the force that keeps you bound to Earth and gives you a sense of up and down. All plants respond to gravity. They also have a sense of up and down—roots grow down and stems grow up. Suppose you place a young seedling on its side, so that its roots and stems stretch out to the side. In a very short time, the tip of the root will begin to turn down, and the tip of the stem will turn up.

Touch

Many plants also respond to touch as a stimulus. Peas, morning glories, tropical vines, and other climbing plants have special stems called tendrils. Tendrils respond to the touch of a nearby object. As the tendrils grow, they wrap around the object. The twining of tendrils around a fence or another plant helps raise a plant into the sunlight.

Teach from Visuals

To help students interpret the photographs of plant responses, ask:

- What is the advantage to a plant to have its roots grow down and its stems grow up? *Roots must take in water, usually found in soil. Leaves on stems must collect sunlight to photosynthesize.*
- What is the advantage of having tendrils? *Plants with weak stems can stretch above the shade of other plants, and get more sunlight.*

Teach Difficult Concepts

Students often have trouble understanding that stems respond to gravity by growing up, although the force of gravity is downward. Explain that roots exhibit a positive response to gravity when they grow down, and stems exhibit a negative response to gravity. You might also try the following demonstration.

Teacher Demo

Soak four corn seeds in water for 24 hours. Crumple up two wet paper towels and place them in the bottom of a petri dish. Arrange the seeds on the paper towels at the 12, 3, 6, and 9 o'clock positions. The pointed ends of the seeds should all face toward the center of the dish. Cover the petri dish, tape it closed, and mark an arrow on the cover. Stand the dish on its edge so the arrow points down. In about two days, the seeds will germinate. Roots will appear first, then green shoots. The roots will grow down, and the shoots will grow up.

DIFFERENTIATE INSTRUCTION

More Reading Support

G What produces a response from an organism? *a stimulus*

H How do plants respond to gravity? *Roots grow down; stems grow up.*

English Learners Have students look up words that have similar roots to *stimulus,* such as *stimulate* and *stimulation.* Have them look up the words *simulate* and *simulation.* Point out that these two groups of words have completely different meanings. Have students use the words in sentences.

Teach from Visuals

To help students interpret the diagram of how plants respond to light, ask:

- What do the dots in the cells in the circle represent? *auxin*
- Why do the cells on the left side of the stem have more auxin than the cells on the right side? *They are on the dark side of the stem.*
- Why does the stem bend to the right? *Because the cells on the left are longer, there is more surface area on the left side of the stem.*

History of Science

In the late 19th century, Charles Darwin and his son Francis conducted some of the first experiments on the response of plants to light. They observed that a grass seedling bends toward light only if the tip of the shoot is present. When they removed the tip or covered it with a tiny opaque cap, the seedling did not bend. They concluded that the tip of the shoot is responsible for sensing light. In 1926 Friedrich Went, a Dutch botanist, identified auxin, the chemical substance that causes plant stems to bend toward light.

Ongoing Assessment

Identify how plants respond to their environment.

Ask: Why do plants grow toward light? *The stimulus of light causes chemical changes that make the plant grow that way.*

How Plants Respond to Light

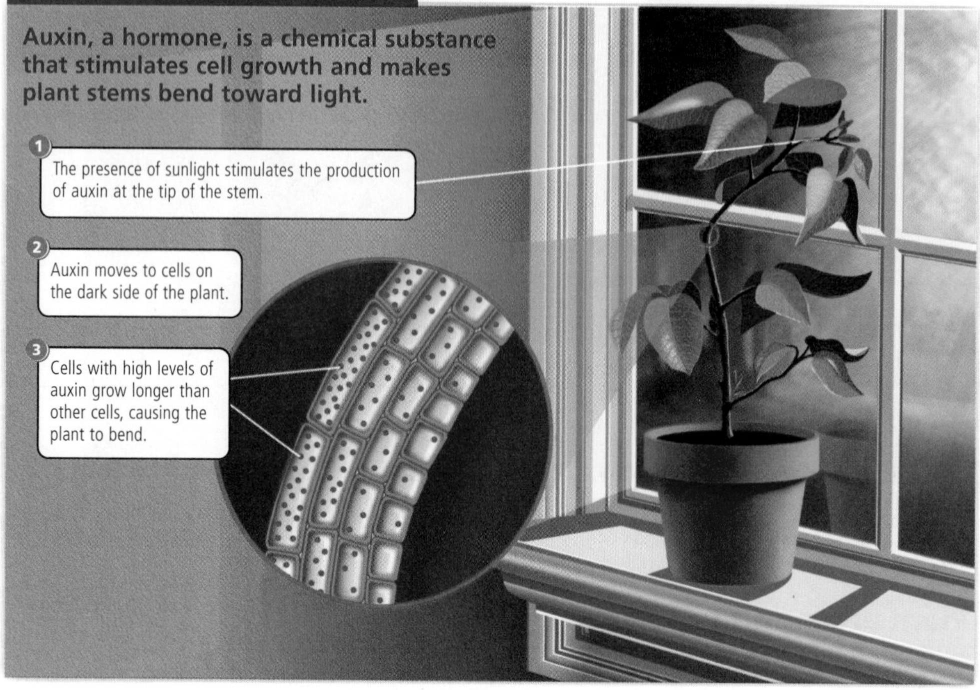
Auxin, a hormone, is a chemical substance that stimulates cell growth and makes plant stems bend toward light.
1 The presence of sunlight stimulates the production of auxin at the tip of the stem.
2 Auxin moves to cells on the dark side of the plant.
3 Cells with high levels of auxin grow longer than other cells, causing the plant to bend.

Light

READING TIP
The words *stimulus, stimuli,* and *stimulate* all have the same root, meaning "to provoke or encourage action."

Light is a powerful stimulus for plants. You can see that stems and leaves grow toward light by placing an indoor plant near a window. After several days, the tips of the stems start to bend toward the window. What happens if you turn the plant around so that those stems reach into the room? The stems will bend as they continue to grow, turning back toward the light.

Plants respond to light with the help of a hormone. A **hormone** is a chemical substance produced in one part of an organism and travels to a different part where it produces a reaction. Hormones act as chemical messengers. They allow an organism to respond to changes in its body or to changes in the environment around it.

VISUALIZATION CLASSZONE.COM
Examine how plants respond to different stimuli.

? I

Auxin (AWK-sihn), a plant hormone that stimulates cell growth, is produced at the tip of a plant stem. Auxin moves away from light. As a result, the cells on the darker side of a plant stem contain more auxin than those on the lighter side. Higher levels of auxin in plant cells on the darker side cause those cells to grow longer. The longer cells cause the plant stem to bend, moving the tip of the stem toward the light.

DIFFERENTIATE INSTRUCTION

More Reading Support

I What is the name of a plant hormone that stimulates cell growth? *auxin*

Alternative Assessment Have students answer the Check Your Reading question by drawing diagrams. Diagrams should include responses to each of the three stimuli.

Advanced Have students design experiments to test plant responses. They should choose one response to test, such as response to gravity, to touch, and so on. They should research the response in a particular plant variety, then write a hypothesis and procedure.

R Challenge and Extension, p. 100

Plants respond to seasonal changes.

Most regions of the world go through seasonal changes every year. For example, during the summer in North America, temperatures rise and the days get longer. As winter approaches, temperatures go down and the days become shorter. These types of seasonal changes have an effect on plants.

For plants, a shorter period of daylight will affect the amount of sunlight available for photosynthesis. Shorter days cause many plants to go into a state of dormancy. When plants are dormant, they temporarily stop growing and so require less energy.

In temperate climates, the approach of winter causes the leaves of deciduous trees to die and drop to the ground. The trees enter a state of dormancy during which their growth is slowed. Other plants, such as wild cornflowers, do not survive the change. New cornflowers will grow the following season, from seeds left behind.

CHECK YOUR READING What stimulus causes a deciduous plant to respond by dropping its leaves?

For many plants, reproduction is also affected by seasonal changes. For some plants, the amount of daylight is a factor. A few plants, such as rice and ragweed, produce flowers only in autumn or winter, when days are short. They are short-day plants. Long-day plants flower in late spring and summer, when days are long. Lettuce, spinach, and irises are long day-plants. You will read more about plants in Chapter 3.

2.2 Review

KEY CONCEPTS

1. What process makes a plant a producer, and what does a plant produce?
2. Name three stimuli that plants respond to, and give examples of how a plant responds.
3. How do seasonal changes affect plants? Give an example.

CRITICAL THINKING

4. **Give Examples** Give three examples of ways that plants are adapted to their environments. How do these adaptations benefit the plant?

CHALLENGE

5. **Apply** Some experiments suggest that the hormone auxin is involved in the twining of tendrils. Use what you know about auxin to explain how it might cause tendrils to twine around anything they touch. Draw a diagram.

EXPLORE the BIG idea

Revisit "Where Does It Come From?" on p. 41. Have students explain how they decided each item's source.

Ongoing Assessment

CHECK YOUR READING *Answer: fewer hours of daylight*

Reinforce the BIG idea

Have students relate the section to the Big Idea.

R Reinforcing Key Concepts, p. 102

2.2 ASSESS & RETEACH

Assess

A Section 2.2 Quiz, p. 24

Reteach

Have students make a poster summarizing plant responses to light, gravity, and touch. Posters should include a drawing or a picture cut out of a magazine, and give the stimulus, the response, and the effect on the plant for each type of response.

Technology Resources

Have students visit **ClassZone.com** for reteaching of Key Concepts.

CONTENT REVIEW

CONTENT REVIEW CD-ROM

ANSWERS

1. photosynthesis; sugar and starch

2. Light: plants grow toward the light; Touch: peas have tendrils that wrap around nearby objects; Gravity: stems grow up and roots grow down.

3. Some plants produce flowers only in autumn or winter. Shortened hours of daylight cause some trees to become dormant.

4. Sample answer: Rapid life cycle gives plants a better chance to reproduce. Odors protect plants from animals that want to eat them. Plants that grow in poor soil get nutrients from insects.

5. Answers should reflect an understanding that a higher level of auxin on one side of a tendril causes the tendril to bend toward the other side.

2.3 FOCUS

Set Learning Goals

Students will

- Describe how animals obtain energy.
- Summarize how animals process food.
- Identify different ways animals respond to their environment.
- Infer what an owl eats and how well it digests its food.

3-Minute Warm-Up

Display Transparency 13 or copy this exercise on the board:

Decide if these statements are true. If not true, correct them.

1. Plant stems grow up and do not respond to gravity. *Plant stems grow up and do respond to gravity.*
2. Plants are producers because they make energy. *Plants are producers because they make sugar and starch.*
3. Carrot plants have roots that are adapted for storing energy-rich carbon compounds. *true*

3-Minute Warm-Up, p. T13

2.3 MOTIVATE

THINK ABOUT

PURPOSE To understand that an animal's diet affects its teeth

DISCUSS Have students brainstorm what different shapes of teeth suggest about an animal's diet. Ask: Why do humans need four kinds of teeth? *We eat animals and plants, so we need teeth that can bite and chew both.*

KEY CONCEPT

2.3 Animals are consumers.

BEFORE, you learned

- Plants are producers
- Plants have adaptations for capturing and storing energy
- Plants respond to different stimuli

NOW, you will learn

- How animals obtain energy
- How animals process food
- About different ways animals respond to their environment

VOCABULARY

THINK ABOUT

What can you tell from teeth?

Many animals have teeth. Teeth bite, grind, crush, and chew. A fox's sharp biting teeth capture small animals that it hunts on the run. A horse's teeth are flat and strong—for breaking down the grasses it eats. Run your tongue over your own teeth. How many different shapes do you notice? What can the shape of teeth suggest about the food an animal eats?

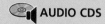

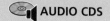

Animals obtain energy and materials from food.

You probably see nonhuman animals every day, whether you live in a rural area or a large city. If the animals are wild animals, not somebody's pet, then chances are that what you see these animals doing is moving about in search of food.

Animals are consumers. A **consumer** is an organism that needs to get energy from another organism. Unlike plants, animals must consume food to get the energy and materials they need to survive. Animals are heterotrophs. A **heterotroph** (HEHT-uhr-uh-TRAHF) is an organism that feeds on, or consumes, other organisms. By definition, animals are, quite simply, multicellular organisms that have adaptations that allow them to take in and process food.

READING TIP

The meaning of *heterotroph* is opposite to that of *autotroph*. The root *hetero-* means "other." *Heterotroph* means "other-feeder," or "feeds on others."

RESOURCES FOR DIFFERENTIATED INSTRUCTION

Below Level

UNIT RESOURCE BOOK

- Reading Study Guide A, pp. 105–106
- Decoding Support, p. 128

AUDIO CDS

Advanced

UNIT RESOURCE BOOK

- Challenge and Extension, p. 111
- Challenge Reading, pp. 124–125

English Learners

UNIT RESOURCE BOOK

Spanish Reading Study Guide, pp. 109–110

AUDIO CDS

- Audio Readings in Spanish
- Audio Readings (English)

Obtaining Food

Food is a source of energy and materials for animals.

Simple feeding Some animals, such as corals, can filter food from their environment.

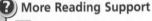

Complex feeding Many animals, such as bats, actively search for and capture food.

Animals need food. And animals have many different ways of getting it. For some animals, feeding is a relatively simple process. An adult coral simply filters food from the water as it moves through the coral's body. Most animals, however, must search for food. Grazing animals, such as horses, move along from one patch of grass to another. Other animals must capture food. Most bats use sound and hearing to detect the motion of insects flying at night. Its wings make the bat able to move through the air quickly and silently.

What Animals Eat

Just about any type of living or once-living material is a source of food for some animal. Animals can be grouped by the type of food they eat.

- A Herbivores (HUR-buh-VAWRS) feed on plants or algae.
- Carnivores (KAHR-nuh-VAWRS) feed on other animals. B
- Omnivores (AWM-nuh-VAWRS) feed on both plants and animals.

MAIN IDEA AND DETAILS Make a chart about the main idea: *Animals obtain energy and materials from food.* Include *herbivore, carnivore,* and *omnivore* in the details.

Another group are those animals that feed on the remains of once-living animals. Many insects do, as do some larger animals, such as vultures. Other animals, such as worms, act as decomposers.

CHECK YOUR READING Describe how herbivores, carnivores, and omnivores get their energy.

Different species of animals have adapted in different ways to take advantage of all the energy-rich material in the environment. To get energy and materials from food, all animals must first break the food down—that is, they must digest it.

2.3 INSTRUCT

Teach from Visuals

To help students interpret the photographs of animals obtaining food, ask:

- How do you think a coral filters food from the water? *Special cells filter tiny organisms out of the water as it flows through the sponge's body.*
- How do you think bats find insects at night? *They use sound and hearing.*

Metacognitive Strategy

Ask students to restate in their own words the definitions of the three groups of animals based on the type of food the animals eat.

Ongoing Assessment

Describe how animals obtain energy.

Ask: Where does an animal get the energy that it needs? *eating and digesting food*

CHECK YOUR READING *Answer: Herbivores feed on plants or algae, carnivores feed on other animals, and omnivores feed on both plants and animals.*

DIFFERENTIATE INSTRUCTION

More Reading Support

A What do herbivores eat? *plants or algae*

B What do carnivores eat? *other animals*

English Learners Place the terms *heterotroph, behavior, predator, prey, migration,* and *hibernation* on the classroom's Science Word Wall. Have students list as many animals as they know in five minutes. Then, have them go down their list and assign *herbivore, carnivore,* or *omnivore* to each animal. If they don't know what a certain animal eats, have them look it up in the encyclopedia.

INVESTIGATE Owl Pellets

PURPOSE To infer what an owl eats and how well it digests its food

TIPS *30 min.* Groupings of two or three students are ideal. Have diagrams of various small rodent skeletal systems available so students can identify the bones. Owl pellets can be ordered from biological supply catalogs. If you order owl pellets, they will come with a bone identification key. Remind students to use caution when handling the needle tool.

WHAT DO YOU THINK? *Students can infer that an owl is a carnivore because they will see fur and bones remaining. They can infer that an owl has no teeth and swallows its prey whole because bones are intact. Owls digest soft tissues but not fur and bones.*

CHALLENGE *Owls eat birds or small rodents such as mice.*

R Datasheet, Owl Pellets, p. 112

Technology Resources

Customize this student lab as needed or look for an alternative. Print rubrics to assess student lab reports.

Lab Generator CD-ROM

Teaching with Technology

If digital cameras are available, students can take photographs of what they find in their owl pellets. If tape recorders are available, students could narrate what they find as they separate the materials in their pellet.

Ongoing Assessment

Summarize how animals process food.

Ask: Why does food have to be processed? *Large, complex compounds in food must be broken down into smaller compounds that a body's cells can use.*

CHECK YOUR READING *Answer: It must be broken into smaller compounds.*

RESOURCE CENTER CLASSZONE.COM

Find out more about animal adaptations.

Processing Food

Energy is stored in complex carbon compounds in food. For the cells in an animal to make use of the energy and materials stored in this food, the large complex compounds must be broken back down into simpler compounds.

CHECK YOUR READING How must food be changed so an animal gets energy?

? C

Digestion is the process that breaks food down into pieces that are small enough to be absorbed by cells. A few animals, such as sponges, are able to take food particles directly into their cells. Most animals, however, take the food into an area of their body where the materials are broken down. Cells absorb the materials they need. Animals such as jellyfish have a single opening in their bodies where food is brought into a central cavity, or gut. The unused materials are released through the same opening.

A digestive system uses both physical and chemical activity to break down food. Many animals have a tubelike digestive system. Food is brought in at one end of the animal, the mouth, and waste is released at the other end. As food moves through the system, it is continually broken down, releasing necessary materials called nutrients to the cells.

INVESTIGATE Owl Pellets

What does an owl eat, and how well does it digest its food?

PROCEDURE

1. Get an owl pellet from your teacher. Open the foil and place the pellet in a tray.
2. Use a needle tool and tweezers to sort through the materials in the pellet and separate them.
3. When you have finished, dispose of the materials according to your teacher's instructions, and wash your hands.

WHAT DO YOU THINK?

- What can you tell about what an owl eats from looking at the remains in the pellet?
- What materials are not digested?

CHALLENGE Use the bone identification key to identify what the owl ate.

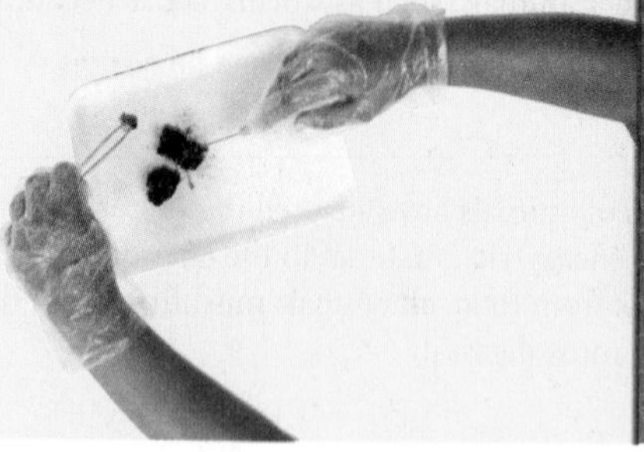

SKILL FOCUS
Inferring

MATERIALS
- owl pellet
- needle tool
- tweezers
- tray
- *for Challenge:* bone identification key

TIME
30 minutes

DIFFERENTIATE INSTRUCTION

More Reading Support

C What is the process that breaks down food? *digestion*

Alternative Assessment Have students give an oral report to the class in which they describe how they made their inferences about what an owl eats and how well it digests its food. Have the class ask questions after the presentation.

C 60 Unit: Diversity of Living Things

Obtaining Oxygen

Animals need oxygen to release the energy in food.

Grasshopper

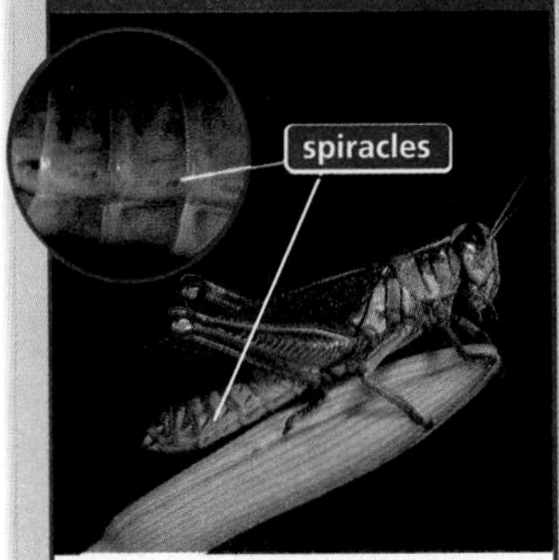

Most insects take in oxygen through body openings called spiracles.

Bass

Fish have gills, which pick up dissolved oxygen as water flows over them.

Tiger

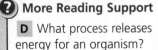

This yawning tiger, like many animals, gets oxygen by inhaling air into its lungs.

Releasing and Storing Energy

Animals obtain energy from sugars and other carbon compounds the same way plants do, through the process of cellular respiration. As you read in Section 2.2, cellular respiration is a process in which energy is released when sugars are broken down inside a cell. The process requires both oxygen and water.

CHECK YOUR READING What is the function of cellular respiration?

Many animals take in water in the same way they take in food, through the digestive system. Oxygen, however, is often taken in through a respiratory system. In many animals, the respiratory system delivers oxygen to the blood, and the blood carries oxygen to the cells.

Animals have different structures for obtaining oxygen. Many insects take in oxygen through spiracles, tiny openings in their bodies. Fish have gills, structures that allow them to pick up oxygen dissolved in the water. Other air-breathing animals take in oxygen through organs called lungs.

Most animals do not feed continuously, so they need to be able to store materials from food in their tissues or organs. Many animals, including humans, take in large amounts of food at one time. This gives an animal time to do other activities, such as caring for young or looking for more food.

Teach from Visuals

To help students interpret the photographs of how animals get oxygen, ask:

- Whose method of getting oxygen do you think provides the most oxygen for the animal? *the tiger's*
- Why do you think a larger animal might need more oxygen than a small one? *They need to use more energy to move and to maintain their body.*

Address Misconceptions

IDENTIFY Ask: How do plants and animals release energy? If students respond that different organisms use different processes, they may hold the misconception that only animals undergo cellular respiration.

CORRECT Demonstrate that plants respire. Put a sprig of *Elodea* in a test tube, cover it with water, and add one milliliter of bromthymol blue. Let the test tube stand for a few minutes. Small bubbles will come from the leaves, and the color will change from blue to green to yellow as the plant produces carbon dioxide during cellular respiration.

REASSESS Ask: What kinds of organisms undergo cellular respiration? *all organisms that use oxygen*

Technology Resources

Visit **ClassZone.com** for background on common student misconceptions.

MISCONCEPTION DATABASE

Ongoing Assessment

CHECK YOUR READING *Answer: to release the energy stored in sugars*

DIFFERENTIATE INSTRUCTION

More Reading Support

D What process releases energy for an organism? *cellular respiration*

E What does a respiratory system do? *It delivers oxygen to the blood.*

Below Level Have students draw a flow chart diagramming the path of food through an animal's body. *Teeth or structures in the mouth grind and cut up the food, which is then digested and absorbed. The processed food moves to cells, where it undergoes cellular respiration, releasing energy.*

Develop Critical Thinking

CONNECT Have students give examples of inherited and learned behavior in humans. *An example of inherited behavior might be jerking one's hand away after touching a hot object. A learned behavior is tying one's shoes.*

Real World Example

Animal behaviorists used to think that birds instinctively knew how to sing the songs of their species. However, young sparrows reared away from other members of their species or in the presence only of other species sang incorrectly. When scientists reared young sparrows in soundproof incubators fitted with microphones and recording devices, the songs the birds sang when they matured were poorly developed. It is now believed that both genetics and learning are necessary if a bird is to sing the common songs of its species.

Ongoing Assessment

CHECK YOUR READING *Answer: A behavior is any observable response to a stimulus.*

Animals interact with the environment and with other organisms.

Animals, as consumers, must obtain food, as well as water, from their environment. An animal's body has many adaptations that allow it to process food. These can include digestive, respiratory, and circulatory systems. Also important are the systems that allow animals to interact with their environment to obtain food. In many animals, muscle and skeletal systems provide movement and support. A nervous system allows the animal to sense and respond to stimuli.

Animals respond to many different types of stimuli. They respond to sights, sounds, odors, light, or a change in temperature. They respond to hunger and thirst. They also respond to other animals. Any observable response to a stimulus is described as a **behavior.** A bird's drinking water from a puddle is a behavior. A lion's chasing an antelope is a behavior, just as the antelope's running to escape the lion is a behavior.

CHECK YOUR READING What is a behavior and how does it relate to a stimulus?

F Some behaviors are inherited, which means they are present at birth. For example, a spider can weave a web without being shown. Other behaviors are learned. For example, the young lion in the photograph learns that a porcupine is not a good source of food.

All behaviors fall into one of three general categories:

- individual behaviors
- interactions between animals of the same species
- interactions between animals of different species

ANALYZE Do you consider the defensive behavior of a porcupine an adaptation?

DIFFERENTIATE INSTRUCTION

More Reading Support

F Is a spider's ability to make a web inherited or learned behavior? *inherited*

Advanced Spiders are predators. They have the adaptation of building webs to catch their prey. The size and design of spider webs can be affected by several factors. Have students each research one type of web-building spider and report on how various stimuli affect the web-building behaviors of the spider.

R Challenge and Extension, p. 111

Have students who are interested in animal behavior read the following article:

R Challenge Reading, pp. 124–125

Individual behaviors often involve meeting basic needs. Animals must find food, water, and shelter. They sleep. They groom themselves. Animals also respond to changes in their environment. A lizard may warm itself in the morning sunlight and then move into the shade when the Sun is high in the sky.

Interactions that occur between animals of the same species are often described as social behaviors. Basic social behaviors include those between parents and offspring and behaviors for attracting a mate. Within a group, animals of the same species may cooperate by working together. Wolves hunt in packs and bees maintain a hive. Behaviors among animals of the same species can also be competitive. Animals often compete for a mate or territory.

For macaques, grooming is both an individual and a social behavior. Here a mother grooms her young.

CHECK YOUR READING What are some ways that animals of the same species cooperate and compete?

Interactions that occur between animals of different species often involve the search for food. A **predator** is an animal that hunts other animals for food. Predators have behaviors that allow them to search for and capture other organisms. A cheetah first stalks an antelope, then chases it down, moving as fast as 110 kilometers per hour.

An animal that is hunted by another animal as a source of food is the **prey.** Behaviors of prey animals often allow them to escape a predator. An antelope may not be able to outrun a cheetah, but antelopes move in herds. This provides protection for the group since a cheetah will kill only one animal. Other animals, such as the pufferfish and porcupine, have defensive behaviors and structures.

VOCABULARY
Create a pair of four squares for the terms *predator* and *prey.*

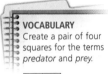

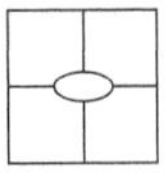

Animals of different species can also interact in cooperative ways. Tickbirds remove ticks from the skin of an impala. This behavior provides food for the bird and provides relief for the impala. Sometimes animals take advantage of the behavior of other animals. Many animals eat the remains of prey left over after a predator has finished feeding.

History of Science

Imprinting is the process of forming a social attachment between parents and offspring. The classic experiments in this field were conducted by Konrad Lorenz in the 1930s. Graylag goslings that hatched from eggs incubated with the mother goose followed and interacted with the mother. However, goslings that hatched from eggs that were isolated in an incubator spent their first few hours with Lorenz. These geese followed Lorenz around and showed no recognition of their mother or of other geese. They behaved as if Lorenz were their mother.

Teach Difficult Concepts

Students often confuse scavengers with decomposers. Decomposers, such as fungi and bacteria, are organisms that break down organic matter into small compounds so they can be taken up and recycled by plants. A scavenger is an animal that eats meat (a carnivore), but the meat it eats is from dead animals and may be decaying or rotted. Vultures, condors, jackals, hyenas, and coyotes are scavengers, as are certain types of insects.

EXPLORE the BIG idea

Revisit "Internet Activity: Bee Dance" on p. 41. Have students explain how bee communication is a type of behavior.

Ongoing Assessment

Identify different ways animals respond to their environment.

Ask: Give an example of cooperative behavior among animals of the same species and of different species. *social behavior in a bee hive, a bird cleaning insects from an impala*

CHECK YOUR READING *Answer: Animals of the same species may cooperate by hunting together or by working together like bees in a hive. They may compete for territory or for a mate.*

DIFFERENTIATE INSTRUCTION

More Reading Support

G What do you call an animal that hunts other animals for food? *a predator*

H What do you call animals that are hunted by others for food? *prey*

English Learners Have English learners write the definitions of *predator* and *prey* in their Science Word Dictionaries. Ask them to think of predator/prey relationships between animals found in their countries of origin or ancestry.

Integrate the Sciences

Many migrating birds navigate by detecting Earth's magnetic field. Studies have shown that placing a powerful magnet near the birds makes them change their direction.

Real World Example

Hibernation is just one example of how organisms escape harsh conditions. Many toads estivate, which means to spend the summer in a dormant state. Toads dig holes in the ground and bury themselves during hot, dry summers to protect themselves from heat and dehydration.

Reinforce the BIG idea

Have students relate the section to the Big Idea.

R Reinforcing Key Concepts, p. 113

2.3 ASSESS & RETEACH

Assess

A Section 2.3 Quiz, p. 25

Reteach

Share with students these sentences explaining the difference between cellular respiration and respiration, a term commonly used to mean breathing:

- Cellular respiration is a chemical process that takes place in cells.
- It involves the chemical breakdown of glucose, releasing energy.
- Respiration, or breathing, is a physical process of taking air into the body.

Then have students work in pairs to create a Venn diagram comparing the two processes. Point out that all multicellular organisms do cellular respiration, but that only animals do respiration.

Technology Resources

Have students visit **ClassZone.com** for reteaching of Key Concepts.

CONTENT REVIEW

CONTENT REVIEW CD-ROM

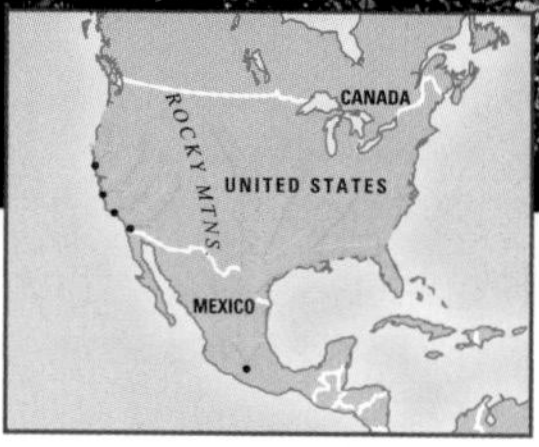

Monarch butterflies migrate each winter to California and Mexico.

Animals respond to seasonal changes.

Animals, like plants, are affected by seasonal changes in their environment. Certain types of food may not be available all year round. A region might go through periods of drought. Some animals do not do well in extreme heat or cold. Unlike plants, animals can respond to seasonal changes by changing their location. **Migration** is the movement of animals to a different region in response to changes in the environment.

Each spring, millions of monarch butterflies begin to fly north from Mexico and parts of southern California. As they migrate, the females lay eggs among milkweed plants. This new generation, when it matures, will continue to travel north, moving into the northern United States and parts of Canada. This second generation of monarchs also lays eggs. Monarchs cannot survive the winter temperatures of the north, so the butterflies make the long journey back to Mexico and California in the fall. These butterflies are a different generation from the butterflies that migrated from the south. This suggests their behavior is inherited, not learned. The fall migration route is shown on the map.

Not all animals migrate in response to seasonal changes. Many nonmigratory animals do change their behaviors, however. For example, when winter cold reduces the food supply, some animals hibernate. **Hibernation** is a sleeplike state that lasts for an extended time period. The body systems of a hibernating animal slow down, so the animal needs less energy to survive. Many animals, including frogs, turtles, fish, and some types of insects, hibernate. You will learn more about different types of animals in Chapters 4 and 5.

2.3 Review

KEY CONCEPTS

1. In what way are animals consumers?
2. Name three body systems that relate to how an animal gets its energy.
3. What is a behavior?

CRITICAL THINKING

4. **Give Examples** Identify three categories of animal behavior and give an example of each.
5. **Analyze** How is migration similar to hibernation? How is it different?

CHALLENGE

6. **Analyze** Scientists often look at feeding patterns as a flow of energy through the living parts of the environment. Describe the flow of energy as it relates to plants, herbivores, and carnivores.

ANSWERS

1. Animals must consume food to live.

2. digestive system, respiratory system, circulatory system

3. an observable response to a stimulus

4. individual: grooming, eating, sleeping; within species: parenting, mating, herding; between species: preying, scavenging

5. Migration and hibernation are both responses to changes in seasons. Migration involves moving between locations, while hibernation involves entering a dormant state.

6. Energy is captured by a plant, then captured by a herbivore that eats the plant, then captured by a carnivore that eats the herbivore. Energy flows from plant to herbivore to carnivore.

SCIENCE on the JOB

WILDLIFE PHOTOGRAPHER

An Animal's World

Have you ever watched a TV documentary, or flipped through the pages of a magazine and wondered, "How did the photographer ever get that picture?" For a wildlife photographer, understanding animal biology is essential.

Cover and Protection

Photographers use camera traps and blinds. A blind, made of branches, is built upwind of an animal gathering place. Steve Winter has built camera traps, where a hidden camera emits a beam of light when an animal steps into it.

Habitat

Certain habitats—such as snow and cold, swamp and mud, or sea water—present challenges to a person with a camera. Underwater photographers need to use scuba gear to swim with animals like the Caribbean reef shark.

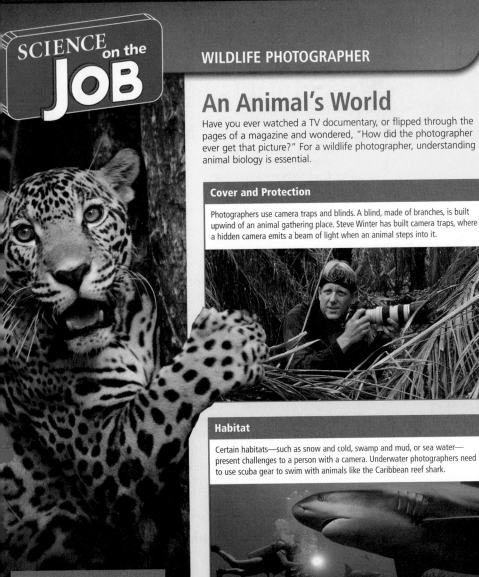

Behavior

To photograph an endangered species like the jaguar, a photographer must learn animal behavior. Steve Winter and a team of scientists used dogs with keen scent-tracking to find jaguars who are active mostly at night. They learned that jaguars have a favorite scratching tree for claw sharpening—perfect photo opportunity!

EXPLORE

1. **OBSERVE** With or without a camera, find a spot where you are likely to find wildlife. Sit as still as possible and wait. What animals do you observe? What do they do?
2. **CHALLENGE** Interview a photographer about digital photography, and ask how technology is changing photography.

EXPLORE

1. ***OBSERVE*** *Many students will observe birds, especially if they set up a feeder or bird bath. Others will observe insects or small animals on the ground.*
2. ***CHALLENGE*** *In digital photography, the photographer has many more options for how the photographic image can be produced.*

SCIENCE ON THE JOB
Relevance of Science to Non-science Jobs

Set Learning Goal

To learn how a wildlife photographer uses knowledge of animal biology

Present the Science

In addition to using scuba gear, underwater photographers use special cameras. Some cameras have all their openings and controls waterproofed by rubber rings. Other cameras are conventional and housed in waterproof cases. Underwater photographers also use strobe lights that must be waterproofed.

Discussion Questions

- Brainstorm what special qualifications or training a wildlife photographer might need.
- Ask: What other people might a wildlife photographer have to work with? *They might need guides who do not speak the same language. They might need to get special permits and government approval to visit certain areas.*
- Ask: What are some pros and cons to this kind of work? How might one person's pro (such as travel excitement, meeting new people) be another person's con (such as travel danger, being away from home and medical care)?

Close

Ask: What science might a wildlife photographer need to understand? *Sample answer: animal behavior, life cycles of various organisms, habitat requirements, optics, ecology, mechanics*

2.4 FOCUS

Set Learning Goals

Students will

- Describe how fungi get energy and materials.
- Identify different types of fungi.
- Recognize how fungi interact with other organisms.
- Conduct an experiment to draw conclusions about how sugar, salt, and sweetener affect the growth of yeast.

3-Minute Warm-Up

Display Transparency 13 or copy this exercise on the board:

Make a chart outlining the relationships between predator and prey animals. Start with one animal (for example, a fly). Next, name a predator of that animal (for example, a frog). Continue your chart by naming an animal that eats the previously named animal. Your chart is finished when you can go no farther.

3-Minute Warm-Up, p. T13

2.4 MOTIVATE

EXPLORE Mushrooms

PURPOSE To examine mushroom spores

TIPS *10 min. on 2 different days*
Fresh mushrooms work better than old mushrooms or dried mushrooms. It may be easier to "pop" the mushroom stem off than to cut it off.

WHAT DO YOU THINK? *The mushroom cap has the same pattern as the print. Spores that were released from the cap made the print.*

KEY CONCEPT

2.4 Most fungi are decomposers.

BEFORE, you learned

- Plants and animals interact with the environment
- Plants transform sunlight into chemical energy
- Animals get energy by eating other organisms

NOW, you will learn

- How fungi get energy and materials
- About different types of fungi
- How fungi interact with other organisms

VOCABULARY

hyphae p. 67
spore p. 67
lichen p. 70

EXPLORE Mushrooms

What does a mushroom cap contain?

PROCEDURE

1. Carefully cut the stem away from the mushroom cap, as near the cap as possible.
2. Place the mushroom cap on white paper and cover it with a plastic cup. Leave overnight.
3. Carefully remove the cup and lift the mushroom cap straight up.
4. Use a hand lens to examine the mushroom cap and the print it leaves behind.

MATERIALS

- fresh store-bought mushrooms
- sharp knife
- clear plastic cup
- paper
- hand lens

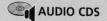

WHAT DO YOU THINK?

- How does the pattern in the mushroom cap compare with the mushroom print?
- What made the print?

Fungi absorb materials from the environment.

MAIN IDEA AND DETAILS Don't forget to make a main idea chart with detail notes on the main idea: *Fungi absorb materials from the environment.*

Plants are producers; they capture energy from the Sun and build complex carbon compounds. Animals are consumers; they take in complex carbon compounds and use them for energy and materials. Fungi (FUHN-jy), at least most fungi, are decomposers. Fungi break down, or decompose, the complex carbon compounds that are part of living matter. They absorb nutrients and leave behind simpler compounds.

Fungi are heterotrophs. They get their energy from living or once-living matter. They, along with bacteria, decompose the bodies of dead plants and animals. They also decompose materials left behind by organisms, such as fallen leaves, shed skin, and animal droppings.

RESOURCES FOR DIFFERENTIATED INSTRUCTION

Below Level

UNIT RESOURCE BOOK

- Reading Study Guide A, pp. 116–117
- Decoding Support, p. 128

AUDIO CDS

Advanced

UNIT RESOURCE BOOK

Challenge and Extension, p. 122

English Learners

UNIT RESOURCE BOOK

Spanish Reading Study Guide, pp. 120–121

AUDIO CDS

- Audio Readings in Spanish
- Audio Readings (English)

Characteristics of Fungi

Except for yeasts, most fungi are multicellular. The cells of a fungus have a nucleus and a thick cell wall, which provides support. Fungi are different from plants and animals in their organization. Plants and animals have specialized cells, which are usually organized into tissues and organs. Multicellular fungi don't have tissues or organs. Instead, a typical fungus is made up of a reproductive body and network of cells that form threadlike structures called **hyphae** (HY-fee).

A mass of hyphae, like the one shown in the diagram below, is called a mycelium (my-SEE-lee-uhm). The hyphae are just one cell thick. This means the cells in the mycelium are close to the soil or whatever substance the fungus is living in. The cells release chemicals that digest the materials around them, and then absorb the nutrients they need. As hyphae grow, openings can form between the older cells and the new ones. This allows nutrients to flow back to the older cells, resulting in what seems like one huge cell with many nuclei.

READING TIP

The root of the word *hyphae* means "web." Look at the diagram below to see their weblike appearance.

Reproduction

Fungi reproduce with spores, which can be produced either asexually or sexually. A **spore** is a single reproductive cell that is capable of growing into a new organism. The mushrooms that you buy at the store are the spore-producing structures of certain types of fungi. These spore-producing structures are reproductive bodies of mushrooms. A single mushroom can produce a billion spores.

Parts of a Fungus

The mycelium makes up a large part of a multicellular fungus.

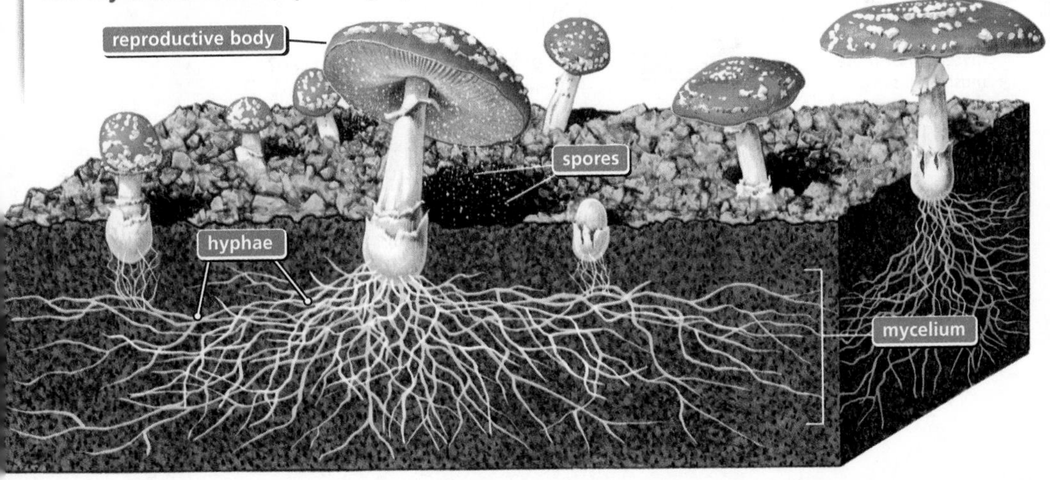

2.4 INSTRUCT

Teach Difficult Concepts

Students may have difficulty understanding how material is digested outside the body and then absorbed. Ask:

- How does digested material get into the fungus? *through the cell walls*
- Why can't a fungus take dead material into its cells? *Cell walls let only tiny things into the cell.*
- What happens to the digested material that the fungus does not need? *Plant roots and other living things absorb it.*

Teach from Visuals

To help students interpret the diagram of fungus growth, ask:

- What is the common name for the reproductive body of the fungus? *a mushroom or toadstool*
- Which comes first, the mycelium or the mushroom? *the mycelium*

Ongoing Assessment

Describe how fungi get energy and materials.

Ask: Where does fungal digestion take place? *in the substance surrounding the hyphae*

DIFFERENTIATE INSTRUCTION

? More Reading Support

A What are hyphae? *threadlike structures that make up a fungus*

B What are spores? *reproductive cells of a fungus*

English Learners Have students write the definitions for *hyphae, spore,* and *lichen* in their Science Word Dictionaries. Have English learners look up the words *hyphae* and *lichen* in a dictionary that provides phonetic spellings. Encourage them to look up words in the dictionary or to practice with an advanced student when they are unsure of pronunciation.

Teach from Visuals

To help students interpret the photograph of mushroom growth, ask:

- What is the mushrooms' food source? *organic matter in the tree trunk*
- Does the photograph show one organism or many organisms? *probably one organism connected by a single mycelium*

Teacher Demo

Demonstrate that fungi need moisture and a food supply to grow. You must use bread without preservatives for this demo. Place a piece of fresh bakery bread in a self-seal plastic bag. Moisten a second piece of the same bread and place it in a second bag. Trap air in the bags so the plastic does not touch the top of the bread, and seal the bags. Incubate the bags in a dark, warm place for several days. Mold will grow on the moistened bread but probably not on the dry bread. Have students examine the mold, but do not allow them to open the bags. Ask what the food source is for the mold.

Integrate the Sciences

There are very few truly poisonous mushrooms. The genus *Amanita* causes 90 percent of deaths by mushroom poisoning. Several members of this group contain amanitin, one of the deadliest poisons found in nature. One cap of a destroying angel mushroom, for example, can kill an adult human. The toxin causes irreparable liver and kidney damage.

Ongoing Assessment

CHECK YOUR READING *Answer: The cap produces and holds the reproductive spores.*

***PHOTO CAPTION** Answer: inside the tree*

RESOURCE CENTER CLASSZONE.COM

Learn more about different types of fungi.

Spores are released into the air and spread by the wind. Because they're so small and light, the wind can carry spores long distances. Scientists have found spores 160 kilometers above Earth's surface. Some spores have a tough outer covering that keeps the reproductive cell from drying out. Such spores can survive for many years. If the parent fungus dies, the spores may remain and grow when conditions are right.

Fungi reproduce in other ways. For example, a multicellular fungi can reproduce asexually when hyphae break off and form a new mycelium. Yeasts, which are single-celled fungi, reproduce asexually by simple cell division or by budding. Yeasts can also produce spores.

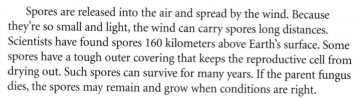

INFER Where is the mycelium of these mushrooms?

Fungi include mushrooms, molds, and yeasts.

A convenient and simple way to study fungi is to look at their forms. They are mushrooms, molds, and yeasts. You are probably familiar with all of them. The mushrooms on your pizza are a fungus. So is the mold that grows if you leave a piece of pizza too long in the refrigerator. The crust of the pizza itself rises because of the activity of yeast.

Mushrooms

What we call a mushroom is only a small part of a fungus. A single mushroom you buy in the store could have grown from a mycelium that fills an area 30 meters across. When you see a patch of mushrooms, they are probably all part of a single fungus.

For humans, some mushrooms are edible and some are poisonous. A toadstool is a poisonous mushroom. The cap of a mushroom is where the spores are produced. Both the cap and the stalk it grows on are filled with hyphae.

CHECK YOUR READING What is produced in a mushroom cap?

Molds

What we call mold, that fuzzy growth we sometimes see on food, is the spore-producing part of another form of fungus. The hyphae of the mold grow into the food, digesting it as they grow. Not all food molds are bad. Different species of the fungus *Penicillium* are used in the production of Brie, Camembert, and blue cheeses. Some species of the *Aspergillus* fungus are used to make soy sauce.

One interesting application of a mold is the use of the fungus *Trichoderma.* This mold grows in soil. The digestive chemicals it produces are used to give blue jeans a stonewashed look.

DIFFERENTIATE INSTRUCTION

More Reading Support

C How do fungi reproduce? *from spores, hyphae, and budding*

Advanced Have interested students research poisonous mushrooms and give a short class presentation on their findings.

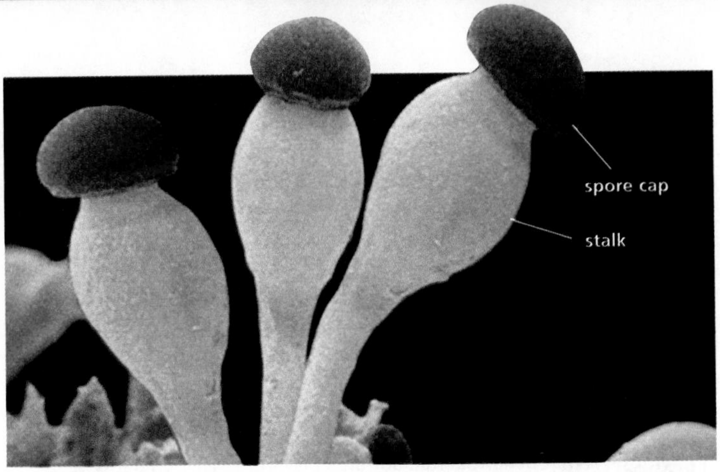

Pilobolus reacts to sunlight as a stimulus. The bend in the stalk will cause the spore cap to fly off.

Many molds cause disease. Fungal molds cause athlete's foot. Molds also affect plants. They are the cause of Dutch elm disease and the powdery white mildews that grow on plants. Compounds made from molds are also used to treat disease. Penicillin is an antibiotic that comes from the *Penicillium* fungus. It is used to fight bacterial diseases, such as pneumonia.

Molds reproduce with spores, which are typically carried by moving air. The "hat thrower" fungus *Pilobolus,* however, has an interesting adaptation for spreading spores. *Pilobolus* grows in animal droppings. It has a spore-containing cap—its hat—that grows on top of a stalk. The stalk responds to light as a stimulus and bends toward the Sun. As the stalk bends, water pressure builds up, causing the spore cap to shoot off, like a tiny cannonball. A spore cap can be thrown up to two meters away. If the spore caps land in grass, then cows and other grazing animals will eat the caps as they graze. A new cycle begins, with more *Pilobolus* being dispersed in the animal's droppings.

Yeasts

Yeasts are single-celled fungi. Some species of fungi exist in both yeast form and as multicellular hyphae. Yeasts grow in many moist environments, including the sap of plants and animal tissues. They also grow on moist surfaces, including shower curtains. Certain yeasts grow naturally on human skin. If the yeast begins to reproduce too rapidly, it can cause disease.

Yeasts are used in many food products. The activity of yeast cells breaking down sugars is what makes bread rise. The genetic material of the yeast *Saccharomyces cerevisiae* has been carefully studied by scientists. The study of this organism has helped scientists understand how genetic material controls the activities of a cell.

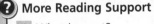

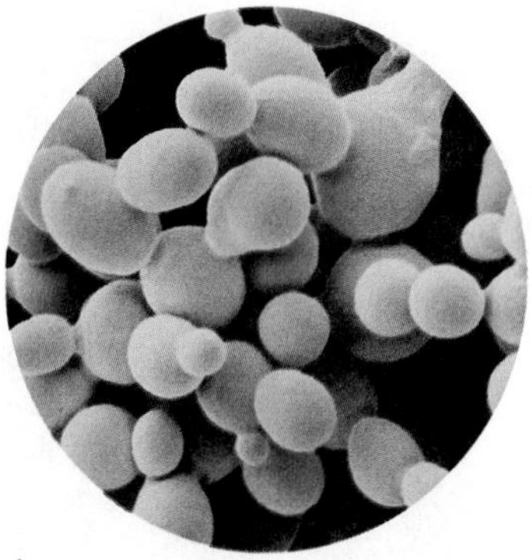

Yeasts are single-celled fungi.

Teach from Visuals

To help students interpret the photograph of *Pilobolus,* ask:

- What is the function of the stalk? *It holds up the spore cap and responds to sunlight.*
- What is the advantage of releasing spores far away from the fungus? *The new organisms that form will not compete with the parent organism for food and moisture.*

Teacher Demo

To demonstrate how yeast cells make bread rise, mix about half an envelope of dry yeast with some warm water in an uncovered jar or beaker. Add a teaspoon of granulated sugar and stir. Set the jar aside for a few minutes. The yeast will begin to ferment the sugar, producing carbon dioxide. The yeast suspension will get foamy and will increase in volume. It is this gas that makes bread dough rise.

Real World Example

Mold-infested office buildings, schools, and homes are being blamed for an increase in serious respiratory illnesses. Coughs, shortness of breath, sinus infections, dizziness, burning eyes, and asthma are common in certain airtight buildings that have water leakage problems. Carpeting and drywall may have to be replaced after the water problem is resolved.

Ongoing Assessment

Identify different types of fungi.

Ask: How are yeasts different from mushrooms and molds? *Yeasts are single-celled.*

DIFFERENTIATE INSTRUCTION

More Reading Support

D What is yeast? *single-celled fungi*

Below Level The reproduction of yeasts can be modeled using modeling clay. Have students make a yeast cell out of clay. They can form a bud on the cell, then pinch it off to form a new yeast organism. This activity will also help students who have impaired vision.

Teach from Visuals

To help students interpret the diagram of a lichen, ask:

- How do the algae provide nutrients for themselves and the fungus? *Algae produce nutrients by photosynthesis.*
- How does the fungus hold the algae in place? *The hyphae wrap around the algae.*

Address Misconceptions

IDENTIFY Ask: What do decomposers decompose? If students cannot answer, they may not know the exact role of decomposers in the environment. If students answer "dead or decaying organic material," ask them where the material comes from and where it goes. If they do not say "from living organisms that die or produce wastes," and "to soil, air, and nutrients," they may hold the misconception that decomposers are the bottom of the food chain.

CORRECT Have students make a series of diagrams showing the steps of the decomposition process.

REASSESS Have students write a short paragraph explaining why decomposers are necessary in the environment. *Dead material would pile up if it were not decomposed. The living world would run out of nutrients if nothing were recycled.*

Technology Resources

Visit **ClassZone.com** for background on common student misconceptions.

MISCONCEPTION DATABASE

Ongoing Assessment

Recognize how fungi interact with other organisms.

Ask: Describe the relationship between algae and fungus in a lichen. *The algae produces food for the fungus, which provides a structure and protection.*

CHECK YOUR READING *Answer: Fungi are decomposers.*

Fungi can be helpful or harmful to other organisms.

Fungi have a close relationship to the environment and all living things in the environment. Fungi, along with bacteria, function as the main decomposers on Earth. The digestive chemicals that fungi release break down the complex compounds that come from living matter, dead matter, or the waste an organism leaves behind. A fungus absorbs what it needs to live and leaves behind simpler compounds and nutrients. These are then picked up again by plants, as producers, to start the cycle over again. Fungi also live in the sea, recycling materials for ocean-living organisms.

CHECK YOUR READING What beneficial role do fungi play in the environment?

The threadlike hyphae of a fungus can grow into and decompose the material produced by another organism. This means that fungi can be helpful, for example, by releasing the nutrients in a dead tree back into the soil. Or fungi can be harmful, for example, attacking the tissues of a plant, such as the Dutch elm.

VOCABULARY Remember to add a four square for *lichen* to your notebook.

Most plants interact with fungi in a way that is helpful. The hyphae surround the plant roots, providing nutrients for the plant. The plant provides food for the fungus. Some fungi live together with single-celled algae, a network referred to as a **lichen** (LY-kuhn). The hyphae form almost a sandwich around the algae, which produce sugars and other nutrients the fungus needs.

Lichen

A lichen is formed by a close association between algae and fungi.

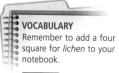

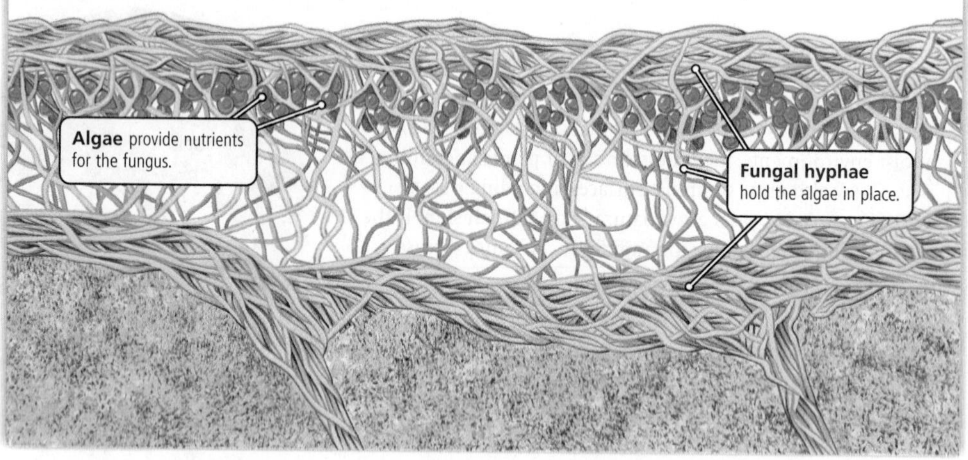

Algae provide nutrients for the fungus.
Fungal hyphae hold the algae in place.

DIFFERENTIATE INSTRUCTION

More Reading Support

E What does algae provide for the fungus in a lichen? *food*

Alternative Assessment Have students make a poster showing the beneficial relationship between a plant and the mycorrhizal fungi that grow around the plant's roots. A small amount of research will give students additional information for their posters and for a class presentation.

Advanced Have students design an experiment to determine the best growing conditions for yeast.

Challenge and Extension, p. 122

Lichens can live just about anywhere. Lichens are found in the Arctic and in the desert. They can even grow on bare rock. The hyphae can break the rock down, slowly, and capture the particles of newly formed soil. This eventually prepares the ground for new plant growth.

On the harmful side, many fungi produce toxins, harmful chemicals. In 1845, a fungus infected Ireland's potato crop, causing the population of Ireland to drop from 8 million to about 4 million. Many people died from disease. Others died from starvation because of the loss of the important food crop. And hundreds of thousands of Irish left Ireland, many emigrating to the United States. Today, several fungal diseases are spreading through the worlds banana crops.

CHECK YOUR READING Name some ways fungi can be harmful to organisms.

The toxic quality of a fungus can be put to good use, as in the case of the antibiotic penicillin. The photographs show what happens when a bacterium comes in contact with penicillin. The antibiotic prevents the bacterial cells from making new cell walls when they divide. This causes the cells to break open and the bacteria to die.

Bacterium before penicillin

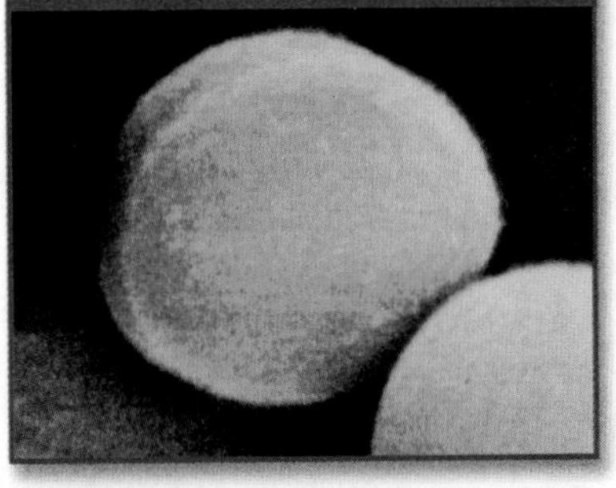

Bacterium after penicillin

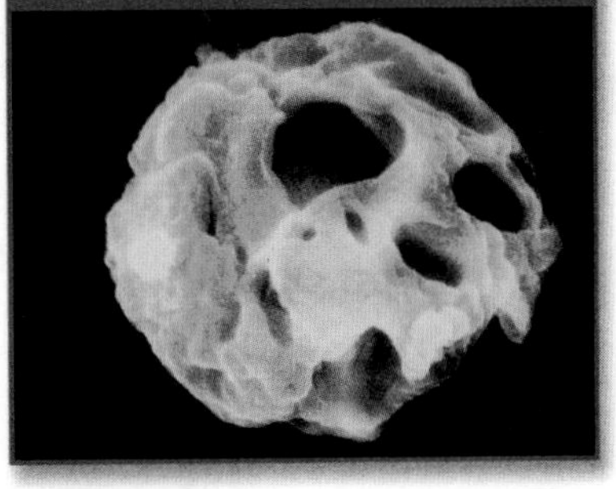

Penicillin is an antibiotic drug made from compounds taken from a species of the *Penicillium* fungus.

2.4 Review

KEY CONCEPTS

1. Describe the structure of a fungus.
2. How do fungi reproduce?
3. Describe two relationships between fungi and other organisms.

CRITICAL THINKING

4. **Analyze** Scientists used to classify fungi as plants. Today, scientists say that fungi are more like animals than plants. How are fungi like plants? How are they like animals?
5. **Predict** What might change in an environment where there were no fungi?

CHALLENGE

6. **Connect** Think of at least one way your life is affected by fungi in each of their three main forms: mushrooms, molds, and yeast. Are these effects beneficial or harmful to you?

ANSWERS

1. A fungus is a network of hyphae that forms reproductive structures above ground to produce spores.

2. from spores, by budding, or from pieces of hyphae

3. Fungi may cause disease, and they act as decomposers.

4. Fungi are like plants because they have a thick cell wall. Fungi are like animals because they get their energy from other organisms.

5. Fungi are decomposers. Without them, organic matter would not be recycled.

6. Sample answer: Eating mushrooms on pizza gives me energy and helps me grow (beneficial). Athlete's foot is a disease caused by a mold (harmful). Yeast makes bread rise for me to eat (beneficial).

History of Science

Penicillin was discovered by chance in 1928 after Alexander Fleming left a petri dish of bacteria uncovered in his lab for a few days. When he returned, the dish had been contaminated by a mold that inhibited bacterial growth. Fleming's discovery revolutionized the practice of medicine when the antibiotic penicillin from the *Penicillium* mold was used to treat infected wounds.

Ongoing Assessment

CHECK YOUR READING *Answer: Fungi spoil food, release toxins that cause disease, and attack food crops and trees.*

Reinforce the BIG idea

Have students relate the section to the Big Idea.

R Reinforcing Key Concepts, p. 123

2.4 ASSESS & RETEACH

Assess

A Section 2.4 Quiz, p. 26

Reteach

Give pairs of students a fresh white mushroom. Have them use a toothpick to pull apart the hyphae that make up the mycelium in the mushroom. Students should also examine the gills on the underside of the cap. After they dissect the mushroom, have them draw it, label the parts, and label the type of organism and its way of feeding.

Technology Resources

Have students visit **ClassZone.com** for reteaching of Key Concepts.

CONTENT REVIEW

CONTENT REVIEW CD-ROM

CHAPTER INVESTIGATION

Focus

PURPOSE To investigate the ability of yeast to use different food sources

OVERVIEW Students will make dough using flour, water, and one of the following additives: sugar, salt, and sweetener. Students will measure how much each dough rises and determine which substances yeast can use as a food source. Students will find the following:

- Yeast use sugar for food and produce carbon dioxide.
- Yeast cannot use artificial sweetener or salt as a food source.

Lab Preparation

- Prior to the investigation, proof the yeast by mixing a sample with warm water and sugar and letting it sit for about 20 minutes. If the yeast is alive, bubbles will form.
- Clothespins must be the spring type to make a tight closure.
- Prior to the investigation, have students read through the investigation and prepare their data tables. Or you may wish to copy and distribute datasheets and rubrics.

UNIT RESOURCE BOOK, pp. 131–139

SCIENCE TOOLKIT, F14

Lab Management

- Have students use a clean spoon to mix each mound or wash their hands between kneading mounds to prevent contamination.
- Remind students not to eat the dough.
- Three teaspoons equals one tablespoon. A tablespoon can be used to measure the sugar, salt, and sweetener.

Teaching with Technology

If probeware is available, a carbon dioxide probe can be inserted into each ball of dough instead of using the straw apparatus.

CHAPTER INVESTIGATION

What Do Yeast Cells Use for Energy?

OVERVIEW AND PURPOSE Yeasts are tiny one-celled fungi that require food, water, and a warm place to grow. When they have a food source, they release carbon dioxide gas as a waste product. Yeast is used to make foods such as bread. In this activity, you will

- observe the activity of yeast
- draw conclusions about the effect of three materials on the activity of yeast

Problem (Write It Up)

How do sugar, salt, and sweetener affect the growth of yeast?

Hypothesize (Write It Up)

Write a hypothesis to explain how sugar, sweetener, and salt affect the activity of yeast in bread dough. Your hypothesis should take the form of an "If . . ., then . . ., because . . ." statement.

MATERIALS
- baking sheet
- flour
- measuring cups
- measuring spoons
- sugar
- artificial sweetener
- salt
- quick-rise yeast
- warm water
- metric ruler
- marker
- clothespins
- clear plastic straws

Procedure

1. Make a data table like the one shown on page 73. Label four sheets of notebook paper *A, B, C,* and *D.*
2. Spread a very thin layer of flour over the baking sheet. Measure $\frac{1}{4}$ cup of flour and place it on the baking sheet as a mound. Repeat three times, forming separate mounds. Label the mounds *A, B, C,* and *D.*

C 72 Unit: Diversity of Living Things

INVESTIGATION RESOURCES

CHAPTER INVESTIGATION, What Do Yeast Cells Use for Energy?
- Level A, pp. 131–134
- Level B, pp. 135–138
- Level C, p. 139

Advanced students should complete Levels B & C.

Writing a Lab Report, D12–13

Technology Resources

Customize this student lab as needed or look for an alternative. Print rubrics to assess student lab reports.

Lab Generator CD-ROM

3 Add 3 teaspoons of sugar to mound A. Add 3 tsp of sweetener to mound B. Add 3 tsp of salt to mound C. Add nothing to mound D.

4 Add $\frac{1}{4}$ tsp of the quick-rise yeast to each of the mounds. Slowly add 1 tsp of warm water to each mound to moisten the mixture. Spread a pinch of flour over your hands and knead the mounds by hand. Add water, 1 tsp at a time until the mixture has the consistency of dough. If the mixture gets too sticky, add more flour. Knead well and form each mound into a ball. Wash your hands thoroughly when you are finished. Do not taste or eat the dough.

5 Push 2 straws into each ball of dough, making sure the dough reaches at least 3 cm into the straws.

step 5

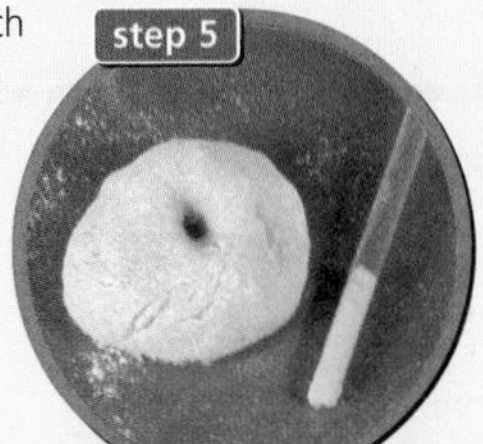

6 Squeeze the end of each straw to push the dough from the ends. Place a clothespin on the end of each straw closest to the dough. Fold and tape the other end. Mark both edges of the dough on the straw. Stand each straw upright on the appropriate piece of paper labeled *A*, *B*, *C*, or *D*.

step 6

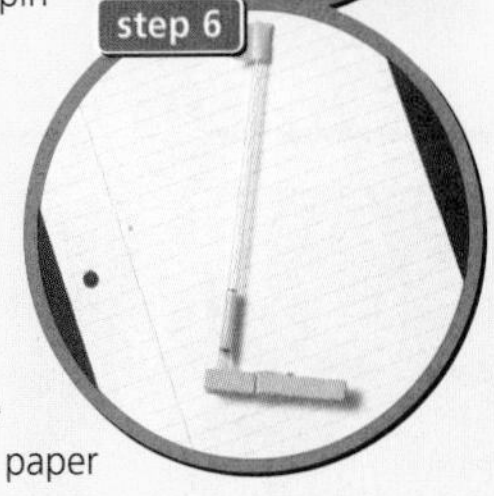

7 Predict which mounds of dough will rise after 30 minutes. Write down your predictions in the data table.

8 After 30 minutes, measure the amount the dough has risen in each straw. Write down the results in the data table.

Observe and Analyze

Write It Up

1. **OBSERVE** In which mounds did the dough rise?
2. **OBSERVE** Did any of the remaining mounds of dough change? Explain.
3. **INFER** What was the purpose of using two straws for each of the mounds?

Conclude

Write It Up

1. **INTERPRET** Which is the most likely source of energy for yeast: salt, sugar, or sweetener? How do you know?
2. **INTERPRET** Compare your results with your hypothesis. How does your data support or disprove your hypothesis?
3. **LIMITATIONS** What limitations or sources of error could have affected your results?
4. **CONNECT** How would you account for the air spaces that are found in some breads?
5. **APPLY** Would you predict that breads made without yeast contain air spaces?

INVESTIGATE Further

CHALLENGE Design an experiment in which you can observe the production of carbon dioxide by yeast.

What Do Yeast Cells Use for Energy?

Table 1. Observations of Dough Rising

Mound	Prediction	Results
A. sugar and yeast		
B. sweetener and yeast		
C. salt and yeast		
D. yeast		

Observe and Analyze

Write It Up

1. *The dough rose in the mixture containing sugar.*
2. *no*
3. *as duplicates to compare results*

Conclude

Write It Up

1. *Sugar; the yeast produced a lot of carbon dioxide in the dough with sugar.*
2. *Check student hypotheses. Students should provide evidence or explanation for describing whether their hypothesis is correct or incorrect.*
3. *Possible sources of error are contamination of mounds with sugar, and inactive yeast.*
4. *The air spaces are pockets of carbon dioxide produced by the yeast in the bread dough.*
5. *No, matzo and tortillas are not made with yeast, so they do not rise.*

INVESTIGATE Further

CHALLENGE One way to observe carbon dioxide production by yeast is to grow yeast in a sugar solution in a test tube that is inverted in a dish containing water. The test tube must be full, with no trapped air pockets. As carbon dioxide is produced, it will rise to the top of the tube (the closed end), displacing some of the liquid.

Post-Lab Discussion

- Ask: What was the purpose of mound D, which did not have anything added except yeast? *It was a control.*
- Ask: Why might the yeast produce carbon dioxide in mound D if no food source was added? *Yeast used the starch in the flour.*
- Have students discuss any new questions that arose from the investigation.
- Have students plan an investigation where the results would help answer one of their questions.

BACK TO

the BIG idea

Ask: How do plants, animals, and fungi work together to maintain a healthy environment? *Plants use the Sun's energy to produce food consumed by animals and fungi. Animals' complex interactions with one another and their environment keep the environment in balance. Fungi act as decomposers.*

KEY CONCEPTS SUMMARY

SECTION 2.1

Ask: How does sexual reproduction contribute to a species' ability to adapt to its environment? *Offspring are genetically different from parents. They may have unique traits that allow them to survive more successfully than their parents.*

SECTION 2.2

Ask: How do types of leaves help plants adapt to different environments? *Leaves of conifers stay green and produce food for the plant in winter. Leaves of deciduous trees drop off in the fall, allowing the trees to rest during winter. Leaves of the Venus flytrap are specially adapted as traps for insects.*

SECTION 2.3

Ask: What function does digestion serve in an animal? *Digestion enables an animal to get energy from food.*

SECTION 2.4

Ask: How does fungi's role as decomposers benefit other organisms? *Fungi's digestive chemicals release useful compounds locked in dead matter and waste materials from living matter so that these compounds can be used again by living things. This process also removes waste from the environment.*

Review Concepts

- Big Idea Flow Chart, p. T9
- Chapter Outline, pp. T15–T16

2 Chapter Review

the BIG idea

Multicellular organisms live in and get energy from a variety of environments.

CONTENT REVIEW
CLASSZONE.COM

KEY CONCEPTS SUMMARY

2.1 Multicellular organisms have many ways of meeting their needs.

- The bodies of multicellular organisms have different levels of organization.
- Multicellular organisms have a wide range of adaptations.
- Multicellular organisms reproduce by sexual reproduction. Some also reproduce asexually.

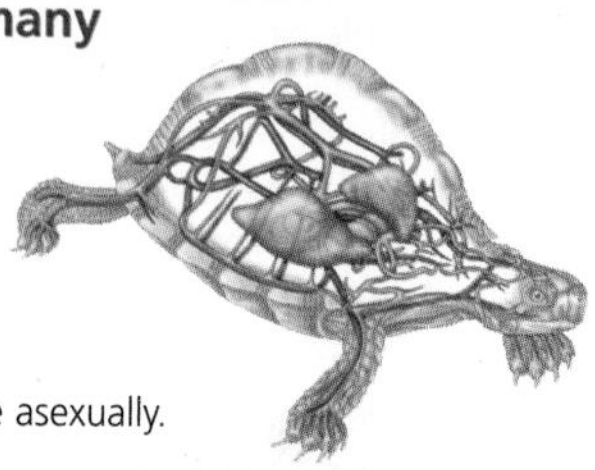

VOCABULARY
tissue p. 44
organ p. 44
sexual reproduction p. 48
meiosis p. 48
fertilization p. 48

2.2 Plants are producers.

Plants capture energy from the Sun and store it as sugar and starch. Plants are adapted to many environments. They respond to stimuli in the environment.

VOCABULARY
autotroph p. 52
photosynthesis p. 52
cellular respiration p. 53
stimulus p. 55

2.3 Animals are consumers.

Animals consume food to get energy and materials. Animals are adapted to many enviroments. They interact with the environment and with other organisms.

VOCABULARY
consumer p. 58
heterotroph p. 58
behavior p. 62
predator p. 63
prey p. 64
migration p. 64
hibernation p. 64

2.4 Most fungi are decomposers.

Fungi absorb energy from their surroundings. Fungi include mushrooms, molds, and yeasts. They affect people and other organisms in both helpful and harmful ways.

VOCABULARY
hyphae p. 67
spore p. 67
lichen p. 70

Technology Resources

Have students visit **ClassZone.com** or use the CD-ROM for a cumulative review of concepts.

CONTENT REVIEW

CONTENT REVIEW CD-ROM

Engage students in a whole-class interactive review of Key Concepts. Edit content as you wish.

POWER PRESENTATIONS

Reviewing Vocabulary

Draw a Venn diagram for each pair of terms. Put at least one shared characteristic in the overlap area, and put at least one difference in the outer circles.

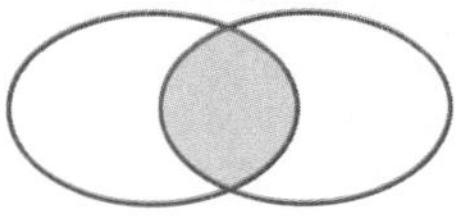

1. tissue, organ
2. autotroph, heterotroph
3. photosynthesis, cellular respiration
4. predator, prey
5. producer, consumer
6. migration, hibernation

Reviewing Key Concepts

Multiple Choice *Choose the letter of the best answer.*

7. Which body system transports materials such as nutrients and oxygen throughout an animal's body?
   - **a.** respiratory system
   - **b.** circulatory system
   - **c.** digestive system
   - **d.** nervous system
8. An example of an adaptation is
   - **a.** a change in climate that increases plant growth
   - **b.** the movement of a group of animals to an area that has more food and water
   - **c.** a change in location of a squirrel's nest
   - **d.** the ability of a plant to resist fungal disease better than other plants
9. Plants capture the Sun's energy through which process?
   - **a.** reproduction
   - **b.** cellular respiration
   - **c.** photosynthesis
   - **d.** digestion
10. Plants produce auxin in response to which stimulus?
    - **a.** light
    - **b.** gravity
    - **c.** temperature
    - **d.** touch
11. A plant is best described as
    - **a.** a herbivore
    - **b.** an omnivore
    - **c.** a carnivore
    - **d.** a producer
12. A carnivore is best described as an animal that
    - **a.** eats plants
    - **b.** eats plants and other animals
    - **c.** eats other animals
    - **d.** makes its own food
13. Mushrooms produce
    - **a.** spores
    - **b.** buds
    - **c.** mold
    - **d.** yeast
14. Fungi and algae together form
    - **a.** hyphae
    - **b.** mushrooms
    - **c.** lichen
    - **d.** mold

Short Answer *Write a short answer to each question.*

15. Write a short paragraph comparing sexual reproduction with asexual reproduction. How are they the same? How are they different?
16. Write a short paragraph to explain how the sugars and starches stored in plant tissue are important to the survival of animals.
17. Write a short paragraph to explain how fungi are dependent on plants and animals for their energy.

ASSESSMENT RESOURCES

UNIT ASSESSMENT BOOK
- Chapter Test A, pp. 27–30
- Chapter Test B, pp. 31–34
- Chapter Test C, pp. 35–38
- Alternative Assessment, pp. 39–40

SPANISH ASSESSMENT BOOK
Spanish Chapter Test, pp. 45–48

Technology Resources

Edit test items and answer choices.

Test Generator CD-ROM

Visit **ClassZone.com** to extend test practice.

Test Practice

Reviewing Vocabulary

Sample answers:

1. same: level of organization in body
tissue: made up of similar cells
organ: made up of different tissues

2. same: organisms
autotroph: gets energy from Sun
heterotroph: gets energy from food

3. same: cellular process
photosynthesis: captures energy
cellular respiration: releases energy

4. same: animal
predator: eats another animal
prey: is eaten

5 same: involves means of getting food
producer: makes own food
consumer: eats or absorbs food

6. same: type of behavior
migration: move with change in environment
hibernation: go into sleeplike state with change in environment

Reviewing Key Concepts

7. b	*11. d*
8. d	*12. c*
9. c	*13. a*
10. a	*14. c*

15. Reproduction is the process by which an organism produces offspring. Both single-celled and some multicellular organisms are capable of both sexual and asexual reproduction. Only one organism is involved in asexual reproduction, and the offspring have the same genetic material as the parent. In sexual reproduction, two parents are involved and the offspring have genetic material from both parents.

16. Most of the food that animals use comes either directly or indirectly from plants. Plants use energy from the Sun to put together complex carbon compounds into sugar and starch. Animals get energy from sugar and starch by eating plants.

17. Fungi absorb nutrients from living material. Many fungi get nutrients from the remains of dead plants and animals or the waste products of living plants and animals. Other fungi grow into the living tissue of organisms.

Thinking Critically

18. *Sun; plants capture energy from Sun; organisms get energy from plants or animals that eat plants.*
19. *insect, worm, woodchuck, robin, owl*
20. *insect: herbivore; worm: decomposer; robin and owl: carnivores; woodchuck: omnivore*
21. *Yes; it has more options because if animal food is not available, it can eat both plants and animals.*
22. *It breaks down plant and animal matter and returns nutrients to soil.*
23. *In winter, less food is available. Some animals migrate; others hibernate.*
24. *Hibernation; it allows an animal to get through winter on little food.*
25. *yes; ensures that animals will survive to reproduce*
26. *Gravity establishes up/down for plants. Roots grow into soil; stems and leaves grow up. Touch allows stems and leaves to find support. Light provides energy.*
27. *All cells support life. Single-celled organisms: functions are done by single cell. Multicellular organisms: cells specialize to do certain jobs.*
28. *Fungal cells can multiply asexually very quickly.*

the BIG idea

29. *Plants capture energy from Sun. Animals get energy from food. Fungi break down complex carbon materials into simple materials.*
30. *Producers, such as plants, use sunlight to make sugar and starch through photosynthesis. Energy flows to consumers when they eat food. Decomposers take materials from living matter and break it down.*

UNIT PROJECTS

Check to make sure students are working on their projects. Check schedules and work in progress.

R Unit Projects, pp. 5–10

Thinking Critically

The diagram below shows a woodland food web. Each arrow starts with a food source and points to a consumer. Use the diagram to answer the next six questions.

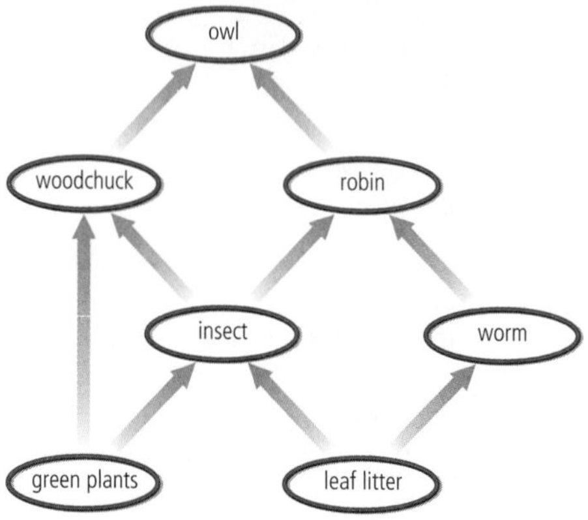

18. ANALYZE What is the original source of energy for all the animals in the food web? Explain your reasoning.

19. CLASSIFY Identify the consumers in this food web.

20. CLASSIFY Identify the animals in the food web as either herbivores, carnivores, omnivores, or decomposers.

21. EVALUATE Does an omnivore have an advantage over carnivores and herbivores in finding food?

22. ANALYZE What role does the worm play in the food web, and why it is important?

23. PREDICT How might this food web change over the course of a year, and how would that affect the feeding activity of animals in the food web?

24. CONNECT A woodchuck is sometimes referred to as a groundhog. Many people celebrate February 2 as groundhog day. The legend is that if a groundhog emerges from its burrow on this day and sees its shadow, then there will be six more weeks of winter. The groundhog is emerging from a long sleeplike state. What is this behavior, and how does it benefit the animal?

25. ANALYZE Do you think the defensive behavior of a porcupine or pufferfish is an adaptation? Explain your reasoning.

26. SYNTHESIZE A plant responds to gravity, touch, and light as stimuli. How does this relate to a plant being a producer?

27. SYNTHESIZE How are the cells of multicellular organisms like those of single-celled organisms? How are they different?

28. ANALYZE What quality of asexual reproduction makes a fungal disease spread so quickly.

the BIG idea

29. SYNTHESIZE Look again at the photograph on pages 40–41. Plants, animals, and fungi are pictured there. How do these organisms get energy and materials from the environment?

30. SUMMARIZE Write a short paragraph to describe how matter and energy move between members of the kingdoms of plants, animals, and fungi. Use the words in the box below. Underline the terms in your answer.

photosysnthesis	consumer
producer	decomposer

UNIT PROJECTS

By now you should have completed the following items for your unit project.

- questions that you have asked about the topic
- schedule showing when you will complete each step of your project
- list of resources including Web sites, print resources, and materials

MONITOR AND RETEACH

If students have trouble with the concepts in Chapter Review item 26, have them review pp. 55–57. Point out the three responses.

Students may benefit from summarizing one or more sections of the chapter.

R Summarizing the Chapter, pp. 149–150

Standardized Test Practice

For practice on your state test, go to . . .
TEST PRACTICE
CLASSZONE.COM

Interpreting Diagrams

The diagram shows the feeding relationships between certain animals and plants in a forest environment. The size of the bars represent the relative numbers of each organism. The arrow shows the flow of energy between these groups of organisms.

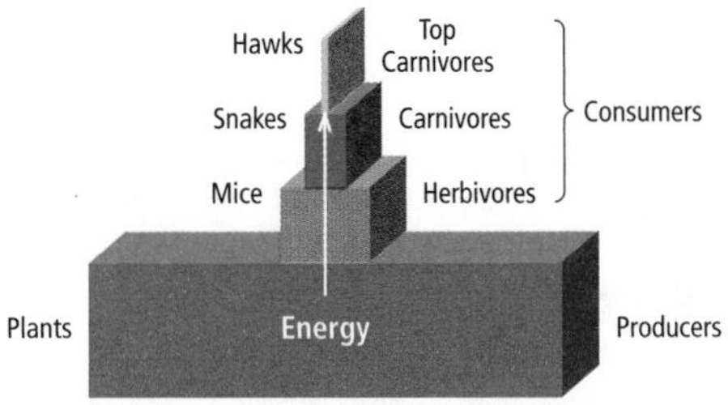

Feeding Relationships Among Plants and Animals

Use the diagram to answer the questions below.

1. Which is the largest group of organisms in the forest?
- **a.** plants
- **b.** mice
- **c.** snakes
- **d.** hawks

2. Most energy in the forest comes from
- **a.** top carnivores
- **b.** carnivores
- **c.** herbivores
- **d.** plants

3. Which description best fits the snake?
- **a.** a producer that feeds upon mice
- **b.** a consumer that is eaten by mice
- **c.** a consumer that feeds upon mice
- **d.** a consumer that feeds upon plants

4. A hawk is a top carnivore that feeds upon both snakes and mice. Which of the following best describes a top carnivore?
- **a.** a carnivore that feeds upon other carnivores
- **b.** a carnivore that feeds upon both carnivores and herbivores
- **c.** a consumer that gets its energy from producers
- **d.** a producer that supplies energy to consumers

5. Which statement best summarizes the diagram?
- **a.** The energy in a forest environment flows one way, from producers to consumers.
- **b.** Consumers don't need as much energy as producers.
- **c.** Energy in a forest environment goes from plants to animals and then back to plants.
- **d.** The number of producers depends on the number of consumers.

Extended Response

6 The diagram above shows the amount of energy available at each level of a forest environment. Describe what happens to the amount of energy going from the producer level into the different levels of consumers. Based on the diagram, not all the energy produced at a given level is available to organisms in the next level. What has happened to that energy? Use the words in the word box in your answer. Underline the words.

producer	energy
consumer	food

7. How would the number of plants, snakes, and hawks be affected if some disease were to reduce the numbers of mice in the forest?

Interpreting Diagrams

1. a   2. d   3. c   4. b   5. a

Extended Response

6. RUBRIC

4 points for a response that correctly answers the question and uses the following terms accurately:
- producer
- consumer
- energy
- food

Sample answer: The diagram starts out large at the base and then gets significantly smaller going from one level to the next. The most energy is with the producers. Consumers get energy from food; some is stored, but some is used by their bodies. Therefore, at each level, less energy is available. The less energy available in a level, the smaller the number of organisms in the level.

3 points correctly answers the question and uses three terms accurately
2 points correctly answers the question and uses two terms accurately
1 point correctly answers the question and uses one term accurately

7. RUBRIC

4 points for a response that correctly answers the question and demonstrates an understanding of a food web.

Sample answer: The numbers of plants might increase if there were fewer mice to feed on plants and plant seeds. The number of snakes and hawks might decrease because both these organisms rely on mice as a food source.

3 points correctly answers the question and demonstrates partial understanding of a food web
2 points correctly answers the question but demonstrates little understanding of a food web
1 point does not correctly answer the question and demonstrates little understanding of a food web

METACOGNITIVE ACTIVITY

Have students answer the following questions in their **Science Notebook:**
1. What concepts about multicellular organisms are still confusing to you?
2. What did you learn about plants that surprised you?
3. How might working with a partner help you with your Unit Project?

FOCUS

Set Learning Goals

Students will

- Examine historical events that contribute to our understanding of biodiversity.
- Learn about some of the technology that makes close observation of different species possible.
- Invent an animal classification system, using criteria such as habitat or diet.

National Science Education Standards

A.9.a–g Understandings About Scientific Inquiry

E.6.a–c Understandings About Science and Technology

F.5.a–g Science and Technology in Society

G.1.a–b Science as a Human Endeavor

G.2.a Nature of Science

G.3.a–c History of Science

INSTRUCT

Point out that the top half of the timeline depicts scientists' progress in understanding the many forms of life. The bottom half provides examples of technology and their application to the study of biodiversity. The gaps represent blocks of time omitted.

Technology

NAVIGATION Any location on Earth can be described using latitude and longitude. Lines of latitude circle Earth in horizontal slices. Lines of longitude divide Earth from pole to pole. Discuss why determining longitude was a harder problem than latitude: You can measure the angle of the Sun above the horizon at noon, to tell latitude. For longitude, you must also measure the time the Sun appeared highest. Keeping time was difficult at sea.

TIMELINES in Science

DISCOVERIES IN Biodiversity

Scientists have discovered new species in the treetops of tropical forests and in the crannies of coral reefs. The quest to catalog the types and numbers of living things, the biodiversity of Earth, began in the late 1600s. A wave of naturalists set sail from Europe to the Americas and to Africa to find specimens of living things.

In the late 19th century, biologists reached agreement on a system for naming and classifying each new species. The mid-20th century brought an understanding of DNA and how it could be used to compare one species with another. Now new organisms could be pinpointed with precision. To this day, millions of undiscovered species lie deep in the unexplored ocean, in tropical forests, and even in heavily trafficked U.S. cities. A large concentration of Earth's known species now live in named and protected biodiversity "hotspots."

1670

Merian Illustrates from Life

In her day, it was typical to work from preserved specimens, but Maria Merian draws from life. Shown below is her illustration of the Legu Lizard she observes in South America. In 1670, she publishes a richly illustrated book of insects, the first to describe the process of metamorphosis.

EVENTS

1670

APPLICATIONS AND TECHNOLOGY

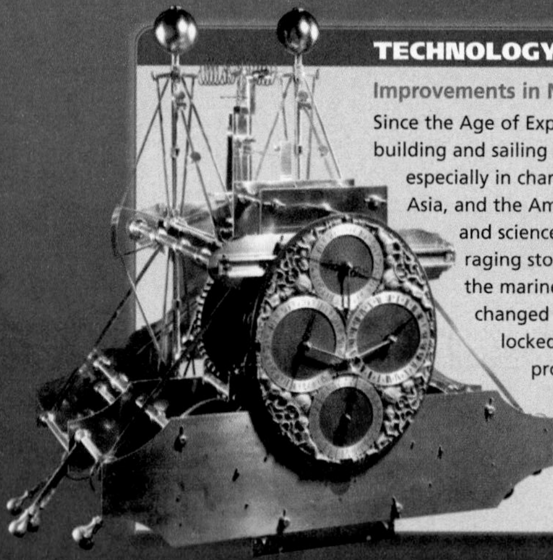

TECHNOLOGY

Improvements in Navigation

Since the Age of Exploration in the 1500s and 1600s, European shipbuilding and sailing had boomed. Vast improvements had been made, especially in charting and mapping. Travel from Europe to Africa, Asia, and the Americas became an important part of business and science. Still, ships often lost their way or wrecked in raging storms. The invention by John Harrison in 1765 of the marine chronometer, a clock that could work at sea, changed navigation forever. Nobody expected a landlocked clockmaker to solve the puzzle, but the sea clock provided an accurate way to record sightings of stars and planets, and thus plot longitude at sea. The ocean remained a dangerous passage, but now, if tossed off course, a captain could still steer clear.

DIFFERENTIATE INSTRUCTION

Below Level Connect the events above the timeline with the technology below it. Ask how each technology helped people make discoveries. *Sample answer: Navigational improvement made it possible to find new places and identify where new organisms were found.*

1775
A Catalog of Living Things
In 1775 Carl Linnaeus completed a book called *Systema Naturae.* His book outlines a system to organize and name plants and animals. The naming system gave scientists a precise and consistent method for sharing discoveries.

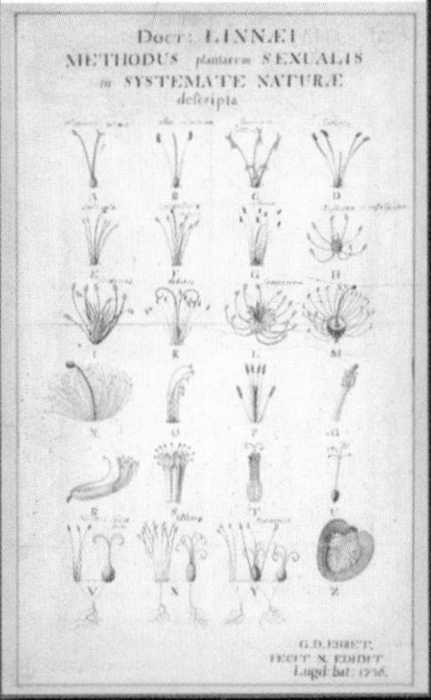

1859
Naturalists in the Amazon
Henry Walter Bates travels from England to the Amazon rainforest in South America with Alfred Russell Wallace. Bates sails home in 1859, bringing over 14,000 specimens, mostly insects. About 8000 of Bates's finds are new discoveries. Wallace loses his collection in a shipwreck.

1889
Naming Discoveries
At the height of discovering new species, conflicts arise over who gets to name living things. In 1889, a conference settles the matter. The first person to publish a description of an organism has the right to claim the discovery and to name it.

1770 1780 1860 1870 1880

TECHNOLOGY
Living Things Too Small to See
Before the 1600s, scientists were unable to see microorganisms. Antony van Leeuwenhoek, a drapery maker who made microscopes in his spare time, was one of the first to observe these tiny organisms. He made the first observation of bacteria, and viewed lake water through a microscope like the one shown to the far right. Others used microscopes to draw detailed close-ups, such as an insect's compound eye shown here.

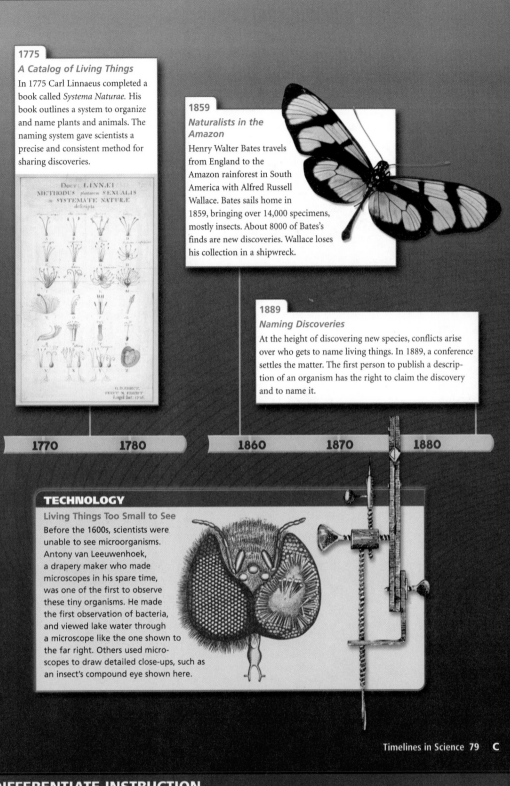

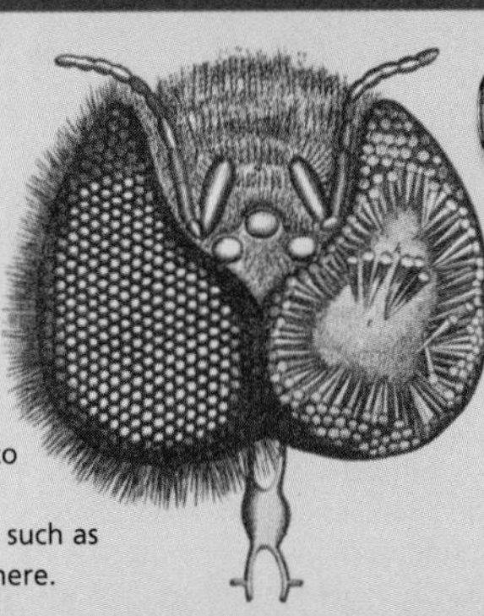

DIFFERENTIATE INSTRUCTION

Advanced Have students choose some plants and animals for which they know the common names, and then research the scientific names. They should write a brief explanation of each name, tracing the name back to its Latin meaning.

Scientific Process

1859 Henry Walter Bates made detailed studies of butterflies, particularly their color patterns. His studies led to the understanding of what is called Batesian mimicry. Some nonpoisonous animals mimic the colors, shape, or behavior of poisonous animals. Bates concluded that the success of organisms with this adaptation was clear evidence for the theory of natural selection.

Technology

TOO SMALL TO SEE The microscope pictured, like all of Leeuwenhoek's microscopes, is a simple one, using only one lens. Though other researchers were using compound microscopes with more than one lens, Leeuwenhoek was so skilled at grinding lenses and adjusting lighting that his simple microscopes actually achieved greater magnification. Ask: Can you think of another example of a scientific tool where simpler is better? *Sample answer: A drawing may tell more than a photograph.*

Integrate the Sciences

1990 Boyle's Law states that the volume of a gas varies inversely with the pressure on it. When a diver goes underwater holding his or her breath, all the air-filled spaces in the body, such as the lungs and ears, will compress. If the diver goes too deep and the spaces compress too much, blood will leave the capillaries to try to fill in the spaces created, causing hemorrhage or ruptured membranes.

Scientific Process

1974 Discuss the practical problems involved in sampling canopy insects. Ask: Can you think of reasons that fogging might not be a good way to collect insects? *Sample answer: Some insects might not be susceptible, and birds might be killed. Also, a scientist might prefer to study living organisms, not dead ones.* Invite students to brainstorm other methods of collecting insects in the forest canopy.

1970

The Bacteria Kingdom Divides

Dr. Carl Woese discovers that bacterialike organisms living in places with extremely high temperatures have genetic material very different from bacteria. Woese suggests a new kingdom or domain called archaebacteria, which is later shortened to Archaea.

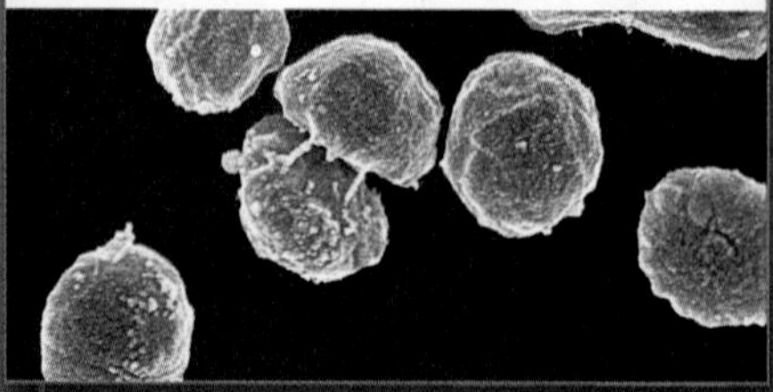

1974

Fogging the Rainforest

Terry Erwin spouts plumes of insecticide into trees and discovers thousands of new insects species in the rainforest canopy. Erwin uses an organism insecticide that does not harm other parts of the ecosystem. Erwin spreads ground sheets or big funnels around a tree, sprays, and, after an hour or so catches the fallen insects, which look like jewels.

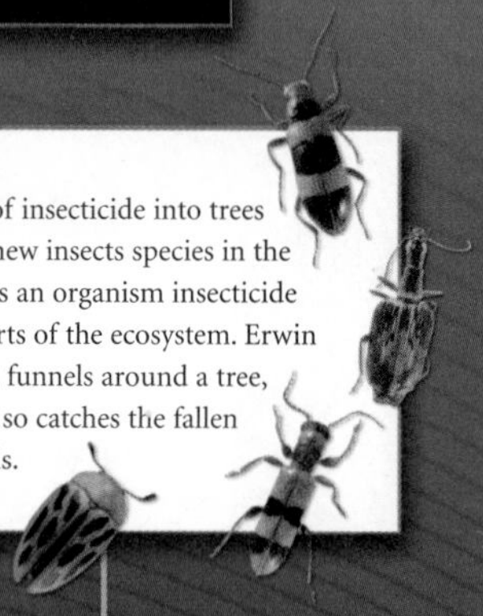

1990

Diver, SCUBA Pioneer, Heads Research

Sylvia Earle is the first woman to serve as chief scientist at the National Oceanic and Atmospheric Administration (NOAA). Earle pioneers SCUBA diving and discovers many new fish and marine species—especially new seaweeds in the Gulf of Mexico. Here she shows a sample to a student in a submersible vehicle.

1970 1980 1990

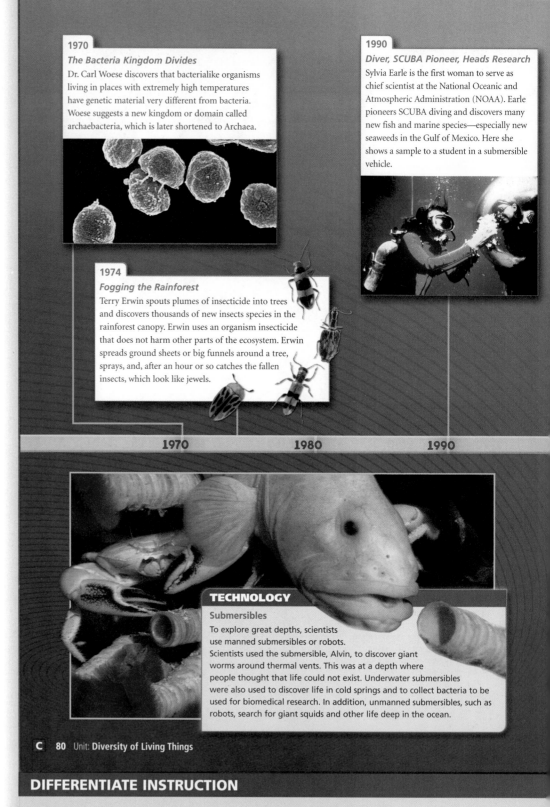

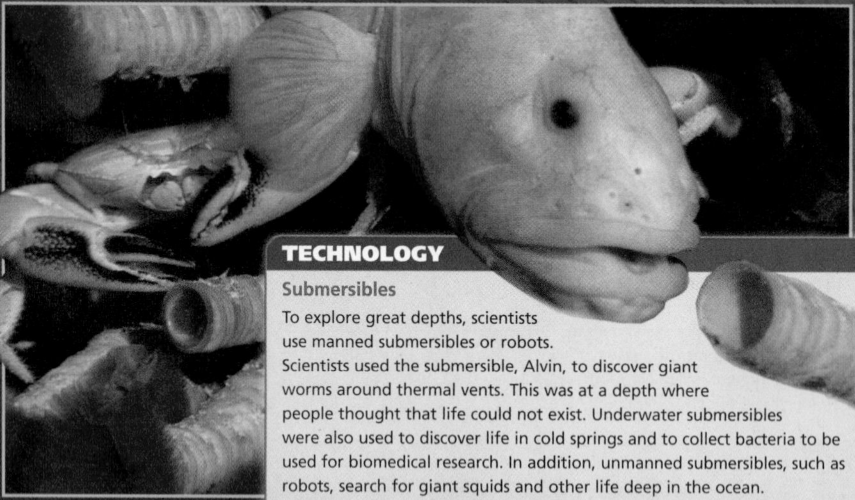

TECHNOLOGY

Submersibles

To explore great depths, scientists use manned submersibles or robots. Scientists used the submersible, Alvin, to discover giant worms around thermal vents. This was at a depth where people thought that life could not exist. Underwater submersibles were also used to discover life in cold springs and to collect bacteria to be used for biomedical research. In addition, unmanned submersibles, such as robots, search for giant squids and other life deep in the ocean.

DIFFERENTIATE INSTRUCTION

Advanced Have students investigate organic insecticides such as the spray used by Terry Erwin. They can compare them to chemical insecticides used in agriculture in rain forests. They should present a detailed comparison of the effects—positive and negative—of each compound on the biodiversity of environments where it is used.

C 80 Unit: Diversity of Living Things

2002 to Present

New Plant Genera

Scientists who study plants and insects in tropical rain forests discover so many new species that they are sometimes backlogged in naming them. In the plant kingdom, it has been estimated that for every 1000 species discovered, one new genus will be found. Occasionally a new plant genus is found in the wild. In 2002 scientists found a new genus of ginger in the wild in Singapore. More and more, genera are identified in the laboratory using DNA.

RESOURCE CENTER
CLASSZONE.COM
Read about current biodiversity discoveries and research.

2000

APPLICATION

Biodiversity Hotspots

In 1988, environmentalist Mike Meyers creates a list of 18 areas that he calls biodiversity "hotspots." In 2000, Myers increases the number of hotspots to 25. How does an area get to be called a hotspot? It has to have a large number of species that exist only in that location and it has to be a region in great danger of habitat loss. Hotspots cover only 1.5 % of the Earth's surface yet they contain 44 percent of all species of higher plants and 35 percent of all land vertebrates. Scientists place most of their effort in discovering new species in these hotspots.

cold ■■■■ hot

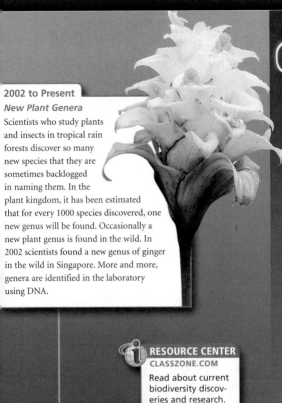

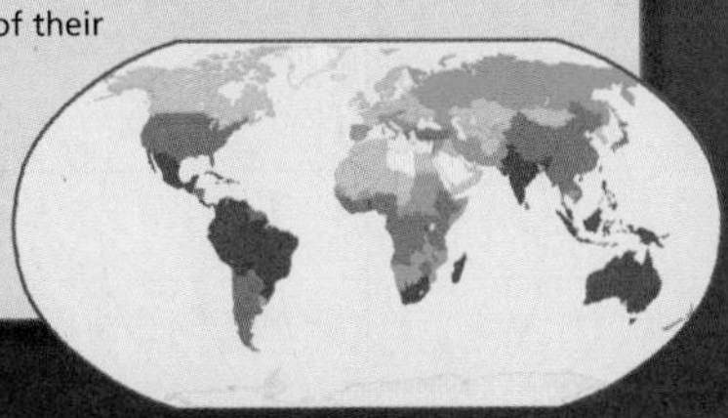

INTO THE FUTURE

Although scientists have explored most of the continents, little is known about what life is like in the more remote areas. Deep-sea exploration, for example, is only just beginning, and big surprises surface with each expedition. Technology will continue to delve deeper toward the floor of deep oceans with underwater robotics, manned submersibles, and better mapping and imaging systems.

Science will increasingly rely on organizing the growing data on biodiversity. Currently, scientists who study one area of life mostly share data with others in the same field. For example, scientists studying plants share their data with other botanists. An effort to create global databases to share information about all species is beginning to bring together this data. Databases that catalogue the genes of living things are being created. The better we know the genetic profile of various species, the better we can identify and compare them.

In addition, the attention to the health of biodiversity hotspots and ecosystems everywhere may help stop extinction, the dying off of a whole species. Extinction decreases the diversity of living things, and scientists have recognized that Earth's biodiversity plays a big role in keeping ecosystems healthy.

ACTIVITIES

Writing About Science: Documentary

Research one hotspot and report on the diverse species living there. Note any legal steps or other efforts to conserve biodiversity in that area.

Reliving History

Devise an animal classification system based on a criterion such as habitat, diet, or behavior. What are the strengths and weaknesses of your system?

Application

BIODIVERSITY HOTSPOTS Over 120 currently used pharmaceutical products are derived from plants. Indigenous people who have a long history of traditional medicine are rich sources of information about medically valuable plants. Ask: Where is a good source of medical plants that are relatively unexplored or unknown? *the rain forest*

INTO THE FUTURE

Have students brainstorm illnesses or problems that might be treated using species that have not yet been discovered. Discuss how to acknowledge and address the needs and rights of indigenous peoples living in biodiversity hotspots, and how to balance their needs with the need to preserve biodiversity.

ACTIVITIES

Writing About Science: Documentary

Have students put their findings into a letter to the editor asking for public support of policies to conserve biodiversity.

Reliving History

One type of classification might be based on diet. Such a classification might be used to analyze how nutrients move in a system.

Technology Resources

Students can visit **ClassZone.com** for current news on biodiversity preservation.

DIFFERENTIATE INSTRUCTION

Below Level Have students choose a biodiversity hotspot and make a collage of some of the species that live there. They can use their own drawings and pictures cut from magazines to highlight as diverse a sample as possible of the types of organisms in each area.

CHAPTER

3 Plants

Life Science
UNIFYING PRINCIPLES

PRINCIPLE 1
All living things share common characteristics.

PRINCIPLE 2
All living things share common needs.

PRINCIPLE 3
Living things meet their needs through interactions with the environment.

PRINCIPLE 4
The types and numbers of living things change over time.

Unit: Diversity of Living Things
BIG IDEAS

CHAPTER 1
Single-Celled Organisms and Viruses
Bacteria and protists have all of the characteristics of living things, while viruses are not alive.

CHAPTER 2
Introduction to Multicellular Organisms
Multicellular organisms live in and get energy from a variety of environments.

CHAPTER 3
Plants
Plants are a diverse group of organisms that live in many land environments.

CHAPTER 4
Invertebrate Animals
Invertebrate animals have a variety of body plans and adaptations.

CHAPTER 5
Vertebrate Animals
Vertebrate animals live in most of Earth's environments.

CHAPTER 3
KEY CONCEPTS

SECTION 3.1

Plants are adapted to living on land.
1. Plants are a diverse group of organisms.
2. Plants share common characteristics.
3. Plant parts have special functions.
4. Plants grow throughout their lifetimes.

SECTION 3.2

Most mosses and ferns live in moist environments.
1. Plant species adapted to life on land.
2. Mosses are nonvascular plants.
3. Mosses reproduce with spores.
4. Ferns are vascular plants.
5. Ferns reproduce with spores.

SECTION 3.3

Seeds and pollen are reproductive adaptations.
1. Seeds are an important adaptation.
2. Some plants reproduce with seeds.
3. Pine trees reproduce with pollen and seeds.
4. Gymnosperms are seed plants.

SECTION 3.4

Many plants reproduce with flowers and fruit.
1. Angiosperms have flowers and fruit.
2. Animals spread both pollen and seeds.
3. Humans depend on plants for their survival.

The Big Idea Flow Chart is available on p. T17 in the **UNIT TRANSPARENCY BOOK.**

Previewing Content

SECTION

3.1 Plants are adapted to living on land. pp. 85–91

1. Plants are a diverse group of organisms.
Plants come in a variety of sizes and shapes, and are found in all environments, from the icy arctic to the tropics.

2. Plants share common characteristics.
All plants are multicellular. Plant cells have a nucleus and cell wall. Plants are producers and have life cycles that are divided into two stages.

3. Plant parts have special functions.
- Plants have two systems—shoot systems that consist of stems and leaves, and root systems.
- **Vascular systems** in plants are made up of two kinds of tissues. Xylem tissue carries water and dissolved nutrients up from the roots. Phloem tissue transports energy-rich materials down from the leaves.
- All plants photosynthesize. Photosynthesis takes place in chloroplasts where sugars and oxygen are produced from the raw materials of carbon dioxide, water, and solar energy. In **transpiration** water vapor moves out of the plant. Stomata regulate the movement of gases and water in and out of the plant.

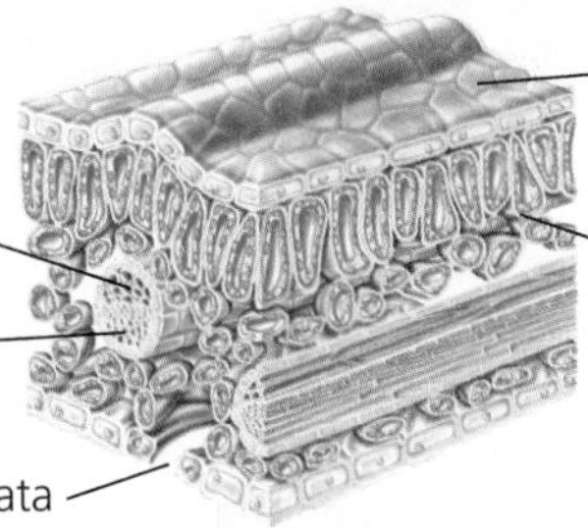

4. Plants grow throughout their lifetimes.
Plants with soft stems survive under the right conditions for growth and photosynthesis. Many of these plants are seasonal. Plants with woody stems have thick, tough stems formed from xylem. These plants survive from year to year.

SECTION

3.2 Most mosses and ferns live in moist environments. pp. 92–97

1. Plant species adapted to life on land.
- Plant life appeared on land about 475 million years ago. Early plants were much like green algae of today.
- As plants moved to land, they adapted structures for support and for taking in nutrients and water.
- The first land plants were similar to present-day mosses and ferns.

2. Mosses are nonvascular plants.
Mosses have simple roots, stems, and leaves, but lack vascular tissues. Mosses are able to reproduce both asexually and sexually.

3. Mosses reproduce with spores.
- In the first generation, moss plants grow from spores. These plants have male and female reproductive parts that can produce sperm and eggs. Fertilization occurs as sperm swim through water to the eggs.
- In the second generation, the fertilized egg grows into a stalk with a capsule. Through meiosis, thousands of spores are produced inside the capsule. The spores are released from the capsule and may grow into new moss plants.

4. Ferns are vascular plants.
The vascular tissues of ferns allow them to grow larger than mosses. Root systems can branch out more and water and nutrients are able to move more efficiently through the plant.

5. Ferns reproduce with spores.
- In the first generation, male and female structures produce eggs. Sperm swim to fertilize the eggs.
- In the second generation, the fertilized eggs grow structures that produce spores. The spores are found in clusters on leafy structures called fronds.

Common Misconceptions

MISCONCEPTION DATABASE
CLASSZONE.COM Background on student misconceptions

PLANT NUTRITION Many studies have shown that students have the misconception that plants get food from the soil by sucking it up through their roots. In fact, plants use carbon dioxide from the air and combine it with the energy from sunlight to synthesize usable energy and materials they need to grow. Plants do not "make food" in photosynthesis, but rather store energy and materials which then become food for any organism that eats the plant.

TE This misconception is addressed on p. 88.

LIVING OR NONLIVING Many students do not classify plants, ferns, and mosses as living or alive. They also often think plant life to be very different than animal life and often do not recognize similarities.

TE This misconception is addressed on p. 95.

SECTION

3.3 Seeds and pollen are reproductive adaptations. pp. 98–106

1. Seeds are an important adaptation.
Seeds are young plants enclosed in protective coatings. Fertilization brings about the growth of the **embryo,** the immature form of the plant inside the seed. An embryo remains dormant until conditions are right for **germination.**

2. Some plants reproduce with seeds.
Seeds contain a food supply for the developing embryo. They can travel long distances, spread by wind, animals, or water. Sperm do not need water to reach the egg, so seed plants can reproduce in drier environments than mosses or ferns.

3. Pine trees reproduce with pollen and seeds.
In seed plants, meiosis and fertilization occur completely within the tissue of the mature plant. Gymnosperms such as pine trees have pollen, small multicellular structures that hold sperm cells. The diagram below summarizes the life cycle of a pine tree.

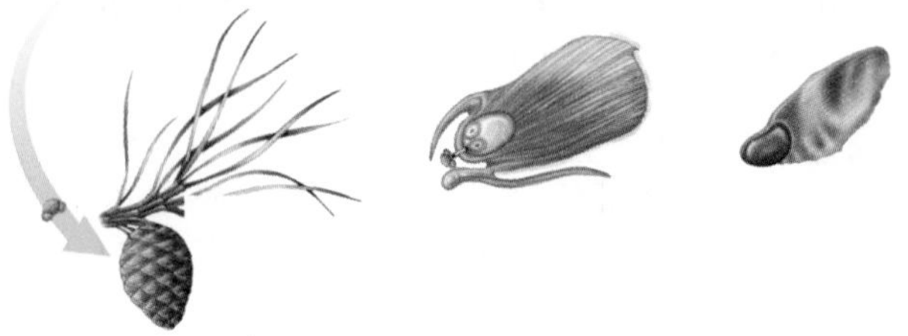

1. Reproductive structures are in pinecones.
2. Male cones: sperm cells are contained in pollen grains. Female cones: contain egg cells.
3. Pollination: pollen grains attach to female cone and release sperm. Fertilization: occurs inside the female structure.
4. Fertilized egg grows into embryo within a seed.

4. Gymnosperms are seed plants.
A gymnosperm is a plant that produces seeds that are not enclosed in flowers or fruit. Four kinds of gymnosperms on Earth today are: conifers, ginkgoes, cycads, and gnetophytes.

SECTION

3.4 Many plants reproduce with flowers and fruit. pp. 107–115

1. Angiosperms have flowers and fruit.
- Most plants on Earth today are angiosperms.
- Angiosperm reproductive cycles are similar to gymnosperms. In angiosperms the sperm and egg cells are contained in a **flower.** When fertilized eggs form seeds, the ovary wall thickens into a **fruit.**
- Male sperm cells are produced in the anther and are released in pollen grains. Female egg cells are produced in the ovary. Pollen grains, when released, attach to the pistil. Fertilization occurs after the pollen tube reaches the ovary and the sperm fertilizes the egg.
- Flowers vary in size, shape, color, and fragrance. The parts of the flower are shown in the diagram.

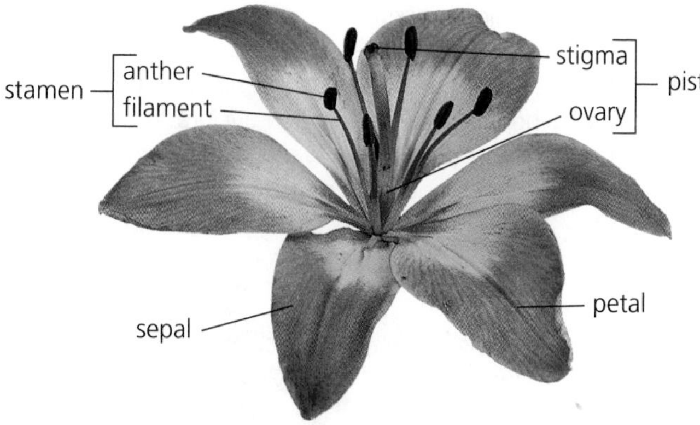

2. Animals spread both pollen and seeds.
- Animals such as bees and hummingbirds feed on the nectar of flowers. These animals pollinate flowers by brushing up against the pollen and carrying it to the reproductive parts of another flower.
- Seeds can be dispersed from fruits by wind, on the fur of animals, and in their digested waste.

3. Humans depend on plants for their survival.
Plants provide oxygen and energy sources for consumers in the form of plant food. Other resources from plants include natural gas, coal, and soil.

Common Misconceptions

SEEDS Only slightly more than half of students above the fourth-grade level classify embryos (seeds and eggs) as part of a living thing. Many students do not think a seed to be a plant. In fact, a seed is a stage in the life cycle of a plant.

MISCONCEPTION DATABASE
CLASSZONE.COM Background on student misconceptions

TE This misconception is addressed on p. 100.

Previewing Labs

Lab Generator CD-ROM
Edit these Pupil Edition labs and generate alternative labs.

EXPLORE the BIG idea

How Are Plants Alike, How Are They Different? p. 83 Students compare and contrast samples of plants to learn about the characteristics of plants.	**TIME** 15 minutes **MATERIALS** samples of local plants, hand lens, notebook
How Are Seeds Dispersed? p. 83 Students make a model of a seed with wing-like structures to examine dispersal by wind.	**TIME** 15 minutes **MATERIALS** paper, scissors, paper clip
Internet Activity: Sprouting Seeds, p. 83 Students learn about plant life cycles by observing seeds sprouting.	**TIME** 20 minutes **MATERIALS** computer with Internet access

SECTION 3.1

EXPLORE Leaf Characteristics, p. 85 Students examine the characteristics of a leaf.	**TIME** 15 minutes **MATERIALS** assorted leaves, hand lens, (optional) colored pencils

SECTION 3.2

EXPLORE Moss Plants, p. 92 Students observe the structures and distinct parts of moss plants.	**TIME** 20 minutes **MATERIALS** live moss plant, hand lens
INVESTIGATE Capturing the Sun's Energy, p. 94 Students measure how much sunlight reaches organisms living in water to learn the advantages for plants of living on land.	**TIME** 15 minutes **MATERIALS** white button, string, empty 1 L clear plastic bottle, water, ruler, kelp granules, tablespoon measure

SECTION 3.3

INVESTIGATE Pinecones, p. 102 Students observe the structure of pinecones to learn about the reproductive structures of gymnosperms.	**TIME** 20 minutes **MATERIALS** dried, open pinecones; beaker of water; paper towels
CHAPTER INVESTIGATION, Which Seeds Will Grow? pp. 104–105 Students compare and contrast germination and growth patterns of various types of seeds.	**TIME** 40 minutes **MATERIALS** assorted seeds, potting soil, paper cups, water, paper towels, labels

SECTION 3.4

EXPLORE Fruit, p. 107 Students explore the inside of different fruits to discover the structures of seeds in angiosperms.	**TIME** 15 minutes **MATERIALS** apple, paper towel, plastic knife, pea pod
INVESTIGATE Flower Parts, p. 111 Students identify the parts of a flower to learn more about the reproductive cycle of angiosperms.	**TIME** 15 minutes **MATERIALS** assorted flowers, hand lens

R **Additional INVESTIGATION,** Fertilization in Angiosperms, A, B, & C, pp. 210–218; Teacher Instructions, pp. 360–361

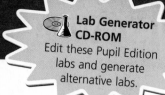

Previewing Chapter Resources

	INTEGRATED TECHNOLOGY	LABS AND ACTIVITIES
CHAPTER 3 **Plants**	**CLASSZONE.COM** • eEdition Plus • EasyPlanner Plus • Misconceptions • Content Review • Test Practice • Simulation • Resource Centers • Internet Activity: Sprouting Seeds • Math Tutorial **SCILINKS.ORG** SCILINKS **CD-ROMS** • eEdition • EasyPlanner • Power Presentations • Content Review • Lab Generator • Test Generator **AUDIO CDS** • Audio Readings • Audio Readings in Spanish	PE EXPLORE the Big Idea, p. 83 • How Are Plants Alike, How Are They Different? • How Are Seeds Dispersed? • Internet Activity: Sprouting Seeds R **UNIT RESOURCE BOOK** Unit Projects, pp. 5–10 **Lab Generator CD-ROM** Generate customized labs.
SECTION 3.1 **Plants are adapted to living on land.** pp. 85–91 Time: 2 periods (1 block) R Lesson Plan, pp. 151–152	**RESOURCE CENTER,** Plant Systems T **UNIT TRANSPARENCY BOOK** • Big Idea Flow Chart, p. T17 • Daily Vocabulary Scaffolding, p. T18 • Note-Taking Model, p. T19 • 3-Minute Warm-Up, p. T20	PE EXPLORE Leaf Characteristics, p. 85
SECTION 3.2 **Most mosses and ferns live in moist environments.** pp. 92–97 Time: 2 periods (1 block) R Lesson Plan, pp. 161–162	**RESOURCE CENTER,** Plant Evolution T **UNIT TRANSPARENCY BOOK** • Daily Vocabulary Scaffolding, p. T18 • 3-Minute Warm-Up, p. T20	PE • EXPLORE Moss Plants, p. 92 • INVESTIGATE Capturing the Sun's Energy, p. 94 R **UNIT RESOURCE BOOK** Datasheet, Capturing the Sun's Energy, p. 170
SECTION 3.3 **Seeds and pollen are reproductive adaptations.** pp. 98–106 Time: 3 periods (1.5 blocks) R Lesson Plan, pp. 172–173	**RESOURCE CENTER,** Seeds, Extreme Seeds T **UNIT TRANSPARENCY BOOK** • Daily Vocabulary Scaffolding, p. T18 • 3-Minute Warm-Up, p. T21 • "Life Cycle of a Pine Tree" Visual, p. T22	PE • INVESTIGATE Pinecones, p. 102 • CHAPTER INVESTIGATION, pp. 104–105 • Extreme Science, p. 106 R **UNIT RESOURCE BOOK** • Datasheet, Pinecones, p. 181 • CHAPTER INVESTIGATION, Which Seeds Will Grow? A, B, & C, pp. 201–209
SECTION 3.4 **Many plants reproduce with flowers and fruit.** pp. 107–115 Time: 3 periods (1.5 blocks) R Lesson Plan, pp. 183–184	• **SIMULATION,** Seed Dispersal • **MATH TUTORIAL** T **UNIT TRANSPARENCY BOOK** • Big Idea Flow Chart, p. T17 • Daily Vocabulary Scaffolding, p. T18 • 3-Minute Warm-Up, p. T21 • Chapter Outline, pp. T23–T24	PE • EXPLORE Fruit, p. 107 • INVESTIGATE Flower Parts, p. 111 • Math in Science, p. 115 R **UNIT RESOURCE BOOK** • Datasheet, Flower Parts, p. 193 • Math Support & Practice, pp. 199–200 • Additional INVESTIGATION, Fertilization in Angiosperms, A, B, & C, pp. 210–218

KEY TO ICONS
- CD/CD-ROM
- Teacher Edition
- UNIT TRANSPARENCY BOOK
- SPANISH ASSESSMENT BOOK
- INTERNET
- Pupil Edition
- UNIT RESOURCE BOOK
- UNIT ASSESSMENT BOOK
- SCIENCE TOOLKIT

READING AND REINFORCEMENT	ASSESSMENT	STANDARDS
• Word Triangle, B18–19 • Mind Map, C40–41 • Daily Vocabulary Scaffolding, H1–8 **UNIT RESOURCE BOOK** • Vocabulary Practice, pp. 196–197 • Decoding Support, p. 198 • Summarizing the Chapter, pp. 219–220 **Audio Readings CD** Listen to Pupil Edition. **Audio Readings in Spanish CD** Listen to Pupil Edition in Spanish.	• Chapter Review, pp.117–118 • Standardized Test Practice, p. 119 **UNIT ASSESSMENT BOOK** • Diagnostic Test, pp. 41–42 • Chapter Test, A, B, & C, pp. 47–58 • Alternative Assessment, pp. 59–60 Spanish Chapter Test, pp. 49–52 **Test Generator CD-ROM** Generate customized tests. **Lab Generator CD-ROM** Rubrics for Labs	National Standards A.2–8, A.9.a–c, A.9.e–f, C.2.b, C.4.c, G.1.b See p. 82 for the standards.
UNIT RESOURCE BOOK • Reading Study Guide, A & B, pp. 153–156 • Spanish Reading Study Guide, pp. 157–158 • Challenge and Extension, p. 159 • Reinforcing Key Concepts, p. 160 • Challenge Reading, pp. 194–195	Ongoing Assessment, pp. 86–90 Section 3.1 Review, p. 91 **UNIT ASSESSMENT BOOK** Section 3.1 Quiz, p. 43	National Standards C.4.c, G.1.b
UNIT RESOURCE BOOK • Reading Study Guide, A & B, pp. 163–166 • Spanish Reading Study Guide, pp. 167–168 • Challenge and Extension, p. 169 • Reinforcing Key Concepts, p. 171	Ongoing Assessment, pp. 93–97 Section 3.2 Review, p. 97 **UNIT ASSESSMENT BOOK** Section 3.2 Quiz, p. 44	National Standards A.2–7, A.9.a–b, A.9.e–f, C.2.b, C.4.c, G.1.b
UNIT RESOURCE BOOK • Reading Study Guide, A & B, pp. 174–177 • Spanish Reading Study Guide, pp. 178–179 • Challenge and Extension, p. 180 • Reinforcing Key Concepts, p. 182	Ongoing Assessment, pp. 99–102 Section 3.3 Review, p. 103 **UNIT ASSESSMENT BOOK** Section 3.3 Quiz, p. 45	National Standards A.2–7, A.9.a–b, A.9.e–f, C.4.c, G.1.b
UNIT RESOURCE BOOK • Reading Study Guide, A & B, pp. 185–188 • Spanish Reading Study Guide, pp. 189–190 • Challenge and Extension, p. 191 • Reinforcing Key Concepts, p. 193	Ongoing Assessment, pp. 108–113 Section 3.4 Review, p. 114 **UNIT ASSESSMENT BOOK** Section 3.4 Quiz, p. 46	National Standards A.2–8, A.9.a–c, A.9.e–f, C.4.c, G.1.b

Previewing Resources for Differentiated Instruction

CHAPTER INVESTIGATION

Leveled resources present the same concepts for different abilities.

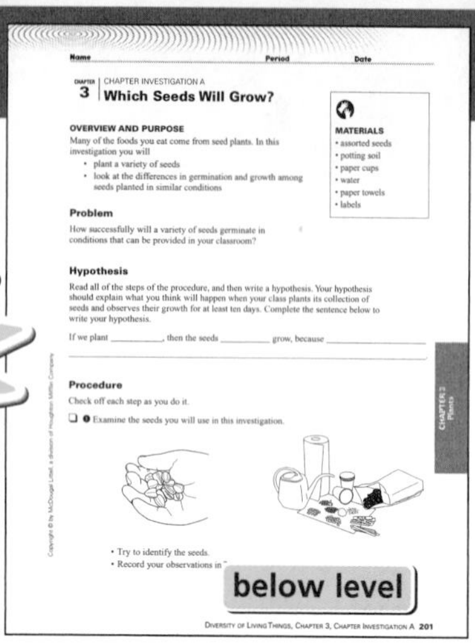

UNIT RESOURCE BOOK, pp. 201–204

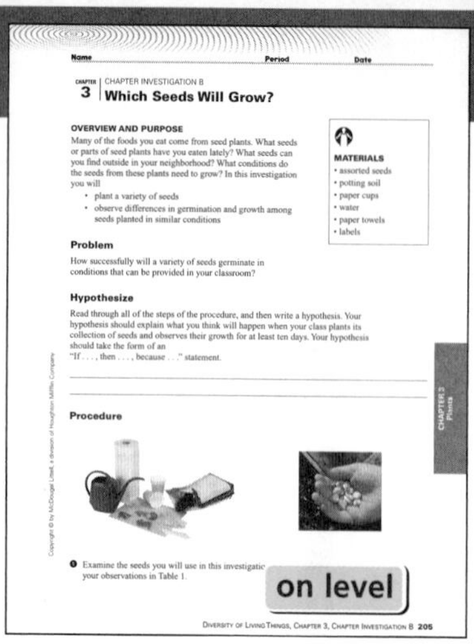

pp. 205–208

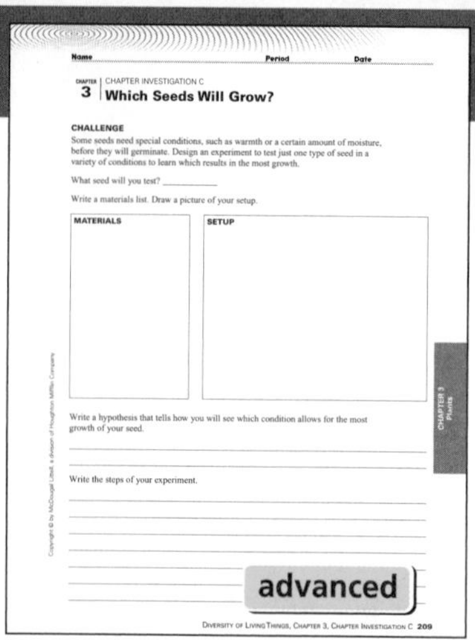

pp. 205–209

READING STUDY GUIDE

Reading Study Guide is also in Spanish.

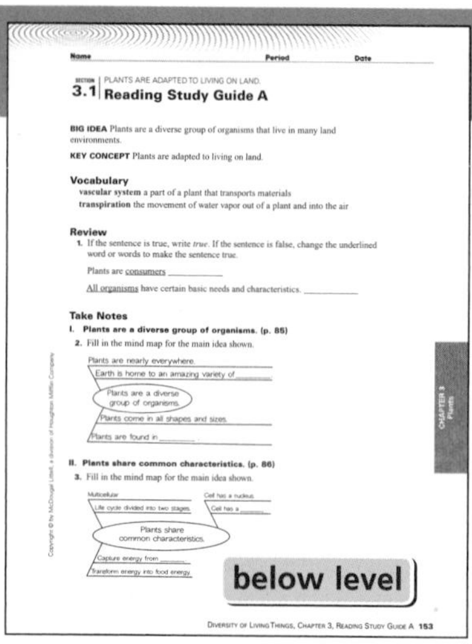

UNIT RESOURCE BOOK, pp. 153–154

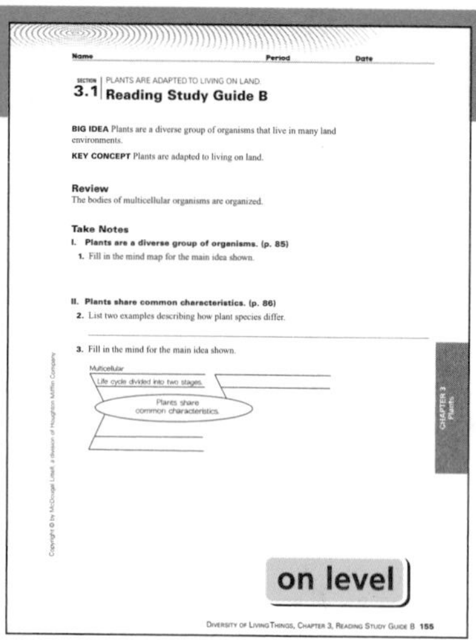

pp. 155–156

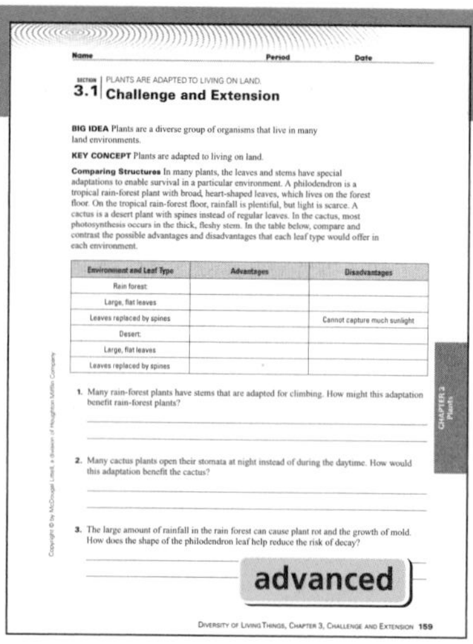

p. 159

CHAPTER TEST

Chapter Test is also in Spanish.

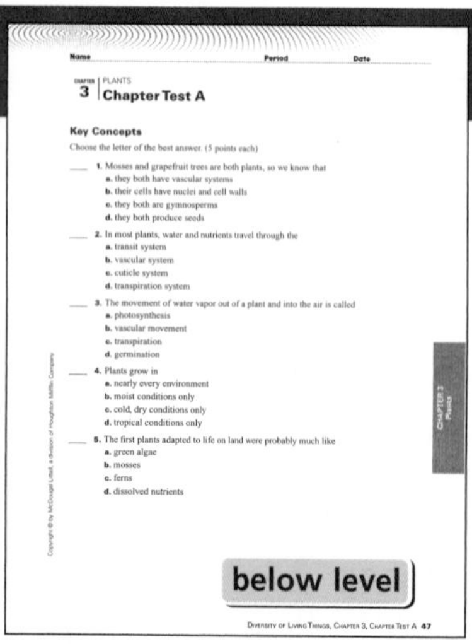

UNIT ASSESSMENT BOOK, pp. 47–50

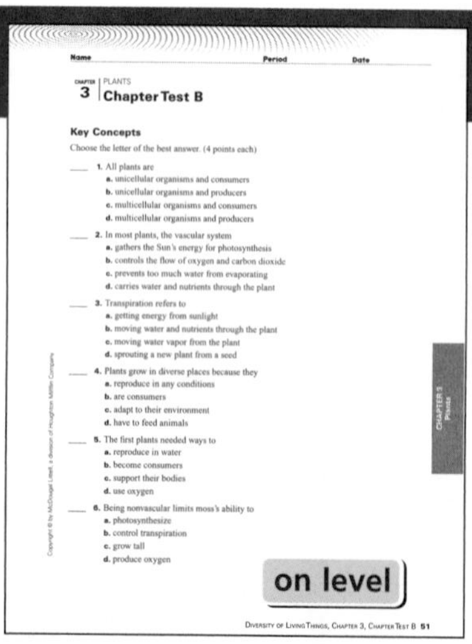

pp. 51–54

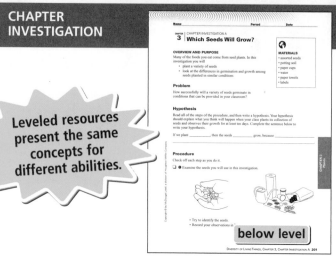

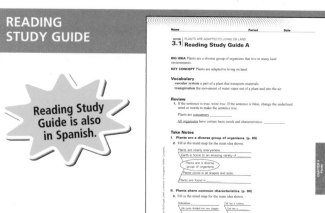

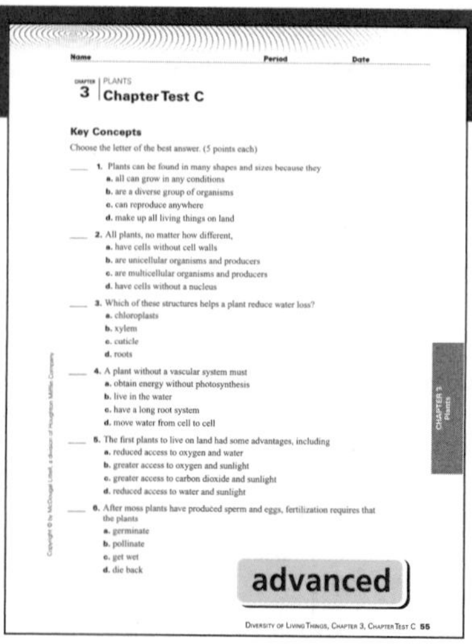

pp. 55–58

TECHNOLOGY

CLASSZONE.COM

CD/CD-ROMS

CLASSZONE.COM

VISUAL CONTENT

T UNIT TRANSPARENCY BOOK, p. T17

T p. T19

T p. T22

MORE SUPPORT

Section 3.1 Plants are adapted to living on land.

Reinforcing Key Concepts

BIG IDEA Plants are a diverse group of organisms that live in many land environments.

KEY CONCEPT Plants are adapted to living on land.

1. Plants are a diverse group of organisms. Plants come in all shapes and sizes, from tiny flowers no bigger than the head of a pin to giant trees taller than a 12-story building. Why do you think plants are such a diverse group?

2. Plants share common characteristics. Despite how different some plants are from others in appearance, they all share certain characteristics. Name two of the characteristics that define a plant as a plant.

3. Plant parts have special functions. Every living cell needs a constant supply of water, nutrients, and food. Draw a picture of a plant, and label the leaves, roots, and stem. Then answer the questions.

4. Plants grow throughout their lifetimes.

160 Diversity of Living Things, Chapter 3, Reinforcing Key Concepts

Chapter 3 Plants

Vocabulary

vascular system	germination	angiosperm
transpiration	pollen	flower
seed	gymnosperm	fruit
embryo		

A. ANSWER AND QUESTION

Diversity of Living Things, Chapter 3, Vocabulary Practice 196

Chapter 3 Plants

Math Support

Using Grids to Estimate

SAMPLE PROBLEM

Diversity of Living Things, Chapter 3, Math Support 199

R UNIT RESOURCE BOOK, p. 160

R pp. 196–197

R p. 199

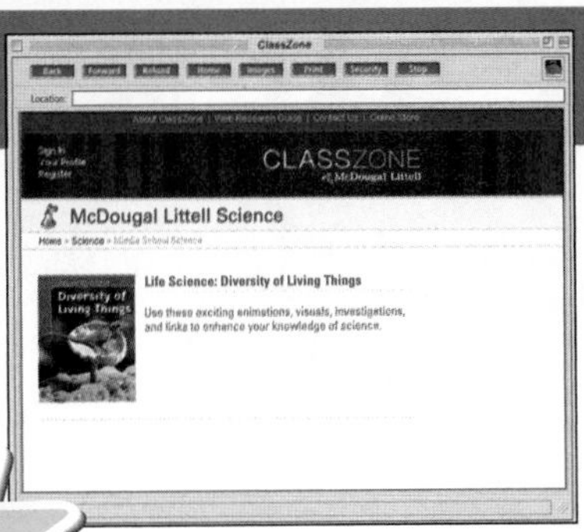

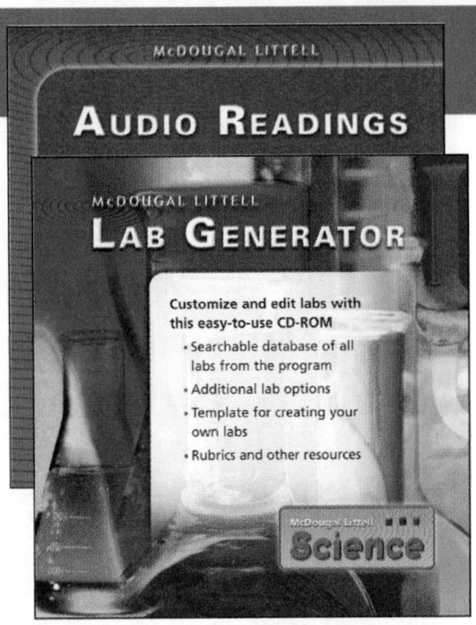

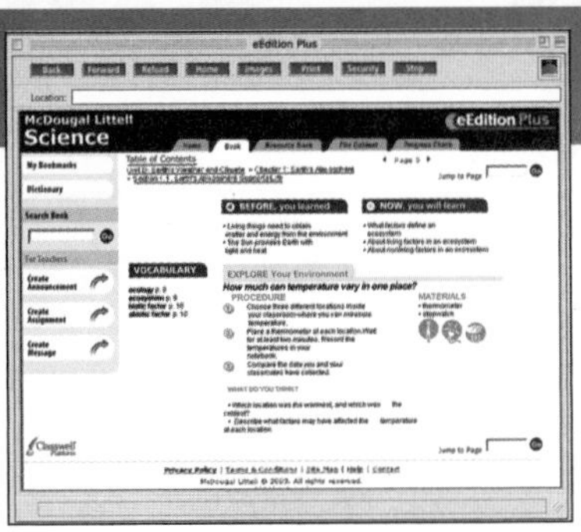

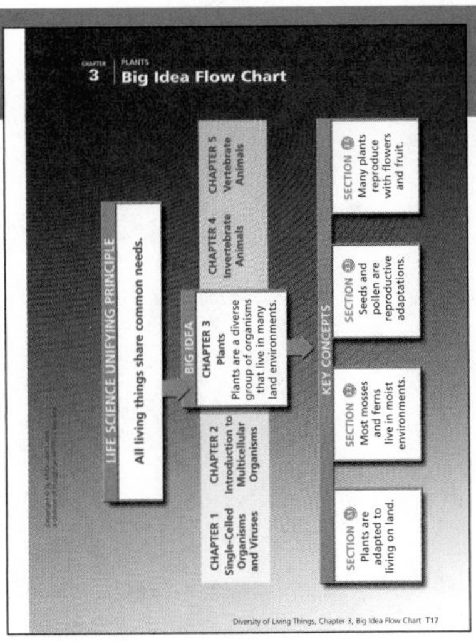

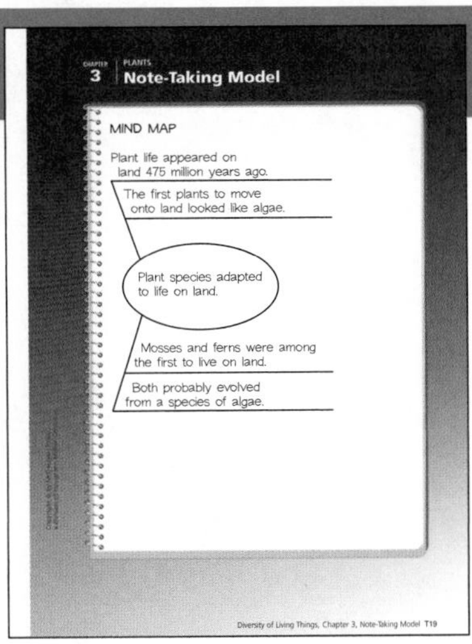

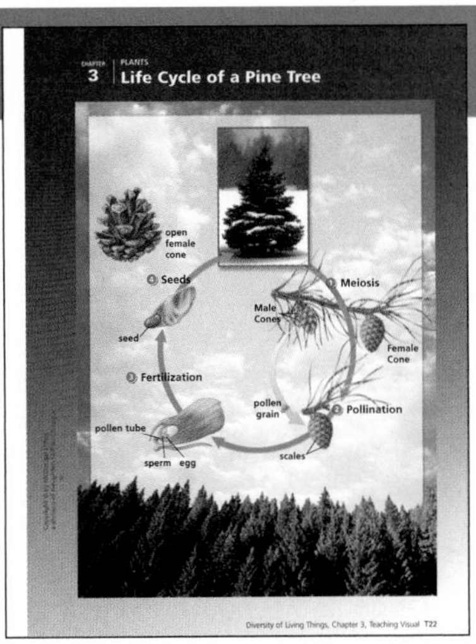

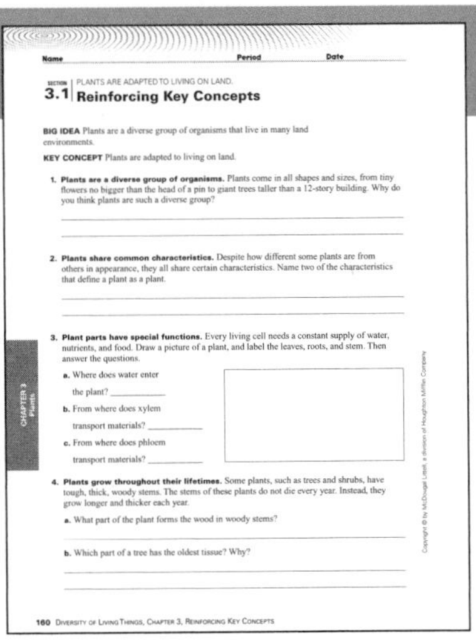

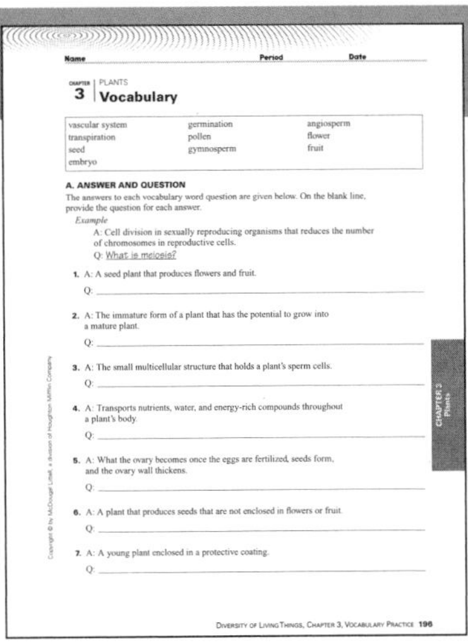

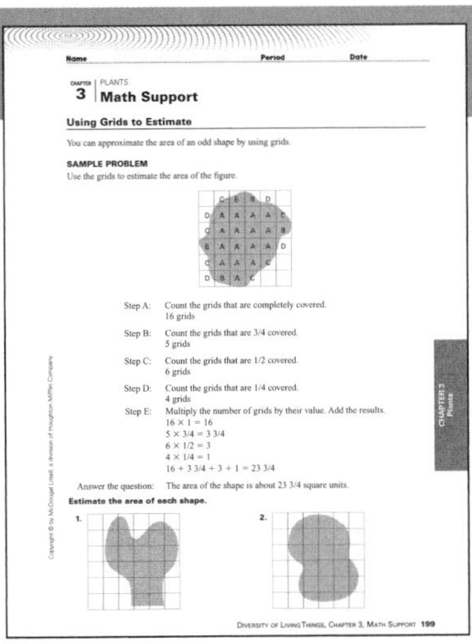

CHAPTER 3 Plants

INTRODUCE

the BIG idea

Have students look at the photograph of plants and discuss how the question in the box links to the Big Idea:

- Why do some plants grow larger than others?
- Why can a tree grow larger and live longer than an animal?

National Science Education Standards

Content

C.2.b Plants reproduce sexually—the egg and sperm are produced in the flowers of flowering plants. An egg and sperm unite to begin development of a new individual.

C.4.c For ecosystems, the major source of energy is sunlight. Energy entering ecosystems as sunlight is transferred by producers into chemical energy through photosynthesis. That energy then passes from organism to organism in food webs.

Process

A.2–8 Design and conduct an investigation; use tools to gather and interpret data; use evidence to describe, predict, explain, model; think critically to make relationships between evidence and explanation; recognize different explanations and predictions; communicate scientific procedures and explanations; use mathematics.

A.9.a–c, A.9.e–f Understand scientific inquiry by using different investigations, methods, mathematics, and explanations based on logic, evidence, and skepticism.

G.1.b Science requires different abilities.

CHAPTER 3 Plants

the BIG idea

Plants are a diverse group of organisms that live in many land environments.

Key Concepts

SECTION 3.1 **Plants are adapted to living on land.** Learn about plant characteristics and structures.

SECTION 3.2 **Most mosses and ferns live in moist environments.** Learn about how mosses and ferns live and reproduce.

SECTION 3.3 **Seeds and pollen are reproductive adaptations.** Learn about seeds, pollen, and gymnosperms.

SECTION 3.4 **Many plants reproduce with flowers and fruit.** Learn about flowers, fruit, and angiosperms.

Internet Preview

CLASSZONE.COM

Chapter 3 online resources: Content Review, Simulation, Visualization, four Resource Centers, Math Tutorial, Test Practice

How does such a large tree get what it needs to survive?

INTERNET PREVIEW

CLASSZONE.COM For student use with the following pages:

Review and Practice

- Content Review, pp. 84, 116
- Math Tutorial: Using Grids to Estimate, p. 115
- Test Practice, p. 119

Activities and Resources

- Internet Activity: p. 83
- Resource Centers: Plant Systems, p. 90; Plant Evolution, p. 92; Seeds, p. 99; Extreme Seeds, p. 106
- Simulation: p. 112

NSTA SCiLINKS scilinks.org

Plant Kingdom
Code: MDL041

EXPLORE the BIG idea

How Are Plants Alike, How Are They Different?

Find samples of plants in your area. Draw or take pictures of them for your science notebook. Note each plant's shape, size, and the environment where you found it.

Observe and Think Do all plants have leaves? Do all plants have flowers?

How Are Seeds Dispersed?

Make a model of a seed with wings, such as a seed from a maple tree. Fold a strip of paper in half the long way and then unfold it. Cut a little less than halfway along the fold. Refold the uncut part of the paper and bend out the two cut strips so they look like wings. Put a paper clip on the bottom of the folded part. Drop your seed.

Observe and Think How does having seeds with wings benefit a plant?

Internet Activity: Sprouting Seeds

Go to **ClassZone.com** to watch a time-lapse video of a seed sprouting.

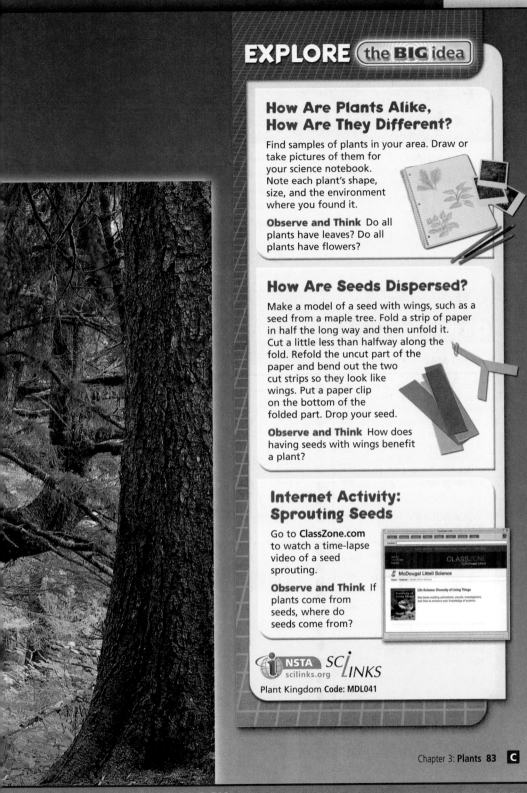

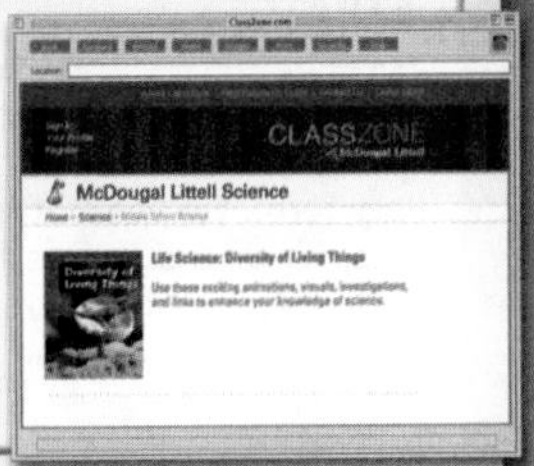

Observe and Think If plants come from seeds, where do seeds come from?

NSTA scilinks.org SCILINKS

Plant Kingdom **Code: MDL041**

TEACHING WITH TECHNOLOGY

Digital Camera Students can use digital cameras to take daily photographs of the germinating seeds, while doing the Chapter Investigation on pp. 104–105. Have students include the photographs in their Science Notebooks.

CBL and Probeware Students might use a pressure sensor to study the movement of water within a plant while studying gas exchange and water loss in plants on p. 89. They could enter and plot the data on a graphing calculator.

EXPLORE the BIG idea

These inquiry-based activities are appropriate for use at home or as a supplement to classroom instruction.

How Are Plants Alike, How Are They Different?

PURPOSE To compare and contrast samples of plants.

TIP *15 min.* Gather plants for students to examine. Use caution to avoid selecting poisonous plants.

Answer: Most plants have leaves, but they vary in shape and size. Most plants have flowers at some point in their life cycle, while some do not.

REVISIT after p. 86.

How Are Seeds Dispersed?

PURPOSE To make a model of a seed with wing-like structures.

TIP *15 min.* Make a sample model for students to follow. Ask students what the paper clip represents. *the seed* Encourage students to try dropping the "seeds" in various wind conditions, or from various heights. They might also experiment with types of paper.

Answer: Wing-like structures allow plants to spread seeds beyond the parent plant. This allows the young plants to grow without competition for sunlight, water, and nutrients.

REVISIT after p. 113.

Internet Activity: Sprouting Seeds

PURPOSE To observe how plants grow from seeds.

TIP *20 min.* Challenge students to create sketches of the sprouting seed.

Answer: Seeds come from the reproductive structures of plants, and are usually found in fruits or pinecones.

REVISIT after p. 99.

PREPARE

CONCEPT REVIEW

Activate Prior Knowledge

- Ask students to name some of the common characteristics that all living organisms share. *All living organisms grow and reproduce and need energy, water, other materials, and living space.*
- What are some parts of plants you have seen? Can you name their function in the plant? *Sample answer: Roots take up water; seeds and flowers help the plant reproduce.*
- Ask students to give an example of an organism that could be called a "producer"—that is, one at the start of a food chain. *a plant or protist*

TAKING NOTES

Mind Map

Students may also use the red headings in the text to form the basis of some of the details in the Mind Map. The maps are useful as study tools when students review the chapter or organize their notes.

Vocabulary Strategy

The word triangle allows students to demonstrate understanding of nouns that can be illustrated with a visual. Illustrating terms and using them in meaningful sentences help students personalize and own the new concepts that challenge them.

Vocabulary and Note-Taking Resources

- Vocabulary Practice, pp. 196–197
- Decoding Support, p. 198

- Daily Vocabulary Scaffolding, p. T18
- Note-Taking Model, p. T19

- Word Triangle, B18–19
- Mind Map, C40–41
- Daily Vocabulary Scaffolding, H1–8

CHAPTER 3

Getting Ready to Learn

CONCEPT REVIEW

- All organisms get water and other materials from the environment.
- Plants have specialized tissues with specific functions.
- Plants are producers.

VOCABULARY REVIEW

meiosis p. 48
fertilization p. 48
photosynthesis p. 52
spore p. 67
cycle *See Glossary.*

CONTENT REVIEW
CLASSZONE.COM
Review concepts and vocabulary.

TAKING NOTES

MIND MAP

Write each main idea, or blue heading, in an oval; then write details that relate to each other and to the main idea. Organize the details so that each spoke of the web has notes about one part of the main idea.

VOCABULARY STRATEGY

Draw a **word triangle** diagram for each new vocabulary term. On the bottom line, write and define the term. Above that, write a sentence that uses the term correctly.
At the top, draw a small picture to remind you of the definition.

See the Note-Taking Handbook on pages R45–R51.

SCIENCE NOTEBOOK

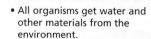

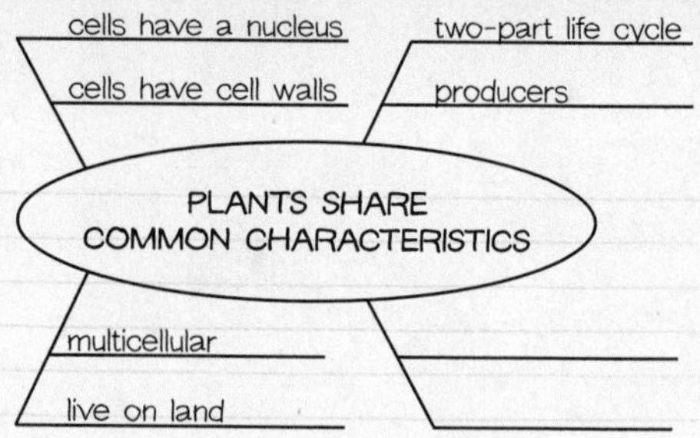

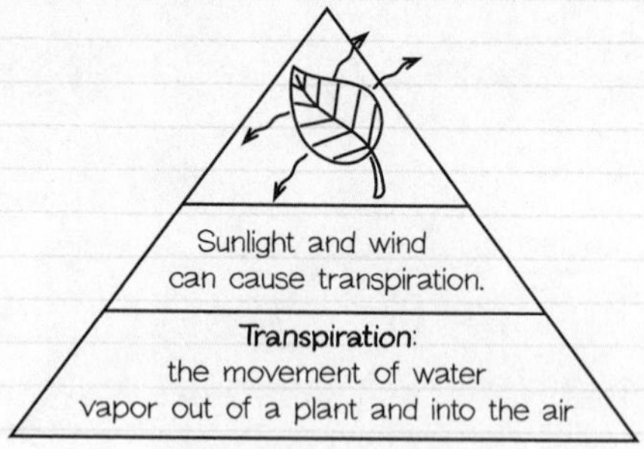

CHECK READINESS

Administer the Diagnostic Test to determine students' readiness for new science content and their mastery of requisite math skills.

Diagnostic Test, pp. 41–42

Technology Resources

Students needing content and math skills should visit **ClassZone.com**.

- **CONTENT REVIEW**
- **MATH TUTORIAL**

CONTENT REVIEW CD-ROM

KEY CONCEPT

3.1 Plants are adapted to living on land.

BEFORE, you learned
- All organisms have certain basic needs and characteristics
- The bodies of multicellular organisms are organized
- Plants are producers

NOW, you will learn
- About plant diversity
- About common characteristics of plants
- How the bodies of plants are organized

VOCABULARY
vascular system p. 87
transpiration p. 88

EXPLORE Leaf Characteristics

What is a leaf?

PROCEDURE

1. Examine the leaf your teacher gives you carefully. Try to notice as many details as you can.
2. Make a drawing of both sides of your leaf in your notebook. Label as many parts as you can and write down your ideas describing each part's function.
3. Compare your diagram and notes with those of your classmates.

MATERIALS
- assorted leaves
- hand lens

WHAT DO YOU THINK?
- What characteristics did most or all of your leaves have?
- How would you describe your leaf to someone who could not see it?

Plants are a diverse group of organisms.

MIND MAP
Make a mind map for the first main idea: *Plants are a diverse group of organisms.*

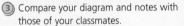

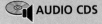

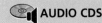

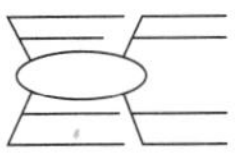

Plants are nearly everywhere. Walk through a forest, and you're surrounded by trees, ferns, and moss. Drive along a country road, and you pass fields planted with crops like cotton or wheat. Even a busy city has tree-lined sidewalks, grass-covered lawns, and weeds growing in vacant lots or poking through cracks in the pavement.

Earth is home to an amazing variety of plant life. Plants come in all shapes and sizes, from tiny flowers no bigger than the head of a pin to giant trees taller than a 12-story building. Plants are found in all types of environments, from the icy Arctic to the steamy tropics.

RESOURCES FOR DIFFERENTIATED INSTRUCTION

Below Level
UNIT RESOURCE BOOK
- Reading Study Guide A, pp. 153–154
- Decoding Support, p. 198

AUDIO CDS

Advanced
UNIT RESOURCE BOOK
- Challenge and Extension, p. 159
- Challenge Reading, pp. 194–195

English Learners
UNIT RESOURCE BOOK
Spanish Reading Study Guide, pp. 157–158

AUDIO CDS
- Audio Readings in Spanish
- Audio Readings (English)

Additional INVESTIGATION,
Fertilization in Angiosperms, A, B, & C, pp. 210–218; Teacher Instructions, pp. 360–361

3.1 FOCUS

Set Learning Goals

Students will
- Make observations about plant diversity.
- Identify common characteristics of plants.
- Describe how the bodies of plants are organized.

3-Minute Warm-Up

Display Transparency 20 or copy this exercise on the board:

Decide if these statements are true. If not true, correct them.

1. Living organisms need only water and air to survive. *Living organisms need to obtain energy, materials, water, and air to survive.*
2. Most multicellular organisms are organized into tissues and organs. *true*
3. Producers are organisms that feed off of other organisms. *Producers are organisms that capture energy to meet their needs.*

3-Minute Warm-Up, p. T20

3.1 MOTIVATE

EXPLORE Leaf Characteristics

PURPOSE To explore the characteristics of a leaf

TIPS *15 min.* Use fallen leaves for this activity if you prefer not to remove leaves from living plants. Students may be able to see stomata on the undersides of certain types of leaves with the aid of a hand lens.

WHAT DO YOU THINK? *Sample answer: Most leaves are green and flat, have veins, are symmetrical, and have stems. Students may describe how big the leaves are and whether the leaves are round or pointed. Pine needles are, in fact, a type of modified leaf.*

3.1 INSTRUCT

Teach from Visuals

To help students interpret the photographs of plant diversity, ask:

- What structures of these plants are you able to observe? *leaves, flowers, stems, branches*
- What structures of these plants are you unable to see? *cells that make up the plants, roots, inside the layers of plant tissues*

EXPLORE the BIG idea

Revisit "How Are Plants Alike, How Are They Different?" on p. 83. Ask: What common features do the plants discussed on this page share with the plants you observed in the activity? *Students should now understand that despite unique or dissimilar appearances all plants are alike in four basic ways. Plants are multicellular, have a two-part life cycle, photosynthesize, and are made up of cells with nuclei and cell walls. They may also mention that the plants they have observed all have leaves, stems, and green color.*

Ongoing Assessment

Make observations about plant diversity.

Ask: What are some ways in which plants are diverse? *They vary in size, shape, color, and the environments in which they live.*

CHECK YOUR READING *Answer: 1) multicellular; 2) cell with nucleus and cell wall; 3) producers that get energy from the Sun; 4) two-part life cycle*

Diversity of Plants

Plant species live in a variety of environments and have a wide range of features.

Orchids

Seed pods of the vanilla orchid are used to flavor food.

Horsetails

Horsetails have a distinctive shape and texture.

Bristlecone Pine

Bristlecone pines are some of the oldest trees on Earth.

Plants share common characteristics.

REMINDER A *species* is a classification used for a group of organisms that are so similar that members of the group can breed and produce offspring that can also reproduce.

Scientists estimate that at least 260,000 different species of plants live on Earth today. The photographs on this page show three examples of plants that are very different from one another.

Orchids are flowering plants that mostly grow in tropical rain forests. To get the sunlight they need, many orchids grow not in the soil but on the trunks of trees. Horsetails are plants that produce tiny grains of a very hard substance called silica. Sometimes called scouring rushes, these plants were once used to scrub clean, or scour, dishes and pots. Bristlecone pine trees live on high mountain slopes in North America where there is little soil and often high winds. These trees grow very slowly and can live for several thousand years.

You can see from these three examples that plant species show great diversity. Despite how different an orchid is from a horsetail and a bristlecone pine, all three plants share certain characteristics. These are the characteristics that define a plant as a plant:

- Plants are multicellular organisms.
- A plant cell has a nucleus and is surrounded by a cell wall. (A)
- Plants are producers. They capture energy from the Sun.
- Plant life cycles are divided into two stages, or generations. (B)

CHECK YOUR READING What characteristics are shared by all plants?

DIFFERENTIATE INSTRUCTION

More Reading Support

A From where do plants obtain energy? *the Sun*

B What is common in all plant life cycles? *There are two stages or generations.*

English Learners Place the terms *vascular system* and *transpiration* on a classroom Science Word Wall with brief definitions of each. Pronounce these words for students.

English learners may not recognize that some words change their spellings when made plural. For example, in this section the words *leaf* and *leaves* are seen repeatedly. Point these out to students and give examples of other word pairs, such as *knife* and *knives*, or *hoof* and *hooves* that change when plural.

Plant parts have special functions.

You could say that a plant lives in two worlds. The roots anchor a plant in the ground. Aboveground, reaching toward the Sun, are stems and leaves. Together, stems and leaves make up a shoot system. These two systems work together to get a plant what it needs to survive.

A plant's root system can be as extensive as the stems and leaves that you see aboveground. Roots absorb water and nutrients from the soil. These materials are transported to the leaves through the stems. The leaves use the materials, along with carbon dioxide from the air, to make sugars and carbohydrates. The stems then deliver these energy-rich compounds back to the rest of the plant.

REMINDER

Plants make sugar through *photosynthesis*. The sugars that are not used immediately are stored as carbohydrates, energy-rich compounds.

CHECK YOUR READING What are two plant systems, and what are their functions?

Transporting Water and Other Materials

Stems serve as the pathway for transporting water, nutrients, and energy-rich compounds from one part of a plant to another. In most plants, the materials move through a **vascular system** (VAS-kyuh-lur) that is made up of long, tubelike cells. These tissues are bundled together and run from the roots to the leaves. A vascular bundle from the stem of a buttercup plant is shown.

Transport is carried out by two types of tissue. Xylem (ZY-luhm) is a tissue that carries water and dissolved nutrients up from the roots. Phloem (FLOH-em) is a tissue that transports energy-rich materials down from the leaves. Xylem cells and phloem cells are long and hollow, like pipes. The xylem cells are a little larger than the phloem cells. Both tissues include long fibers that help support the plant body, as well as cells that can store extra carbohydrates for energy.

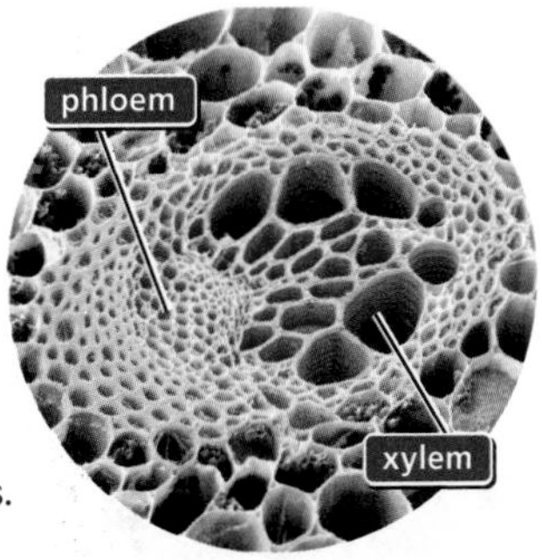

This vascular bundle has been magnified 113×.

Vascular System

The vascular system transports materials throughout a plant's body.

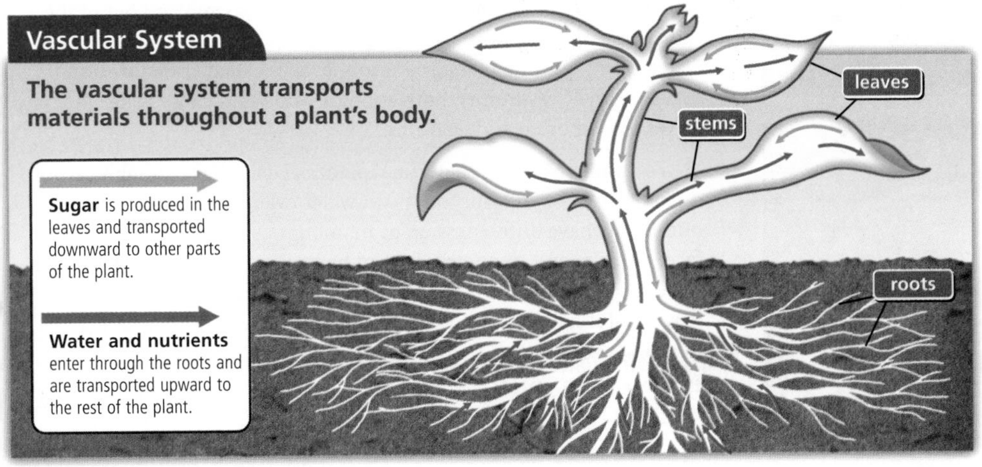

Sugar is produced in the leaves and transported downward to other parts of the plant.

Water and nutrients enter through the roots and are transported upward to the rest of the plant.

Real World Example

The oldest trees on Earth are the bristlecone pines. Some bristlecones are over 4000 years old. The oldest known tree is nicknamed Methuselah, estimated to be over 4600 years old. The tree's exact location in the Inyo National Forest in California is kept secret by the U.S. Forest Service. The oldest trees are not necessarily the largest. Bristlecone pines grow very slowly. Their age can be determined by the number of growth rings they have.

Teach from Visuals

To help students interpret the diagram of vascular systems in plants, ask:

- In what parts of a plant are sugars made? *the green parts of a plant, its leaves and stems*
- Through which parts of a plant do water and some trace nutrients enter? *roots*

Teach Difficult Concepts

Tell students that plants need sugar for survival. Sugar, which is a molecule that contains carbon, is an essential nutrient for plant growth and function. Some other nutrients necessary for plants, such as nitrogen, enter through the roots. The bulk of material needed for a plant's growth is from water (absorbed from roots) and carbon dioxide (taken in from the air).

Ongoing Assessment

Identify common characteristics of plants.

Ask: What are two ways in which plant cells are all alike? *They each have a nucleus and a cell wall.*

CHECK YOUR READING *Answer: A root system brings in water and nutrients; a shoot system made up of leaves and stems makes energy-rich materials and moves material up and down through the plant.*

DIFFERENTIATE INSTRUCTION

More Reading Support

C What compounds are made in the leaves of plants? *sugars, carbohydrates*

D What tissues are in a plant's vascular system? *xylem, phloem*

Advanced Provide students with a sampling of leaves and a field guide for plants in your area. Challenge students to use the field guide to identify the plants. Ask students to describe the various criteria used to identify the plants.

R Challenge and Extension, p. 159

Have students who are interested in how plants use water read the following article:

R Challenge Reading, pp. 194–195

Address Misconceptions

IDENTIFY Ask students about where plants get sugars. Students may hold the misconception that plants get food from the soil.

CORRECT Ask students to study the formula for photosynthesis. Explain that the Sun provides energy for a series of chemical reactions that take place in plant cells. Plants do not eat "food," instead they make their own sugar. Ask students to name the raw materials and products of photosynthesis. *raw materials: carbon dioxide, water, and sunlight; products: sugars and oxygen*

REASSESS Sketch a tiny seedling and a tall tree on the board. Ask: Where did this plant get the materials to grow from so tiny to so big? *It absorbed water with its roots and used the Sun's energy to synthesize sugar in its leaves.*

Technology Resources

Visit **ClassZone.com** for background on common student misconceptions.

MISCONCEPTION DATABASE

History of Science

Many scientists have contributed to the understanding of how plants get the materials they need. It began in the mid 1600s. Belgian chemist Jan Baptista van Helmont grew a tree in a carefully measured amount of soil and gave it only water over a five-year period. Since the weight of the soil remained relatively unchanged, he concluded that plants do not "eat" soil.

Ongoing Assessment

CHECK YOUR READING *Answer: Photosynthesis is a series of chemical reactions that use energy from the Sun to produce sugars. It is important since this is how plants get their energy.*

CHECK YOUR READING *Answer: The plant's cuticle and stomata keep it from drying out.*

Making Sugars

Plants produce sugars through the process of photosynthesis. Photosynthesis is a series of chemical reactions that capture light energy from the Sun and convert it into chemical energy stored in sugar molecules. The starting materials needed are carbon dioxide, water, and light. The end products are sugars and oxygen. The chemical reaction for photosynthesis can be written like this:

carbon dioxide + water + sunlight → sugars + oxygen

?E Photosynthesis takes place in chloroplasts, structures that contain chlorophyll. Chlorophyll, the chemical that gives plants their green color, produces the reaction. Most chloroplasts in a plant are located in leaf cells. As you can see from the illustration on page 89, the structure of the leaf is specialized for capturing light energy and producing sugar.

CHECK YOUR READING What is photosynthesis, and why is it important?

?F The upper surface of the leaf, which is turned toward the Sun, has layers of cells filled with chloroplasts. Vascular tissue located toward the center of the leaf brings in water and nutrients and carries away sugars and carbohydrates. Tiny openings at the bottom of the leaf, called stomata (STOH-muh-tuh), lead to a network of tiny spaces where gases are stored. The carbon dioxide gas needed for photosynthesis comes in through the stomata, and oxygen gas moves out. This process is called gas exchange.

Open Stomata

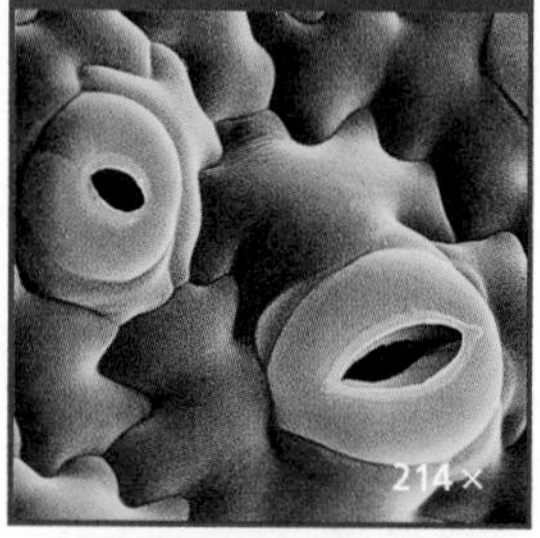

Closed Stomata

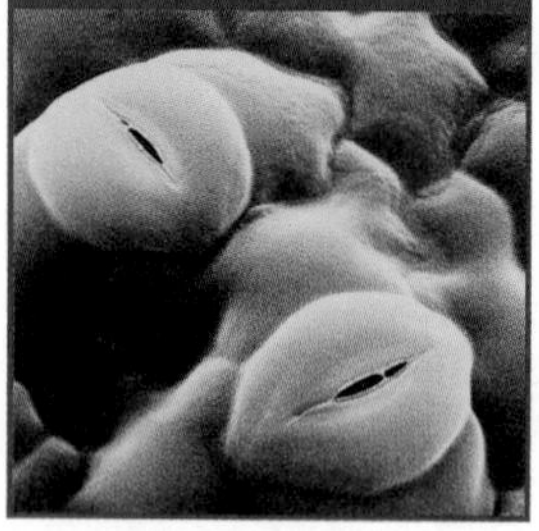

Controlling Gas Exchange and Water Loss

For photosynthesis to occur, a plant must maintain the balance of carbon dioxide and water in its body. Carbon dioxide gas from the air surrounding a plant enters through the stomata in its leaves. Open stomata allow carbon dioxide and oxygen to move into and out of the leaf. These openings also allow water to evaporate. The movement of water vapor out of a plant and into the air is called **transpiration** (TRAN-spuh-RAY-shuhn). Both sunlight and wind cause water in leaves to evaporate and transpire.

For photosynthesis to occur, a plant needs to have enough carbon dioxide come in without too much water evaporating and moving out. Plants have different ways of maintaining this balance. The surface of leaves and stems are covered by a waxy protective layer, called a cuticle. The cuticle keeps water from evaporating. Also, when the air is dry, the stomata can close. This can help to prevent water loss.

CHECK YOUR READING What are two ways plants have to keep from losing too much water?

DIFFERENTIATE INSTRUCTION

More Reading Support

E Where does photosynthesis take place? *chloroplasts*

F Through which structures in the leaf do gases enter and leave the plant? *stomata*

Below Level To help students remember the process of photosynthesis, have them write a recipe. Students can list the "main ingredients" (carbon dioxide, water, and sunlight). Make sure students include process steps as well as ingredients in their recipe: the main ingredients are "combined," or "mixed," inside a chloroplast. The results yield sugars and oxygen.

Advanced Have a small group of advanced students design an experiment to recreate Priestley's experiment to show that plants release gas. They can perform the experiment with adult supervision.

C 88 Unit: Diversity of Living Things

Inside a Leaf

The leaf is an organ that produces sugars. It is made up of different types of cells and tissues.

Cells at the surface produce a waxy cuticle that keeps the leaf from losing water.
Most chloroplasts are located in cells of the upper layer of the leaf.
Xylem transports water and nutrients up from the roots.
Phloem transports energy-rich compounds made in the leaf down to other parts of the plant.
Carbon dioxide, oxygen, and water vapor move into and out of the leaf through stomata.

READING VISUALS How is the top of a leaf different from the bottom of a leaf?

DIFFERENTIATE INSTRUCTION

Below Level Ask students to describe in their own words the processes of photosynthesis and transpiration, using the main concepts presented in the visual.

English Learners Have students use a dictionary to find the origins of the word parts in *chloroplast* and *chlorophyll.* You might also write related words, such as *plastic, phyllo dough,* and *chlorine,* on the board. Chloro- *comes from a Greek word meaning* green; phyll- *comes from the Greek word for* leaf, *and* plast- *comes from the Greek word for* molded.

Teach from Visuals

To help students interpret the diagram of leaf structure, ask:

- Through which structure of the leaf does transpiration take place? *stomata*
- Where are most of the chloroplasts in the leaf located? Why do you think this is so? *In the upper layer; the chloroplasts need to be exposed to sunlight for photosynthesis to take place.*

Teaching with Technology

If pressure sensor probeware is available, have students use it to measure pressure in stems and leaves of a classroom plant. They can use a graphing calculator to show the data visually. Then ask them to use their data to infer which times had faster or slower transpiration rates.

Teacher Demo

Obtain two long balloons and inflate them. Use the balloons to demonstrate the opening and closing of stomata. Ask: Under what conditions do stomata open and close? *Stomata open to allow oxygen in and carbon dioxide and water vapor out; stomata close under dry conditions to prevent water loss.*

Language Arts Connection

Students can write or narrate the story of imaginary travel through the structure of a leaf. Have them describe what cells are at the surface, and what structures such as xylem, phloem, and stomata look like and do. Assign a choice of formats for the story such as poem, newscast, letters, diary, or comic strip.

Ongoing Assessment

Describe how the bodies of plants are organized.

Ask: Which tissue in a leaf transports sugars and runs through all parts of the plant? *phloem*

READING VISUALS *Answer: The top of a leaf has a waxy covering (cuticle) and more cells with chloroplasts; the bottom of the leaf has stomata.*

Develop Critical Thinking

INFER Direct attention to the photograph of the prickly pear cactus. Ask: How might the spines be a benefit to the cactus other than being a way to minimize water loss? ***The spines keep it from being eaten by consumers.***

Integrate the Sciences

Colors of the visible spectrum include red, orange, yellow, green, blue, and violet. Most objects reflect some of the colors of the spectrum while they absorb other colors. When light strikes the leaves of a plant, the green waves of the spectrum are reflected, giving the leaves their green color. All of the other colors are absorbed.

Ongoing Assessment

CHECK YOUR READING *Answer: Plant stems can 1) store food; 2) provide support for the plant body; and 3) transport materials via the vascular system. Some plants, such as cacti, perform photosynthesis in their stems.*

APPLY Plant stems branch as they grow. Do cactus plants have branches?

The stomata are an adaptation that allows a plant to adjust to daily changes in its environment. Plants can respond to hot, dry weather by keeping their stomata closed. Stomata can be open during the night, when evaporation is less likely to occur. Most plants have stomata. G

Some species of plants have special adaptations for survival in a particular environment. For example, a cactus plant has adaptations that allow it to survive in a desert. The spines of a cactus are actually modified leaves. A plant with regular leaves would lose too much water through transpiration. In cacti, most photosynthesis occurs in the thick fleshy stem, where the cactus also stores water and carbon dioxide gas.

Plants grow throughout their lifetimes.

Plants grow for as long as they live. This is true for plants that live for only one season, such as sunflowers, and for plants that can live for many years, such as trees. Plants grow bigger when cells at the tips of their roots and stems divide and multiply more rapidly than other plant cells do. A plant's roots and stems can grow longer and thicker and can branch, or divide. However, only stems grow leaves. Leaves grow from buds produced by growth tissue in a plant's stems. The bud of an oak tree is shown on page 91.

Plant stems are structures with more than one function. You have read that a plant's stem includes its vascular system, which allows the plant to transport materials between its leaves and roots. Long stiff fibers in the tissues of the vascular system provide support and give the plant shape. Plant stems can also store the sugars produced by photosynthesis. Many plants, including broccoli, celery, and carrots, convert sugars into carbohydrates and then store this energy-rich material in their stems or roots.

A CHECK YOUR READING What are three functions of plant stems?

Plants with Soft Stems

RESOURCE CENTER
CLASSZONE.COM
Learn more about plant systems.

The soft stems and leaves of many weeds, wildflowers, garden flowers, and vegetables die when the environment they live in becomes too cold or too dry. This type of plant survives by using the carbohydrates stored in its roots. Then, when the environment provides it with enough warmth, water, and sunlight, the plant will grow new, soft, green stems and leaves.

DIFFERENTIATE INSTRUCTION

More Reading Support

G Describe how stomata respond to hot, dry weather. *They stay closed during the day and may open at night when it is cooler.*

English Learners Pair students who are new to English with partners who are advanced. Have them use web sites, botany books, or field guides to prepare a booklet of Unusual Plant Adaptations. Encourage each pair to present a picture, diagram, or illustration of the plant's unusual structure(s). They should include a sentence on how the structure benefits the plant or suits it to its environment. Suggest students research one of the following plants to begin: staghorn, rafflesia, mangrove, Venus flytrap, and bristlecone pine.

Plant Growth

Plants, such as these oak trees, grow most when there is enough warmth, water, and sunlight.

The tips of the shoots produce buds, which become new leaves and stems.

Plants with Woody Stems

Some plants, such as trees and shrubs, have tough, thick stems that do not die each year. These stems keep growing longer and thicker. As the stems grow, they develop a type of tough xylem tissue that is not found in soft stems. This tough xylem tissue is called wood. The growing tissues in woody stems are located near the outer surface of the stem, right under the bark. This means that, for a tree like one of the oaks in the photograph above, the center of the trunk is the oldest part of the plant.

3.1 Review

KEY CONCEPTS

1. What characteristics do all plants have in common?
2. How does the structure of a leaf relate to its function?
3. What tissues move materials throughout a plant?

CRITICAL THINKING

4. **Summarize** Describe how the structure of a leaf allows a plant to control the materials involved in photosynthesis.
5. **Analyze** Do you think the stems of soft-stemmed plants have chloroplasts? How about woody-stemmed plants? Explain your reasoning. Hint: Think about the color of each.

CHALLENGE

6. **Evaluate** Scientists who study the natural world say that there is unity in diversity. How does this idea apply to plants?

ANSWERS

1. Plants are multicellular with cells that have a nucleus and a cell wall. They are producers and have a two-stage life cycle.

2. A leaf is suited for photosynthesis, having cells with many chloroplasts at its surface. Xylem brings in water. Stomata let in carbon dioxide. The cuticle keeps water in. Phloem transports sugars.

3. xylem and phloem

4. Stomata regulate the amount of carbon dioxide, oxygen, and water vapor that enters and exits the plant.

5. Soft-stemmed plants have green stems, with cells that have chloroplasts. Woody stems are brown or grey; the cells do not have chloroplasts.

6. Plants differ in height. Some are soft-stemmed. Some are woody. They live in many different environments. Yet the basic structure is the same: roots, stems, leaves. All rely on photosynthesis to get needed energy and materials.

Reinforce the BIG idea

Have students relate the section to the Big Idea.

R Reinforcing Key Concepts, p. 160

3.1 ASSESS & RETEACH

Assess

A Section 3.1 Quiz, p. 43

Reteach

Have students create comic strips that describe the process of photosynthesis. Tell students to include in their comic strips the following key points, which can be provided before they begin:

- where photosynthesis occurs: *in chloroplasts*
- where raw materials come from: *sunlight and air*
- what the raw materials are: *carbon dioxide, water, and light energy*
- how the raw materials get into the leaf: *CO_2 through stomata, light absorbed by chloroplasts, water by roots*
- what the end products are: *sugar and oxygen*
- how the end products move from the leaf to other parts of the plant or out into the air: *Sugars are carried by phloem to all parts of plant; oxygen is released through stomata.*

Technology Resources

Have students visit **ClassZone.com** for reteaching of Key Concepts.

CONTENT REVIEW

CONTENT REVIEW CD-ROM

3.2 FOCUS

Set Learning Goals

Students will

- Explain how plants are adapted to live on land.
- Describe the reproductive cycle of mosses.
- Describe the reproductive cycles of ferns.
- Measure distance to infer how much sunlight reaches an organism.

3-Minute Warm-Up

Display Transparency 20 or copy this exercise on the board:

Complete the chart to describe the characteristics of plants.

Plant Adaptations	
Characteristics	*multicellular, cell walls, nuclei, two-stage life cycles*
Functions	*photosynthesis, transpiration*
Structures	*roots, stems, leaves*

3-Minute Warm-Up, p. T20

3.2 MOTIVATE

EXPLORE Moss Plants

PURPOSE To observe the structures of moss plants

TIP *15 min.* Growing moss plants can be purchased from biological supply companies. If possible, obtain "sporulating" moss so that students can observe the spore-making stalk-and-capsule structures (sporophytes) as well as the more familiar leafy green parts (gametophytes).

WHAT DO YOU THINK? *Sample answer: Students observations may include noticing that typical moss has leaf-like structures that branch off a main stem. These structures are often arranged in a spiral around the central stem. Moss plants are not very tall and the plant bodies are only a few millimeters thick.*

KEY CONCEPT

3.2 Most mosses and ferns live in moist environments.

BEFORE, you learned

- All plants share common characteristics
- The body of a plant has specialized parts
- Plants grow throughout their lifetimes

NOW, you will learn

- About the first plants
- About reproduction in nonvascular plants, such as mosses
- About reproduction in vascular plants such as ferns

EXPLORE Moss Plants

What do moss plants look like?

PROCEDURE

1. Use a hand lens to examine a moss plant. Look for different structures and parts you can identify.
2. Draw a diagram of the moss plant in your notebook. Label parts you identified and parts you would like to identify.
3. Write a brief description of each parts function.

MATERIALS

- live moss plant
- hand lens

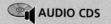

WHAT DO YOU THINK?

- How would you describe a moss plant to someone who had never seen one?
- How does a moss plant compare with the other plants you are familiar with?

Plant species adapted to life on land.

Life first appeared on Earth about 3.8 billion years ago. Tiny single-celled and multicellular organisms lived in watery environments such as warm, shallow seas, deep ocean vents, and ponds. Fossil evidence suggests that plant life did not appear on land until about 475 million years ago. The ancestors of the first plants were among the first organisms to move onto land.

What did these plantlike organisms look like? Scientists think they looked much like the green algae you can find growing in watery ditches or shallow ponds today. Both green algae and plants are autotrophs, or producers. Their cells contain chloroplasts that enable them to convert the Sun's light energy into the chemical energy stored in sugars.

RESOURCE CENTER
CLASSZONE.COM
Explore plant evolution.

RESOURCES FOR DIFFERENTIATED INSTRUCTION

Below Level

UNIT RESOURCE BOOK

- Reading Study Guide A, pp. 163–164
- Decoding Support, p. 198

AUDIO CDS

Advanced

UNIT RESOURCE BOOK

Challenge and Extension, p. 169

English Learners

UNIT RESOURCE BOOK

Spanish Reading Study Guide, pp. 167–168

AUDIO CDS

- Audio Readings in Spanish
- Audio Readings (English)

The First Plants

Suppose that hundreds of millions of years ago, the area now occupied by your school was a shallow pond full of tiny, floating organisms that could photosynthesize. The Sun overhead provided energy. The pond water was full of dissolved nutrients. The organisms thrived and reproduced, and over time the pond became crowded. Some were pushed to the very edges of the water. Then, after a period of dry weather, the pond shrank. Some organisms at the edge were no longer in the water. The ones that were able to survive were now living on land.

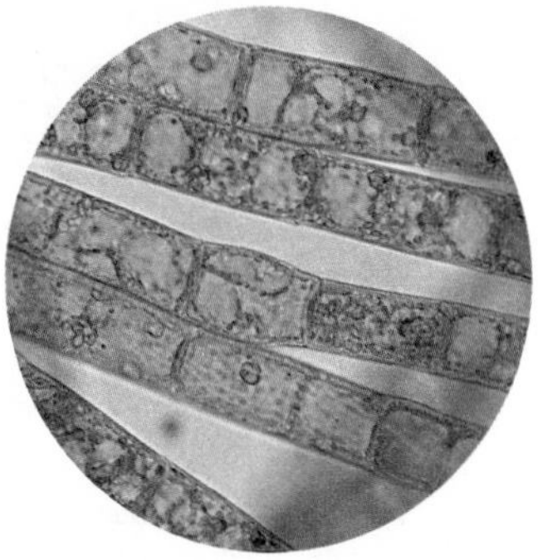

Scientists think the first plants shared a common ancestor with green algae, shown here magnified about 80×.

Scientists think that something like this took place in millions of watery environments over millions of years. Those few organisms that were stranded and were able to survive became ancestors to the first plants. Life on land is very different from life in water. The first plants needed to be able to get both nutrients and water from the land. There is no surrounding water to provide support for the body or to keep body tissues from drying out. However, for organisms that survived, life on land had many advantages. There is plenty carbon dioxide in the air and plenty of direct sunlight.

CHECK YOUR READING Why is having plenty of sunlight and water an advantage for a plant? Your answer should mention photosynthesis.

Mosses and Ferns

A Among the first plants to live on Earth were the ancestors of the mosses and ferns you see today. Both probably evolved from species of algae that lived in the sea and freshwater. Mosses are simpler in structure than ferns. Mosses, and two closely related groups of plants known as liverworts and hornworts, are descended from the first **B** plants to spread onto the bare rock and soil of Earth. Ferns and their relatives appeared later.

This diorama shows what a forest on Earth might have looked like about 350 million years ago.

DIFFERENTIATE INSTRUCTION

More Reading Support

A Which plants most likely first appeared on land? *mosses*

B Which plants appeared after mosses? *ferns*

Advanced Have students make their own dioramas similar to the Field Museum diorama of a swamp forest in the prehistoric Carboniferous Era shown on this page. They can work from the photographs on pp. 93–97 to plan materials that resemble the mosses, ferns, liverworts, and hornworts of 350 million years ago. They should add a caption to their display that summarizes their understanding of "Earth's First Plants."

R Challenge and Extension, p. 169

3.2 INSTRUCT

Integrate the Sciences

The first plants to invade the land probably appeared during the Ordovician Period of the Paleozoic Era (505 to 438 million years ago). The oldest fossil evidence of plants is of liverworts from the rocks of the Devonian Period, about 400 million years ago.

Teach from Visuals

To help students interpret the diorama of forests growing 350 million years ago, ask: How might an artist know what plants looked like millions of years ago? ***Fossil evidence would tell scientists about the structure and size of plants.***

Real World Example

Point out that the illustration shows a diorama from the Field Museum of Natural History in Chicago. It is an artist's depiction of a swamp forest of the Carboniferous Era (about 350 million years ago). The plants that look like trees are giant club mosses, which are a type of fern. The plants in the foreground in front of the fallen log are ancestors of horsetails, similar to the ones in the photograph on p. 86.

Ongoing Assessment

Describe how the first plants appeared on land.

Ask: From which organisms did the first land plants evolve? *algae*

CHECK YOUR READING *Answer: The more sunlight and water available to the plant, the more energy-rich sugars the plant can produce through photosynthesis.*

INVESTIGATE Capturing the Sun's Energy

PURPOSE To explore how much sunlight reaches an organism in water

TIPS *20 min.* Before the investigation, attach the button to the string and lower it into the water. The button should not float. When the kelp granules are added, part of the water will be obscured. Explain that the kelp represents organisms floating in the ocean.

WHAT DO YOU THINK? *The second distance was shorter. On land, all organisms are "on the surface" of Earth and so have generally better access to sunlight.*

CHALLENGE *The kelp granules represent organisms for resources. On land, where there is generally better access to resources such as sunlight, there is less competition for such resources.*

R Datasheet, Capturing the Sun's Energy, p. 170

Technology Resources

Customize this student lab as needed or look for an alternative. Print rubrics to assess student lab reports.

Lab Generator CD-ROM

Develop Critical Thinking

SYNTHESIS Write the term *osmosis* on the board and explain to students that this is the process by which water moves through a membrane such as a cell membrane. Ask: If there is a higher concentration of water molecules outside a cell membrane than on the inside of the cell membrane, what will happen? *The water will move through the membrane to the area of lesser concentration.* How does osmosis relate to nonvascular plants? *Nonvascular plants such as mosses transport water from one part of the plant to another by the process of osmosis.*

Ongoing Assessment

CHECK YOUR READING *Answer: Mosses are nonvascular plants. Water passes through their bodies cell by cell. This limits their size because their stems can be only a few cells thick.*

INVESTIGATE Capturing the Sun's Energy

How much sunlight reaches an organism living in water?

SKILL FOCUS
Measuring

MATERIALS
- white button
- string
- empty clear plastic bottle
- water
- ruler
- kelp granules
- tablespoon

TIME
20 minutes

PROCEDURE

1. Thread the string through the holes in the button so that the button hangs flat. Fill the empty bottle with clean water.
2. Look down through the top of the bottle. Lower the button into the water until it either disappears from view or reaches the bottom. Have a classmate measure how far the button is from the surface of the water.
3. Add two spoonfuls of kelp granules to the water. Repeat step 2.

WHAT DO YOU THINK?

- How did the distance measured the second time compare with your first measurement?
- Why might a photosynthetic organism living on land get more sunlight than one living in water?

CHALLENGE What do the kelp granules in this experiment represent? What does that suggest about the advantages of living on land?

Mosses are nonvascular plants.

Moss plants have adaptations for life on land. Mosses have simple roots, stems, and leaves. Moss cells also have special areas for storing water and nutrients. Each moss cell, like all plant cells, is surrounded by a thick wall that provides it with support. Mosses do not grow very large, but there are many species of them, and over millions of years they have adapted to survive in many environments.

C If you look closely at a clump of moss, you will see that it is actually made up of many tiny, dark green plants. Mosses belong to a group called the nonvascular plants. Nonvascular plants do not have vascular tissue. Water and nutrients simply move through nonvascular plants' bodies cell by cell. A plant can get enough water this way as long as its body is no more than a few cells thick.

CHECK YOUR READING What limits the size of moss plants?

D Water also plays a part in the reproductive cycle of a moss plant. In the first part of the cycle, the moss grows and maintains itself, producing the male and female structures needed for sexual reproduction. If conditions are right and there is enough water, the plant enters a spore-producing stage, the second part of the cycle.

DIFFERENTIATE INSTRUCTION

More Reading Support

C Why are mosses described as nonvascular plants? *They do not have vascular tissues.*

D Which reproduction involves male and female structures? *sexual*

English Learners The investigation on p. 94 uses imperative sentences. The word "you" is eliminated in the directions. Make sure English learners know how to read and follow these directions. For example, the direction "Look down through the top of the bottle," is to be understood as "You look down through the top of the bottle."

Mosses reproduce with spores.

Mosses, ferns, and fungi all reproduce with spores. Spores are an important adaptation that allowed the ancestors of these organisms to reproduce on land. A spore is a single reproductive cell that is protected by a hard, watertight covering. The covering prevents the cell from drying out. Spores are small and can be transported through the air. This means offspring from spores can grow in places that are distant from the parent organisms.

REMINDER
Sexual reproduction involves two processes: fertilization and meiosis.

The green moss plants you are familiar with have grown from spores. They represent the first generation. Within a clump of moss are both male and female reproductive structures. When conditions are right, these structures produce sperm and eggs. Fertilization can occur only if water is present because the tiny moss sperm move by swimming. A layer of water left by rain is one way sperm can move to the eggs on another part of the plant.

The fertilized egg grows into a stalk with a capsule on the end—the second generation of the plant. The stalk and capsule grow from the female moss plant. Inside the capsule, the process of meiosis produces thousands of tiny spores. When the spores are released, as shown in the photograph, the cycle can begin again.

Moss Releasing Spores

IDENTIFY Point out the two generations of the moss plant shown here.

Address Misconceptions

IDENTIFY Ask students if a moss is more like a rock or an animal. Students may hold the misconception that plants are not living things. They may say plants have more in common with a rock because neither can move around. However plants have much more in common with other living things, such as animals, than they do with nonliving things, such as rocks.

CORRECT Have students recall what characteristics and needs are shared by all living things. *characteristics: grow, reproduce, have an organized structure, use energy; needs: water, oxygen, nutrients, energy, living space*

REASSESS Ask students if mosses fit this description. *Mosses reproduce, grow, and need water and nutrients.*

Technology Resources

Visit **ClassZone.com** for background on common student misconceptions.

MISCONCEPTION DATABASE

Teach Difficult Concepts

Students may not understand the concept of two generations, the idea that a plant takes two different forms during its reproductive cycle. Ask:

- What does a moss plant grow from? *spores*
- How are spores produced? *Fertilization must occur in the moss plant, producing a stalk and capsule. Spores are produced by meiosis in the capsule.*
- What two forms does a moss take during its reproductive cycle? *moss plant and stalk with capsule*

DIFFERENTIATE INSTRUCTION

More Reading Support

E What do male moss plants produce in order to reproduce sexually? *sperm*

F What does the fertilized egg of a moss plant grow into? *a stalk with a capsule on the end*

Advanced Pair students to create a diagram to show the two-part life cycle of mosses. Students should label stages and identify the generation to which each stage belongs. They should use the last two paragraphs and photograph on p. 95 to break out steps of the cycle and relate them in a repeating sequence. *FIRST GENERATION: 1. Capsule releases spores. 2. Spores grow into green moss plant with male and female structures. 3. Sperm fertilizes egg. SECOND GENERATION: 4. Stalk and capsule grow from egg. 5. Inside capsule meiosis produces spores.*

Ongoing Assessment

PHOTO CAPTION Answer: The green clumps are the first generation; the stalk and capsule are the second generation.

Teacher Demo

You can demonstrate how mosses can grow by asexual reproduction. Take moss and crumble it in a blender. Mix 1 part moss to 2 parts buttermilk (for nutrients). Spread the mixture on some rocks in a terrarium and keep moist and shady (3–6 hours of sunlight per day). The moss cells in the mix will grow into new moss plants in about 8 weeks.

Develop Critical Thinking

INFER Direct attention to the photographs of nonvascular and vascular plants. Ask: What is different about the arrangement of cells in the fern compared to those in a moss? *The fern has vascular tissue, that is groups of cells working together to move nutrients and water throughout the plant. The moss does not.*

Teach from Visuals

To help students understand the photograph on p. 97 of the fern and its spores, ask: What part of the fern life cycle is shown? How do you know? *The second part of the life cycle is shown because the leafy frond has developed spore-producing clusters.*

Ongoing Assessment

Describe the function of vascular tissue.

Ask: What are two functions of vascular tissue that enable vascular plants to grow larger than nonvascular plants? *Vascular tissue moves nutrients and water more quickly and also provides support.*

CHECK YOUR READING *Answer: In sexual reproduction the genetic material from two parents comes together producing offspring with a mix of genetic material. In asexual reproduction, the genetic material of the offspring is the same as the parent's.*

READING VISUALS *Answer: The penny and the person in the photographs show the difference in scale between the two plants.*

G Mosses, like other plants, can also reproduce asexually. A small piece of a moss plant can separate and can grow into a new plant, or new plants can branch off from old ones. Asexual reproduction allows plants to spread more easily than sexual reproduction. However, the genetic material of the new plants is the same as that of the parent. Sexual reproduction increases genetic diversity and the possibility of new adaptations.

CHECK YOUR READING Compare and contrast sexual and asexual reproduction. Your answer should mention *genetic material.*

Ferns are vascular plants.

H Ferns, and two closely related groups of plants known as horsetails and club mosses, were the first plants on Earth with vascular systems. The tubelike tissue of a vascular system moves water through a plant's body more quickly than when water moves cell by cell. Because of this, vascular plants can grow much larger than nonvascular plants. Vascular tissue also provides support for the weight of a larger plant.

The presence of vascular tissue has an effect on the development of roots, stems, and leaves. The root system can branch out more, anchoring a larger plant as well as providing water and nutrients. Vascular tissue moves materials more efficiently and gives extra support. The stems can branch out and more leaves can grow. This results in more sugars and other materials needed for energy and growth.

Nonvascular

Liverworts are tiny nonvascular plants. The liverworts shown here are life-size.

Vascular

Ferns are vascular plants. Notice how tall a tree fern can grow.

READING VISUALS COMPARE AND CONTRAST How do the penny and the person help to show that the tree ferns are much larger than the liverworts?

DIFFERENTIATE INSTRUCTION

More Reading Support

G How are mosses able to reproduce? *asexually and sexually*

H What were the first plants with vascular systems? *ferns*

Below Level Have pairs of students create a two-column chart to compare mosses and ferns. Each student should complete one column of the chart for each plant. Students then highlight in red how the plants are similar. Then they highlight in blue how the plants are different.

Ferns reproduce with spores.

You may have seen ferns growing in the woods or in a garden. The leaves of ferns, called fronds, are often included in a flower bouquet. The next time you have a chance to look at a fern frond, take a look at the back. You will probably see many small clusters similar to those shown to the right. The clusters are full of spores.

As fern fronds grow, they produce clusters of spores on the back of the fronds.

Ferns, like mosses, have a two-part life cycle. In ferns, spores grow into tiny structures that lie very close to the ground. You would have to look closely to find these structures on the ground, for they are usually smaller than the size of your thumbnail. Within these structures are the sperm- and egg-producing parts of the fern plant. This is the first generation of the plant. Like mosses, the sperm of a fern plant need water to swim to the egg. So fertilization occurs only when plenty of moisture is present.

The second part of a fern life cycle is the plant with fronds that grows from the fertilized egg. As the fronds grow, the small egg-bearing part of the plant dies away. The fronds produce clusters, and cells within those clusters undergo meiosis and produce spores. The more the fern grows, the more clusters and spores it produces. The spores, when they are released, spread through the air. If conditions are right where the spores land, they grow into new fern plants and a new cycle begins. This is sexual reproduction. Ferns, like mosses, also reproduce asexually. New ferns branch off old ones, or pieces separate from the plant and grow.

CHECK YOUR READING Explain one way that sexual reproduction in ferns is similar to reproduction in mosses. Explain one way it is different.

3.2 Review

KEY CONCEPTS

1. For the ancestors of the first plants, what were some advantages to living on land?
2. What are three adaptations that make mosses able to live on land?
3. What are two characteristics you can observe that distinguish vascular plants from nonvascular plants?

CRITICAL THINKING

4. **Synthesize** Vascular plants such as ferns can grow bigger and taller than nonvascular plants such as mosses. Does this mean they can also capture more sunlight? Explain your answer.
5. **Compare** Make a chart that shows how the life cycles of mosses and ferns are different, and how they are similar.

CHALLENGE

6. **Evaluate** Consider the conditions that are needed for mosses and ferns to reproduce sexually. Sexual reproduction increases genetic diversity within a group of plants, while asexual reproduction does not. Explain why asexual reproduction is still important for both moss and fern plants.

ANSWERS

1. They had easier access to carbon dioxide and sunlight.

2. Mosses have simple roots, stems, leaves; can store water and nutrients; and produce spores.

3. Vascular plants can grow taller and have larger stems and leaves.

4. The larger the plant, the more chloroplasts it has to capture energy.

5. Differences: main moss plant produces eggs and sperm, main fern plant produces spores; similarities: two-part life cycle, have spores, need water

6. Asexual reproduction allows plants to spread more quickly. Plants can reproduce under conditions that are not favorable to sexual reproduction.

Ongoing Assessment

Describe the life cycles of ferns.

Ask: Where are the spore-producing clusters found on ferns, and what do the spores do? *On the underside of the leafy part of the fern, known as a frond; the spores grow into tiny structures as one stage of the fern life cycle.*

CHECK YOUR READING *Sample answer: Similarities: two-part life cycles, sperm that swim through water, special spore making structures, spores grow into new plants. Difference: the fern's dominant stage is its spore-producing stage.*

Reinforce the BIG idea

Have students relate the section to the Big Idea.

R Reinforcing Key Concepts, p. 171

3.2 ASSESS & RETEACH

Assess

A Section 3.2 Quiz, p. 44

Reteach

Write the terms on the board. For each term, have students tell if it relates to moss reproduction, fern reproduction, or both. Then have them tell if the term occurs in the first or second generation.

1. spore *both, second generation*
2. sperm *both, first generation*
3. egg *both, first generation*
4. fertilization *both, first generation*
5. capsule *moss, second generation*
6. stalk *moss, second generation*
7. frond *fern, second generation*

Technology Resources

Have students visit **ClassZone.com** for reteaching of Key Concepts.

CONTENT REVIEW

CONTENT REVIEW CD-ROM

3.3 FOCUS

Set Learning Goals

Students will

- Discuss the advantages that plants with pollen and seeds have over non-seed plants.
- Describe how some plants reproduce with pollen and seeds.
- Observe conditions that make a pinecone open.

3-Minute Warm-Up

Display Transparency 21 or copy this exercise on the board:

Predict what will happen:

A group of moss plants live in a moist area. The area experiences a period of drought. How does this affect the moss plants? Explain.

The moss plants will not be able to reproduce because the sperm need water in which to reach the eggs of the female plant. The plants may die.

3-Minute Warm-Up, p. T21

3.3 MOTIVATE

THINK ABOUT

PURPOSE To understand how seeds are adapted for survival and reproduction

DISCUSS Invite students to describe what they see in the photograph. Explain that the seed and the plant embryo it contains can become a living organism. The thousand-year-old lotus seeds were able to survive because the structure of the seed provides a protective coating and food for the embryo. This allows the embryo to remain dormant (capable of living but not growing or acting) until the conditions for growth or activity are right.

KEY CONCEPT

3.3 Seeds and pollen are reproductive adaptations.

BEFORE, you learned

- Plant species evolved from algaelike ancestors
- Mosses are nonvascular plants that reproduce with spores
- Ferns are vascular plants that reproduce with spores

NOW, you will learn

- How some plants reproduce with pollen and seeds
- About the advantages of pollen and seeds

VOCABULARY

THINK ABOUT

Is a seed alive?

A lotus is a type of pond lily that is commonly found in water gardens. The plants take root in the bottom of a pond. A plant scientist in California experimented with lotus seeds from China that were over 1000 years old. The scientist made a small opening in the hard covering of each seed and planted the seeds in wet soil. Some of the seeds sprouted and grew. What made it possible for these seeds to survive for such a long time? Is a seed alive?

Seeds are an important adaptation.

Spores are one adaptation that made it possible for plants to reproduce on land. Seeds are another. A **seed** is a young plant that is enclosed in a protective coating. Within the coating are enough nutrients to enable the plant to grow. Seeds and spores can both withstand harsh conditions. Seeds, however, have several survival advantages over spores. These advantages make it possible for seed plants to spread into environments where seedless plants are less likely to survive.

VOCABULARY
Remember to make word triangles for *seed* and *embryo* in your notebook.

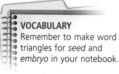

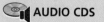

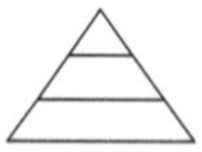

In seedless plants, such as mosses and ferns, fertilization brings about the growth of the next generation of the plant. In seed plants, there is a step in between. Fertilization brings about the growth of an embryo. An **embryo** (EHM-bree-OH) is the immature form of an organism that has the potential to grow and develop. The seed protects the plant embryo until conditions are right for it to grow.

RESOURCES FOR DIFFERENTIATED INSTRUCTION

Below Level

UNIT RESOURCE BOOK

- Reading Study Guide A, pp. 174–175
- Decoding Support, p. 198

AUDIO CDS

Advanced

UNIT RESOURCE BOOK

Challenge and Extension, p. 180

English Learners

UNIT RESOURCE BOOK

Spanish Reading Study Guide, pp. 178–179

AUDIO CDS

- Audio Readings in Spanish
- Audio Readings (English)

An embryo can remain inside a seed for a long time without growing. When moisture, temperature, and other conditions are right, a seed will start to grow. **Germination** (JUR-muh-NAY-shuhn) is the beginning of growth of a new plant from a spore or a seed. If you've ever planted a seed that sprouted, you've observed germination.

CHECK YOUR READING What is germination?

When a seed germinates, it takes in water from its surroundings. As the embryo begins to grow, it uses the stored nutrients in the seed for energy and materials. The nutrients need to last until the new plant's roots and shoots can start to function.

Inside a Seed

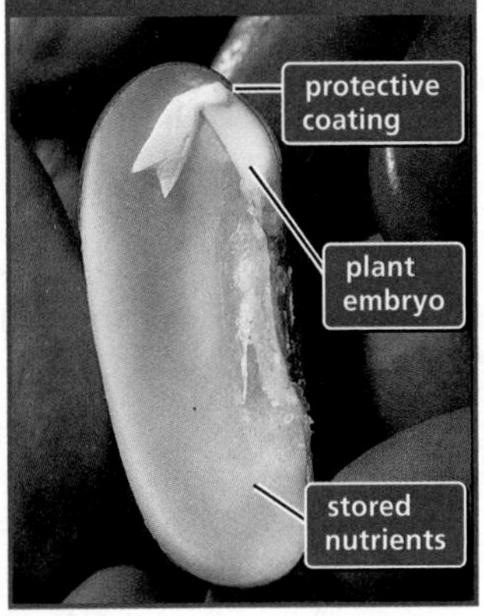
protective coating
plant embryo
stored nutrients

Some plants reproduce with seeds.

In most places, plants that reproduce with seeds are common and easy to see. Trees, bushes, flowers, and grasses are all seed plants. It's a bit harder to find plants that reproduce with spores, such as mosses or ferns. Why are there so many more seed plants in the world? The diagram below shows some of the differences and similarities between seeds and spores.

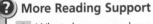

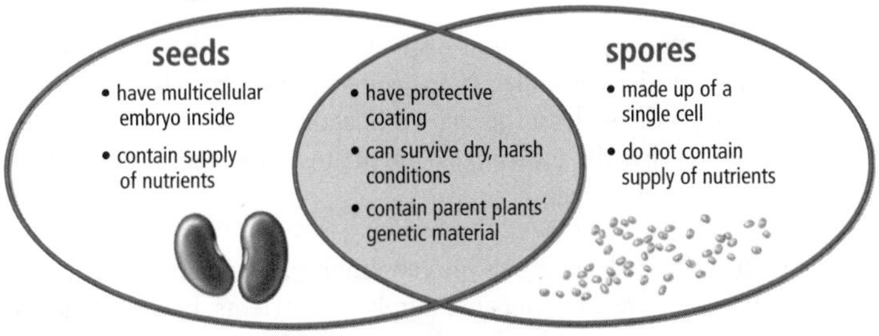
seeds
• have multicellular embryo inside
• contain supply of nutrients
• have protective coating
• can survive dry, harsh conditions
• contain parent plants' genetic material
spores
• made up of a single cell
• do not contain supply of nutrients

One important difference between a seed and a spore is that the seed contains a multicellular organism. If you look closely at the photograph above you can see the tiny leaves at the top of the embryo, with the root below. Spores are just a single cell. Seeds can be spread by wind, animals, or water. Spores are mostly carried by the wind. The sperm of seed plants, unlike the sperm of mosses and ferns, do not need water to reach the egg. One important similarity between a seed and a spore is that both can grow into a new plant.

CHECK YOUR READING Name three ways seeds are different from spores.

RESOURCE CENTER
CLASSZONE.COM
Learn more about seeds.

DIFFERENTIATE INSTRUCTION

More Reading Support

A What does an embryo use stored nutrients for? *energy and materials*

B What are some examples of seed plants? *trees, bushes, flowers, and grasses*

English Learners Venn diagrams, such as the one on this page, can be especially helpful in summarizing content in a visual way. Be sure to explain the significance of each section of the diagram, especially the center. Have students make their own version of this Venn diagram. They should write in their own words the separate and the shared qualities of seeds and spores. You might simplify language for them as: Seeds—new plant inside, food supply; Both—protected, do not dry out, can grow into a new plant; Spores—one cell, no food supply.

3.3 INSTRUCT

Teach from Visuals

To help students interpret both the photograph of the seed and the Venn diagram, ask the following questions that relate the visuals to the text:

- What is the function of the stored food supply? *to provide the growing embryo with nutrients*
- What is the function of the seed coat? *to protect the embryo from harsh conditions in the environment*
- What advantages do seeds have over spores? *Seeds have an embryo, a tiny plant ready to grow. Seeds also have a food supply for the plant once it starts to grow.*

EXPLORE the BIG idea

Revisit "Internet Activity: Sprouting Seeds" on p. 83. Ask students if their ideas of what seeds are have changed based on their observations and what they have learned from the text.

Ongoing Assessment

Discuss the advantages that plants with pollen and seeds have over nonseed plants.

Ask students to list two advantages that seeds provide in a plant life cycle. *Seeds provide a supply of nutrients to the developing plant. Seeds can be spread by wind, animals, or water.*

CHECK YOUR READING *Answer: Germination is the growth of a new plant from a seed or a spore under the right conditions of moisture and temperature.*

CHECK YOUR READING *Answer: Seeds contain a multicellular embryo; they have their own food supply; they can be transported by animals or water as well as by wind.*

Address Misconceptions

IDENTIFY Ask students if a seed is alive. Students may not realize that a seed is capable of becoming a living plant. A seed does not show the characteristics of a living thing until it germinates, when it begins to grow.

CORRECT Have students describe or sketch what happens to a plant as it grows: roots and shoots develop so a plant can get water and nutrients from its environment. (Remind them of the Internet Activity on p. 83.). Ask: What are the characteristics of life? *organization, growth, reproduction, and response*

REASSESS Ask students to draw a diagram of the life cycle of a pine tree. Remind them that the cycle should include two generations, and also pollen and seeds. Ask them to indicate where meiosis and fertilization occur.

Technology Resources

Visit **ClassZone.com** for background on common student misconceptions.

MISCONCEPTION DATABASE

Metacognitive Strategy

Students can use the heads in the caption boxes in the illustration on p. 101 as a starting point to create memory aids for life cycles of pine trees. They could choose to write, draw, or create manipulative cards, or even to create a song. Ask them to explain why their strategies work well for remembering.

Ongoing Assessment

Describe how some plants reproduce with pollen and seeds.

Ask: What is pollen? What is pollination? *Pollen contains sperm cells. Pollination occurs when a pollen grain attaches to the part of the plant that holds the egg. The sperm are released.*

CHECK YOUR READING *Answer: Pollen grains contain sperm cells, and female cone scales contain egg cells.*

Pine trees reproduce with pollen and seeds.

Seed plants, such as pine trees, do not have swimming sperm. Instead, they have pollen. A **pollen** grain is a small multicellular structure that holds a sperm cell. It has a hard outer covering to keep the sperm from drying out. Pollen grains can be carried from one plant to another by wind, water, or by animals such as insects, bats, or birds. The process of pollination is completed when a pollen grain attaches to the part of a plant that contains the egg and releases the sperm.

? C

REMINDER A *cycle* is a series of events that repeat regularly.

The life cycle of a pine tree provides an example of how seed plants reproduce. As you read the numbered paragraphs on this page, follow the numbers on the labels of the diagram on page 101.

1. The reproductive structures of a pine tree are the pinecones. Meiosis occurs in the pine cones, producing sperm and egg cells. Each tree has separate male and female cones.

Pinecones release lots of pollen into the air.

2. In male cones, the sperm cells are contained in pollen grains, which are released into the air. In female cones, the egg cells are enclosed in protective compartments within the cone scales.
3. The female cone produces a sticky substance. When a pollen grain lands, it sticks. A pollen tube begins to grow from the pollen grain through the scale to the egg. Fertilization occurs when the pollen tube reaches the protective compartment and sperm travel through it to one of the eggs.
4. The fertilized egg grows into an embryo. The compartment, with its protective covering and supply of nutrients, becomes the seed. The pinecone eventually releases its seeds. The winged seeds can float through the air and may be carried long distances by the wind. If the seed lands on the ground and germinates, it can become a new pine tree.

In the life cycle of a pine tree, meiosis and fertilization occur completely within the tissue of the mature plant. Fertilization doesn't require an outside source of water. The sperm cells in a pollen grain and the egg cells in a cone scale represent the first generation of the plant. The seed and the tree that grows from the seed represent the second generation.

CHECK YOUR READING What do pollen grains and cone scales contain?

DIFFERENTIATE INSTRUCTION

More Reading Support

C How is pollen carried from plant to plant? *by wind, water, and animals*

Below Level Have pairs of students create a flow chart to illustrate the life cycle of the pine tree. Students should include the following terms in their flow charts: *meiosis, spore cells, pollen grains, sperm cells, egg cells, fertilization, embryo, seed.* Allow time for pairs to share their flow charts with other students and make necessary adjustments or changes.

Life Cycle of a Pine Tree
1 Meiosis
A mature pine tree has both male and female cones. Meiosis occurs inside the cones, producing sperm and egg cells.
Male Cones
Female Cone
2 Pollination
Male cones release pollen into the air. A pollen grain sticks to the scales of a female cone, and a pollen tube begins to grow.
pollen grain
scales
3 Fertilization
Sperm move through the pollen tube to fertilize one of the egg cells at the base of the female cone scale.
pollen tube
sperm
fertilized egg
4 Seeds
Each fertilized egg becomes an embryo in a seed. A female cone can contain many seeds. The cone opens and the seeds are released.
seed
open female cone

READING VISUALS Use the definition of a cycle to predict what happens after seeds are released from pinecones.

DIFFERENTIATE INSTRUCTION

Below Level Invite students to bring in any pinecones they might have collected to share with the class. Have students note the various shapes and sizes of the cones. You may use the cones for the investigation on p. 102.

Advanced Challenge students to find pinecones or parts that match each of the stages and structures shown in the diagram. If pine trees are not local, have students find images of the parts and trace, draw, or clip-and-paste copies with captions.

R Challenge and Extension, p. 180

Teach from Visuals

To help students interpret the diagram of the life cycle of pine trees, ask:

- Where are the reproductive structures of the tree found? *in the pinecones*
- What is produced by the male and female cones? *sperm and eggs*
- When does fertilization occur? *when the sperm reaches one of the egg cells inside the female cone scale*

T The "Life Cycle of a Pine Tree" visual is also available as T22 in the Unit Transparency Book.

Develop Critical Thinking

ANALYZE Direct attention to the illustration of the pine seed. Ask: How do you think the shape of the seed affects how it travels? *The shape of the seed allows the wind to carry the seed and prevent it from falling directly under the tree.*

Real World Example

Because of the many uses of conifers, large forests are grown in many areas of the United States. A method called clearcutting is used to obtain the lumber in which all of the trees are cut down in a forest at once. An alternative method, called sustained yield, allows foresters to harvest lumber without destroying entire forests of trees. In this method, only a percentage of lumber is cut every year, allowing a constant supply of trees to be maintained in the forest.

Ongoing Assessment

READING VISUALS *Answer: The seed will be carried away by the wind, germinate if conditions are right, and grow into a pine tree similar to the one in the top picture, allowing the cycle to continue.*

INVESTIGATE Pinecones

PURPOSE To observe the structure of pinecones

TIPS *20 min.* Ask students to collect fallen pinecones and allow them to dry in the classroom. Cones fall from conifers at varying stages of ripeness, so students may find seeds inside the pinecones. You may wish to call "time" after 10 minutes in water to be sure students observe drying as well as moistening.

WHAT DO YOU THINK? *The cone floated in the water and gradually the scales closed. When the cone dried out, the cone opened up again.*

CHALLENGE *Students will likely have similar results with other pine species.*

R Datasheet, Pinecones, p. 181

Technology Resources

Customize this student lab as needed or look for an alternative. Print rubrics to assess student lab reports.

Lab Generator CD-ROM

Social Studies Connection

Pine trees, or conifers, are a common lumber source for the papermaking industry. Students can do research into the current and past lumber harvesting practices for papermaking. Have them write a for *and* against one of these practices, such as clearcutting, forced burning, or sustained yield harvesting.

Ongoing Assessment

CHECK YOUR READING *Answer: Gymnosperm seeds are naked because they are not contained in fruit.*

INVESTIGATE Pinecones

What conditions make a pinecone open?

SKILL FOCUS
Observing

MATERIALS
- dried, open pinecones
- beaker of water
- paper towels

TIME
20 minutes

PROCEDURE

1. Place your pinecone in the beaker of water. Observe any changes that take place. Leave the cone in the water until the changes stop.
2. Remove the cone from the water and place it on a paper towel to dry. Observe any changes that take place as the cone dries.

WHAT DO YOU THINK?

- What did you observe when the cone was in the water?
- What happened when the cone dried out?

CHALLENGE Try this procedure on cones from different plant species.

Gymnosperms are seed plants.

Pollen and seeds are reproductive adaptations. They did not appear in plants until millions of years after seedless plants such as mosses and ferns had already begun to live on land. Today, however, most of the plant species on Earth reproduce with seeds, and many species of seedless plants have become extinct. Some scientists think this is because over time Earth's climate has become drier and cooler. Seed plants are generally better adapted for reproducing in dry, cool environments than seedless plants are.

MIND MAP
Make a mind map for the main idea that *gymnosperms are seed plants.* Don't forget to include the definition of a gymnosperm in your mind map.

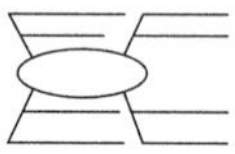

D Fossil evidence shows that species of seed plants in the **gymnosperm** (JIHM-nuh-SPURM) group have existed on Earth for more than 250 million years. Plants classified as gymnosperms produce seeds, but the seeds are not enclosed in fruit. The word *gymnosperm* comes from the Greek words for "naked seed." There are four types of gymnosperms living on Earth today.

CHECK YOUR READING What is distinctive about gymnosperm seeds?

Conifers

E The conifers, or cone-bearing trees, are the type of gymnosperm you are probably most familiar with. The conifers include pine, fir, spruce, hemlock, cypress, and redwood trees. Many conifers are adapted for living in cold climates, where there is relatively little water available. Their leaves are needle-shaped and have a thick cuticle. This prevents the plant from losing much water to transpiration. Conifers can also keep their needles for several years, which means the plants can produce sugars all year.

C 102 Unit: Diversity of Living Things

DIFFERENTIATE INSTRUCTION

More Reading Support

D Which kinds of seed plants do scientists think were the first to inhabit Earth? *gymnosperms*

E What are conifers? *cone-bearing trees*

Inclusion For students who have vision impairments, the "Investigate Pinecones" on this page can be adapted so that "observations" are based on touch rather than visual image. Students can place their fingers on the pinecone during the time it is in water, and again while it is drying. Time them for about 10 minutes for each step.

Ginkgo trees like this one are gymnosperms that do not produce cones. Their seeds are exposed to the environment.

Other Gymnosperms

The other types of living gymnosperms are cycads, gnetophytes, and ginkgoes. These three types of gymnosperms appear to be quite different from one another. Cycads are palmlike plants that are found in tropical areas. They produce cones for seeds. Many cycads produce poisonous compounds. Gnetophytes are another type of tropical gymnosperm that produces cones. Chemicals taken from certain gnetophyte plants have been used to treat cold symptoms for thousands of years.

Ginkgoes are gymnosperms with fleshy seeds that hang from their branches. Ginkgoes are often grown in parks and along streets, but you will not often see them with seeds. People avoid putting male and female ginkgo trees together because the seed coat of a ginkgo produces a particularly foul smell.

3.3 Review

KEY CONCEPTS

1. How are seeds different from spores?
2. How does fertilization in seed plants differ from fertilization in seedless plants?
3. Seed plants are found in environments where seedless plants are not. List at least two reasons why.

CRITICAL THINKING

4. **Compare and Contrast** What is the difference between a spore and a pollen grain? In what ways are spores and pollen grains similar?
5. **Hypothesize** Gymnosperms produce a lot of pollen, and most of it blows away, never fertilizing an egg. Why might this characteristic help a plant species survive?

CHALLENGE

6. **Analyze** Like all plants, a pine tree has a two-part life cycle. Look again at the diagram on page 101, and then make a new version of it in your notebook. Your version should show where in the life cycle each generation begins and ends. Hint: reread the text on page 100.

ANSWERS

1. Seeds: multicellular embryo, food supply. Spores: single cells

2. The sperm of seed plants do not need water to carry them to the egg.

3. Fertilization does not require water. Seeds are spread by animals.

4. Spores are single cells that grow into plants that produce sperm and eggs. Pollen are multicellular and carry only sperm cells. Both are protected by a hard covering.

5. A lot of pollen increases the chances of a pollen grain reaching the female cone of another plant.

6. Diagrams should show that a new generation begins in the pinecones, with meiosis producing sperm and egg cells. The seed produced after fertilization represents the second generation, the pine tree.

Develop Critical Thinking

SYNTHESIZE Compare the needles of conifers with the leaves of deciduous trees (trees that lose their leaves in winter). Ask:

- How are conifer's needles adapted to surviving winter conditions? *The shape of the needles allows the tree to conserve water when little is available.*
- Since there is less surface area on the needles for photosynthesis to take place, how do you think these plants get energy and materials they need? *The needles have less surface area but they are green all year and do photosynthesize. There may be more needles than leaves. They keep the needles all year.*

Reinforce the BIG idea

Have students relate the section to the Big Idea.

R Reinforcing Key Concepts, p. 182

3.3 ASSESS & RETEACH

Assess

A Section 3.3 Quiz, p. 45

Reteach

Reproduce the photograph on p. 95 of Moss Releasing Spores and the illustration of the pine tree life cycle on p. 101. (Or use the transparency "Life Cycle of the Pine Tree," in the Unit Transparency Book.) Have student volunteers use these two diagrams to explain examples of the following concepts.

- Plants reproduce well on land.
- Pollen and seeds have made some plants spread out more than others.

Technology Resources

Have students visit **ClassZone.com** for reteaching of Key Concepts.

CONTENT REVIEW

CONTENT REVIEW CD-ROM

CHAPTER INVESTIGATION

Focus

PURPOSE To compare and contrast germination and growth patterns of various types of seeds

OVERVIEW Students will set up a classroom greenhouse and plant a variety of seeds. Students will determine:

- differences in germination.
- differences in growth patterns.

Lab Preparation

- Have students collect seeds ahead of time and make predictions as to which seeds will grow. They may use seeds they collect from the outdoors or from foods. Dried beans and mixed bird seed work. Popcorn, rice, and grocery-store fruit seeds are unlikely to grow.
- Point out that the data tables will vary, depending on the seeds used.
- Prior to the investigation have students read through the investigation and prepare their data tables. Or you may wish to copy and distribute datasheets and rubrics.

UNIT RESOURCE BOOK, pp. 201–209

SCIENCE TOOLKIT, F14

Lab Management

- Prompt students to control the following: containers, planting medium, the depth at which seeds are planted, the amounts of light and warmth, and the amount of water.

INCLUSION For students with attention disorders or learning disabilities, prepare a checklist of plant care. Have students use the checklist each day that they care for and observe their plants.

Teaching with Technology

If digital cameras or stop-frame video are available, students may make close-up photographs of their plants. They can use time-lapse slides or videos to show the pattern of growth.

CHAPTER INVESTIGATION

Which Seeds Will Grow?

OVERVIEW AND PURPOSE Many of the foods you eat come from seed plants. What seeds or parts of seed plants have you eaten lately? What seeds can you find outside in your neighborhood? What conditions do the seeds from these plants need to grow? In this investigation, you will

- plant a variety of seeds
- observe differences in germination and growth among seeds planted in similar conditions

Problem

Write It Up

How successfully will a variety of seeds germinate in conditions that can be provided in your classroom?

Hypothesize

Write It Up

Read through all of the steps of the procedure and then write a hypothesis. Your hypothesis should explain what you think will happen when your class plants its collection of seeds and observes their growth for at least ten days. Your hypothesis should take the form of an "If . . . , then . . . , because . . ." statement.

MATERIALS
- assorted seeds
- potting soil
- paper cups
- water
- paper towels
- labels

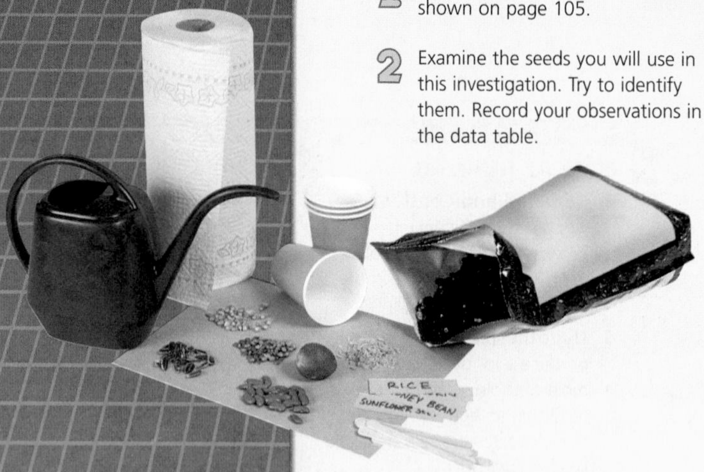

Procedure

1. Make a data table in your **Science Notebook** like the one shown on page 105.
2. Examine the seeds you will use in this investigation. Try to identify them. Record your observations in the data table.

step 2

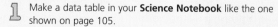

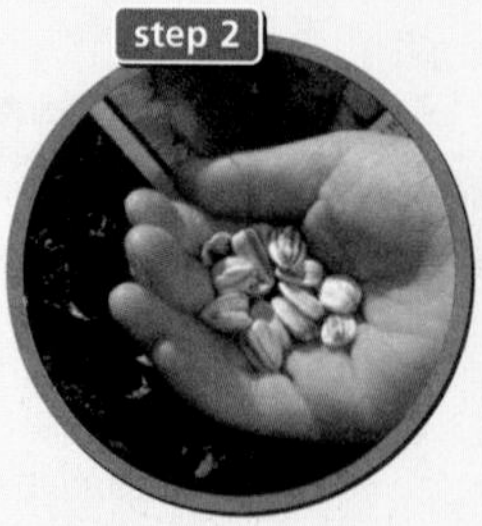

INVESTIGATION RESOURCES

CHAPTER INVESTIGATION, Which Seeds Will Grow?
- Level A, pp. 210–204
- Level B, pp. 205–208
- Level C, p. 209

Advanced students should complete Levels B & C.

Writing a Lab Report, D12–13

Technology Resources

Customize this student lab as needed or look for an alternative. Print rubrics to assess student lab reports.

Lab Generator CD-ROM

3 Use the materials provided to plant the seeds. Remember that the planting conditions should be the same for each of the seeds. Label each container.

4 Decide where you will keep your seeds while they are growing and how often you will check and water them. Be sure to keep the growing conditions for all of the seeds the same. Wash your hands.

5 Observe your seeds for at least ten days. Check and water them according to the plan you made.

step 5

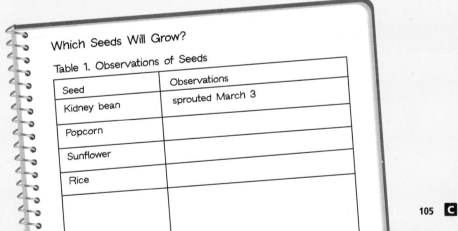

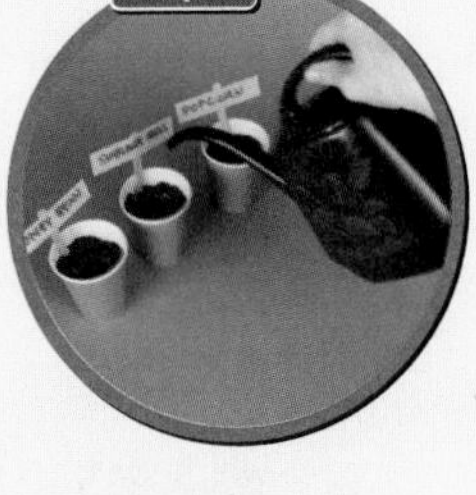

Observe and Analyze

1. In your **Science Notebook**, draw and label a diagram showing how you planted the seeds, the materials you used, and the place where they are being kept.
2. Each time you check on your seeds, record your observations in your data table.
3. **IDENTIFY** Which seeds germinated? What differences in growth and development did you observe in the different types of seeds?

Conclude

1. **INTERPRET** Compare your results with your hypothesis. Did your data support your hypothesis?
2. **INFER** What patterns or similarities did you notice in the seeds that grew most successfully?
3. **IDENTIFY LIMITS** What unexpected factors or problems might have affected your results?
4. **APPLY** Use your experience to tell a young child how to grow a plant from a seed. What type of seeds would you suggest? Write directions for planting and caring for seeds that a younger child could understand.

INVESTIGATE Further

CHALLENGE Some seeds need special conditions, such as warmth or a certain amount of moisture, before they will germinate. Design an experiment in which you test just one type of seed in a variety of conditions to learn which results in the most growth. Include a hypothesis, a materials list, and a procedure when you write up your experiment.

Which Seeds Will Grow?

Table 1. Observations of Seeds

Seed	Observations
Kidney bean	sprouted March 3
Popcorn	
Sunflower	
Rice	

Observe and Analyze

1. *Students should make sure that all plants are grown under the same conditions.*
2. *Check that students' data tables are complete and correspond to the actual seeds they chose to plant.*
3. *Answers will depend on the seeds selected for the lab.*

Conclude

1. *Students should provide evidence or explanation for describing whether their hypotheses are correct or incorrect.*
2. *Sample answer: Seeds that grew most successfully were planted near the top of the soil and not too close together.*
3. *Sample answers: environment of classroom, limitations of container, soil quality, type of seeds*
4. *Answers should be based on experience gained from this investigation.*

INVESTIGATE Further

CHALLENGE Collect student procedures. Procedures should allow student to test at least one variable (amount of water, temperature, soil type, soil depth, etc.) to determine what conditions are best for the seed.

Post-Lab Discussion

- Discuss why some of the seeds students planted grew better than others, or why some of the seeds did not grow. *Sample answer: Some seeds may take longer to germinate, or may require conditions that are not possible to duplicate in a classroom.*
- Why is it important to have all seeds grown under the same conditions? *If there is more than one variable, it will be difficult to draw conclusions related to the types of seeds.*

EXTREME SCIENCE
Fun and Motivating Science

Set Learning Goal
To understand the ability of seeds to survive under extreme conditions

Present the Science
Many seeds remain dormant, or inactive, after they have been dispersed. Under the right conditions of temperature, moisture, and oxygen, seeds will germinate. Seeds with thick or waxy coats may block the entry of water and oxygen. These seeds may remain dormant for extended periods of time. Most seeds, however, lose the ability to germinate within several years.

Discussion Questions
Review with students the various ways that seeds are dispersed.

Ask: By what methods were the seeds discussed on the page dispersed:

- Jackpine seeds? *were released by cones and spread through air*
- Sea rocket seed? *traveled by water*
- Calvaria seeds? *were spread through the wastes of an animal*

Close
Ask: Why do you think seeds are adapted to germinate under extreme conditions? *Without these adaptations, the seeds might not germinate and the plant species would be in danger of becoming extinct.*

Technology Resources

Students can visit **ClassZone.com** for more links about extreme seeds.

RESOURCE CENTER

EXTREME SCIENCE

AMAZING SEEDS

Forest Fires release jackpine seeds from their pinecones to sprout after the fire stops.

Seed Survivors

Seeds can not only survive some harsh conditions, but often, harsh conditions make it possible for seeds to grow into new plants.

Forest Fires
Jackpine seeds are locked inside pinecones by a sticky resin. The pinecones survive hot forest fires that melt the resin and release the seeds. In fact, without the high temperatures from fires, the seeds would never get the chance to sprout and grow into jackpine trees.

Bomb Damage
In 1940, during World War II, The British Museum was firebombed. People poured water over the burning museum and its contents. In the museum were silk-tree seeds collected in China and brought to the museum in 1793. After the firebombing 147 years later, the seeds sprouted.

Dodo Digestion

Seeds of Calvaria trees, which live on the island of Mauritius have very hard outer shells. The outer shell must be softened before the seeds can sprout. Hundreds of years ago, dodo birds ate the Calvaria fruits. Stones and acids in the birds' digestive tract helped soften the seed. After the dodo birds deposited them, the softened seeds would sprout. Dodo birds went extinct in 1681, and no young Calvaria trees grew. In 1975, only about 13 Calvaria trees remained. Recently, scientists have used artificial means to grind and break down the Calvaria seed cover and foster new tree growth.

EXPLORE

1. **INFER** Why do you think the silk tree seeds sprouted? Think about how the seeds' environment changed.
2. **CHALLENGE** If you were a scientist who wanted to help the Calvaria trees make a comeback, what methods might you try for softening the seeds?

RESOURCE CENTER
CLASSZONE.COM
Find out more about extreme seeds.

EXPLORE

1. ***INFER*** *Heat and water from the firebombing most likely gave the right conditions of moisture and temperature for the seeds to germinate.*
2. ***CHALLENGE*** *Accept all reasonable responses. Students might suggest soaking the seeds in a moist environment containing acids to simulate the digestive tract of the dodo birds.*

KEY CONCEPT

3.4 Many plants reproduce with flowers and fruit.

BEFORE, you learned
- Seed plants do not have swimming sperm
- Gymnosperms reproduce with pollen and seeds

NOW, you will learn
- About flowers and fruit
- About the relationship between animals and flowering plants
- How humans need plants

VOCABULARY
angiosperm p. 107
flower p. 108
fruit p. 108

EXPLORE Fruit

What do you find inside fruit?

PROCEDURE

1. Place the apple on a paper towel. Carefully cut the apple in half. Find the seeds.
2. Place the pea pod on a paper towel. Carefully split open the pea pod. Find the seeds.
3. Both the apple and the pea pod are examples of fruits. In your notebook, draw a diagram of the two fruits you examined. Label the fruit and the seeds.

MATERIALS
- apple
- paper towel
- plastic knife
- pea pod

WHAT DO YOU THINK?
- How many seeds did you find?
- What part of an apple do you eat? What part of a pea?

Angiosperms have flowers and fruit.

MIND MAP
Make a mind map diagram for the main idea: *Angiosperms have flowers and fruit.*

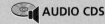

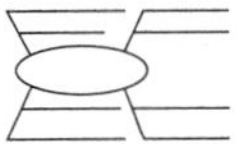

Have you ever eaten peanuts, grapes, strawberries, or squash? Do you like the way roses smell, or how spider plants look? All of these plants are angiosperms, or flowering plants. An **angiosperm** (AN-jee-uh-SPURM) is a seed plant that produces flowers and fruit. Most of the species of plants living now are angiosperms. The grasses at your local park are angiosperms. Most trees whose leaves change color in the fall are angiosperms.

The sperm of a flowering plant are protected in a pollen grain and do not need an outside source of water to reach the eggs. The eggs develop into embryos that are enclosed within seeds. Both generations of angiosperms and gymnosperms occur within a single plant.

RESOURCES FOR DIFFERENTIATED INSTRUCTION

Below Level

UNIT RESOURCE BOOK
- Reading Study Guide A, pp. 185–186
- Decoding Support, p. 198

AUDIO CDS

Additional INVESTIGATION, Fertilization in Angiosperms, A, B, & C, pp. 210–218; Teacher Instructions, pp. 360–361

Advanced

UNIT RESOURCE BOOK
Challenge and Extension, p. 191

English Learners

UNIT RESOURCE BOOK
Spanish Reading Study Guide, pp. 189–190

AUDIO CDS
- Audio Readings in Spanish
- Audio Readings (English)

3.4 FOCUS

Set Learning Goals

Students will
- Describe the characteristics of flowers and fruit.
- Analyze the relationship between animals and flowering plants.
- Discuss human uses of plants.
- Observe and identify parts of a flower.

3-Minute Warm-Up

Display Transparency 21 or copy this exercise on the board:

Match each definition with the correct term.

Definitions
1. the immature form of a plant that has the potential to grow and develop *b*
2. the multicellular structure that holds the sperm cell *e*
3. the beginning of growth from a spore or seed *c*

Terms

a. seed	d. pollination
b. embryo	e. pollen
c. germination	

T 3-Minute Warm-Up, p. T21

3.4 MOTIVATE

EXPLORE Fruit

PURPOSE To explore the inside of different fruits

TIPS *15 min.* Apples may be cut ahead of time. Put lemon juice on cut surfaces to reduce browning. Students may not think that foods such as pea pods are actually fruits. Encourage students to determine why these foods are considered fruits as they examine them.

WHAT DO YOU THINK? *Number of seeds found will vary, but students should find several in each fruit. In the apple, you would eat the fruit (the ripened ovary). In pea pods, you would eat the seeds, and in some cases, the fruit (pod) as well.*

3.4 INSTRUCT

Teach Difficult Concepts

Students may confuse angiosperms and gymnosperms and the differences in their life cycles. To help them understand the differences, discuss the following questions:

- Compare the reproductive structures of angiosperms with those of a pine tree. How are they different? *In angiosperms, the reproductive structures are in flowers. In pine trees, they are in pine cones.*
- How is pollination similar in both types of plants? *In both plants, the pollen grain goes through a pollen tube to reach the egg.*
- Compare the seeds of angiosperms with those of a pine tree. How are they different? *In pine trees, seeds are not enclosed inside a fruit, as in angiosperms.*

Language Arts Connection

Many of the structures and processes in an angiosperm life cycle are used in English as metaphors for ideas and human actions. Some such as seed, fruit, germ, and kernel have multiple definitions relating to plants or to ideas. Some may be used in poetry or a story to create similes (sprout like a weed, fresh as a flower) or metaphors, implied comparisons that do not use "like" or "as." Assign students an exercise of creating their own original similes and metaphors based on words describing angiosperm structures or processes.

Ongoing Assessment

CHECK YOUR READING *Answer: Angiosperms have flowers and fruit.*

CHECK YOUR READING *Answer: Flowers produce egg and sperm cells. Once the egg in the ovary is fertilized, the seeds grow and the ovary becomes a fruit.*

The reproductive cycles of angiosperms and gymnosperms are alike in many ways. Both angiosperms and gymnosperms have separate male and female reproductive structures. In some species, male and female parts grow on the same plant, but in others there are separate male and female plants.

VOCABULARY Remember to add word triangles for *flower* and *fruit* to your notebook.

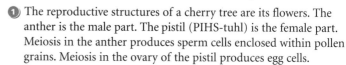

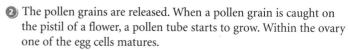

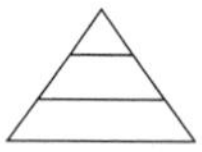

An important difference between angiosperms and gymnosperms is that in angiosperms, the sperm and egg cells are contained in a flower. The **flower** is the reproductive structure of an angiosperm. Egg cells develop in a part of the flower called an ovary. Once the eggs are fertilized and the seed or seeds form, the ovary wall thickens and the ovary becomes a **fruit.** (A)

CHECK YOUR READING What reproductive structures do angiosperms have that gymnosperms do not?

The diagram on page 109 shows the life cycle of one type of angiosperm, a cherry tree. As you read the numbered paragraphs below, follow the numbers on the labels in the diagram.

(B)

1. The reproductive structures of a cherry tree are its flowers. The anther is the male part. The pistil (PIHS-tuhl) is the female part. Meiosis in the anther produces sperm cells enclosed within pollen grains. Meiosis in the ovary of the pistil produces egg cells.
2. The pollen grains are released. When a pollen grain is caught on the pistil of a flower, a pollen tube starts to grow. Within the ovary one of the egg cells matures.
3. Fertilization occurs when the pollen tube reaches the ovary and sperm fertilizes the egg. The fertilized egg grows into an embryo and develops a seed coat. The ovary develops into a fruit.
4. The fruit may fall to the ground or it may be eaten by animals. If the seed inside lands in a place where it can germinate and survive, it will grow into a new cherry tree.

CHECK YOUR READING What is the flower's role in the sexual reproduction of an angiosperm?

Many flowering plants also reproduce asexually. New shoots can grow out from the parent plant. For example, strawberries and spider plants can reproduce by sending out shoots called runners. New plants grow from the runners, getting nutrients from the parent until the roots of the new plant are established. Plants can spread quickly this way. This form of asexual reproduction allows plants to reproduce even when conditions are not right for the germination of seeds.

DIFFERENTIATE INSTRUCTION

More Reading Support

A Which structure of the flower develops into the fruit? *the ovary*

B Which part of the flower is the male part? The female part? *male—anther, female—pistil*

English Learners As students read the text and study the diagram on pp. 108–109, they will encounter many unfamiliar words, such as *anther, pistil, ovary,* and *angiosperm.* Encourage students to make a picture dictionary for "Flowering Plants" by defining unknown words and drawing pictures for as many examples as they can. Students new to English may have difficulty with words such as *shoots, grain, embryo, pollen, runners,* and *germinate,* all of which are important to the life cycle of flowering plants.

Life Cycle of a Cherry Tree

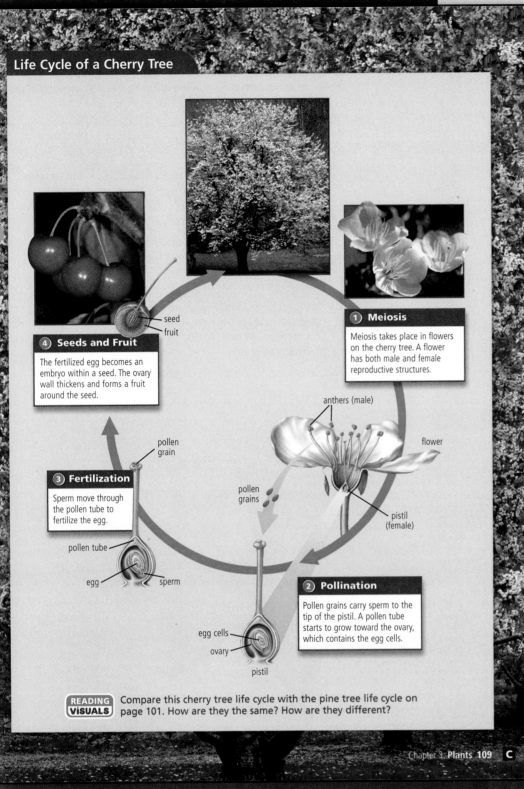

seed
fruit
4 Seeds and Fruit
The fertilized egg becomes an embryo within a seed. The ovary wall thickens and forms a fruit around the seed.
1 Meiosis
Meiosis takes place in flowers on the cherry tree. A flower has both male and female reproductive structures.
anthers (male)
flower
pollen grain
3 Fertilization
Sperm move through the pollen tube to fertilize the egg.
pollen grains
pistil (female)
pollen tube
egg
sperm
2 Pollination
Pollen grains carry sperm to the tip of the pistil. A pollen tube starts to grow toward the ovary, which contains the egg cells.
egg cells
ovary
pistil

READING VISUALS Compare this cherry tree life cycle with the pine tree life cycle on page 101. How are they the same? How are they different?

DIFFERENTIATE INSTRUCTION

Below Level Ask students to create a flow chart illustrating the stages for angiosperm reproduction. Students can compare the flow charts to the ones they made for pine tree reproduction in Section 3.3 (p. 100). Ask students to compare their flow charts and describe how they are alike and how they differ.

Teach from Visuals

To help students interpret the diagram of the life cycle of a cherry tree, ask:

- Where does meiosis occur? *in the flowers*
- What happens during pollination? *Pollen grains are released and land on the pistil.*
- Where does fertilization take place? *in the ovary*
- How do the pollen grains reach the ovary? *by producing a pollen tube*

Metacognitive Strategy

Ask students to restate the steps for the life cycle of a cherry tree to a partner or group using their own words. Encourage them to move in order from one stage to the next. Have them discuss which steps are easy or difficult to recall. Next have them imagine a vivid image for each step. They can discuss how this may or may not increase their ability to remember.

Arts Connection

Since flowers have been an inspiration to painters in nearly all cultures and in all periods of art history, the cherry blossom is a common motif. Display Japanese prints and watercolors, as well as Dutch still life paintings and Victorian botanical drawings. Students can apply their science learning to identifying structures of flowers in the artworks. They may also try sketching or painting their own likenesses of live, or cut, flowers.

Ongoing Assessment

READING VISUALS *Answer: Both diagrams show tree life cycles. Both life cycles have meiosis, pollen, pollination, and fertilization steps. Both result in seeds being produced. Differences include: angiosperms have male and female structures within a single flower, and an ovary that will become a fruit.*

Arts Connection

Allow students to create models of flowers, using materials they find at home or in the classroom. For example, they could use colorful wrapping paper or tissue paper for sepals and petals, toothpicks for filaments, clay for anthers, a pipe cleaner for the stigma, and a foam ball for the ovary. Allow students class time to share their models and identify the parts of the flower. They might also create a caption describing parts of the model.

Teach from Visuals

To help students interpret the diagram of flower parts, ask:

- Which two structures make up the stamen? *anther and filament*
- Which two structures make up the pistil? *stigma and ovary*

Ongoing Assessment

Describe the characteristics of flowers and fruit.

Ask: What part of a flower later becomes a fruit? What did that part contain in the flower, and what does it contain as it becomes fruit? *An ovary becomes a fruit. The ovary contained an egg which when fertilized became a seed in the fruit.*

CHECK YOUR READING *Answer: The stamen is the male reproductive structure of the plant. The pistil is the female reproductive structure.*

Flowers

Flowers vary in size, shape, color, and fragrance. They all have some similar structures, although they are not always as easy to see as in the lily pictured below.

- Sepals are leafy structures that enclose the flower before it opens. When the flower blooms, the sepals fall open and form the base of the flower.
- Petals are leafy structures arranged in a circle around the pistil. The petals open as the reproductive structures of the plant mature. Petals are often the most colorful part of a flower. The petals help to attract animal pollinators.
- C The stamen (STAY-muhn) is the male reproductive structure of a flower. It includes a stalk called a filament and the anther. The anther produces sperm cells, which are contained in pollen grains.
- D The pistil is the female reproductive structure of the flower. The ovary is located at the base of the pistil and contains the egg cells that mature into eggs. At the top of the pistil is the stigma, where pollen grains attach.

CHECK YOUR READING What are the stamen and pistil?

Parts of a Flower

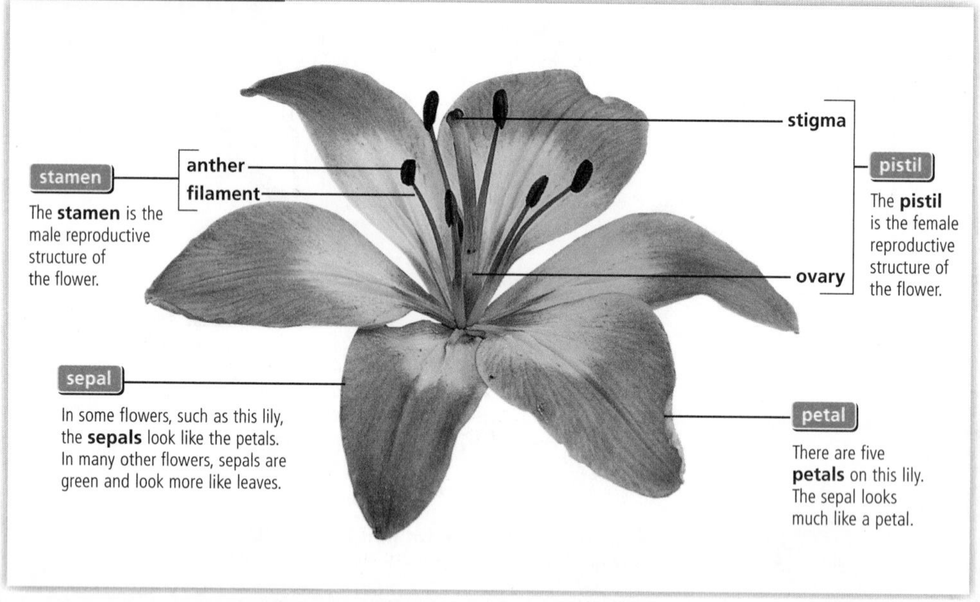
stamen
The **stamen** is the male reproductive structure of the flower.
anther
filament
stigma
pistil
The **pistil** is the female reproductive structure of the flower.
ovary
sepal
In some flowers, such as this lily, the **sepals** look like the petals. In many other flowers, sepals are green and look more like leaves.
petal
There are five **petals** on this lily. The sepal looks much like a petal.

DIFFERENTIATE INSTRUCTION

More Reading Support

C Where in the stamen is the pollen produced? *in the anther*

D Where in the pistil are the eggs produced? *in the ovary*

Advanced Invite interested students to research and report on carnivorous plants, such as the Venus flytrap, pitcher plants, and sundew plants. Ask students to report on how the structure of the flowers and the plant are adapted to trap food.

R Challenge and Extension, p. 191

INVESTIGATE Flower Parts

What parts of a flower can you identify?

PROCEDURE

1. Examine the flower you are given. Try to notice as many details as you can. Draw a diagram of the flower in your notebook and label its parts.
2. Carefully take your flower apart. Sort the parts. Draw and label one example of each part in your notebook.

WHAT DO YOU THINK?

- Which of the parts of a flower labeled in the diagram on page 110 did you find in your flower?
- Based on your experience, what would you look for if you were trying to decide whether a structure on an unfamiliar plant was a flower?

SKILL FOCUS
Observing

MATERIALS
- assorted flowers
- hand lens

TIME
15 minutes

Fruit

A fruit is a ripened plant ovary. Some ovaries contain more than one seed, such as an apple. Some contain only one seed, like a cherry. Apples and cherries are called fleshy fruits, because they have juicy flesh. The corn you eat as corn on the cob is a fleshy fruit. There are also dry fruits. The shells of peanuts, walnuts, and sunflowers are dry fruits. The structures of a dry fruit help protect the seed. Some dry fruits, like the winged fruit of a maple tree or the feathery tip of a dandelion seed, have structures that allow the seeds to be carried by the wind.

E

Animals spread both pollen and seeds.

Reproduction in many types of flowering plants includes interactions between plants and animals. The plants are a source of food for the animal. The animals provide a way to transport pollen and seeds. As they eat, animals move pollen from flower to flower and seeds from place to place.

Have you ever watched a honeybee collect nectar from a flower? Nectar is a sweet sugary liquid located at the bottom of the flower. As the bee crawls around in the flower, reaching down for the nectar, it rubs against the anthers and picks up pollen grains. When the bee travels to another flower, some of that pollen rubs off onto the pistil of the second flower.

F

CHECK YOUR READING How do bees benefit from the flowers they pollinate?

INVESTIGATE Flower Parts

PURPOSE To identify the parts of a flower

TIPS *15 min.* Flowers with anatomies that are easy to identify (both male and female parts) include lilies, tulips, and gladiolus. Flowers that are "daisy-like" will be more difficult to identify.

WHAT DO YOU THINK? *Part of the flower that students identify will vary depending on the selection of flowers. Students might say that they would look for brightly colored petal-like parts, pollen, parts that look like anthers, and stamens and pistils in the center of the bloom.*

R Datasheet, Flower Parts, p. 199

Technology Resources

Customize this student lab as needed or look for an alternative. Print rubrics to assess student lab reports.

Lab Generator CD-ROM

Teacher Demo

Provide examples of fleshy and dry fruits for students to examine. (Dry fruits include acorns, walnuts, and other nuts; fleshy fruits include mangoes, oranges, pears, and many similar fruits.) Allow students time to examine the fruits with a hand lens and note the color, texture, and shape. Cut several of the fleshy fruits in half and have students observe the ovaries and seeds.

Ongoing Assessment

CHECK YOUR READING *Answer: Bees get energy and material from the nectar they drink.*

DIFFERENTIATE INSTRUCTION

More Reading Support

E Give an example of a fleshy fruit and a dry fruit. *Sample answer: cherry, peanut*

F What is the sugary liquid located at the bottom of a flower? *nectar*

Advanced Write the terms *monocot* and *dicot* on the board and tell students that angiosperms come in these two distinct types. Have students find out the characteristics of each and give examples of each. Encourage students to illustrate their findings with diagrams, photographs, and illustrations.

Inclusion "Investigate Flower Parts" can be tailored for students of differing abilities by providing flowers that are less challenging (such as a lily) or more challenging (such as a daisy or sunflower) to dissect and identify.

Integrate the Sciences

Scientists have discovered that honeybees can distinguish at least four different colors of the spectrum: yellow, blue green, blue, and ultraviolet. Their ability to see ultraviolet allows bees to follow ultraviolet patterns in flowers that are not visible to the human eye.

Bats, on the other hand, are able to locate sources of food, such as insects or flowers through echolocation. They emit high-frequency pulses of sound that are not audible to human ears and listen for the echoes that return. The echoes enable the bats to determine position and direction of food objects.

Teach Difficult Concepts

Some students think that pollination and fertilization are the same process. Point out that pollination is necessary for fertilization to occur. Refer students back to the "Life Cycle of a Cherry Tree" diagram on p. 109. Ask: What is the difference between pollination and fertilization? *Pollination moves sperm from the male part of the flower (anther) to the female part of the flower (stigma). Fertilization occurs when the egg and sperm are united (when sperm travels down a pollen tube to reach an egg in the ovary).*

Ongoing Assessment

Analyze the relationship between animals and flowering plants.

Ask: How do animals help plants grow and reproduce? *Animals pollinate plants by spreading pollen from one plant to another, and help plants grow by dispersing their seeds.*

CHECK YOUR READING *Answer: Animal pollinators move pollen from flower to flower, often of the same species. Wind blows pollen around and there is only a small chance it will land on the female part of a plant of the right species.*

An animal that pollinates a flower is called a pollinator. Bees and other insects are among the most important pollinators. Bees depend on nectar for food, and they collect pollen to feed their young. Bees recognize the colors, odors, and shapes of flowers. Bee-pollinated plants include sunflowers, rosemary, lavender, and thousands of other species.

G The relationship between angiosperms and their pollinators can be highly specialized. Sometimes the nectar is located in a tube-shaped flower. Only certain animals, for example hummingbirds with long, slender beaks, can pollinate those flowers. Some flowers bloom at night. These flowers attract moths and bats as pollinators. Night-blooming flowers are usually pale, which means they are visible at night. Also, they may give off a strong scent to attract animal pollinators.

The advantage of animal pollination is that the pollen goes to where it is needed most. The pollen collected by a bee has a much better chance of being brought to another flower. By comparison, pollen grains that are spread by the wind are blown in all directions. Each grain has only a small chance of landing on another flower. Wind-pollinated plants produce a lot more pollen than plants that are pollinated by animals.

CHECK YOUR READING What is the advantage of animal pollination over wind pollination?

SIMULATION CLASSZONE.COM
Compare the different ways seeds are dispersed.

The fruits produced by angiosperms help to spread the seeds they contain. Some seeds, like dandelion and maple seeds, are carried by the wind. Many seeds are scattered by fruit-eating animals. The seeds go through the animal's digestive system and are eventually deposited on the ground with the animal's waste.

Animals eat fleshy fruit and distribute the seeds with their waste.

The burrs in this horse's mane are dry fruit that contain seeds.

DIFFERENTIATE INSTRUCTION

More Reading Support

G Nectar found in a long slender flower is likely pollinated by which animal? *hummingbird*

H How are dandelion seeds dispersed? *by wind*

Below Level Ask students to draw a comic strip of animals pollinating flowers or dispersing seeds. Students should include captions of the action in their strips to explain how the processes occur.

Additional Investigation To reinforce Section 3.4 learning goals, use the following full-period investigation:

R **Additional INVESTIGATION,** Fertilization in Angiosperms, A, B, & C; pp. 210–218; 360–361
(Advanced students should complete Levels B and C.)

Animals also help to scatter some types of dry fruits—not by eating them, but by catching them on their fur. Have you ever tried to pet a dog that has run through a grassy field? You might have noticed burrs stuck in the animal's fur. The seeds of many grasses and wildflowers produce dry fruits that are covered with spines or have pointed barbs. Seeds protected by these types of dry fruits stick to fur. The seeds travel along with the animal until the animal rubs them off.

Humans depend on plants for their survival.

Without plants, humans and all other animals would not be able to live on Earth. After plants adapted to life on land, it become possible for animals to live on land as well. Land animals rely on plants for food and oxygen. Many animals live in or near plants. Plants also supply materials humans use every day.

Food and Oxygen

All organisms must have energy to live. For animals, that energy comes from food. Plants, especially angiosperms, are the main source of food for all land animals. Plants capture energy from the Sun to make sugars and carbohydrates. Those same energy-rich materials are then consumed by animals as food. Even animals that eat other animals depend on plants for survival, because plants may provide food for the animals they eat.

Photosynthesis, the process that plants use to produce sugars and carbohydrates, also produces oxygen. The oxygen in the air you breathe is the product of the photosynthetic activity of plants and algae. Animals, including humans, need oxygen to release the energy stored in food.

Plants capture light energy from the Sun and store it in sugars and carbohydrates.

Energy Resources and Soil

Plants are an important source of many natural resources. Natural gas and coal are energy resources that formed deep underground from the remains of plants and other organisms. Natural gas and coal are important fuels for many purposes, including the generation of electricity.

Even the soil under your feet is a natural resource associated with plants. Plant roots can break down rock into smaller and smaller particles to form soil. When plants die, their bodies decay and add richness to the soil.

History of Science

Humans began the transition from a hunter-gatherer society to cultivating the land and domestication of animals around 9000 B.C. to 7000 B.C. Scientists think that agricultural techniques developed somewhat simultaneously throughout the world. In the Middle East, people began to harvest types of grain and barley. The domestication of cattle and sheep allowed for the cultivation of crops. In China, millet and rice were both grown by 4000 B.C. The Americas saw cultivation of peppers, avocados, and amaranths around 7000 B.C. to 5000 B.C.

EXPLORE the BIG idea

Revisit "How Are Seeds Dispersed?" on p. 83. Ask: What features of seeds enable them to be dispersed by both animals and wind? *Sample answers: Seeds have burrs that stick to animals, seeds have wings that allow them to be blown in the wind. Animals eat the fruit surrounding seeds and deposit the seeds in wastes.*

Ongoing Assessment

Discuss human uses of plants.

Ask: Other than food, what are some resources for humans that come from plants? *Sample answer: gas and coal obtained from long-decayed plants, soil and fertilizer, oxygen in air, clothing fibers, paper, wood for building*

DIFFERENTIATE INSTRUCTION

More Reading Support

I What is the main source of food for land animals? *plants*

Alternative Assessment Ask students to create a mind map showing the importance of plants to human survival. Students should include subheads of food and oxygen and energy resources and soil in their graphics.

Teach from Visuals

To help students interpret the photographs of plants and products ask: About how many pounds of cotton are needed to produce a single pair of jeans? *about two and one-half pounds*

Reinforce the BIG idea

Have students relate the section to the Big Idea.

R Reinforcing Key Concepts, p. 193

3.4 ASSESS & RETEACH

Assess

A Section 3.4 Quiz, p. 46

Reteach

Ask students to identify an angiosperm they are familiar with. Prompt volunteers to discuss:

- parts of the plant and their functions.
- the stages the plant goes through when it reproduces.
- how pollen and seeds from the plant are dispersed.

As you guide them through their identification, students may wish to sketch or diagram the answers. Then pairs can review the notes. *Sample answer: A flower contains male and female structures: sperm cells are produced in anther and egg cells in ovary. Pollen grains with sperm cells are released and attach to pistil. When pollen tube reaches ovary, sperm can fertilize egg, seeds and fruit form. Pollen and seeds are spread by animals, wind, and water.*

Technology Resources

Have students visit **ClassZone.com** for reteaching of Key Concepts.

CONTENT REVIEW

CONTENT REVIEW CD-ROM

From Plants to Products

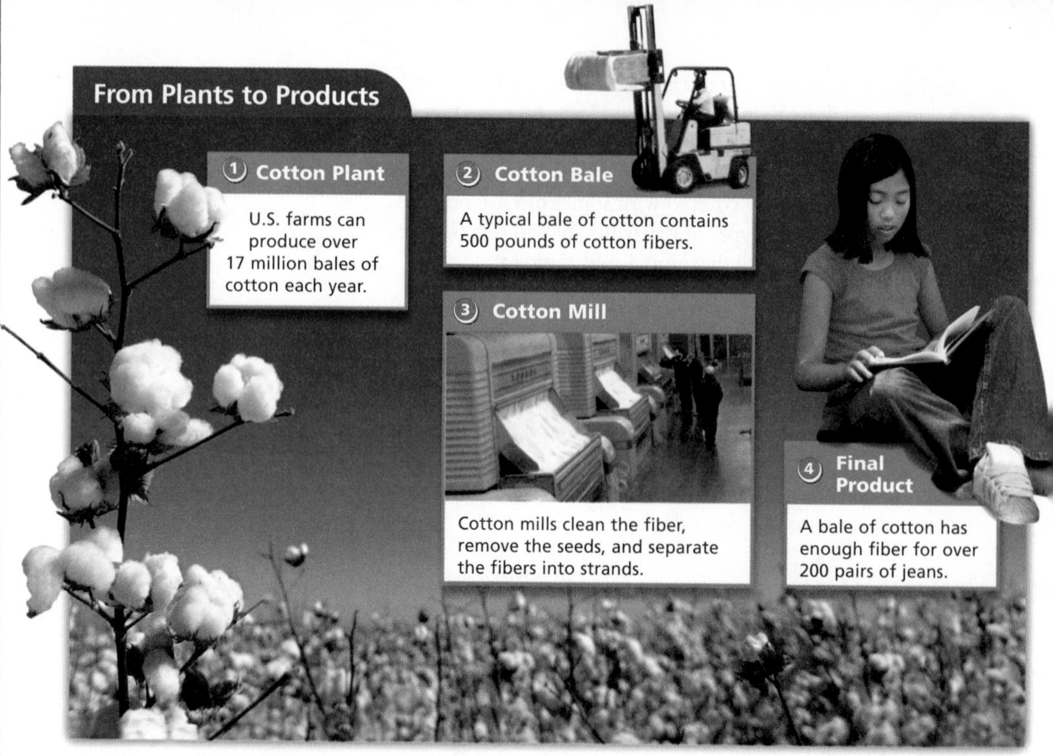

1 Cotton Plant

U.S. farms can produce over 17 million bales of cotton each year.

2 Cotton Bale

A typical bale of cotton contains 500 pounds of cotton fibers.

3 Cotton Mill

Cotton mills clean the fiber, remove the seeds, and separate the fibers into strands.

4 Final Product

A bale of cotton has enough fiber for over 200 pairs of jeans.

Other Products

Plant materials are part of many products people use every day. Plants provide the wood used to build houses and the wood pulp used to make paper for books like the one you are reading. The cotton in blue jeans comes from plants. So do many dyes that are used to add color to fabrics. Aspirin and many other medicines made by drug companies today are based on chemicals originally found in plants.

3.4 Review

KEY CONCEPTS

1. How do flowers relate to fruit?
2. How are animals involved in the life cycles of some flowering plants?
3. List three ways that humans depend upon plants.

CRITICAL THINKING

4. **Predict** If you observed three plants in a forest—a moss, a fern, and a flowering plant—which do you think would have the most insects nearby?
5. **Connect** Draw an apple like the one shown on page 107. Label three parts of the fruit and explain from which part of an apple flower each part grew.

CHALLENGE

6. **Synthesize** There are more species of flowering plants on Earth than species of mosses, ferns, or cone-bearing plants such as pine trees. How do you think the different ways spores, pollen, and seeds are spread affect the genetic diversity of different types of plants? Explain your reasoning.

ANSWERS

1. Ovary of flower becomes fruit.

2. Some animals feed on nectar, carrying pollen from flower to flower. Others eat fruit and disperse seeds as waste.

3. Sample answers: oxygen, food, material, coal and gas, wood, paper, fabric, medicine

4. flowering plants because insects feed on nectar and fruit

5. Upper part of pistil becomes stem, fertilized eggs become seeds, ovary becomes fruit.

6. Fruits enable plant species to spread seeds into different environments because seeds are spread more readily. This increases genetic diversity. Spores and pollen are more limited in ways they can spread.

MATH in SCIENCE

SKILL: USING GRIDS TO ESTIMATE

Chloroplast Math

MATH TUTORIAL
CLASSZONE.COM
Click on Math Tutorial for more help with perimeter and area.

You can't count the number of chloroplasts in a leaf very easily, but you can estimate the number. For example, if you know the number of chloroplasts in a small area, you can estimate the number of chloroplasts in a whole leaf.

Example

Suppose you are studying how lilacs make food from sunlight. You read that there are 50 million chloroplasts for every square centimeter of a leaf. You want to know the number in a whole leaf.

(1) Cover the leaf with centimeter grid paper.

(2) Count the number of whole squares covering the leaf.
7

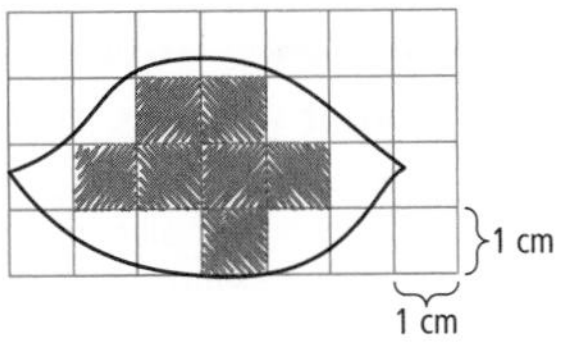

(3) Match pairs or sets of partly covered squares, that add up to a whole square.
7 + 5 = 12

(4) Add on any remaining 0.5 (half), 0.025 (quarter), or 0.75 (three-quarters) of a square.
12 + .5 = 12.5

(5) Finally, multiply the number of squares by the number of chloroplasts in one square.

ANSWER 50,000,000 × 12.5 = 625,000,000.

Give estimates for the following amounts.

1. Trace the beech leaf shown on this page onto a sheet of centimeter grid paper. What is the leaf's approximate area in cm^2?

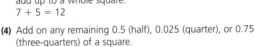

2. About how many chloroplasts are in this beech leaf?
3. A eucalyptus leaf is long and thin. Suppose a healthy leaf is 1.5 centimeters wide and 6 centimeters long. Estimate its area. Hint: Make a sketch.
4. What is the approximate number of chloroplasts in the eucalyptus leaf described above?

CHALLENGE Collect two leaves. Trace the leaves on centimeter grid paper. Label each tracing with its name, estimated area, and its approximate number of chloroplasts.

ANSWERS

1. about 11 cm^2

2. about 550,000,000 chloroplasts (11 × 50,000,000 = 550,000 000)

3. about $8\frac{1}{2}$ cm^2

4. about 425,000,000 chloroplasts ($8\frac{1}{2}$ × 50,000,000 = 425,000 000)

***5. CHALLENGE** Check students' leaf samples and sketches.*

MATH IN SCIENCE
Math Skills Practice for Science

Set Learning Goal

To use area concepts to estimate the number of chloroplasts on a leaf

Present the Science

Chloroplasts are mostly found on the surface of the leaf where they are in full view of sunlight. A single leaf can contain millions of chloroplasts.

Develop Measurement Skills

Areas of irregular objects can be found by using a grid and estimating square units that cover the area.

As students work through the example, make sure they accurately combine parts of squares to equal one whole square. For example, combine 3/4 of a square with 1/4 to get a full square.

DIFFERENTIATION TIP Enlarge a copy of the leaf by about 60% and have students use one-inch grid paper to make their tracings.

Close

Ask students how using grids can help scientists study plants in other ways. *Sample answer: Scientists might estimate numbers of stomata, of cells, of xylem or phloem, etc., in leaves or stems.*

Point out that scientists can use grids for sampling purposes. If a square foot of a field contains a certain number of plants, scientists can estimate the number in the whole field.

R
- Math Support, p. 199
- Math Practice, p. 200

Technology Resources

Students can visit **ClassZone.com** to practice using grids to estimate.

MATH TUTORIAL

BACK TO

the BIG idea

Ask students to choose one of the groups of plants in Chapter 3 (mosses, ferns, gymnosperms, and angiosperms) and to name an adaptation that allows the plant to survive in its environment. *Sample answer: Ferns and mosses have sperm that swim through water, enabling them to reproduce. In gymnosperms, sperm are contained in pollen that travels through air. Angiosperms have seeds that are enclosed in fruits.*

KEY CONCEPTS SUMMARY

SECTION 3.1
Ask: How is the tree able to get essential water and nutrients from the soil to its leaves and energy from its leaves to the rest of tree? *Bundles of vascular tissues called* xylem *and* phloem *transport these materials.*

SECTION 3.2
Ask students to explain how water is transported throughout the bodies of mosses. *Mosses are nonvascular plants and do not have tissue for transporting water. Instead water travels from one cell to the next.*

SECTION 3.3
Ask students to describe the advantages gymnosperms have over mosses and ferns in reproduction. *Gymnosperm have seeds that contain food for the embryo; sperm does not need water to reach an egg.*

SECTION 3.4
Ask: What is the main difference between reproduction in angiosperms and gymnosperms? *Male and female reproductive parts are found in a flower in angiosperms. Seeds are surrounded by fruit.*

Review Concepts

- Big Idea Flow Chart, p. T17
- Chapter Outline, pp. T23–T24

3 Chapter Review

the BIG idea

Plants are a diverse group of organisms that live in many land environments.

CONTENT REVIEW
CLASSZONE.COM

KEY CONCEPTS SUMMARY

3.1 Plants are adapted to living on land.

All plants share common characteristics. The parts of a plant are specialized to get water and nutrients from the soil, gases from the air, and energy from the Sun. Plants have tissues, organs, and organ systems.

VOCABULARY
vascular system p. 87
transpiration p. 88

3.2 Most mosses and ferns live in moist environments.

The ancestors of present-day mosses and ferns were among the first land plants. Mosses are small nonvascular plants. Ferns are larger vascular plants. Both reproduce with spores and need moisture for a sperm to reach an egg.

3.3 Seeds and pollen are reproductive adaptations.

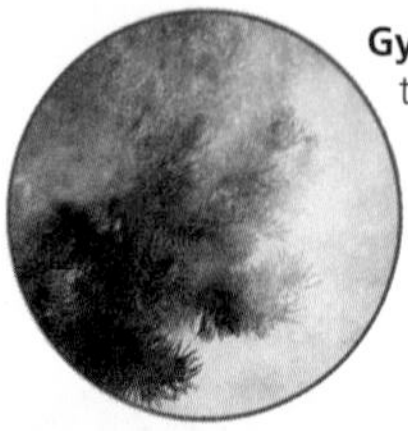

Gymnosperms, such as the pine tree, reproduce with **pollen** and seeds.

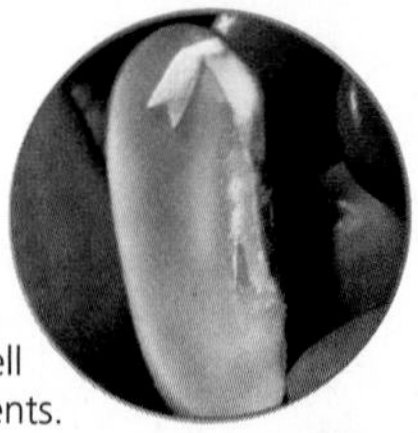

Seeds provide protection for the young plant as well as a supply of nutrients.

VOCABULARY
seed p. 98
embryo p. 98
germination p. 99
pollen p. 100
gymnosperm p. 102

3.4 Many plants reproduce with flowers and fruit.

Angiosperms use flowers and fruit to reproduce. **Flowers** produce pollen and contain the plant's reproductive structures.

Fruit develops after pollination and contains seeds. Animals eat fruit and transport seeds to new locations.

VOCABULARY
angiosperm p. 107
flower p. 108
fruit p. 108

C 116 Unit: Diversity of Living Things

Technology Resources

Have students visit **ClassZone.com** or use the CD-ROM for a cumulative review of concepts.

CONTENT REVIEW

CONTENT REVIEW CD-ROM

Engage students in a whole-class interactive review of Key Concepts. Edit content as you wish.

POWER PRESENTATIONS

Reviewing Vocabulary

Write a statement describing how the terms in each pair are related to each other.

1. transpiration, cuticle
2. spores, seeds
3. seed, embryo
4. flower, fruit

The table below shows Greek (G.) and Latin (L.) words that are roots of words used in this chapter.

angeion (G.)	vessel or holder
gymnos (G.)	naked
sperma (G.)	seed
germinatus (L.)	sprout
pollen (L.)	fine flour

Describe how these word roots relate to the definitions of the following words.

5. angiosperm
6. gymnosperm
7. germination
8. pollen

Reviewing Key Concepts

Multiple Choice *Choose the letter of the best answer.*

9. Which of these is a characteristic of only some plants?
   - **a.** They produce sugars.
   - **b.** They are multicellular.
   - **c.** They have a vascular system.
   - **d.** They have alternation of generations.

10. Which part of a plant anchors it in the soil?
    - **a.** shoot system
    - **b.** root system
    - **c.** vascular system
    - **d.** growth tissue

11. Which plant tissue transports water to different parts of the plant?
    - **a.** vascular
    - **b.** leaf
    - **c.** stomata
    - **d.** cuticle

12. Which part of a leaf does *not* allow transpiration?
    - **a.** stomata
    - **b.** cell wall
    - **c.** stem
    - **d.** cuticle

13. Which of these structures do mosses and ferns reproduce with?
    - **a.** seeds
    - **b.** growth tissue
    - **c.** spores
    - **d.** pollen

14. How are mosses different from ferns, pine trees, and flowering plants?
    - **a.** Mosses reproduce through sexual reproduction.
    - **b.** Mosses need moisture to reproduce.
    - **c.** Mosses produce sugar through photosynthesis.
    - **d.** Mosses have no vascular tissue.

15. What do seeds have that spores and pollen do not?
    - **a.** a supply of nutrients
    - **b.** a reproductive cell
    - **c.** a protective covering
    - **d.** a way to be transported

Short Answer *Write a short answer to each question.*

16. What are three ways plants are important to humans?

17. Explain how the bat in these photographs is interacting with the flower. How might the activity benefit the plant?

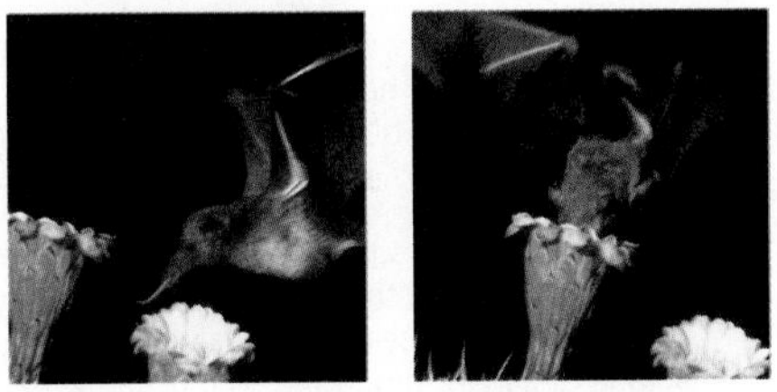

ASSESSMENT RESOURCES

UNIT ASSESSMENT BOOK
- Chapter Test, Level A, pp. 47–50
- Chapter Test, Level B, pp. 51–54
- Chapter Test, Level C, pp. 55–58
- Alternative Assessment, pp. 59–60

SPANISH ASSESSMENT BOOK
Spanish Chapter Test, pp. 49–52

Technology Resources

Edit test items and answer choices.

Test Generator CD-ROM

Visit **ClassZone.com** to extend test practice.

Test Practice

Reviewing Vocabulary

1. Transpiration is movement of water vapor out of a plant through the leaves. The cuticle is a waxy layer on the surface of leaves that limits transpiration.

2. Both grow into new plants. Spores are single cells released from a moss or fern. Seeds contain multi-celled embryos and are produced by angiosperms or gymnosperms.

3. The embryo is a new plant. It is contained in the seed, which provides protection and nutrients.

4. The flower is the reproductive part of the plant. Egg cells inside the flower are fertilized, forming seeds. The fruit grows around the seeds.

5. Angiosperms are plants with seeds enclosed in fruit. The fruit is like a vessel or holder for the seeds.

6. Gymnosperm seeds are not enclosed in fruit. In this way they are naked seeds.

7. Germination is the process by which a seed or spore begins to sprout into a plant.

8. Pollen is like fine flour in that it is made up of small particles.

Reviewing Key Concepts

9. c

10. b

11. a

12. d

13. c

14. d

15. a

16. Answers should mention three of the following: Plants supply oxygen, materials for shelter, food; contribute to energy resources; and are used to manufacture cloth, paper, medicines, and other products.

17. As the bat feeds on the nectar, it picks up pollen. The bat brings pollen to the next plant it visits. This enables the plant to be pollinated. It also increases genetic diversity in plants.

Thinking Critically

18. The cuticle is a waxy covering that keeps water needed for photosynthesis from evaporating.

19. Xylem and phloem are in the vascular system. Xylem brings water and nutrients in; phloem moves energy-rich materials out.

20. Stomata allow exchange of gases. When air is dry, the stomata close so water will not evaporate.

21. Chloroplasts produce sugars using energy from sunlight. More are in the upper leaf, where there is plenty of sunlight.

22. Photosynthesis is the main function. A large surface allows for a lot of energy from sunlight. Stomata allow CO_2 gas in and let O_2 gas out. Xylem brings water and nutrients in, phloem transports sugars out.

23. Pine trees—gymnosperms; cherry trees—angiosperms; angiosperms are more widespread.

24. A large amount of pollen increases the chances of fertilization.

25. egg: all; pollen and seed: pine tree, cherry tree; cone: pine tree; flower and fruit: cherry tree

26. Plants provide food, O_2, shelter. Animals transport pollen and seeds.

27. Plants can produce offshoots, such as runners. Plants can spread more quickly, but genetic material is the same as parent's.

the BIG idea

28. The tree has a large root system, a vascular system to move materials, and lots of leaves for photosynthesis.

29. Answers should refer to at least three different types of plants.

UNIT PROJECTS

Collect schedules, materials lists, and questions. Be sure dates and materials are obtainable, and questions are focused.

R Unit Projects, pp. 5–10

Thinking Critically

The next four questions refer to the labeled parts of the diagram of the leaf below.

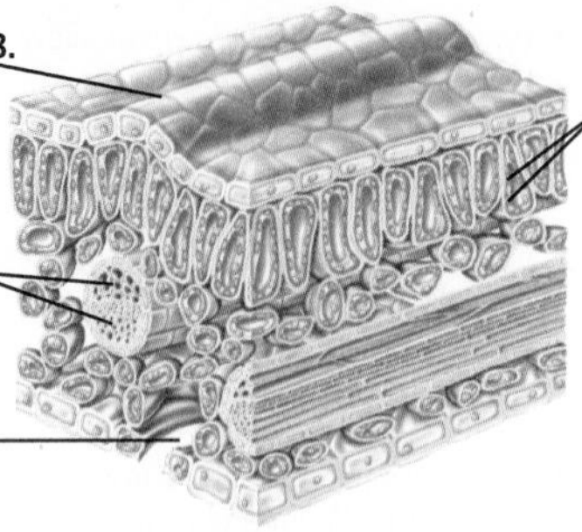

18. CLASSIFY Identify the layer that covers a leaf. What function does it serve, and why is that function important?

19. CLASSIFY What are the two tissues shown here, how do they function, and what system do they belong to?

20. HYPOTHESIZE Identify the structure and write a hypothesis about what would happen to this structure if the air around the plant was very hot and dry.

21. ANALYZE What are these structures? What is their function? Why are there more of them at the top of the leaf than at the bottom?

22. COMMUNICATE A leaf is made up of different tissues and specialized cells. What is the main function of a leaf? How does the organization of tissues and cells in a leaf help it to carry out this function? Include the following terms in your answer: *sunlight*, *energy*, *oxygen gas*, *carbon dioxide gas*, *water*, and *sugar*.

23. CLASSIFY What is the name of the group of plants pine trees belong to? What group of plants do cherry trees belong to? Which of the two groups is more widespread?

24. CONNECT You or someone you know may be allergic to pollen that is in the air during some plants' growing seasons. What is the advantage of a plant producing so much pollen?

25. SUMMARIZE Copy and complete the table below, indicating which reproductive structures are part of the life cycles of mosses, ferns, pine trees, and cherry trees. The first row is done.

	moss	fern	pine tree	cherry tree
sperm	✓	✓	✓	✓
egg				
pollen				
seed				
cone				
flower				
fruit				

26. SYNTHESIZE Plants were among the first organisms to live on land. Name at least two ways that plants made land habitable for animals. In what way did animals help plants to spread farther onto the land?

27. PROVIDE EXAMPLES Describe how a plant might reproduce asexually. Explain one advantage and one disadvantage asexual reproduction might have for the plant you described.

the BIG idea

28. SUMMARIZE Look again at the photograph on pages 82–83. Now that you have finished the chapter, how would you change or add details to your answer to the question on the photograph?

29. SYNTHESIZE Think of three different types of plants that you have seen. Use what you know about those plants as supporting evidence in a paragraph you write on one of these topics:

Plants are a diverse group of organisms.

Plants share common characteristics.

UNIT PROJECTS

Check your schedule for your unit project. How are you doing? Be sure that you have placed data or notes from your research in your project folder.

MONITOR AND RETEACH

If students are having trouble applying the concepts in items 18–21, suggest that they review the visual on p. 89. Have them create a word frame diagram for each of the following words: *stomata, chloroplast, xylem, phloem, photosynthesis,* and *leaf*. Tell them to leave the center blank and write the word very small on the back of the page. They can then trade frames with a partner and guess the words.

Students may benefit from summarizing one or more sections of the chapter.

R Summarizing the Chapter, pp. 219–220

C 118 Unit: **Diversity of Living Things**

Standardized Test Practice

For practice on your state test, go to . . .
TEST PRACTICE
CLASSZONE.COM

Analyzing Data

A pesticide is a material that kills pests, such as insects. A plant specialist wants to know if a new pesticide has any effect on the production of oranges. Grove A is planted using the same pesticide that was used in previous years. The new pesticide is used in grove B. Both groves have the same number of orange trees. The bar graph shows the data for one season.

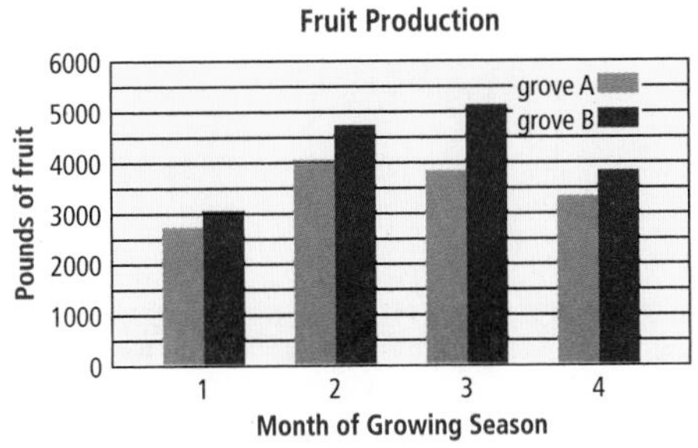

Choose the letter of the best response.

1. About how many pounds of fruit were grown in grove A during month 1 of the growing season?
   a. about 2000 lb
   b. about 2700 lb
   c. about 3000 lb
   d. about 3500 lb

2. During which month in the growing season was the most fruit produced?
   a. month 1
   b. month 2
   c. month 3
   d. month 4

3. About how many pounds of fruit were produced altogether by both groves in month 2?
   a. about 4000 lb
   b. about 4700 lb
   c. about 5000 lb
   d. about 8700 lb

4. Comparing the production in both groves, during which month was there the least amount of difference in the number of pounds of fruit produced?
   a. month 1
   b. month 2
   c. month 3
   d. month 4

5. Based on the data in the graph, what might the plant specialist conclude about the effectiveness of the new pesticide used in grove B?
   a. The pesticide is effective only after three months of growth.
   b. The pesticide does not have any effect on orange production.
   c. The pesticide used in grove A is more effective than the pesticide used in grove B.
   d. The pesticide increases the overall production of oranges throughout the growing season.

Extended Response

6. Pesticides are intended to kill insects that harm the growth of a plant. However, not all insects are harmful. Insects often pollinate the flowers of fruit trees. Pollination leads to fertilization of the flower, and fruit grows from the fertilized flower. Describe what might happen if a pesticide kills insects that are pollinators. What effect might such a pesticide have on the flowers of the plant and on its fruit?

7. Pesticides are only one factor that affect the growth of plants and the number of flowers or fruit the plant produces. Write a paragraph that names some other factors that may affect plant growth. Include environmental factors and factors that are controlled by humans.

METACOGNITIVE ACTIVITY

Have students answer the following questions in their **Science Notebook:**

1. What plants were familiar to you before you read the chapter? What were some unfamiliar plants or new plants you learned about?
2. What most surprised you about the relationships between plants and animals, including humans?
3. What activity or question helped you most in learning about plants?

Analyzing Data

1. b *3. d* *5. d*
2. c *4. a*

Extended Response

6. RUBRIC

4 points for a response that correctly identifies four ways that plants will be affected.

Sample answer: If the insecticide kills insects that pollinate the flowers, the flowers will still grow. With fewer pollinating insects around, there is less chance that the insects will feed on nectar and pollinate the flowers. If flowers are not pollinated, the eggs will not be fertilized and produce fruit. Fruit production will decline and the plant population will be in danger of declining as well.

3 points correctly identifies three ways that plants will be affected
2 points correctly identifies two ways that plants will be affected
1 point correctly identifies one way that plants will be affected

7. RUBRIC

4 points for a response that correctly describes environmental factors and those that are controlled by humans.

Sample answer: The amount of precipitation affects growth, either too little or too much. Temperature can also affect growth—too high and the plants dry out; too low and the plants freeze. The quality of soil is also a factor. Humans can alter the amount of water by irrigating fields. They can improve the quality of soil by fertilizing it. They can breed plants that are better able to grow and produce fruit.

3 points describes at least two effects of environment and one effect of humans
2 points describes at least one effect of environment and one effect of humans
1 point describes at least one effect of environment or one effect of humans

CHAPTER

4 Invertebrate Animals

Life Science UNIFYING PRINCIPLES

PRINCIPLE 1
All living things share common characteristics.

PRINCIPLE 2
All living things share common needs.

PRINCIPLE 3
Living things meet their needs through interactions with the environment.

PRINCIPLE 4
The types and numbers of living things change over time.

Unit: Diversity of Living Things BIG IDEAS

CHAPTER 1
Single-Celled Organisms and Viruses
Bacteria and protists have the characteristics of living things, while viruses are not alive.

CHAPTER 2
Introduction to Multicellular Organisms
Multicellular organisms live in and get energy from a variety of environments.

CHAPTER 3
Plants
Plants are a diverse group of organisms that live in many land environments.

CHAPTER 4
Invertebrate Animals
Invertebrate animals have a variety of body plans and adaptations.

CHAPTER 5
Vertebrate Animals
Vertebrate animals live in most of Earth's environments.

CHAPTER 4 KEY CONCEPTS

SECTION 4.1

Most animals are invertebrates.

1. Invertebrates are a diverse group of organisms.
2. Sponges are simple animals.

SECTION 4.2

Cnidarians and worms have different body plans.

1. Cnidarians have simple body systems.
2. Animals have different body plans.
3. Most worms have complex body systems.

SECTION 4.3

Most mollusks have shells, and echinoderms have spiny skeletons.

1. Mollusks are soft-bodied animals.
2. Mollusks show a range of adaptations.
3. Echinoderms have unusual adaptations.

SECTION 4.4

Arthropods have exoskeletons and joints.

1. Most invertebrates are arthropods.
2. Insects are six-legged arthropods.
3. Crustaceans live in water and on land.
4. Arachnids are eight-legged arthropods.
5. Millipedes and centipedes are arthropods.

The Big Idea Flow Chart is available on p. T25 in the **UNIT TRANSPARENCY BOOK.**

Previewing Content

SECTION

4.1 Most animals are invertebrates. pp. 123–127

1. Invertebrates are a diverse group of organisms.
Invertebrates include all groups of animals that do not have backbones. The six groups of **invertebrates** include sponges, cnidarians, worms, mollusks, echinoderms, and arthropods. Invertebrates live in a variety of environments on both land and in water.

2. Sponges are simple animals.

- A **sponge** is made up of a body wall with inner and outer layers.
- Flagella that line the body cavity of the sponge move water through the sponge, as shown. Specialized food cells digest plankton and other tiny organisms that sponges feed upon.

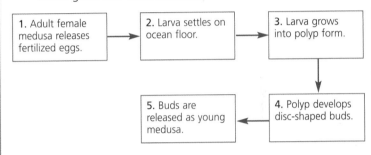

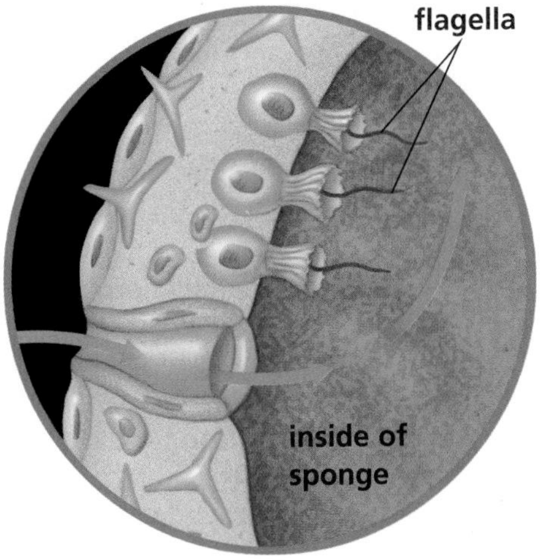

- Sponges reproduce asexually by budding. The new sponges are genetically identical to the parent sponge. Sponges also reproduce sexually by releasing sperm into the water. Sperm from one sponge then fertilizes the eggs of another sponge. **Immature** sponges, or **larvae,** are then released into the water where they attach to an underwater surface and remain **sessile.**

SECTION

4.2 Cnidarians and worms have different body plans. pp. 128–135

1. Cnidarians have simple body systems.

- **Cnidarians** include jellyfish, corals, sea anemones, and hydras. All cnidarians have **tentacles** with specialized stinging cells. These cells contain nematocysts, capsules that hold poisonous barbed filaments. Tissues in the cnidarians body include muscle and nerve cells that allow them to sense and respond to stimuli in the environment.
- Cnidarians reproduce both asexually and sexually. The jellyfish life cycle includes a polyp stage that is sessile and a medusa stage that is mobile. The polyps release disk-shaped buds, which grow into adult medusas.

1. Adult female medusa releases fertilized eggs. →
2. Larva settles on ocean floor. →
3. Larva grows into polyp form. →
4. Polyp develops disc-shaped buds. →
5. Buds are released as young medusa.

2. Animals have different body plans.

- A radial body plan is organized around a central point, usually a mouth that leads into a cavity where food is digested.
- A bilateral body plan has two mirror images of each other. Bilaterally shaped animals have front and back ends. The front end usually includes a head and mouth that consumes food, while the back end releases wastes.

3. Most worms have complex body systems.

- Segmented worms, called annelids, are divided into segments. Annelids have digestive, circulatory, muscular, and nervous systems. Annelids reproduce both asexually and sexually.
- Flatworms are the simplest worms, often living as parasites. Roundworms are more complex than flatworms.

Common Misconceptions

MISCONCEPTION DATABASE
CLASSZONE.COM Background on student misconceptions

IS IT AN ANIMAL? Many studies have shown that students often think that animals only include large land animals, such as pets, farm animals, and those seen in zoos. When it comes to invertebrates, fewer than half of elementary students will identify an animal as such. In fact, any multicellular organism that does not photosynthesize and can move on its own is an animal.

TE This misconception is addressed on p. 124.

REPRODUCTION Students commonly hold these misconceptions about reproduction: 1) Animals plan their reproductive strategies; 2) Sexual reproduction must involve a separate male and female organism; and 3) Sexual reproduction must involve mating. In fact, none of the above are necessarily true of all animals.

TE This misconception is addressed on p. 130.

Previewing Content

SECTION

4.3 Most mollusks have shells, and echinoderms have spiny skeletons. pp. 136–141

1. Mollusks are soft-bodied animals.
Mollusks include bivalves, gastropods, and cephalopods. Most mollusks have a hard, outer shell and relatively well-developed organ systems. They also have a muscular foot and a mantle, a layer of folded skin that protects internal organs. The table shows the mollusk groups and their characteristics.

Mollusks			
	Bivalves	**Gastropods**	**Cephalopods**
Characteristics	made up of two matching shell halves	most are protected by a spiral-shaped shell	most have eyes and tentacles for capturing prey
Examples	clams, scallops, mussels, oysters	snails, slugs, conches, whelks, periwinkles	squids, chambered nautiluses, octopuses

2. Mollusks show a range of adaptations.
The muscular foot is a feature that has been adapted to function differently for each group of mollusks.

3. Echinoderms have unusual adaptations.
- **Echinoderms** include radially symmetric sea stars, sea urchins, and sand dollars. These animals have mouths that are located in the center of the body and feed upon the ocean floor.
- Echinoderms have spines and internal skeletons. Water vascular systems and tube feet enable echinoderms to move and feed.

SECTION

4.4 Arthropods have exoskeletons and joints. pp. 142–149

1. Most invertebrates are arthropods.
Arthropods have a segmented body covered with an **exoskeleton,** an outer covering that is made up of chitin. This exoskeleton is shed as the animal grows in a process called **molting.** Arthropods have complex body systems. The three main groups of arthropods are insects, crustaceans, and arachnids.

2. Insects are six-legged arthropods.
- All adult **insects** have three distinct body segments and six legs.

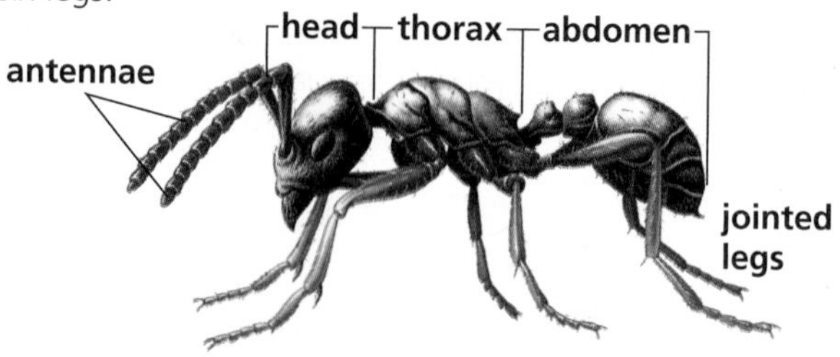

- Insects have a variety of adaptations such as specialized mouth parts, wings, and color and shape to blend into the environment.
- Insects undergo a process called **metamorphosis** in which the insect's appearance and body systems change at each life stage. Insects may go through simple or complete metamorphosis.

3. Crustaceans live in water and on land.
Crustaceans include lobsters, shrimp, crabs, barnacles, krill, and copepods. They have complex body systems that include a circulatory system.

4. Arachnids are eight-legged arthropods.
Arachnids include spiders, mites, ticks, and scorpions. They have two body segments as opposed to the three body segments in insects. Spiders have a unique adaptation for capturing prey by spinning webs made of silk.

5. Millipedes and centipedes are arthropods.
Millipedes have two pairs of walking legs on multiple body segments. Centipedes have one pair of walking legs per body segment.

Common Misconceptions

ANIMAL TRAITS Among middle school students, reasons for identifying something as an animal commonly include: has four legs, makes sound, and has fur. Most of the animals in this chapter would not fit any of those criteria.

TE This misconception is addressed on p. 144.

MISCONCEPTION DATABASE
CLASSZONE.COM Background on student misconceptions

Previewing Labs

Lab Generator CD-ROM
Edit these Pupil Edition labs and generate alternative labs.

EXPLORE (the BIG idea)

Worm-Watching, p. 121
Students observe worms in order to describe the worms' characteristics, activities, and behavior.

TIME 10 minutes
MATERIALS notebook, outdoor area, hand lens

Insects and You, p. 121
Students list the types of insects seen during a week in order to discover diversity and characteristics.

TIME 10 minutes
MATERIALS notebook, outdoor area, hand lens

Internet Activity: Invertebrate Diversity, p. 121
Students observe invertebrates to learn about the different types.

TIME 20 minutes
MATERIALS computer with Internet access

SECTION 4.1

INVESTIGATE Invertebrates, p. 124
Students make a potato trap to determine the kinds of invertebrates that live in local environments.

TIME 20 minutes
MATERIALS potato, plastic knife, spoon, masking tape

SECTION 4.2

EXPLORE Worm Movement, p. 128
Students observe a worm to explain how its body shape affects its movement.

TIME 20 minutes
MATERIALS worm, tray, soil to cover bottom of tray, spray bottle filled with distilled water

---

CHAPTER INVESTIGATION, Worm Behavior, pp. 134–135
Students observe earthworm behavior and predict how they will respond to environmental stimuli.

TIME 40 minutes
MATERIALS potting soil, coarse sand, aquarium, filter paper, spray bottle, small beaker, distilled water, 5 or more worms, 2 containers (one for untested worms and one for tested worms), stopwatch, (for Challenge: flashlight)

SECTION 4.3

INVESTIGATE Mollusks and Echinoderms, p. 138
Students examine mollusk and echinoderm shells to compare and contrast their characteristics.

TIME 15 minutes
MATERIALS selection of mollusk shells, sea stars, and sand dollars

---

SECTION 4.4

EXPLORE Arthropods, p. 142
Students observe pillbugs to describe their characteristics.

TIME 20 minutes
MATERIALS clear container, shoebox (half of cover removed), pillbugs, hand lens

---

INVESTIGATE Insect Metamorphosis, p. 146
Students observe mealworms to discover how often they molt.

TIME 20 minutes
MATERIALS glass jar with lid with air holes, oat bran, potato, carrot, mealworms, (for Challenge: petri dish, tweezers)

R **Additional INVESTIGATION,** Exploring How a Squid Moves, A, B, & C, pp. 280–288; Teacher Instructions, pp. 360–361

Previewing Chapter Resources

	INTEGRATED TECHNOLOGY	LABS AND ACTIVITIES
CHAPTER 4 **Invertebrate Animals**	**CLASSZONE.COM** • eEdition Plus • EasyPlanner Plus • Misconception Database • Content Review • Test Practice • Resource Centers • Visualization • Internet Activity • Math Tutorial **SCILINKS.ORG** SCILINKS **CD-ROMS** • eEdition • EasyPlanner Plus • Power Presentations • Content Review • Lab Generator • Test Generator **AUDIO CDS** • Audio Readings • Audio Readings in Spanish	PE EXPLORE the Big Idea, p. 121 • Worm-Watching • Insects and You • Internet Activity: Invertebrate Diversity R **UNIT RESOURCE BOOK** Unit Projects, pp. 5–10 **Lab Generator CD-ROM** Generate customized labs.
SECTION 4.1 **Most animals are invertebrates.** pp. 123–127 Time: 2 periods (1 block) R Lesson Plan, pp. 221–222	• **MATH TUTORIAL** T **UNIT TRANSPARENCY BOOK** • Big Idea Flow Chart, p. T25 • Daily Vocabulary Scaffolding, p. T26 • Note-Taking Model, p. T27 • 3-Minute Warm-Up, p. T28	PE • INVESTIGATE Invertebrates, p. 124 • Math in Science, p. 127 R **UNIT RESOURCE BOOK** • Datasheet, Invertebrates, p. 230 • Math Support, p. 269 • Math Practice, p. 270
SECTION 4.2 **Cnidarians and worms have different body plans.** pp. 128–135 Time: 3 periods (1.5 blocks) R Lesson Plan, pp. 232–233	**RESOURCE CENTER,** Types of Worms T **UNIT TRANSPARENCY BOOK** • Daily Vocabulary Scaffolding, p. T26 • 3-Minute Warm-Up, p. T28 • "Inside an Earthworm" Visual, p. T30	PE • EXPLORE Worm Movement, p. 128 • CHAPTER INVESTIGATION, Worm Behavior, pp. 134–135 R **UNIT RESOURCE BOOK** CHAPTER INVESTIGATION, Worm Behavior, A, B, & C, pp. 271–279
SECTION 4.3 **Most mollusks have shells, and echinoderms have spiny skeletons.** pp. 136–141 Time: 2 periods (1 block) R Lesson Plan, pp. 242–243	**RESOURCE CENTER,** Mollusks **VISUALIZATION,** Cephalopods T **UNIT TRANSPARENCY BOOK** • Daily Vocabulary Scaffolding, p. T26 • 3-Minute Warm-Up, p. T29	PE • INVESTIGATE Mollusks and Echinoderms, p. 138 • Think Science, p. 141 R **UNIT RESOURCE BOOK** • Datasheet, Mollusks and Echinoderms, p. 251 • Additional INVESTIGATION, Exploring How a Squid Moves, A, B, & C, pp. 280–288
SECTION 4.4 **Arthropods have exoskeletons and joints.** pp. 142–149 Time: 3 periods (1.5 blocks) R Lesson Plan, pp. 253–254	**RESOURCE CENTER,** Diversity of Arthropods T **UNIT TRANSPARENCY BOOK** • Big Idea Flow Chart, p. T25 • Daily Vocabulary Scaffolding, p. T26 • 3-Minute Warm-Up, p. T29 • Chapter Outline, pp. T31–T32	PE • EXPLORE Arthropods, p. 142 • INVESTIGATE Insect Metamorphosis, p. 146 R **UNIT RESOURCE BOOK** Datasheet, Insect Metamorphosis, p. 262

KEY TO ICONS: CD/CD-ROM · Teacher Edition · UNIT TRANSPARENCY BOOK · SPANISH ASSESSMENT BOOK · INTERNET · Pupil Edition · UNIT RESOURCE BOOK · UNIT ASSESSMENT BOOK · SCIENCE TOOLKIT

READING AND REINFORCEMENT	ASSESSMENT	STANDARDS
• Choose Your Own Strategy, B18–27 • Combination Notes, C36 • Daily Vocabulary Scaffolding, H1–8 **UNIT RESOURCE BOOK** • Vocabulary Practice, pp. 266–267 • Decoding Support, p. 268 • Summarizing the Chapter, pp. 289–290 **Audio Readings CD** Listen to Pupil Edition. **Audio Readings in Spanish CD** Listen to Pupil Edition in Spanish.	• Chapter Review, pp.150–151 • Standardized Test Practice, p. 152 **UNIT ASSESSMENT BOOK** • Diagnostic Test, pp. 61–62 • Chapter Test, A, B, & C, pp. 67–78 • Alternative Assessment, pp. 79–80 Spanish Chapter Test, pp. 53–56 **Test Generator CD-ROM** Generate customized tests. **Lab Generator CD-ROM** Rubrics for Labs	National Standards A.1–8, A.9.a–c, A.9.e–g, C.1.d, C.5.a, G.1.b See p. 120 for the standards.
UNIT RESOURCE BOOK • Reading Study Guide, A & B, pp. 223–226 • Spanish Reading Study Guide, pp. 227–228 • Challenge and Extension, p. 229 • Reinforcing Key Concepts, p. 231	Ongoing Assessment, pp. 123–125 Section 4.1 Review, p. 126 **UNIT ASSESSMENT BOOK** Section 4.1 Quiz, p. 63	National Standards A.2–8, A.9.a–c, A.9.e–f, C.1.d, C.5.a, G.1.b
UNIT RESOURCE BOOK • Reading Study Guide, A & B, pp. 234–237 • Spanish Reading Study Guide, pp. 238–239 • Challenge and Extension, p. 240 • Reinforcing Key Concepts, p. 241 • Challenge Reading, pp. 264–265	Ongoing Assessment, pp. 129, 131–133 Section 4.2 Review, p. 133 **UNIT ASSESSMENT BOOK** Section 4.2 Quiz, p. 64	National Standards A.1–7, A.9.a–b, A.9.e–g, C.1.d, C.5.a, G.1.b
UNIT RESOURCE BOOK • Reading Study Guide, A & B, pp. 244–247 • Spanish Reading Study Guide, pp. 248–249 • Challenge and Extension, p. 250 • Reinforcing Key Concepts, p. 252	Ongoing Assessment, pp. 137–140 Section 4.3 Review, p. 140 **UNIT ASSESSMENT BOOK** Section 4.3 Quiz, p. 65	National Standards A.2–7, A.9.a–b, A.9.e–f, C.1.d, C.5.a, G.1.b
UNIT RESOURCE BOOK • Reading Study Guide, A & B, pp. 255–258 • Spanish Reading Study Guide, pp. 259–260 • Challenge and Extension, p. 261 • Reinforcing Key Concepts, p. 263	Ongoing Assessment, pp. 142–149 Section 4.4 Review, p. 149 **UNIT ASSESSMENT BOOK** Section 4.4 Quiz, p. 66	National Standards A.2–7, A.9.a–b, A.9.e–f, C.1.d, C.5.a, G.1.b

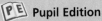

Previewing Resources for Differentiated Instruction

CHAPTER INVESTIGATION

Leveled resources present the same concepts for different abilities.

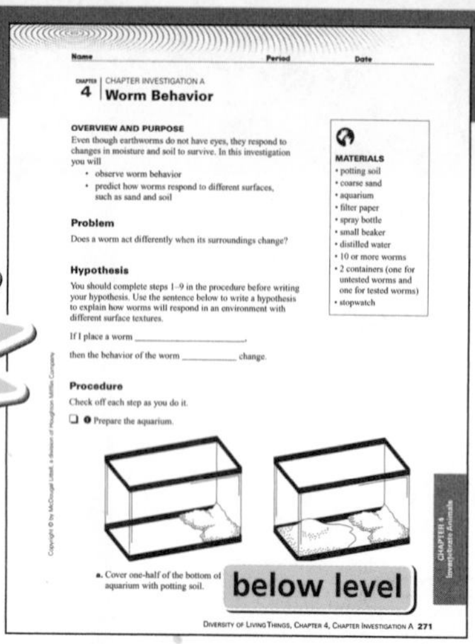

R **UNIT RESOURCE BOOK,** pp. 271–274

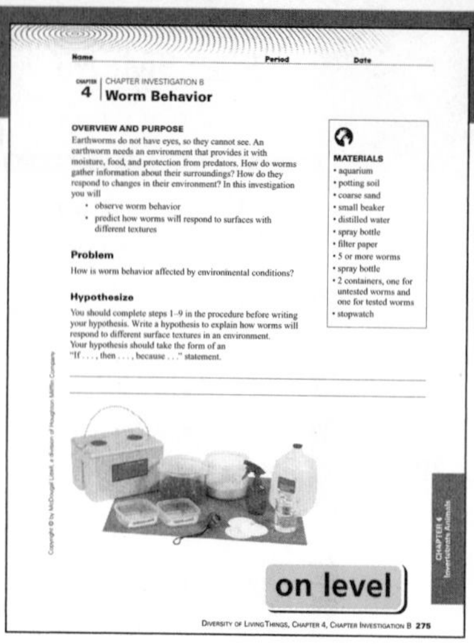

R pp. 275–278

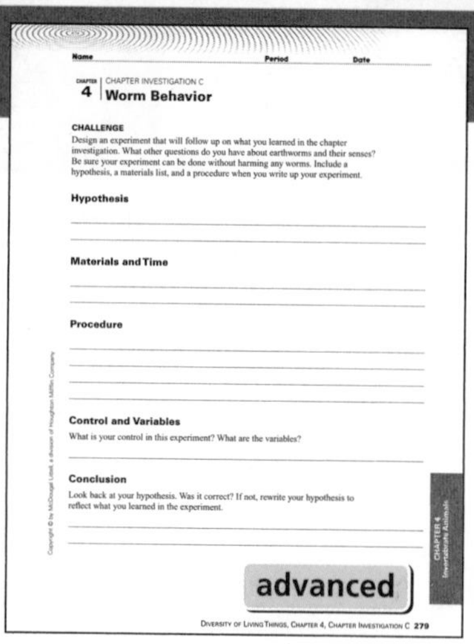

R pp. 275–279

READING STUDY GUIDE

Reading Study Guide is also in Spanish.

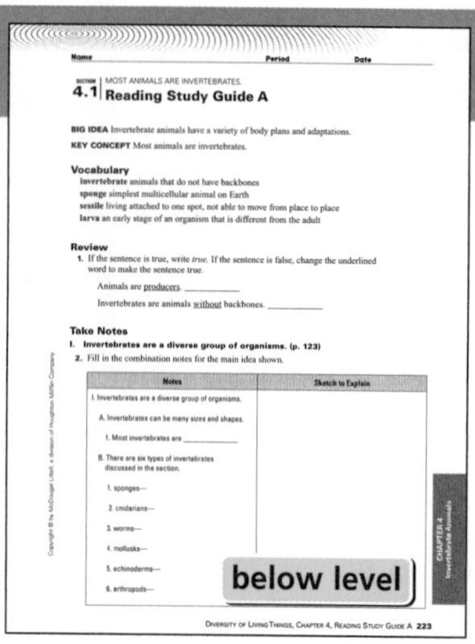

R **UNIT RESOURCE BOOK,** pp. 223–224

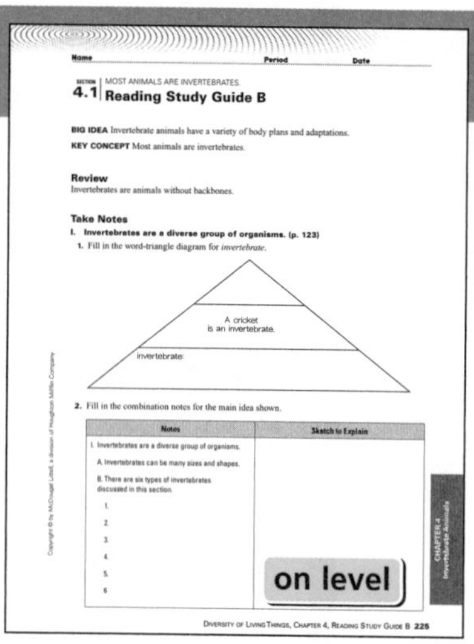

R pp. 225–222

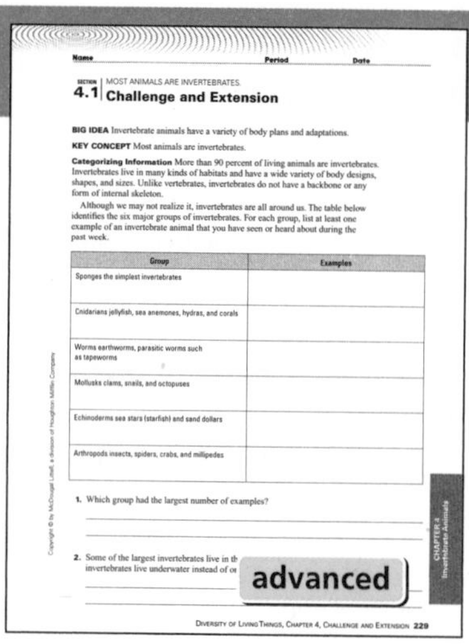

R p. 229

CHAPTER TEST

Chapter Test is also in Spanish.

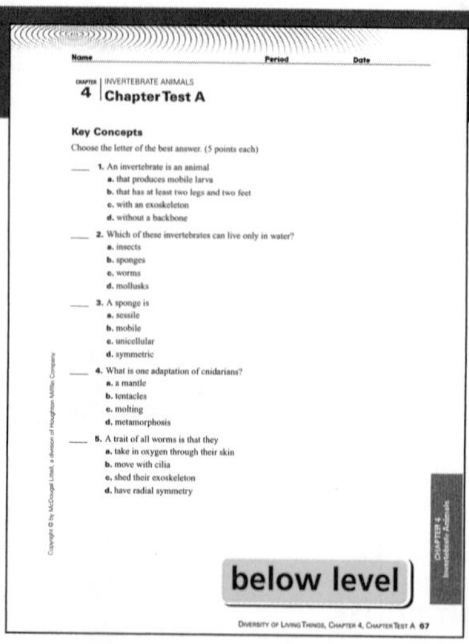

A **UNIT ASSESSMENT BOOK,** pp. 67–70

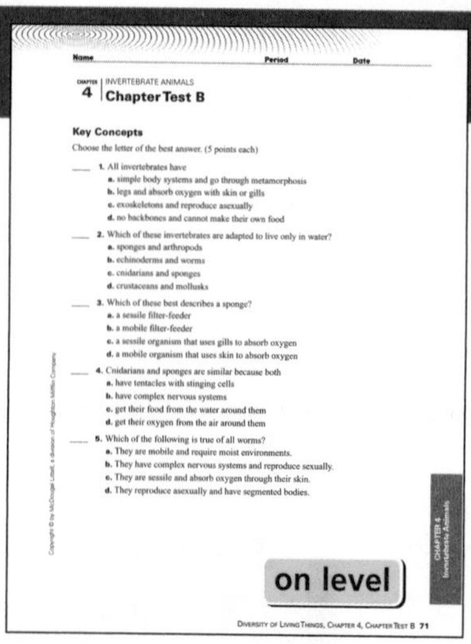

A pp. 71–74

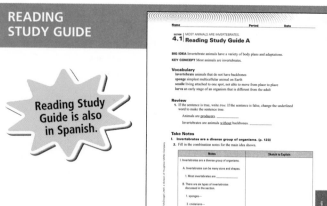

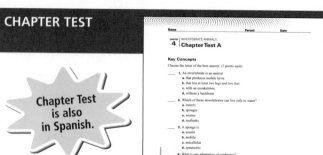

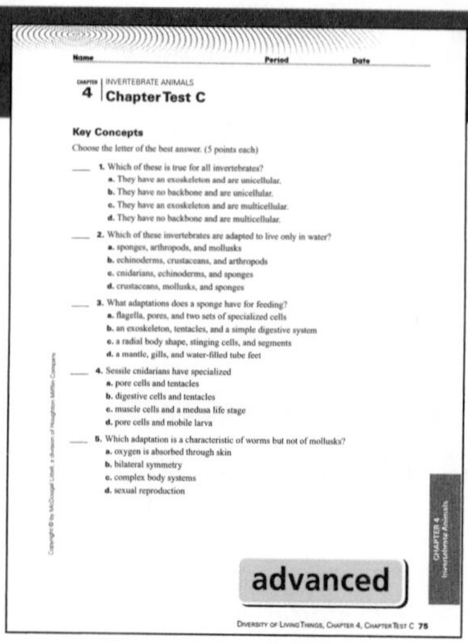

A pp. 75–78

TECHNOLOGY

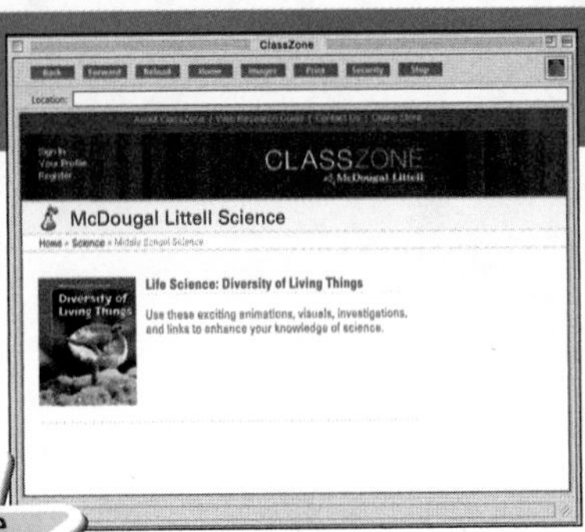

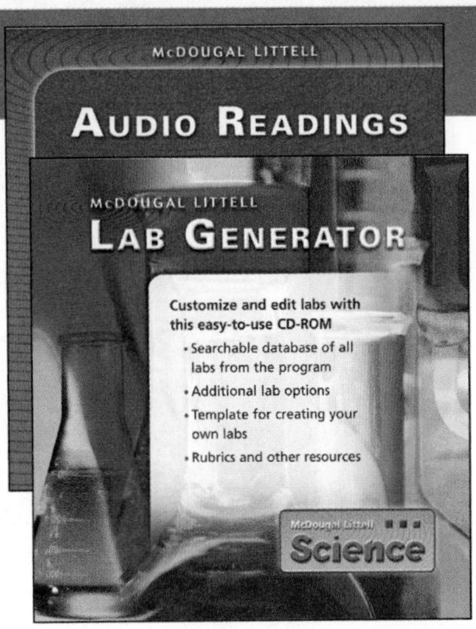

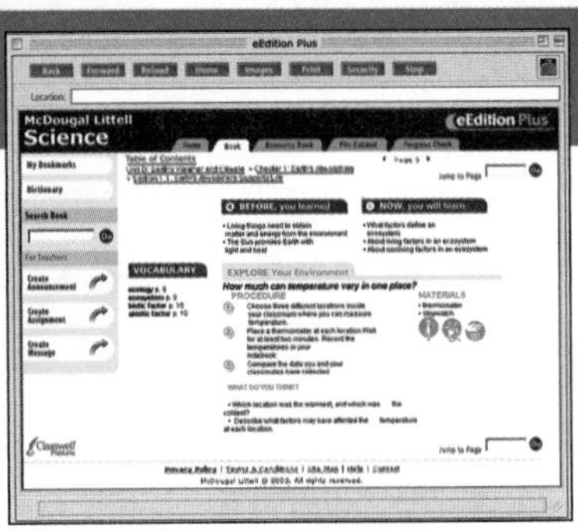

There are three Resource Centers for this chapter.

CLASSZONE.COM

CD/CD-ROMS

CLASSZONE.COM

VISUAL CONTENT

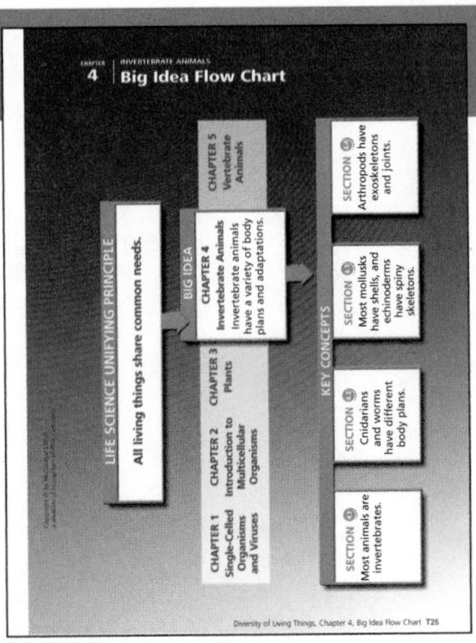

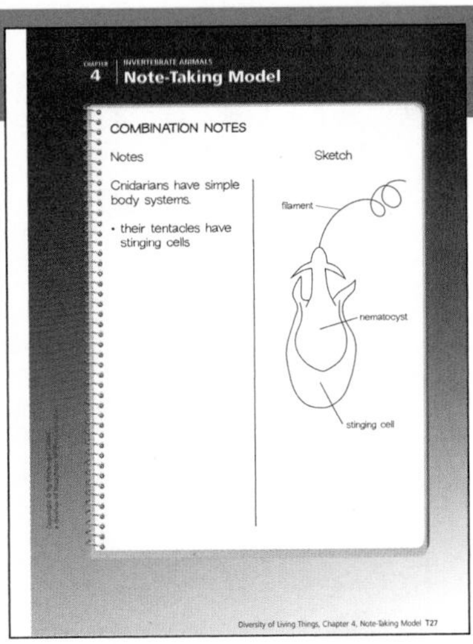

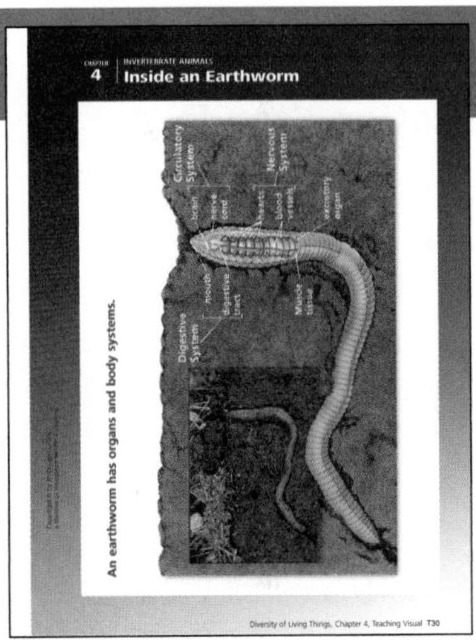

UNIT TRANSPARENCY BOOK, p. T25

p. T27

p. T30

MORE SUPPORT

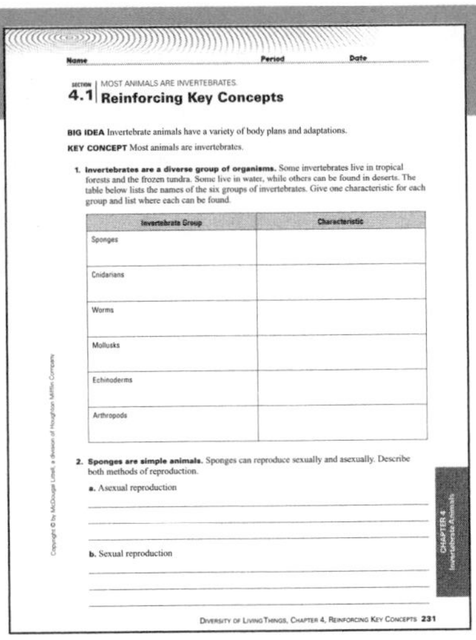

Name Period Date

SECTION 4.1 | MOST ANIMALS ARE INVERTEBRATES.
Reinforcing Key Concepts

BIG IDEA Invertebrate animals have a variety of body plans and adaptations.

KEY CONCEPT Most animals are invertebrates.

1. **Invertebrates are a diverse group of organisms.** Some invertebrates live in tropical forests and the frozen tundra. Some live in water, while others can be found in deserts. The table below lists the names of the six groups of invertebrates. Give one characteristic for each group and list where each can be found.

Invertebrate Group	Characteristic
Sponges	
Cnidarians	
Worms	
Mollusks	
Echinoderms	
Arthropods	

2. **Sponges are simple animals.** Sponges can reproduce sexually and asexually. Describe both methods of reproduction.
   a. Asexual reproduction
   b. Sexual reproduction

DIVERSITY OF LIVING THINGS, CHAPTER 4, REINFORCING KEY CONCEPTS 231

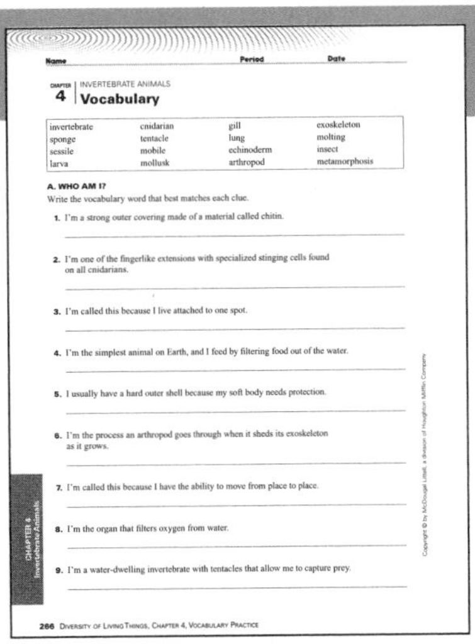

Name Period Date

CHAPTER 4 | INVERTEBRATE ANIMALS
Vocabulary

invertebrate	cnidarian	gill	exoskeleton
sponge	tentacle	lung	molting
sessile	mobile	echinoderm	insect
larva	mollusk	arthropod	metamorphosis

A. WHO AM I?
Write the vocabulary word that best matches each clue.

1. I'm a strong outer covering made of a material called chitin.
2. I'm one of the fingerlike extensions with specialized stinging cells found on all cnidarians.
3. I'm called this because I live attached to one spot.
4. I'm the simplest animal on Earth, and I feed by filtering food out of the water.
5. I usually have a hard outer shell because my soft body needs protection.
6. I'm the process an arthropod goes through when it sheds its exoskeleton as it grows.
7. I'm called this because I have the ability to move from place to place.
8. I'm the organ that filters oxygen from water.
9. I'm a water-dwelling invertebrate with tentacles that allow me to capture prey.

266 DIVERSITY OF LIVING THINGS, CHAPTER 4, VOCABULARY PRACTICE

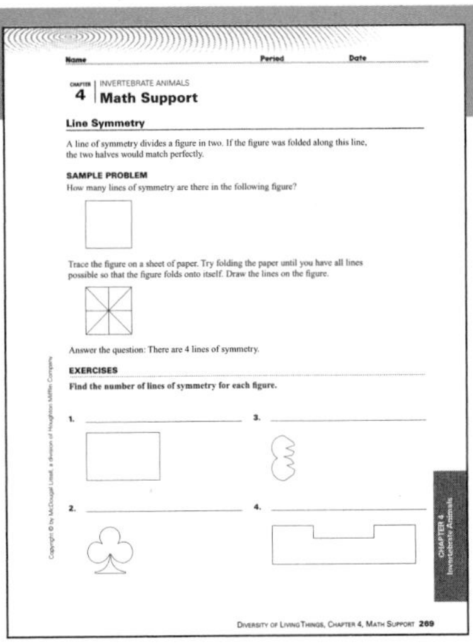

Name Period Date

CHAPTER 4 | INVERTEBRATE ANIMALS
Math Support

Line Symmetry

A line of symmetry divides a figure in two. If the figure was folded along this line, the two halves would match perfectly.

SAMPLE PROBLEM
How many lines of symmetry are there in the following figure?

Trace the figure on a sheet of paper. Try folding the paper until you have all lines possible so that the figure folds onto itself. Draw the lines on the figure.

Answer the question: There are 4 lines of symmetry.

EXERCISES
Find the number of lines of symmetry for each figure.

1.
2.
3.
4.

DIVERSITY OF LIVING THINGS, CHAPTER 4, MATH SUPPORT 269

Reinforcing Key Concepts for each section

UNIT RESOURCE BOOK, p. 231

pp. 266–267

p. 269

CHAPTER 4

Invertebrate Animals

INTRODUCE

the BIG idea

Have students look at the photograph of the jellyfish and the diver and discuss how the question in the box links to the Big Idea:

- How is the body of the jellyfish adapted to finding and eating food?
- How are our bodies adapted to taking in food?
- With students, brainstorm ways that organisms find and eat food.

National Science Education Standards

Content

C.1.d Specialized cells perform specialized functions in multicellular organisms.

C.5.a Although different species might look dissimilar, the unity among organisms becomes apparent from an analysis of internal structures.

Process

A.1–8 Identify questions that can be answered through scientific investigations; design and conduct an investigation; use tools to gather and interpret data; use evidence to describe, predict, explain, model; think critically to make relationships between evidence and explanation; recognize different explanations and predictions; communicate scientific procedures and explanations; use mathematics.

A.9.a–c, A.9.e–g Understand scientific inquiry by using different investigations, methods, mathematics, and explanations based on logic, evidence, and skepticism. Data often results in new investigations.

G.1.b Science requires different abilities.

CHAPTER 4

Invertebrate Animals

the BIG idea

Invertebrate animals have a variety of body plans and adaptations.

Key Concepts

SECTION 4.1 **Most animals are invertebrates.**
Learn about sponges and other invertebrates.

SECTION 4.2 **Cnidarians and worms have different body plans.**
Learn how the body plans of cnidarians are different from those of worms.

SECTION 4.3 **Most mollusks have shells, and echinoderms have spiny skeletons.**
Learn about how mollusks and echinoderms meet their needs.

SECTION 4.4 **Arthropods have exoskeletons and joints.**
Learn about insects, crustaceans, and arachnids.

Internet Preview

CLASSZONE.COM

Chapter 4 online resources: Content Review, Visualization, four Resource Centers, Math Tutorial, Test Practice

How does this jellyfish find food and eat?

INTERNET PREVIEW

CLASSZONE.COM For student use with the following pages:

Review and Practice
- Content Review, pp. 122, 150
- Math Tutorial: Line Symmetry, p. 127
- Test Practice, p. 153

Activities and Resources
- Internet Activity: Invertebrate Diversity, p. 121
- Resource Centers: Worms, p. 132; Mollusks, p. 138; Arthropods, p. 143
- Visualization: Cephalopods, p. 138

NSTA scilinks.org SCILINKS

Sponges **Code: MDL042**

EXPLORE the BIG idea

Worm-Watching

Find a worm and observe it for a while in its natural environment. How much can you learn about how a worm meets its needs by watching it for a short time? Record your observations in your notebook.

Observe and Think What body parts can you identify on a worm? How would you describe a worm's activities and behavior?

Insects and You

Think about times when you have seen or noticed insects. Make a list of your experiences in your notebook. Then, start keeping a list of all the insects you see for the next week. How long do you think this list will be?

Observe and Think How many different types of insects did you see? Where did you see them? What characteristics do you think insects share?

Internet Activity: Invertebrate Diversity

Go to **ClassZone.com** to learn more about different types of invertebrates.

Observe and Think Of every 20 animal species on Earth, 19 are invertebrates. How many types of invertebrates can you name?

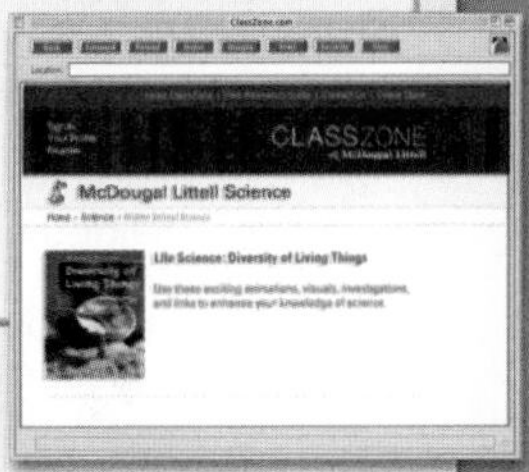

NSTA scilinks.org SCILINKS
Sponges **Code: MDL042**

TEACHING WITH TECHNOLOGY

Electronic Encyclopedias Students can use electronic encyclopedias to investigate the body structures of echinoderms and mollusks, while doing the Investigation on p. 138. Have students print pictures and write their own captions to create a display.

Video Camera Students might want to use a video camera to record close-up observations of worms while investigating worm behavior on pp. 134–135.

Electronic Databases Students can compile a "Fun Facts" database about arthropods as they read about them on pp. 148–149. They can print or post the database to share it.

EXPLORE the BIG idea

These inquiry-based activities are appropriate for use at home or as a supplement to classroom instruction.

Worm-Watching

PURPOSE To observe how a worm meets it needs in its environment.

TIPS *10 min.* If worms are unavailable in an outdoor environment, set up a classroom observation tank with worms and moist soil. Remind students not to harm the worms. Students might gently move the worm they observe from pavement to a grassy area to observe in both environments.

Answer: Students might identify the "head," "mouth," and "tail." They might also notice body segmentation. Students will notice a variety of worm activities.

REVISIT after p. 132.

Insects and You

PURPOSE To discover the variety of insect characteristics.

TIP *10 min.* Some students may name animals that are not insects. Revisit the activity after discussing the characteristics of insects in section 4.4.

Sample answer: ten different types; students will observe most insects outdoors but in many places; common characteristics might include wings, legs, antennae.

REVISIT after p. 145.

Internet Activity: Invertebrate Diversity

PURPOSE To observe the variety of invertebrates.

TIP *20 min.* Challenge students to learn to recognize invertebrates, especially local species.

Answer: Have students make a classroom list of invertebrates and add to it as they study the chapter.

REVISIT after p. 149.

PREPARE

CONCEPT REVIEW

Activate Prior Knowledge

- Ask students to name some of the common characteristics that all living organisms share. *growth, response, reproduction, organization*
- Discuss how organisms meet their needs such as obtaining food, water, and oxygen. *Students will describe various methods, including respiration, eating, photosynthesis.*
- Ask students to name some invertebrate animals. *Students should not list fish, reptiles, amphibians, birds, or mammals, nor microorganisms.*

TAKING NOTES

Combination Notes

Combining a sketch with notes will help students to visualize a new concept and connect it with a concrete example. Later, they can fold the paper and study from the sketches, quizzing each other about written concepts.

Vocabulary Strategy

Allowing students to choose their own strategy helps them to assess their own learning styles and to tailor note-taking to their personal needs. Students might trade their description wheels, four squares, or word triangles with others when they are complete. They can use the diagrams as study aids.

Vocabulary and Note-Taking Resources

R
- Vocabulary Practice, pp. 266–267
- Decoding Support, p. 268

T
- Daily Vocabulary Scaffolding, p. T26
- Note-Taking Model, p. T27

- Choose Your Own Strategy, B18–27
- Combination Notes, C36
- Daily Vocabulary Scaffolding, H1–8

CHAPTER 4

Getting Ready to Learn

CONCEPT REVIEW

- All animals need energy, materials, and living space.
- Animals get energy and materials from food.
- Animals have different adaptations and behaviors for meeting their needs.

VOCABULARY REVIEW

predator p. 63

prey p. 63

adaptation *See Glossary.*

CONTENT REVIEW
CLASSZONE.COM
Review concepts and vocabulary.

TAKING NOTES

COMBINATION NOTES

To take notes about a new concept, first make an informal outline of the information. Then make a sketch of the concept and label it so you can study it later. Use arrows to connect parts of the concept when appropriate.

SCIENCE NOTEBOOK

Notes
Sponges are simple animals
- no organs
- attached to one place
- remove food from water

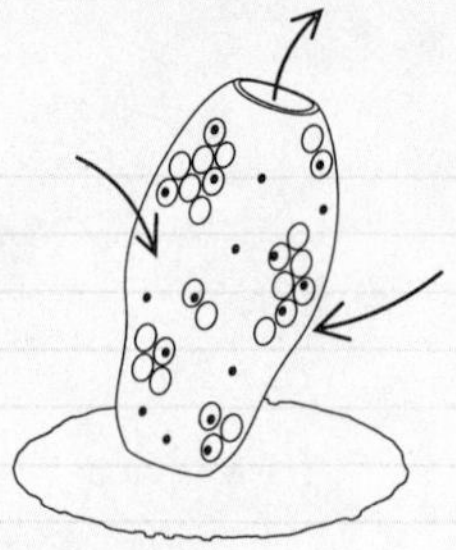

CHOOSE YOUR OWN STRATEGY

For each new vocabulary term, take notes by choosing one of the strategies from earlier chapters—**description wheel, four square,** or **word triangle.** You can also use other vocabulary strategies that you might already know.

Description Wheel

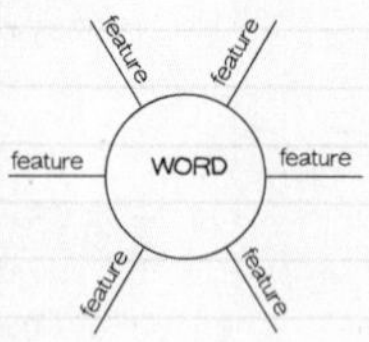

Four Square

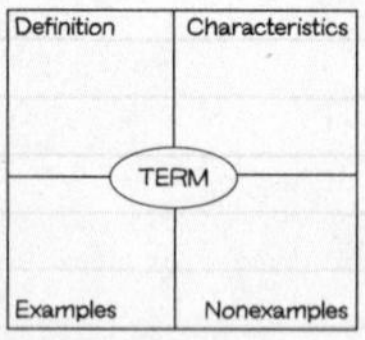

Word Triangle

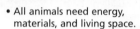

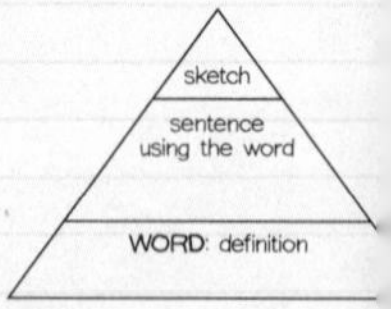

See the Note-Taking Handbook on pages R45–R51.

CHECK READINESS

Administer the Diagnostic Test to determine students' readiness for new science content and their mastery of requisite math skills.

A Diagnostic Test, pp. 61–62

Technology Resources

Students needing content and math skills should visit **ClassZone.com**.

- **CONTENT REVIEW**
- **MATH TUTORIAL**

CONTENT REVIEW CD-ROM

KEY CONCEPT

4.1 Most animals are invertebrates.

BEFORE, you learned

- Animals are consumers; they get food from the environment
- Most animals have body systems, including tissues and organs
- Animals interact with the environment and other animals

NOW, you will learn

- About the diversity of invertebrates
- About six groups of invertebrates
- How sponges get energy

VOCABULARY

invertebrate p. 123
sponge p. 125
sessile p. 125
larva p. 126

THINK ABOUT

What makes an animal an animal?

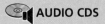

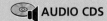

A sponge is an animal. It has no head, eyes, ears, arms, or legs. A sponge doesn't have a heart or a brain or a mouth. It doesn't move. Typically, it spends its life attached to the ocean floor. Many people used to think that sponges were plants that had adapted to life in the water. Scientists, however, classify them as animals. How might you decide if the organism in the photograph is an animal?

Invertebrates are a diverse group of organisms.

COMBINATION NOTES Make notes and diagrams for the first main idea: *Invertebrates are a diverse group of organisms.* Include a sketch of a member of each group.

About one million invertebrate species live on Earth. **Invertebrates** are animals that do not have backbones. In fact, invertebrates do not have any bone tissue at all. Invertebrates can be found just about everywhere, from frozen tundra to tropical forests. Some invertebrates live in water, while others survive in deserts where there is almost no water. Many invertebrates live inside other organisms.

Most invertebrate animals are small. Crickets, oysters, sea stars, earthworms, ants, and spiders are some examples of invertebrates. The fact that invertebrates do not have backbones for support tends to limit their size. However, some ocean-dwelling invertebrates can be quite large. For example, the giant squid can grow to 18 meters (59 ft) in length and can weigh over 450 kilograms (992 lb).

RESOURCES FOR DIFFERENTIATED INSTRUCTION

Below Level

UNIT RESOURCE BOOK

- Reading Study Guide A, pp. 223–224
- Decoding Support, p. 268

AUDIO CDS

Advanced

UNIT RESOURCE BOOK

Challenge and Extension, p. 229

English Learners

UNIT RESOURCE BOOK

Spanish Reading Study Guide, pp. 227–228

AUDIO CDS

- Audio Readings in Spanish
- Audio Readings (English)

4.1 FOCUS

Set Learning Goals

Students will

- Discuss the diversity of invertebrates.
- Describe the six groups of invertebrates.
- Explain how sponges get energy.
- Observe invertebrates that live in local environments.

3-Minute Warm-Up

Display Transparency 28 or copy this exercise on the board:

Decide which of these statements are true. If not true, correct them.

1. Invertebrates are animals that lack internal skeletons. *true*
2. Animals are producers; they make their own food through photosynthesis. *Animals are consumers; they must eat food.*
3. Tissues are made up of groups of similar organs. *Tissues are made up of groups of similar cells.*

3-Minute Warm-Up, p. T28

4.1 MOTIVATE

THINK ABOUT

PURPOSE To understand the characteristics of an animal

DISCUSS Brainstorm with students a list of characteristics of animals and write students' responses on the board.

Sample answer: Because the animal is not green and lives underwater, it is not a plant. You might observe if the organism consumes food. You could also take a cell sample and examine the cell for the presence of cell walls.

Ongoing Assessment

Discuss the diversity of invertebrates.

Ask: Give an example of how invertebrates are a diverse group of organisms. *Sample answer: They can range in size from an 18 m giant squid to a tiny ant.*

4.1 INSTRUCT

INVESTIGATE Invertebrates

PURPOSE Observe the kinds of invertebrates that live in local environments

TIPS *20 min.* Potato traps buried in moist soil will have the best results. If students bury traps in a variety of locations, they will collect a larger sample. The activity works best in warm weather.

WHAT DO YOU THINK? *Answers depend on location, season, climate.*

CHALLENGE *Students should note that they may find different types of invertebrates in the potato trap.*

R Datasheet, Invertebrates, p. 230

Technology Resources

Customize this student lab as needed or look for an alternative. Print rubrics to assess student lab reports.

Lab Generator CD-ROM

Address Misconceptions

IDENTIFY Ask students for examples of animals. If students name only vertebrates, they may hold the misconception that animals include only large land animals.

CORRECT Have students make a chart of the six types of invertebrates. Include a drawing of an example for each type.

REASSESS Ask students to describe some invertebrate animals.

Technology Resources

Visit **ClassZone.com** for background on common student misconceptions.

MISCONCEPTION DATABASE

Ongoing Assessment

Describe the six groups of invertebrates.

Ask: Which three groups of invertebrates contain animals that live on land?
worms, mollusks, and arthropods

INVESTIGATE Invertebrates

Which types of invertebrates live near you?

SKILL FOCUS
Observing

MATERIALS
- potato
- knife
- spoon
- masking tape

TIME
20 minutes

PROCEDURE

1. Cut the potato in half lengthwise. Scoop out a hole and carve a channel so it looks like the photograph below.
2. Put the two halves back together and wrap them with masking tape. Leave the channel uncovered. It is the entrance hole.
3. Take the potato trap outside and bury it upright in soil, with the entrance hole sticking out of the ground. Wash your hands.
4. Collect the potato the next day. Remove the masking tape and look inside.

WHAT DO YOU THINK?

- Observe the contents of the potato. Record your observations.
- Would you classify the contents of the potato as living or nonliving? Do you think they are animals or plants?

CHALLENGE Predict how your observations would be different if you buried the potato in a different place.

In this chapter, you will learn about six groups of invertebrates:

- **Sponges** are the simplest invertebrates. They live in water. They filter food from the water that surrounds them.
- **Cnidarians** also live in water. Animals in this group have a central opening surrounded by tentacles. They take in food and eliminate waste through this opening. Jellyfish, sea anemones, hydras, and corals are cnidarians.
- **Worms** are animals with soft, tube-shaped bodies and a distinct head. Some worms live inside other animals. Others live in the water or on land.
- **Mollusks** have a muscular foot that allows them to move and hunt for food. Some mollusks live on land. Others live in water. Clams, snails, and octopuses are mollusks.
- **Echinoderms** are water animals that have a central opening for taking in food. Sea stars and sand dollars are echinoderms.
- **Arthropods** are invertebrates that are found on land, in the water, and in the air. They have legs. Some have wings. Insects, spiders, crabs, and millipedes are arthropods.

C 124 Unit: Diversity of Living Things

DIFFERENTIATE INSTRUCTION

More Reading Support

A Which group of invertebrates are the simplest? *sponges*

B What body structure allows mollusks to move? *a muscular foot*

English Learners To better understand the process of filter feeding in sponges, let English learners who speak the same first language discuss it together in their common language. Encourage them to use the visual on p. 125 for help. Then, have each student write out or repeat aloud the process in English.

Below Level Have students write a song or a comedy sketch that introduces some sponges and their features to the class. Include details about how they get food, move, and reproduce, as well as behavior and environment.

Sponges are simple animals.

Sponges are the simplest multicellular animals on Earth. These invertebrates are **sessile** (SEHS-EEL) organisms, which means they live attached to one spot and do not move from place to place. Most live in the ocean, although some live in fresh water. Sponges have no tissues or organs. The body of a sponge is made up of a collection of cells. The cells are organized into a body wall, with an outside and an inside. Sponges are adapted to feed continuously. They feed on plankton and other tiny organisms that live in the water.

VOCABULARY
Make a description wheel for *sponge*. Include information on its specialized cells.

Specialized Cells

A sponge meets its needs with cells specialized for different functions. Pore cells along the body wall create tiny openings throughout the body. The pores lead into larger canals and sometimes a central opening, where cells with tiny hairs, or flagella, move water through the sponge. As water moves out, more water enters, as shown in the diagram below. Specialized cells filter out food particles and oxygen. Other specialized cells digest the food.

CHECK YOUR READING What adaptations does a sponge have for obtaining food?

Feeding in Sponges

Structures in a sponge's body function to remove food from water.

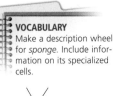

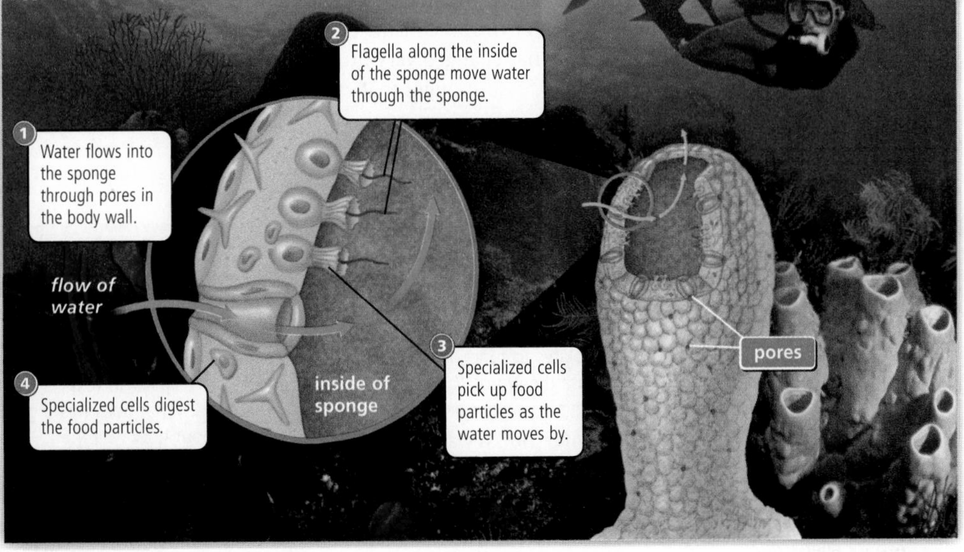

1 Water flows into the sponge through pores in the body wall.
2 Flagella along the inside of the sponge move water through the sponge.
3 Specialized cells pick up food particles as the water moves by.
4 Specialized cells digest the food particles.
flow of water
inside of sponge
pores

Teacher Demo

Bring to class samples of natural sponges for students to examine. Provide a hand lens for students to observe the surface of the sponge and its pores. Explain that natural sponges are not the same as the synthetic sponges often used in the kitchen.

Language Arts Connection

Write the term *Porifera* on the board and tell students that this is the name of the phylum to which sponges belong. Ask students to use a dictionary to find out the derivation of the term. ***Porifera comes from the Latin word,* porus, *which means "pore."*** A common word with this derivation is *porous,* meaning full of pores, or holes.

Teach from Visuals

To help students interpret the diagram of how sponges feed, ask:

- What structures help pump water and food particles through the body cavity of the sponge? *flagella*
- Why must the sponge depend on water flowing through its body to feed? *Sponges are attached to one spot and do not move.*

Ongoing Assessment

Explain how sponges get energy.

Ask: How do sponges feed and digest food? *Specialized cells pick up food particles as water flows through the sponge's body cavity. These cells digest the food particles.*

CHECK YOUR READING *Answer: The sponge has specialized cells for moving water as well as filtering food and digesting it. It also has a body shape adapted for moving water through it.*

DIFFERENTIATE INSTRUCTION

More Reading Support

C What term describes how sponges are attached to one spot? *sessile*

D What do sponges feed on? *plankton, tiny water organisms*

Advanced Challenge students to design an experiment to see whether natural or common household sponges are more absorbent. Give the following guidelines:

- Always start with a dry sponge.
- Make sure the methods are scientifically sound.
- Use a graph to show your results.

R • Challenge and Extension, p. 229

Teach Difficult Concepts

In addition to spicules, the body walls of many sponges contain a substance called spongin, which makes them softer to touch than species that have a lot of spicules. Interlocking spicules of sponges can be considered a kind of exoskeleton, or outer skeleton. Ask students to identify the spicules in the diagram of the sponge on p. 125.

Reinforce the BIG idea

Have students relate the section to the Big Idea.

R Reinforcing Key Concepts, p. 231

4.1 ASSESS & RETEACH

Assess

A Section 4.1 Quiz, p. 63

Reteach

Ask students to explain why some people might think a sponge is a plant. *Sponges are attached to one spot.*

Then write the following on the board: *Characteristics of Sponges.* Encourage students to name facts about sponges that enable us to classify it as an animal. *Sample answers: It feeds on plankton and tiny organisms; it gets water from its environment; it reproduces asexually and sexually; it has a means of protecting itself.*

Technology Resources

Have students visit **ClassZone.com** for reteaching of Key Concepts.

CONTENT REVIEW

CONTENT REVIEW CD-ROM

Another adaptation sponges have are structures that make the body stiff. Most sponges have spicules (SPIHK-yoolz), which are needlelike spines made of hard minerals such as calcium or silicon. Spicules help give the sponge its shape and provide support. In some sponges, spicules stick out from the body. This may make the sponge less likely to become a source of food for other animals.

This basket sponge is releasing microscopic larvas into the water.

Reproduction

Sponges can reproduce asexually. Buds form alongside the parent sponge or the buds break off and float away. Tiny sponges can float quite a distance before they attach to the ocean floor or some underwater object and start to grow.

Sponges also reproduce sexually, as most multicellular organisms do. In sponges, sperm are released into the water. In some sponges, the eggs are released too. In this case, fertilization occurs in the water. In other sponges, the eggs are contained in specialized cells in the body wall. Sperm enter the sponge to fertilize the egg.

A fertilized egg becomes a larva. A **larva** is an immature form—an early stage—of an organism that is different from the parent. Sponge larvas are able to swim. They move away from the parent and will grow into a sponge once they attach to some underwater surface. Then they become sessile, like their parents.

Sponges provide a good starting point for studying other invertebrates. There are many different types of invertebrates, with a wide range of body structures and behaviors. Invertebrates have adapted to many different environments. By comparison, the sponge is a simple organism that has changed very little over time. Sponges today look very similar to fossil sponges that are millions of years old.

4.1 Review

KEY CONCEPTS

1. Make a table with six columns. Write the name of an invertebrate group above each column. Fill in the table with a characteristic and an example for each group.
2. What does it mean that sponges are sessile?
3. How do sponges meet their need for energy?

CRITICAL THINKING

4. **Apply** Give two examples of how structure in a sponge relates to function. You should use the words *flagella* and *spicule* in your answer.
5. **Infer** How is water involved in the reproductive cycle of a sponge?

CHALLENGE

6. **Analyze** Sponges have lived on Earth for hundreds of millions of years. Sponges today look very similar to fossil sponges. What does this suggest about how well the simple structure of a sponge meets its needs? Do species always change over time?

ANSWERS

1. Check students' tables with the information given on p. 124.

2. Sessile means that sponges live their lives attached to one spot.

3. They filter food from the surrounding water.

4. The sponge's pores and flagella allow water to move into and around the sponge's body, where specialized cells pick up and digest food that floats in the water. Spicules are needlelike structures that provide protection from predators.

5. Water allows sperm to travel to eggs, and larvae or buds to float far away.

6. Sponges evolved everything they need to meet their needs millions of years ago and have not been forced to evolve since. No.

MATH in SCIENCE

SKILL: LINE SYMMETRY

Mirror, Mirror

A pattern or shape that has *line symmetry* contains a mirror image of its parts on either side of a straight line. Think about the shapes of a starfish or a butterfly.

MATH TUTORIAL
CLASSZONE.COM
Click on Math Tutorial for more help with line symmetry.

Example

If you were to cut these shapes out of flat paper, you could fold the paper along a line of symmetry. The two halves would match.

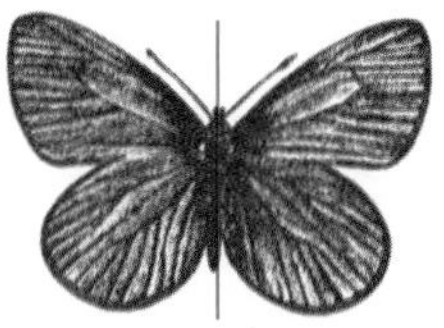

Some shapes have more than one line of symmetry. The shapes of most anemones, and starfish, like flowers, can fold along two, three, or more lines of symmetry.

Sketch each shape. Then draw any lines of symmetry. If there are none, write "zero."

1)

2)

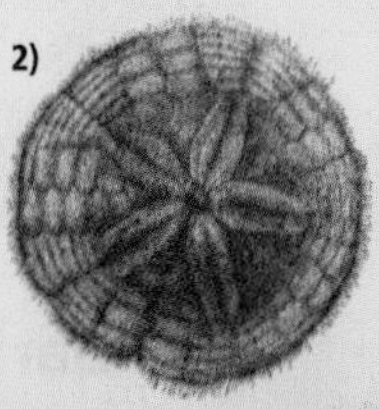

3)

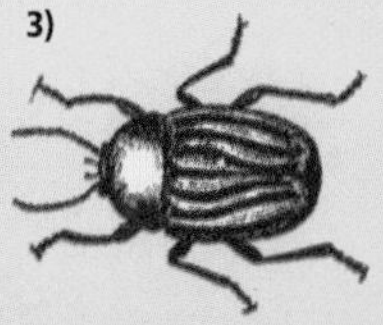

4)

5)

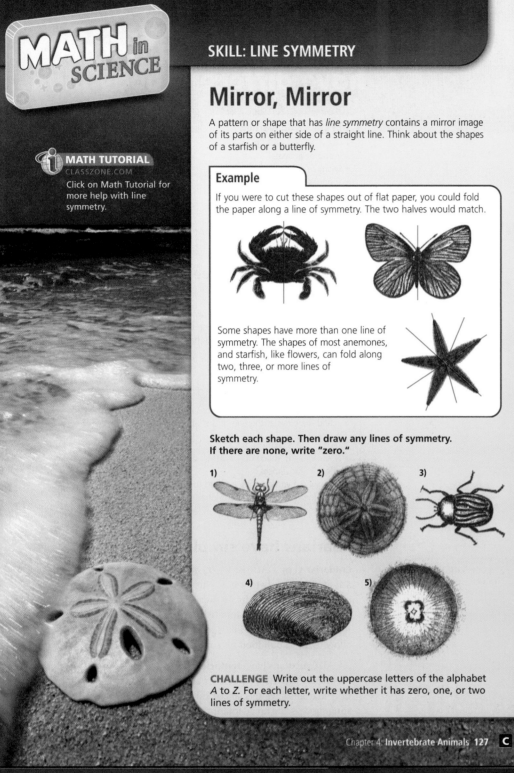

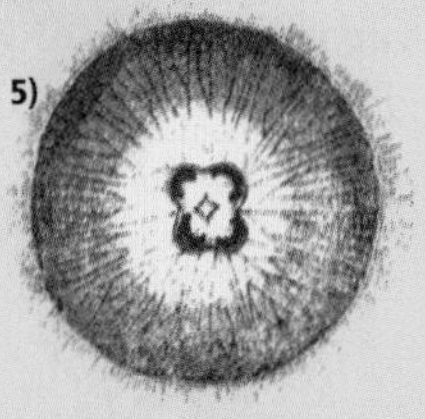

CHALLENGE Write out the uppercase letters of the alphabet *A* to *Z*. For each letter, write whether it has zero, one, or two lines of symmetry.

ANSWERS

1. one line of symmetry
2. five lines of symmetry
3. one line of symmetry
4. zero
5. four lines of symmetry
CHALLENGE *zero: F, G, J, L, N, P, Q, R, S, Z; one: A, B, C, D, E, K, M, T, U, V, W, Y; two: H, I, O, X*

MATH IN SCIENCE
Math Skills Practice for Science

Set Learning Goal

To apply the mathematics of line symmetry to the body plans of invertebrates

Present the Science

Organisms that have radial or bilateral body shapes are symmetric. That is, they have parts that form mirror images. An organism that has a radial body shape will have more than one line of symmetry. Some organisms with symmetrical body plans will appear asymmetrical (having no lines of symmetry) when viewed from different angles. The symmetry in nature is never an exact match, just as circles and squares in nature are never as exact as they are in a mathematical ideal.

Develop Geometry Skills

Review symmetry with students by drawing shapes of circles, squares, triangles, and irregular polygons on the board. Have students note which shapes have one line of symmetry, multiple lines of symmetry, or no lines of symmetry.

DIFFERENTIATION TIP Have students fold papers in half and cut out shapes along the folded parts. Students can unfold the shapes to see the line of symmetry.

Close

Ask students to discuss the difference between the symmetry of a flat shape on paper and symmetry in three dimensions. Why might the clam, shown in figure 4, have a symmetrical body plan, though it appears not to? *It has two shells that are identical. If held up so you see the line between the shells, it will present a symmetrical shape.*

- Math Support, p. 269
- Math Practice, p. 270

Technology Resources

Students can visit **ClassZone.com** to practice determining symmetry.

MATH TUTORIAL

4.2 FOCUS

Set Learning Goals

Students will

- Describe the body systems of cnidarians.
- Analyze body symmetry and feeding patterns.
- Describe body systems in worms.

3-Minute Warm-Up

Display Transparency 28 or copy this exercise on the board:

Fill in the circles surrounding the word web to describe the characteristics of animals.

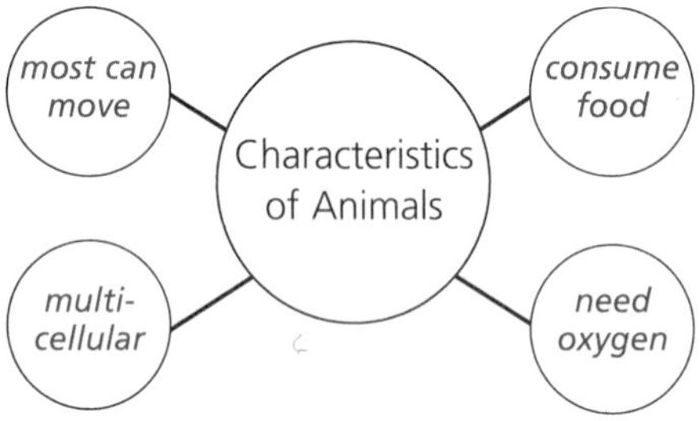

3-Minute Warm-Up, p. T28

4.2 MOTIVATE

EXPLORE Worm Movement

PURPOSE To understand how the body shape of a worm affects its movement

TIPS *20 min.* Ensure that all surfaces the worms will be contacting are kept moist; the worms will die if they dry out. Remind students to treat worms humanely, which means paying attention to what the worms need to thrive. See the notes for the Chapter Investigation, pp. 134–135, for more information on obtaining, handling, and caring for worms in the classroom.

WHAT DO YOU THINK? *Students should note that segmented worms move by contracting and relaxing their muscles. Worms also have very small bristles that anchor them in the soil.*

KEY CONCEPT

4.2 Cnidarians and worms have different body plans.

BEFORE, you learned

- Invertebrates are a diverse group of animals
- Sponges are sessile organisms
- Sponges meet their needs with simple bodies and specialized cells

NOW, you will learn

- About body systems in cnidarians
- About body symmetry and feeding patterns
- About body systems in worms

VOCABULARY

cnidarian p. 128
tentacle p. 128
mobile p. 130

EXPLORE Worm Movement

How does body shape affect movement?

PROCEDURE

1. Put a thin layer of soil on the tray and gently place the worm on it. Use the spray bottle to keep the soil and your hands moist.
2. Draw a sketch of the worm and try to identify the parts of its body.
3. Record your observations of its movement.
4. Follow your teacher's instructions in handling the worm and materials at the end of the lab. Wash your hands.

MATERIALS

- worm
- tray
- soil
- spray bottle filled with distilled water

WHAT DO YOU THINK?

- How does the shape of a worm's body affect its movement?
- How would you describe a worm to someone who has never seen one?

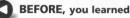

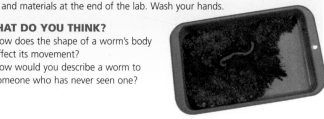

Cnidarians have simple body systems.

COMBINATION NOTES
Make notes and diagrams for the main idea: *Cnidarians have simple body systems.*

Cnidarians (ny-DAIR-ee-uhnz) are invertebrates. Like sponges, cnidarians are found only in water. This group includes jellyfish, corals, sea anemones, and small freshwater organisms called hydras. Most cnidarians feed on small plankton, fish, and clams. Many cnidarians are sessile for most of their lives. Like sponges, cnidarians have adaptations that allow them to pull food in from the water that surrounds them.

All cnidarians have **tentacles,** fingerlike extensions of their body that reach into the water. Other animals have tentacles, but the tentacles of cnidarians have specialized stinging cells. The tentacles, with their stinging cells, are an adaptation that enables cnidarians to capture prey.

RESOURCES FOR DIFFERENTIATED INSTRUCTION

Below Level

UNIT RESOURCE BOOK

- Reading Study Guide A, pp. 234–235
- Decoding Support, p. 268

AUDIO CDS

Advanced

UNIT RESOURCE BOOK

- Challenge and Extension, p. 240
- Challenge Reading, pp. 264–265

English Learners

UNIT RESOURCE BOOK

Spanish Reading Study Guide, pp. 238–239

AUDIO CDS

- Audio Readings in Spanish
- Audio Readings (English)

Stinging Cells in Cnidarians

Jellyfish have specialized cells on their tentacles.

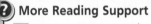

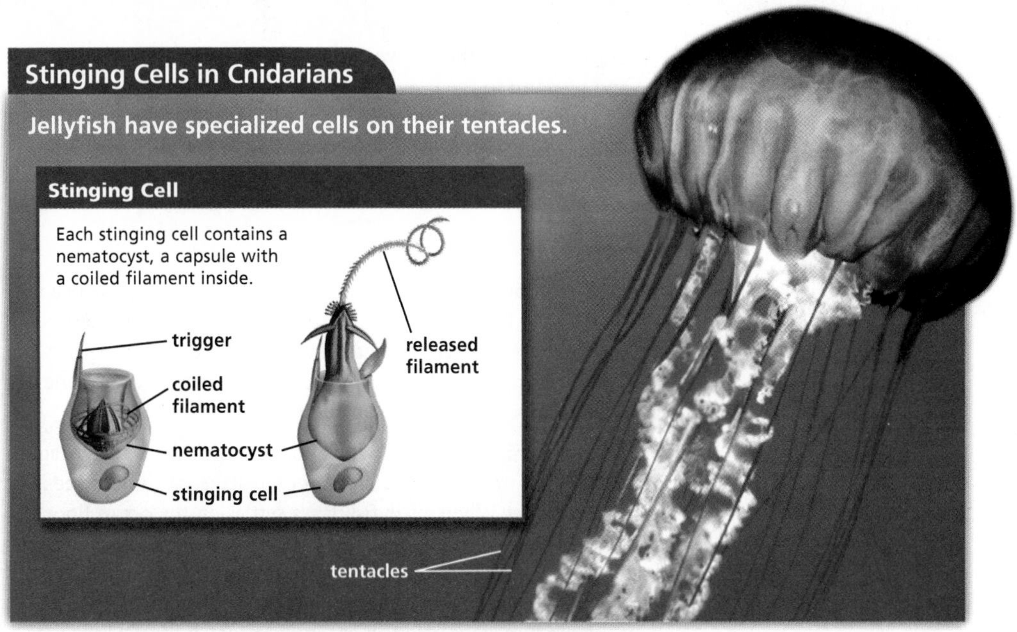

Each stinging cell has a nematocyst (NEHM-uh-tuh-SIHST), a capsule that holds a barbed filament. The filament is like a tiny hollow tube coiled up inside the capsule. When prey comes into contact with the stinger cell, the filament is released. Sometimes this stinger wraps itself around the prey. In most species of cnidarians, the stinger stabs the prey and releases a poison from its tip. These stingers are what produce the sting of a jellyfish. Stinging cells have a second function. They protect cnidarians from predators.

CHECK YOUR READING Describe how the structure of a nematocyst allows it to function in capturing food and providing protection.

Tissues and Body Systems

A cnidarian's body is made up of flexible layers of tissue. These tissues, along with specialized cells, make up its body systems. The tissues are organized around a central opening where food is taken in and wastes are released. The tentacles bring the prey into this opening. The opening leads into a cavity, a gut, where the food is digested.

Cnidarians have a simple muscle system for movement. Even though cnidarians are sessile during most of their lives, they still move their bodies. Cnidarians bend from side to side and extend their tentacles. Adult jellyfish swim. The movement is produced by muscle cells that run around and along the sides of its body. When the muscle cells shorten, or contract, they produce movement.

4.2 INSTRUCT

Teach from Visuals

To help students interpret the diagram of stinging cells in cnidarians, ask:

- What structure is found inside the stinging cell? *the nematocyst*
- What is a nematocyst? What is contained inside the nematocyst? *The nematocyst is a capsule that contains a coiled filament.*
- What is the purpose of the coiled filament? *It is a stinger that releases poison to capture prey or ward off predators.*

Integrate the Sciences

The Great Barrier Reef off the coast of Australia is the world's largest collection of coral reefs, shoals, and islets that extends for about 1,250 miles (2000 km) and has an area of 135,000 square miles (350,000 square km). The Great Barrier Reef is made up of corals dating back to the Miocene Epoch (23.7 to 5.3 million years ago). The reef is home to at least 300 species of coral, as well as anemones, sponges, worms, mollusks, arthropods, and a variety of fish. Reef-building coral are generally found in the tropical areas between the Tropic of Cancer and the Tropic of Capricorn.

Ongoing Assessment

Describe the body systems of cnidarians.

Ask: What body systems enable cnidarians to respond to their environment? *muscle cells that run along the side of the body and a network of nerve cells*

CHECK YOUR READING *Answer: When the trigger comes into contact with a prey, the nematocyst opens and a poisonous filament stabs the prey.*

DIFFERENTIATE INSTRUCTION

More Reading Support

A What happens when the trigger of the nematocyst comes in contact with prey? *Stinger is released.*

B How do cnidarians move? *by contracting cells in their bodies*

English Learners Point out that the "c" in *cnidarian* is silent. Pair English learners with native English speakers and have them create a list of words in which the first consonant is silent. Examples include: *know, knot, who, gnat, gnome, whose, psychology,* and *pseudonym.*

Address Misconceptions

IDENTIFY Ask students how reproduction occurs in cnidarians. If students respond that two cnidarians must find a mate to produce offspring, then they may hold the misconception that animal reproduction must involve separate male and female organisms, that it must involve mating, and/or that animal reproduction involves choice and planning.

CORRECT Have students create a flow chart that shows the jellyfish life cycle. Students should use the following terms in their flow charts: *larva, polyp,* and *medusa.* Ask student to indicate in their charts in which stages the jellyfish are sessile and in which stage they are mobile.

REASSESS Ask students to describe how water is involved in the reproduction of cnidarians. *Buds that are produced by a parent organism are carried away by water and form a new individual. Eggs are fertilized by sperm carried to them in water.*

Technology Resources

Visit **ClassZone.com** for background on common student misconceptions.

MISCONCEPTION DATABASE

Teach from Visuals

To help students understand the life cycle diagram of jellyfish, ask:

- During which stage of the life cycle are the organisms sessile? *during the polyp stage*
- During which stage of the life cycle are the jellyfish considered to be adults? *during the medusa stage*

Language Arts Connection

The term *medusa* is taken from Greek mythology. Have students read versions of the myth and give a short retelling to the class. Then ask: Why is the adult jellyfish named after this mythical being? *A ball shape with tentacles looks like a head with snakes for hair.*

Cnidarians, when they move, interact with the environment. They sense and respond to the prey that come in contact with their tentacles. This behavior is the result of a simple nervous system. Cnidarians have a network of nerve cells, a nerve net that extends throughout their bodies.

Reproduction

VOCABULARY
Choose a strategy from earlier chapters, such as a four square, or one of your own to take notes on the term *mobile.*

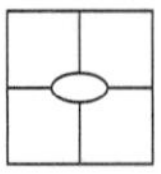

Cnidarians reproduce both sexually and asexually, and water plays a role in both processes. Buds produced by asexual reproduction are carried away from the sessile parent by water. In sexual reproduction, sperm are carried to the egg. Fertilization results in a free-swimming larva. The larva, if it survives, develops into an adult. (C)

Jellyfish are cnidarians with a life cycle that includes several stages. A jellyfish's body, or form, is different at each stage, as you can see in the diagram below. When a jellyfish larva settles on the ocean floor, it grows into a form called a polyp. The polyp, which is sessile, develops disk-shaped buds that stack up like a pile of plates. The buds, once they are released, are called medusas. Each medusa is an adult jellyfish. In the medusa stage, jellyfish are **mobile,** which means they can move their bodies from place to place. (D)

Jellyfish Life Cycle

Jellyfish have a two-part life cycle that includes both a mobile and a sessile form.

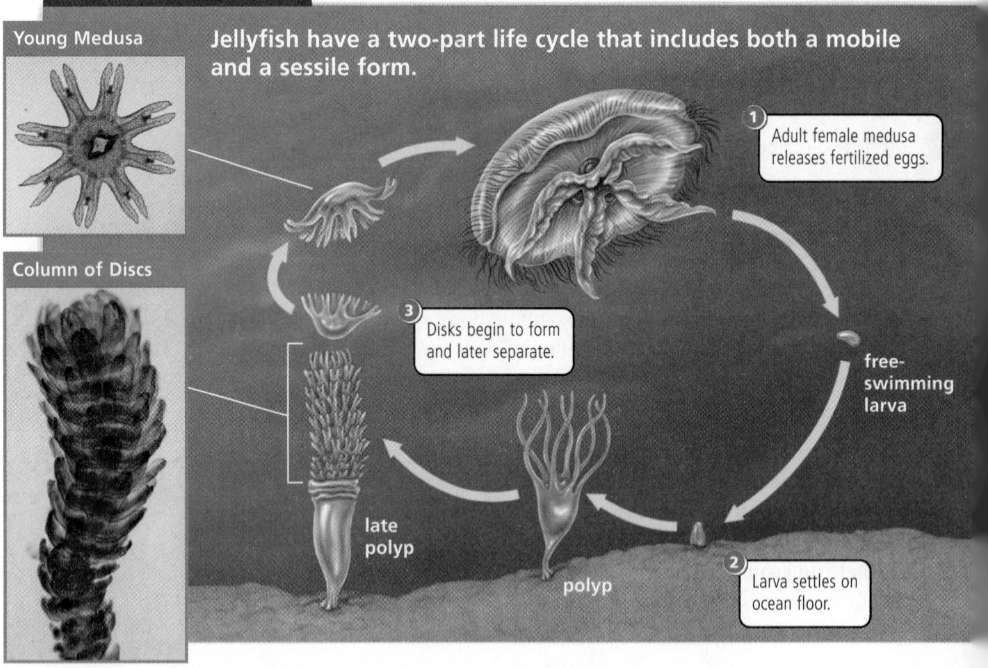
Young Medusa
Column of Discs
1 Adult female medusa releases fertilized eggs.
free-swimming larva
2 Larva settles on ocean floor.
polyp
late polyp
3 Disks begin to form and later separate.

C 130 Unit: Diversity of Living Things

DIFFERENTIATE INSTRUCTION

More Reading Support

C What is the early stage of an organism called? *the larva*

D When buds are released from a polyp, what are they called? *medusas*

Below Level Have pairs of students work together to create a science fiction comic book story about a jellyfish. The details of the story should use scientifically accurate knowledge. Have them include facts about the jellyfish life cycle and habitat and how jellyfish move and feed.

Radial Symmetry

Animals with a circular body shape, such as this sea anemone, have radial symmetry.

Bilateral Symmetry

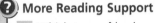

Animals with identical right and left sides, such as this butterfly, have bilateral symmetry.

Animals have different body plans.

Scientists sometimes use the term body plan to describe the shape of an animal's body. Most cnidarians have a body plan with radial symmetry. This means the body is organized around a central point, a mouthlike opening that leads into a gut. You can see from the diagram of the jellyfish life cycle on page 30 that both the polyp and medusa have radial symmetry.

A radial body plan allows a sessile organism, such as the sea anenome shown in the photograph above, to capture food from any direction. A radial body plan also affects how a mobile animal moves. A jellyfish medusa moves forward by pushing down on the water. It has to stop moving to change direction.

Most animals, including worms, butterflies, birds, and humans, have a body plan with bilateral symmetry. One half looks just like the other, as you can see in the photograph of the butterfly above. You can recognize a bilaterally symmetrical shape because there is only one way to draw a line dividing it into two equal halves.

Animals with bilateral symmetry have a forward end where the mouth is located. This is the animal's head. The animal moves forward, head first, in search of food. A bilateral body shape works well in animals that are mobile. Food enters at one end, and is processed as it moves through the body. Once all the nutrients have been absorbed, the remaining wastes are released at the other end.

READING TIP The root of the word *radial* means "ray," like the spoke of a wheel. The roots of the word *bilateral* are *bi-*, meaning "two," and *lateral*, meaning "side."

CHECK YOUR READING Describe how radial symmetry and bilateral symmetry affect an animal's feeding behaviors.

DIFFERENTIATE INSTRUCTION

More Reading Support

E Which type of body shape do cnidarians have? *radial symmetry*

F Which type of body shape do humans have? *bilateral symmetry*

Advanced Challenge students to choose a cnidarian not discussed in detail in the text, such as hydra, sea anemone, or coral and create a poster showing the body systems of the organism and how it reproduces and obtains food.

R Challenge and Extension, p. 240

Have students who are interested in venomous jellyfish read the following article:

R Challenge Reading, pp. 264–265

Teach from Visuals

To help students understand the concept of radial and bilateral body shape shown in the photographs, ask:

- What is different about the body shapes you see here from the body shape of a sponge? *A sponge does not have a regular shape; it does not have a mouth or other body structures such as antennae or wings.*
- Do you think that most animals have radial or bilateral symmetry? *bilateral*
- What are some examples of animals with bilateral symmetry? *Answers may include fish, frogs, turtles, snakes, birds, humans, insects, dogs, and so on.*

Teach Difficult Concepts

Students may not fully understand that some invertebrates do not have distinct digestive and excretory systems. Direct students' attention to the photograph of the sea anemone. Point out that organisms with radial symmetry often have only one body cavity. It is through this cavity that food is ingested and waste is excreted.

Ongoing Assessment

Analyze body symmetry and feeding patterns.

What is the difference between radial and bilateral body symmetry in relation to an animal's way of feeding? *In radial symmetry, the body is organized around a central point. The mouth is in the center. In bilateral symmetry, the body forms around a center line. The mouth is at one end of the line, and the other end is where the waste is released.*

CHECK YOUR READING *Answer: An animal with radial symmetry can capture food from any direction. An animal with bilateral symmetry has one end that is its front. The animal moves itself forward, in search of food.*

Teach Difficult Concepts

To help students understand the increasing complexity of organisms, have them compare the body systems of cnidarians with those of the more complex annelids. The body shapes of the less complex organisms tend to be radially symmetric as opposed to bilaterally symmetric.

EXPLORE the BIG idea

Revisit "Worm-Watching" on p. 121. Now that students have studied the body systems in worms, ask them if their understandings about worms have changed and if they would like to add to their observations.

Develop Critical Thinking

COMPARE AND CONTRAST Have students work together to create Venn diagrams to compare worms and jellyfish. Challenge teams of students to find as many similarities and as many differences as they can.

Teach from Visuals

To help students interpret the diagram of the internal structure of an earthworm, ask them to identify the body system of each of the organs labeled in the diagram. For example, the brain belongs to the nervous system.

This visual is also available as T30 in the Unit Transparency Book.

Arts Connection

Ask students to stage a mock interview between an earthworm and a lifestyle reporter for a celebrity magazine or celebrity news show. The questions and answers should address how the worm moves and gets food as well as how it interacts with its environment.

Ongoing Assessment

CHECK YOUR READING *Sample answer: Two body systems found in worms are the digestive system and excretory system.*

RESOURCE CENTER
CLASSZONE.COM
Learn more about the many types of worms.

Most worms have complex body systems.

Some worms have simple bodies. Others have well-developed body systems. Worms have a tube-shaped body, with bilateral symmetry. In many worms, food enters at one end and is processed as it moves through a digestive tract. Worms take in oxygen, dissolved in water, through their skin. Because of this, worms must live moist environments. Many live in water.

Segmented Worms

Segmented worms have bodies that are divided into individual compartments, or segments. These worms are referred to as annelids (AN-uh-lihdz), which means "ringed animals." One annelid you might be familiar with is the earthworm. As the diagram below shows, an earthworm's segments can be seen on the outside of its body.

An earthworm has organs that are organized into body systems. The digestive system of an earthworm includes organs for digestion and food storage. It connects to the excretory system, which removes waste. Earthworms pass soil through their digestive system. They digest decayed pieces of plant and animal matter from the soil and excrete what's left over. A worm's feeding and burrowing activity adds nutrients and oxygen to the soil.

CHECK YOUR READING Name two body systems found in earthworms.

Inside an Earthworm

An earthworm has organs and body systems.

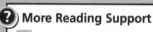

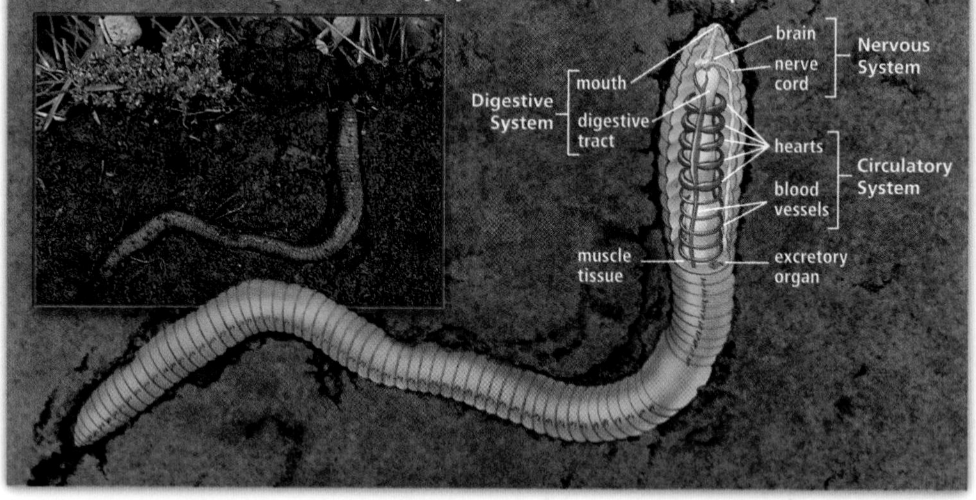

DIFFERENTIATE INSTRUCTION

More Reading Support

G What term describes segmented worms? *annelids*

Advanced Have students observe the unsegmented section about one-third of the way down the earthworm's body in the illustration on p. 132. Explain that this structure is the *clitellum*, the reproductive structure of the worm. It encloses fertilized eggs and then slides off the worm's body. Young worms hatch in the resulting cocoon from eggs encased within it. Ask: How can you tell that earthworms reproduce sexually? *Two worms exchange sperm that fertilize the eggs inside the clitellum. Since sperm cells and egg cells are involved, it is sexual reproduction.*

Earthworms have several layers of muscle tissue in their body wall. Hairlike bristles on the segments help to anchor a worm in the soil as it moves. The nervous system includes a brain and a nerve cord that runs through the body. An earthworm can detect strong light and vibrations in the soil. These stimuli signal danger to a worm. An earthworm also has a circulatory system. It is made up of several hearts that pump blood through blood vessels.

Some annelids reproduce asexually, while others reproduce sexually. There are no distinct male or female worms. Earthworms, for example, carry both male and female reproductive structures. To reproduce, two worms exchange sperm. The sperm fertilize eggs the worms carry in their bodies. The eggs are laid and later hatch into larvas.

Flatworms and Roundworms

Flatworms are the worms with the simplest bodies. Some are so small and flat that they move with cilia, not muscles. These flatworms absorb nutrients directly through the skin. Many flatworms live as parasites, feeding off other organisms. For example, tapeworms are flatworms that infect humans and other animals. The tapeworm has no need for a digestive system because it gets digested nutrients from its host.

Roundworms are found just about everywhere on Earth. The bodies of roundworms are more complex than those of flatworms. They have muscles to move with, and a nervous system and digestive system. Some roundworms are important decomposers on land and in the water.

Three Types of Worms

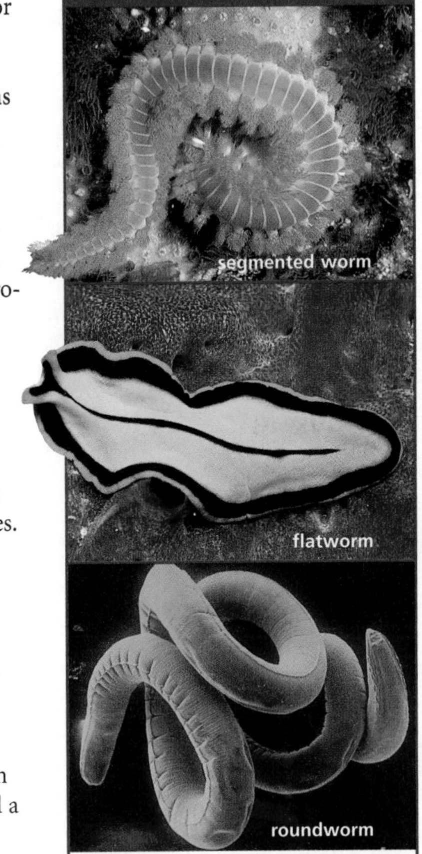

Segmented worms, flatworms, and roundworms are the most common worms on Earth.

4.2 Review

KEY CONCEPTS

1. What adaptation do cnidarians have for capturing prey?
2. What is the difference between radial symmetry and bilateral symmetry?
3. Pick two systems found in an earthworm and describe how they work together.

CRITICAL THINKING

4. **Predict** How might having sense organs located at the front end of the body benefit an animal?
5. **Infer** Would the food of a jellyfish medusa be different from the food of a polyp? Support your answer.

CHALLENGE

6 **Compare and Contrast** Describe how the different body symmetries of cnidarians and segmented worms affect their movement and feeding behaviors.

ANSWERS

1. tentacles with stinging cells

2. Radial symmetry is symmetry around a center; bilateral symmetry has two identical halves.

3. Sample answer: Wastes from digestion pass through the excretory system.

4. Sense organs allow an animal to collect information about the environment. Eyes located in the front allow an animal to seek out food and determine where to move.

5. No, both forms have tentacles and nematocysts. Medusas may eat larger prey since they are usually larger.

6. Cnidarians use the same opening for eating and propulsion. Annelids have a head and move forward toward food.

Ongoing Assessment

Describe body systems in worms.

Ask: How is the digestive system of the earthworm similar to that of humans? *The earthworm has organs that break down, digest, and store food. The food enters through a mouth and moves through the body out of the other end.*

Reinforce the BIG idea

Have students relate the section to the Big Idea.

R Reinforcing Key Concepts, p. 241

4.2 ASSESS & RETEACH

Assess

A Section 4.2 Quiz, p. 64

Reteach

Write the following organisms on the board: *jellyfish, sea anemone, earthworm, flatworm.* For each organism, have students tell if it is has radial body shape or bilateral body shape. Then have students name the body systems for each.

jellyfish and sea anemone: radial; nervous systems

earthworm and flatworm: bilateral; circulatory, digestive, nervous systems

For in-depth reteaching, have small groups of students develop larger-than-life models or diagrams for jellyfish and for earthworms and their environments. They should base the models on the text and on the illustrations on pp. 130 and 132.

Technology Resources

Have students visit **ClassZone.com** for reteaching of Key Concepts.

CONTENT REVIEW

CONTENT REVIEW CD-ROM

CHAPTER INVESTIGATION

Focus

PURPOSE To observe the behavior of worms and hypothesize about how worms will respond to stimuli, and to test hypotheses

OVERVIEW Students will create a worm farm, using different types of surfaces to test the following:

- how earthworms respond to sand and potting soil surfaces.
- how earthworms behave and respond to changes in their environment.

Lab Preparation

- This investigation can be done with earthworms collected from outdoors, or with worms from a kit available at biological supply houses.
- Vermiculite (worm composting) kits tend to use red "wiggler" worms *(Eisenia foetida)* rather than the familiar earthworm *(Lumbricus terrestrial)*. Both species are safe choices. Redworms are not native to the U.S. and should not be released.
- The set-up for the investigation can be prepared in advance.
- Explain how to distinguish between the anterior and posterior of a worm.
- Prior to the investigation, have students read through the investigation and prepare their data tables. Or you may wish to copy and distribute datasheets and rubrics.

UNIT RESOURCE BOOK, pp. 271–279

SCIENCE TOOLKIT, F14

Lab Management

- All surfaces that worms come in contact with should be moist.
- Students should care for and treat worms humanely, which also means knowing what conditions the animals need in order to thrive.

SAFETY Make sure students wash their hands thoroughly after the lab.

INCLUSION Students can draw pictures of their observations rather than write out results, or can share results in a discussion.

CHAPTER INVESTIGATION

Worm Behavior

OVERVIEW AND PURPOSE Earthworms do not have eyes, so they cannot see. An earthworm needs an environment that provides it with moisture, food, and protection from predators. How do worms gather information about their surroundings? How do they respond to changes in their environment? In this investigation you will

- observe worm behavior
- predict how worms will respond to surfaces with different textures

Problem

Write It Up

How is worm behavior affected by environmental conditions?

Hypothesize

Write It Up

You should complete steps 1–9 in the procedure before writing your hypothesis. Write a hypothesis to explain how worms will respond to different surface textures in an environment. Your hypothesis should take the form of an "If . . . , then . . . , because . . ." statement.

MATERIALS

- aquarium
- potting soil
- coarse sand
- small beaker
- distilled water
- spray bottle
- filter paper
- 5 or more worms
- 2 containers, one for untested worms and one for tested worms
- stopwatch
- *for Challenge* flashlight

Procedure

1. Make a data table in your **Science Notebook** like the one shown on page 135.
2. Cover one half of the bottom of the aquarium with potting soil and the other half with sand.
3. Fill the beaker with 250 mL of distilled water and use it to fill the spray bottle. Spray all the water over the potting soil so it is evenly moistened.

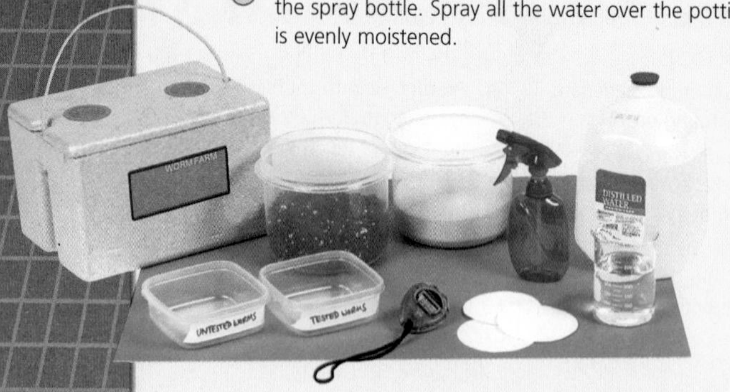

INVESTIGATION RESOURCES

CHAPTER INVESTIGATION, Title
- Level A, pp. 271–274
- Level B, pp. 275–278
- Level C, p. 279

Advanced students should complete Levels B & C.

Writing a Lab Report, D12–13

Technology Resources

Customize this student lab as needed or look for an alternative. Print rubrics to assess student lab reports.

Lab Generator CD-ROM

4 Repeat step 3, but this time moisten the sand. Refill the spray bottle.

5 Place a piece of filter paper in the middle of the aquarium so it is half on the soil and half on the sand, as shown.

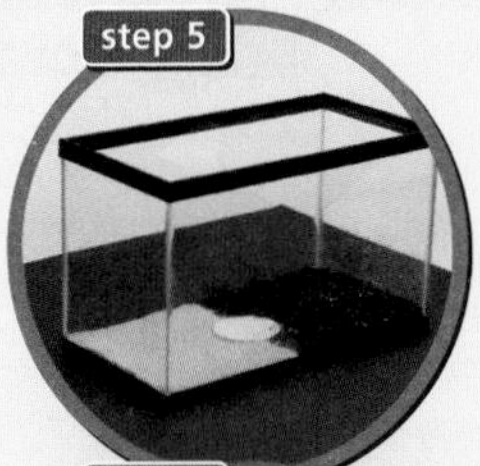

6 Put on your gloves. Spray your hands with water. Gently remove one worm and observe it until you can tell which end is its head.

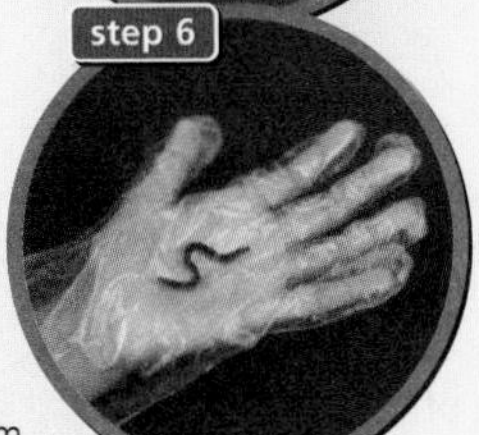

7 Start the stopwatch as you place the worm on the middle of the filter paper. Note which part of the aquarium the worm's head points toward.

8 Observe the worm's behavior for two minutes and then remove it carefully from the aquarium and place it in the container for tested worms.

9 Write your observations in your data table. State your hypothesis.

10 Fix the sand, soil, and paper in the aquarium so they are arranged as they were for the first worm. Then repeat steps 6–9 with at least four more worms.

11 Return the worms to their original living place. Wash your hands.

Observe and Analyze (Write It Up)

1. **OBSERVE** What behaviors suggest that worms gather information about their surroundings?
2. **OBSERVE** What evidence did you see to suggest that worms respond to information they get about their surroundings?
3. **INTERPRET DATA** What patterns did you notice in the behavior of the worms you tested?

Conclude (Write It Up)

1. **INTERPRET** Compare your results with your hypothesis. Does your data support your hypothesis?
2. **IDENTIFY LIMITS** What sources of error could have affected your investigation?
3. **EVALUATE** Based on your observations and evidence, what conclusions can you draw about the connection between worm behavior and environmental conditions?

INVESTIGATE Further

CHALLENGE Worms respond to light as a stimulus. Design an experiment to test the reaction of worms to the presence of light.

Worm Behavior

Hypothesis

Table 1. Observations of Tested Worms

	Starting Position	Ending Position	Description of Behavior
Worm 1			
Worm 2			
Worm 3			
Worm 4			
Worm 5			

Observe and Analyze (Write It Up)

1. *The worms' behavior will vary depending on the conditions of the experimental setting. Worms generally proceed off the filter paper in whatever direction they are pointing until they encounter an obstacle such as the aquarium wall.*
2. *Sample answer: Worms turn in the direction they are pointing and hug the wall of the aquarium until they locate a surface to burrow into.*
3. *Sample answer: Some worms explore and reject certain environments depending on moisture and ability to burrow into the soil or potting soil.*

Conclude (Write It Up)

1. *Answers should give evidence from results to support or to counter student hypotheses.*
2. *Sample answer: How the worms were placed, the type of soil and food used, and the condition of the worms could have affected the results.*
3. *Sample answer: Worms will usually move into environments more suitable to them: a moist, moderately cool area with food.*

INVESTIGATE Further

CHALLENGE Check student designs. Experiments should use a control and test only one variable (amount of light).

Teaching with Technology

If video cameras are available, students may make close-up video shots of their earthworm observations.

Post-Lab Discussion

- Discuss with students any new questions that arose from the investigation. What do their results suggest about their hypotheses? What new hypotheses might be inferred from their observations? How might they conduct a new experiment to test these ideas?
- Ask students to describe how a worm is able to move. *An earthworm moves by contracting muscles, then relaxing them.*

4.3 FOCUS

Set Learning Goals

Students will

- Describe different types of mollusks and their features.
- Describe different types of echinoderms and their features.
- Observe details of mollusk and echinoderm shells.

3-Minute Warm-Up

Display Transparency 29 or copy this exercise on the board:

Match the definitions to the terms.

Definitions

1. attached to one spot *a*
2. fingerlike extensions of the body *d*
3. early immature stage of an organism *e*

Terms

a. sessile
b. mobile
c. invertebrate
d. tentacle
e. larva

3-Minute Warm-Up, p. T29

4.3 MOTIVATE

THINK ABOUT

PURPOSE To understand how a snail uses its muscular foot to move itself

DISCUSS If possible, have students observe a snail through the glass of an aquarium. While observing the action of the snail's foot, students may also be able to see the snail's mouth and its radula. Discuss with students how the snail is adapted for life in an aquatic environment. *The snail glides along the glass surface through muscular contractions of its foot. It has a hard shell for protection, and gills to breathe underwater.*

KEY CONCEPT

4.3 Most mollusks have shells, and echinoderms have spiny skeletons.

BEFORE, you learned

- Body shape affects how animals move and behave
- Cnidarians have radial symmetry and simple body systems
- Worms have bilateral symmetry and complex body systems

NOW, you will learn

- About different types of mollusks and their features
- About different types of echinoderms and their features

VOCABULARY

mollusk p. 136
gill p. 137
lung p. 137
echinoderm p. 139

THINK ABOUT

How does a snail move?

Snails belong to a group of mollusks called gastropods. The name means "belly foot." Snails are often put into aquariums to clean up the algae that can build up along the walls of the tank. If you get a chance, look at a snail moving along the glass walls of an aquarium, observe how it uses its foot to move. How would you describe the action of the snail's foot?

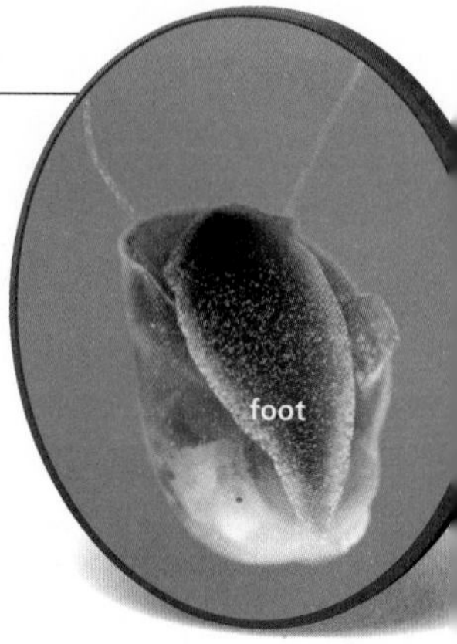

Mollusks are soft-bodied animals.

VOCABULARY
Choose a strategy from earlier chapters, such as a word triangle, or one of your own to take notes on the term *mollusk*.

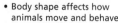

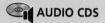

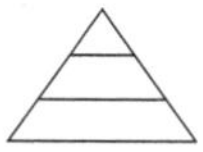

One characteristic that is shared by all **mollusks** is a soft body. Many of these invertebrate animals also have an outer shell to protect their body. Oysters, clams, snails, and mussels are all mollusks. So are octopuses, squids, and slugs. Mollusks live on land and in freshwater and saltwater environments. You will read about three groups of mollusks: bivalves, gastropods, and cephalopods.

Most mollusks have well-developed organ systems. They have muscles, a digestive system, a respiratory system, a circulatory system, and a nervous system with sensory organs. Mollusks reproduce sexually, and in most species there are distinct male and female organisms. Two adaptations distinguish mollusks as a group. First, all mollusks have a muscular foot. A mollusk's head is actually attached to its foot. Second, all mollusks have a mantle, a layer of folded skin that protects its internal organs.

RESOURCES FOR DIFFERENTIATED INSTRUCTION

Below Level

UNIT RESOURCE BOOK

- Reading Study Guide A, pp. 244–245
- Decoding Support, p. 268

AUDIO CDS

Additional INVESTIGATION, Exploring How a Squid Moves, A, B, & C, pp. 280–288; Teacher Instructions, pp. 360–361

Advanced

UNIT RESOURCE BOOK

Challenge and Extension, p. 250

English Learners

UNIT RESOURCE BOOK

Spanish Reading Study Guide, pp. 248–249

AUDIO CDS

- Audio Readings in Spanish
- Audio Readings (English)

Bivalves

Bivalves are named for a hard shell that is made up of two matching halves. Clams, mussels, scallops, and oysters are all bivalves. The shell, when it is closed, completely encloses the body. If you've ever seen a raw oyster, you know that a bivalve's body looks like a mass of tissue. Bivalves do not have a distinct head, but they do have a mouth and sensory organs. The scallop shown in the photograph has light-sensitive organs that look like tiny eyes.

Bivalve

Most of this blue-eyed scallop's body is inside its two-part shell.

Bivalves are filter feeders, they filter food from the surrounding water. To move, a bivalve balances upright, opens its shell, and extends its foot. The animal moves by pushing the foot in and out. The foot is also used for burrowing, digging down into the sand.

The invertebrates you've studied so far—sponges, cnidarians, and worms—take in oxygen all along the surface of their bodies. A bivalve takes in oxygen through a pair of gills. A **gill** is an organ that filters dissolved oxygen from water. The gill is an adaptation that allows an organism to take in a lot of oxygen in just one area of its body. It is made up of many folds of tissue that create a large surface area. Blood picks up the oxygen and moves it to the rest of the animal's body. In most bivalves, the gills also filter food from the water.

RESOURCE CENTER
CLASSZONE.COM
Discover more about mollusks.

CHECK YOUR READING What are the two functions of gills and how do those functions relate to where bivalves live?

Gastropods

Gastropods are the most diverse group of mollusks. Some, such as snails and slugs, live on land. Many live in water, for example, conches, whelks, and periwinkles. Many gastropods are protected by a spiral-shaped shell. To protect itself, a gastropod withdraws into the shell.

Gastropod

This brown-lipped snail extends most of its body out of its shell when it moves.

The gastropod's head is located at the end of its foot. The head has eyes and specialized tentacles for sensing. Many gastropods have a cutting mouth part, called a radula, that shreds their food. Some gastropods eat animals, but most feed on plants and algae. Gastropods that live in water have gills. Some gastropods that live on land have lungs. A **lung** is an organ that absorbs oxygen from the air. Like gills, lungs have a large surface area.

? A

4.3 INSTRUCT

Teach from Visuals

- To help students interpret the photographs of the bivalve and the gastropod, ask: What type of body symmetry do these organisms have? *bilateral*
- How do the shells of bivalves differ from those of the gastropods? *Bivalves have shells that are made up of two matching parts, while gastropods have one large domed shell that encloses the body.*

Develop Critical Thinking

APPLY Have students to discuss the reasons why a large surface area in gills and lungs is an adaptation that is an advantage for mollusks. *The larger surface area enables more water or air to pass over the gills and lungs. Organisms with gills and lungs do not need to absorb oxygen through the cells of their bodies as do sponges and cnidarians.*

Real World Example

Mollusks such as mussels, oysters, squid, octopuses, and snails are considered a delicacy in the United States and around the world. Invite students to research the nutrients found in these foods and to make inferences about why certain nutrients are abundant.

Ongoing Assessment

CHECK YOUR READING *Answer: Bivalve gills filter oxygen and food from water. They are specialized to function in water.*

DIFFERENTIATE INSTRUCTION

More Reading Support

A What adaptations do gastropods have for taking in oxygen? *lungs and gills*

English Learners Have students write definitions and/or descriptions for *mollusk, clam, snail, octopus,* and *squid* in their Science Word Dictionaries. Students can draw pictures next to each description. As they read through section 4.3, they can add to their sketches with labels for body parts, such as *gill, radula, mantle, tentacle,* and so on.

INVESTIGATE Mollusks and Echinoderms

PURPOSE To observe details of mollusk and echinoderm shells

TIPS *15 min.* Invite students to bring in samples of mollusk shells, sand dollars, and sea stars that they may have collected from the beach. Samples may also be obtained through biological supply houses.

WHAT DO YOU THINK? *Answers will depend on the samples that students observe. Be sure that both similarities and differences are listed.*

CHALLENGE *Inferences should mention that the mollusks' bodies are inside their shells with a foot that could be extended or taken into the shell, while the echinoderms' bodies were once covered with a layer of skin, and that echinoderms' tube feet were on the outside of their bodies.*

R Datasheet, Mollusks and Echinoderms, p. 251

Technology Resources

Customize this student lab as needed or look for an alternative. Print rubrics to assess student lab reports.

Lab Generator CD-ROM

Teaching with Technology

If computers and Internet access are available, have students use electronic encyclopedias to investigate the body structures of mollusks and echinoderms. Have students print out pictures and use them in conjunction with the investigation. They should caption the pictures using their own words and observations.

Ongoing Assessment

CHECK YOUR READING *Answer: Cephalopods have vision and tentacles adapted for hunting. They can move quickly because of a mantle that is adapted for producing a jet of water that moves the animal.*

INVESTIGATE Mollusks and Echinoderms

How do mollusk shells compare with echinoderm skeletons?

PROCEDURE

1. Closely observe the mollusk shells and skeletons of sea stars and sand dollars you are given.
2. Examine the shape and texture of each. Sort them by their characteristics.

WHAT DO YOU THINK?

How are the shells and skeletons the same? How are they different?

CHALLENGE Based on your observations, what can you infer about the bodies of living mollusks, sea stars, and sand dollars?

SKILL FOCUS
Observing

MATERIALS
Selection of mollusk shells, sea stars, and sand dollars

TIME
15 minutes

Cephalopods

Cephalopods (SEHF-uh-luh-PAHDZ) live in saltwater environments. Octopuses, squids, and chambered nautiluses are cephalopods. Among mollusks, cephalopods have the most well-developed body systems.

Cephalopods have a brain and well-developed nerves. They have a pair of eyes near their mouth. The foot, which surrounds the mouth, has tentacles for capturing prey. The mantle is adapted to push water forcefully through a tube-shaped structure called a siphon. This produces a jet of water that moves the animal. Gills take in oxygen, which is picked up by blood vessels and pumped through the body by three hearts. (? B)

Cephalopods

This Maori octopus has a well-developed head attached to a foot with eight tentacles.

Octopuses and squids do not have protective shells. They do have protective behaviors, however. Some can change body color to match their surroundings. Some release dark clouds of inklike fluid into the water, to confuse their predators. The lack of a shell lets them move freely through the water.

The nautilus is the only cephalopod that has a shell. The shell is made up of separate compartments, or chambers. The nautilus itself lives in the outermost chamber. The inner chambers are filled with gas, which makes the animal better able to float. The chambered shell also provides the soft-bodied nautilus with protection from predators. (? C)

VISUALIZATION CLASSZONE.COM
Watch how different cephalopods move.

CHECK YOUR READING How is the foot of a cephalopod adapted for hunting?

DIFFERENTIATE INSTRUCTION

More Reading Support

B What two structures do cephalopods have that other bivalves do not? *gills and a siphon*

C Which cephalopod has an exterior shell? *the nautilus*

Below Level Use the table below to help students compare and contrast the differences among the three types of mollusks.

	Bivalves	Gastropods	Cephalopods
Outer covering	matching shells	spiral shells	none except nautilus
Body features	muscular foot	head, eyes, tentacles	siphon, gills, eyes, tentacles
Movement	muscular foot	muscular foot	mantle

Mollusks show a range of adaptations.

You might not think that a clam would belong to the same group as an octopus. These organisms look very different from one another. They also interact with the environment in different ways. The great variety of mollusks on Earth today provides a good example of how adaptations within a group can lead to great diversity. A good example of this is the range of adaptations shown in the shape and function of a mollusk's foot.

The foot of the bivalve is a simple muscular structure that moves in and out of its shell. The foot allows a bivalve to crawl along the ocean floor and to bury itself in the sand. Gastropods have a head at the end of the foot, which runs the length of the body. Muscles in the foot produce ripples that allow the gastropod to glide over a surface as it searches for food. In cephalopods, the foot has tentacles to pull food into its mouth. The tentacles also help some cephalopods move along the ocean floor.

COMPARE How does the foot of this clam compare with the foot of an octopus?

Echinoderms have unusual adaptations.

Echinoderms are a group of invertebrates that live in the ocean. In their adult form, their bodies have radial symmetry. Sea stars, sea urchins, sea cucumbers, and sand dollars belong to this group. Echinoderms feed off the ocean floor as they move along. An echinoderm's mouth is located at the center of the body, on the underside. Some echinoderms, such as sea urchins and sand dollars, filter food from their surroundings. Others, such as sea stars, are active predators that feed on clams, snails, and even other echinoderms.

COMBINATION NOTES
Remember to take notes and make sketches for the main idea: *Echinoderms have unusual adaptations.*

Spines and Skeletons

Echinoderm means "spiny-skinned." Some of the more familiar echinoderms have long, sharp spines, like the sea urchin in the photograph at the bottom of this page. However, some echinoderm species, such as sea cucumbers, have spines that are very small.

One unusual adaptation that echinoderms have is a type of skeleton. Remember that echinoderms are invertebrates, they have no bone tissue. The echinoderm skeleton is made up of a network of stiff, hard plates. The plates lie just under the surface of the echinoderm's skin. Some echinoderms, such as sea stars, have skeletons with loosely connected plates and flexible arms. In other echinoderms, such as sand dollars,the plates grow close together, so the skeleton does not allow for much flexibility.

This purple sea urchin has very obvious spines.

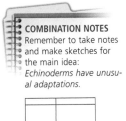

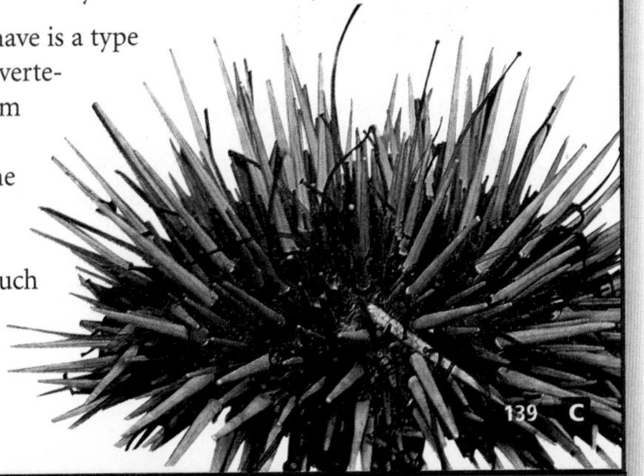

139 C

DIFFERENTIATE INSTRUCTION

More Reading Support

D Which type of body shape do echinoderms have? *radial symmetry*

E What makes up an echinoderm's skeleton? *stiff, hard plates*

Additional Investigation To reinforce Section 4.3 learning goals, use the following full-period investigation:

R **Additional INVESTIGATION,** Exploring How a Squid Moves, A, B, & C, pp. 280–288, 360–361
(Advanced students should complete Levels B and C.)

Advanced Have students look up roots and origins of *cephalopod*, *gastropod*, *arthropod*, and *echinoderm*. Challenge them to find related words and write context sentences.

R Challenge and Extension, p. 250

Real World Example

A new species of deep sea squid has been spotted in the waters of the Gulf of Mexico, Pacific, Atlantic, and Indian Oceans at depths ranging from 1900 meters (6200 feet) to 4800 meters (16,000 feet). The newly discovered species has been spotted only eight times, with the first sighting in 1995. Unlike more familiar squids, this species has 10 long arm-like appendages, which radiate from the main axis of the body, and two large fins. Sizes of the squid are estimated from 1.5 to 7 meters (5 to 23 feet).

Develop Critical Thinking

INFER Direct attention to the photograph of the purple sea urchin, pointing out that you can see its tube feet among its spines. Ask: What do you think is the purpose of the spines? *to ward off predators*

Teacher Demo

Many grocery stores stock fresh, frozen, and dried octopus and squid. Since many students will not have seen these animals, you may wish to show them to students.

Language Arts Connection

Interested students might read *2000 Leagues Under the Sea,* by Jules Verne or read excerpts from it and create a two-column chart showing the "Scientifically Accurate" and "Scientifically Inaccurate" information about giant squids.

Ongoing Assessment

Describe different types of mollusks and their features.

Ask: What adaptations of echinoderms enable them to live in a water environment? *Echinoderms have mouths located on the underside of their bodies, enabling them to feed off ocean floors. Spines ward off predators. They have a skeleton.*

Teach from Visuals

To help students interpret the photograph of the sea stars, ask: How does a sea star use its tube feet to feed? *It spreads its arms over its prey and attaches its tube feet to the two shells. The suction tube feet gradually pull the shell open.*

Ongoing Assessment

Describe the different types of echinoderms and their features.

Ask: How does the water vascular system enable the sea star to move? *Tiny openings along the surface of the body allow water into the tubes of the system. When the openings close, suction is produced at the base of the tubes allowing the sea star to pull itself.*

Reinforce the BIG idea

Have students relate the section to the Big Idea.

R Reinforcing Key Concepts, p. 252

4.3 ASSESS & RETEACH

Assess

A Section 4.3 Quiz, p. 65

Reteach

Write the following terms on the board. Have students tell whether each is a mollusk or an echinoderm. Then have students describe a characteristic for each.

- bivalve *mollusk, two shells*
- gastropod *mollusk, spiral shell*
- cephalopod *mollusk, tentacles*
- sea urchin *echinoderm, spines*
- sea star *echinoderm, radial symmetry*

Technology Resources

Have students visit **ClassZone.com** for reteaching of Key Concepts.

CONTENT REVIEW

CONTENT REVIEW CD-ROM

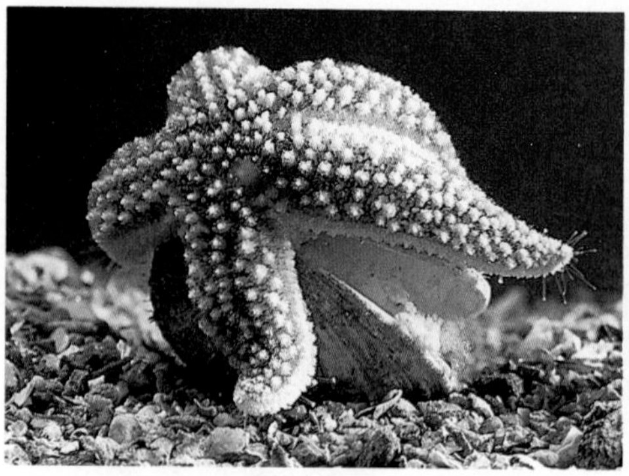

This sea star has captured a bivalve and is using its tube feet to open the shell.

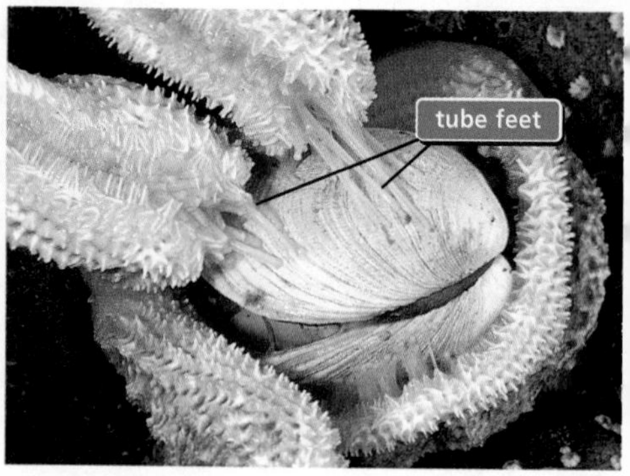

This sea star's arms have been pulled up to show how its tube feet are attached to the bivalve's shell.

Water Vascular System and Tube Feet

Another adaptation that is unique to echinoderms is a water vascular system. This system is made up of water-filled tubes that radiate out from the center of the echinoderm's body. Tiny openings along the upper surface of the echinoderm's body feed water into these tubes. At the base of the tubes is a series of tube feet.

Muscles attached to the top of each tube can close the tube off, producing suction at the base of the tube. The tube feet stick to the ocean floor, allowing the echinoderm to pull itself along. The tube feet can also be used for hunting prey. For example, a sea star can surround a clam or oyster with its body, as shown in the photograph on the left. The tube feet pull the shell open. Then, the sea star's stomach is pushed out through its mouth and into the bivalve's shell, where it begins to digest the bivalve's body. Not all echinoderms eat other animals. Some, like sea urchins, feed off algae on the ocean floor.

4.3 Review

KEY CONCEPTS

1. What two features do all mollusks have?
2. What are two features all echinoderms have?
3. What are two functions of tube feet in echinoderms?

CRITICAL THINKING

4. **Analyze** For mollusks and echinoderms, what are the advantages and disadvantages of having a shell or spiny skeleton?
5. **Compare and Contrast** Compare the foot of mollusks with the tube feet of echinoderms.

CHALLENGE

6. **Analyze** Animals with lungs or gills can be larger than animals that take in oxygen through their skin. What feature do both lungs and gills have that affect the amount of oxygen they can absorb? What role does the circulatory system play?

ANSWERS

1. muscular foot, mantle

2. spiny skeleton, water vascular system

3. movement, getting food

4. Shells and spiny skeletons provide protection from predators. Shells are heavy and not flexible. Spiny skeletons may inhibit movement and cannot protect tube feet.

5. Both use water. In cephalopods, water is drawn in and forcefully pushed out, moving the animal through the water. In echinoderms, the animals use water to create suction and move along the ocean floor.

6. Blood is able to move oxygen to different parts of the body quickly. Oxygen that enters the body across an outer surface must move into the body cell by cell.

Think SCIENCE

SKILL: EVALUATING CONCLUSIONS

Eating Well

Common sea stars have five arms. When one arm is missing, sea stars are, amazingly, able to grow another. Scientists interested in this amazing ability designed an experiment to see how having four arms instead of five affect a sea star's ability to consume prey. The sea stars use tube feet on each arm to pry open the shells of mussels and eat the animal's insides.

Scientists, working in 1999 and 2000, examined common sea stars caught in fishing gear. The study divided them into two groups: one group with five arms and one group with only four arms. Each sea star was tested to see how well it could open and eat a mussel.

Observations

Scientists made these observations:

a. Most of the sea stars with all five arms opened and ate a mussel.
b. Fewer than half of the sea stars with four arms opened and ate a mussel.
c. All the sea stars that opened a mussel took about 13 hours to finish eating it.

Conclusions

Here are some conclusions about sea stars eating mussels:

a. Common sea stars are most likely to feed successfully if they have all their arms.
b. A common sea star with only four arms will starve.
c. Common sea stars eat slowly regardless of how many arms they have.
d. Common sea stars with only four arms choose mussels that are difficult to open.

Evaluate the Conclusions

On Your Own Think about each observation that the scientists noted. Do they support the conclusions? Do some observations support one conclusion but not another? If a conclusion is not supported, what extra observations would you need to make?

With A Partner Compare your thinking with your partner' s thinking. Do you both agree with the conclusions?

CHALLENGE Why do you think the sea stars with only four arms were less likely to open and eat a mussel?

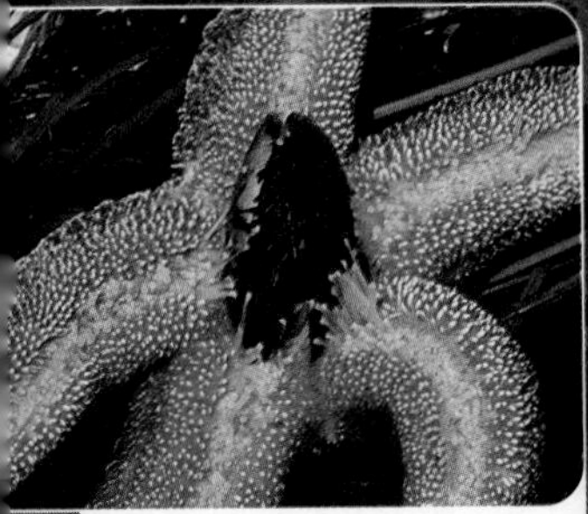

The common sea star *(Asterias rubeus)* can regrow a missing arm. Its main prey is the mussel.

ANSWERS

Conclusion a: Reasonable. Observations a. and b. support this conclusion. Observation c. does not have any bearing on this conclusion.
Conclusion b: Not reasonable. Observation b. does not support this conclusion since some of the four-armed sea stars were able to open the mussels.
Conclusion c: Reasonable. Observation c. supports this conclusion.
Conclusion d: Not reasonable. Neither observations a., b., or c. support this conclusion as the difficulty of opening the mussels is not a factor.
CHALLENGE *These sea stars most likely have less strength and leverage than those with five arms.*

THINK SCIENCE
Scientific Methods of Thinking

Set Learning Goal

To evaluate conclusions by deciding whether or not they are based on observations and prior knowledge

Present the Science

Common sea stars have the ability to regenerate—to grow lost parts. A sea star can grow a new arm if it loses one. Some species of sea stars can regenerate completely from one arm and a piece of the central disc.

Guide the Activity

- Remind students that conclusions must be based upon observations and data collected from experiments.
- Students should check that each conclusion is supported by observations. Point out that the observations in the notebook are quantitative. That is, data can be counted, measured, or calculated.
- Remind students that to evaluate means to judge a statement based on criteria, such as observations.

COOPERATIVE LEARNING STRATEGY

- Group students in pairs to complete the activity.
- Have one student explain what makes each statement an "observation" that is listed as such.
- The other student can explain why each conclusion statement qualifies as a conclusion.
- Lastly, have partners work through the questions by discussing the validity of each conclusion.

Close

Ask: Why is it important to check each conclusion against each observation? *One observation may disprove a conclusion.*

4.4 FOCUS

Set Learning Goals

Students will

- Explore arthropods, the largest group of invertebrates.
- Recognize that all arthropods have exoskeletons.
- Describe the process of metamorphosis that insects undergo.
- Observe the development process of mealworms, including molting.

3-Minute Warm-Up

Display Transparency 29 or copy this exercise on the board:

Complete the Venn diagram. List unique features in the separate circles. Then list similarities in the overlap.

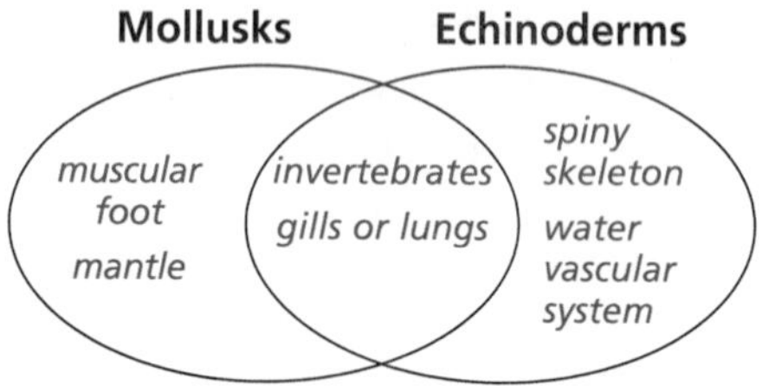

3-Minute Warm-Up, p. T29

4.4 MOTIVATE

EXPLORE Arthropods

PURPOSE To describe some characteristics of a pill bug, a land invertebrate

TIP *20 min.* Allow students to share and discuss their sketches.

WHAT DO YOU THINK? *Features might include mouth, head, eyes, tentacles, antennae, and jointed legs or other appendages. Pill bugs have bilateral symmetry.*

Ongoing Assessment

Explore arthropods, the largest group of invertebrates.

Ask: What types of organisms are included in the group of arthropods? *insects, shrimp, and spiders*

KEY CONCEPT

4.4 Arthropods have exoskeletons and joints.

BEFORE, you learned

- Mollusks are invertebrates with soft bodies, some have shells
- Echinoderms have spiny skeletons
- Different species adapt to the same environment in different ways

NOW, you will learn

- About different groups of arthropods
- About exoskeletons in arthropods
- About metamorphosis in arthropods

VOCABULARY

arthropod p. 142
exoskeleton p. 143
molting p. 143
insect p. 145
metamorphosis p. 146

EXPLORE Arthropods

What are some characteristics of arthropods?

PROCEDURE

1. Observe the pillbugs in their container. Draw a sketch of a pillbug.
2. Gently remove the pillbugs from their container and place them in the open end of the box. Observe and make notes on their behavior for several minutes.
3. Return the pillbugs to their container.

WHAT DO YOU THINK?

- Describe some of the characteristics you noticed about pillbugs.
- Are pillbugs radially or bilaterally symmetrical?

MATERIALS

- clear container
- shoebox with half of cover removed
- pillbugs
- hand lens

Most invertebrates are arthropods.

COMBINATION NOTES
Make notes and diagrams for the main idea: *Most invertebrates are arthropods.*

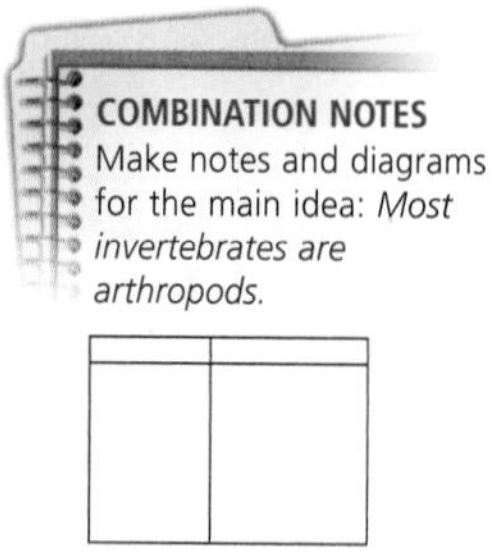

There are more species of arthropods than there are any other type of invertebrate. In fact, of all the animal species classified by scientists, over three-quarters are arthropods. An **arthropod** is an invertebrate that has a segmented body covered with a hard outer skeleton. Arthropods can have many pairs of legs and other parts that extend from their body. Insects are arthropods, so are crustaceans such as the shrimp, and arachnids such as the spider.

Fossil evidence shows that arthropods first appeared on land about 420 million years ago, around the same time as plants. Arthropods are active animals that feed on all types of food. Many arthropods live in water, but most live on land.

RESOURCES FOR DIFFERENTIATED INSTRUCTION

Below Level

UNIT RESOURCE BOOK

- Reading Study Guide A, pp. 255–256
- Decoding Support, p. 268

AUDIO CDS

Advanced

UNIT RESOURCE BOOK
Challenge and Extension, p. 262

English Learners

UNIT RESOURCE BOOK
Spanish Reading Study Guide, pp. 259–260

AUDIO CDS

- Audio Readings in Spanish
- Audio Readings (English)

RESOURCE CENTER
CLASSZONE.COM

Find out more about the diversity of arthropods.

The exoskeleton of this crayfish completely covers its body.

Exoskeletons and Jointed Parts

One adaptation that gives arthropods the ability to live in many different environments is the exoskeleton. An **exoskeleton** is a strong outer covering, made of a material called chitin. The exoskeleton completely covers the body of an arthropod. In a sense, an exoskeleton is like a suit of armor that protects the animal's soft body. For arthropods living on land, the exoskeleton keeps cells, tissues, and organs from drying out.

CHECK YOUR READING What are two functions of an exoskeleton?

A suit of armor is not much good unless you can move around in it. The arthropod's exoskeleton has joints, places where the exoskeleton is thin and flexible. There are joints along the different segments of the animal's body. An arthropod body typically has three sections: a head at one end, a thorax in the middle, and an abdomen at the other end. Legs are jointed, as are other parts attached to the body, such as antennae and claws. Muscles attach to the exoskeleton around the joints, enabling the arthropod to move.

The exoskeleton is like a suit of armor in one other way. It doesn't grow. An arthropod must shed its exoskeleton as it grows. This process is called **molting.** For an arthropod, the times when it molts are dangerous because its soft body is exposed to predators.

COMPARE How does the shape of this cicada's molted exoskeleton compare to the shape of its body?

Complex Body Systems

Arthropods have well-developed body systems. They have a nervous system with a brain and many different sensory organs. Their digestive system includes a stomach and intestines. Arthropods have an open circulatory system, which means the heart moves blood into the body directly. There are no blood vessels. Arthropods reproduce sexually. An arthropod has either a male or a female reproductive system.

4.4 INSTRUCT

Teach from Visuals

To help students interpret the photograph of the crayfish, ask: What body features of this arthropod help it to move around on land? *segmented body and jointed legs*

Integrate the Sciences

The fossil record shows that the most primitive insects lived during the Middle Devonian period about 350 million years ago. Most of the early insect fossils, however, are from the Carboniferous Period (about 360 to about 280 million years ago). There are many gaps in the insect fossil record, but there have been epochs of "explosive" evolution during which many species appeared. After arthropods became firmly established, fossil evidence indicates that many insects and plants co-evolved due to their interdependence.

Ongoing Assessment

Recognize that all arthropods have exoskeletons.

Ask: What is an exoskeleton? *It is a strong outer covering made of chitin, which completely covers the body of an arthropod.*

CHECK YOUR READING *Answer: The exoskeleton protects the animal and keeps land-dwelling arthropods from drying out.*

DIFFERENTIATE INSTRUCTION

More Reading Support

A How is the exoskeleton attached to the joints of the arthropod? *by muscles*

B What term describes the process of shedding an exoskeleton? *molting*

English Learners Have students write the definitions for the terms *exoskeleton, molting,* and *metamorphosis* in their Science Word Dictionaries. Pronounce these words for students. Use the Chapter Summary page to help students focus on key concepts and vocabulary.

Teach from Visuals

To help students compare the three photographs showing characteristics of the three major groups of arthropods, ask:

- How do arachnids differ from the insect and crustacean groups? *They have 4 pairs of legs instead of 3 pairs. They have no antennae.*
- How are the insect and arachnid groups alike? *Most live on land.*

Address Misconceptions

IDENTIFY Ask students which characteristics they would use to identify an organism as an animal. Many students will hold the misconception that animals have four legs, make sounds, and have fur. Most of the animals in this chapter would not fit any of those criteria.

CORRECT Construct a bulletin board with labels *Fur, Makes Sound,* and *Four Legs*. Have students put the names of invertebrate species on scraps of paper and tack them under one of the headings, or in a separate area.

REASSESS Based on the bulletin board above, ask students to re-think their criteria for characteristics of animals. Student answers should include the many characteristics of invertebrates.

Technology Resources

Visit **ClassZone.com** for background on common student misconceptions.

MISCONCEPTION DATABASE

Ongoing Assessment

READING VISUALS *Answer: All have legs and segmented bodies, eyes, and heads.*

Three Major Groups of Arthropods

Scientists have named at least ten groups of arthropods, but most arthropod species belong to one of three groups: insects, crustaceans, or arachnids.

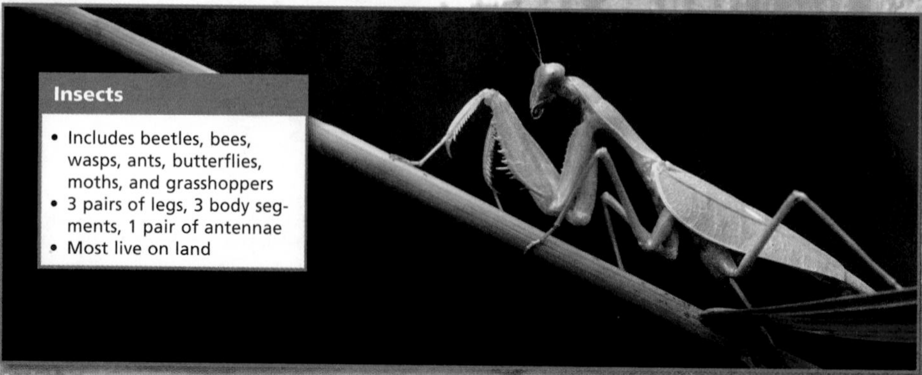

Insects

- Includes beetles, bees, wasps, ants, butterflies, moths, and grasshoppers
- 3 pairs of legs, 3 body segments, 1 pair of antennae
- Most live on land

Crustaceans

- Includes shrimp, crabs, lobsters, barnacles, and pill bugs
- Number of body segments and pairs of legs varies, 2 pairs of antennae
- Most live in water; some live on land

Arachnids

- Includes spiders, ticks, mites, and scorpions
- 4 pairs of legs, 2 body segments, no antennae
- Most live on land

READING VISUALS What body features can you see that are shared by all of these arthropods?

DIFFERENTIATE INSTRUCTION

Below Level Provide students with close-up photographs obtained from nature magazines of the major groups of arthropods. Have students examine the photographs for structures such as number of pairs of legs, presence and number of antennae, and number of body segments. Have students fill in a chart with the characteristics pictured, then classify each arthropod as an insect, crustacean, or arachnid.

Advanced Challenge students to develop a classification key using "yes" and "no" questions to classify the major groups of arthropods.

C 144 Unit: Diversity of Living Things

Parts of an Insect's Body

Adult insects have three body segments and six jointed legs.

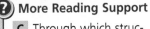

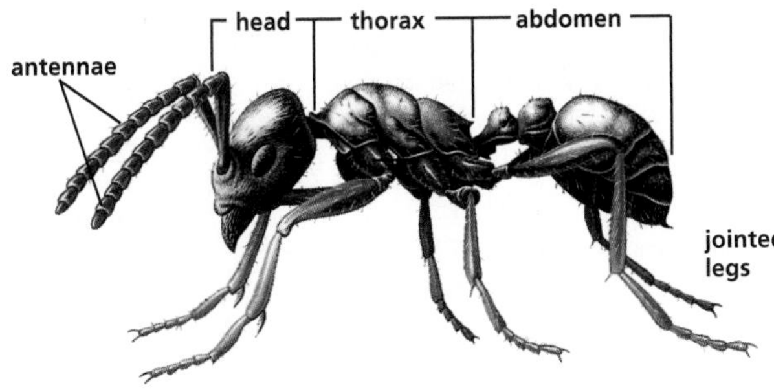

READING TIP

The word *insect* relates to its body being **in sect**ions. Note the three sections in the diagram of the ant.

Insects are six-legged arthropods.

Scientists have so far identified over 700,000 insect species. **Insects** are arthropods that as adults have three body segments, a pair of antenna, and six legs attached to the middle segment, the thorax. Insect species have adapted to all sorts of environments and live on every continent. Most insects live on land. These insects obtain oxygen through spiracles, small openings in their exoskeleton. C

VOCABULARY
Don't forget to take notes on the term *insect,* using a strategy from an earlier chapter or one that you already know.

CHECK YOUR READING What are two characteristics all adult insects share?

Insects show great diversity in appearance. Many species have adaptations in color and shape that allow them to blend into their environments. For example, a stick insect is the same color and shape as a twig. Insect bodies also have different adaptations. Many insects have compound eyes and antennae, which are sensory organs. Many insects fly, having one or two pairs of wings.

Many insects are herbivores. And many insect species have mouth parts adapted for feeding on specific plants. A butterfly, for example, has a tubelike mouth that can reach into a flower to get nectar. Insects that feed on flowers often help the plants reproduce because the insects carry pollen from flower to flower. Other insects harm the plants they feed on. A grasshopper has jawlike mouth parts that crush parts of a plant. Many plants have defensive adaptations, such as poisons in leaves and stems, to keep insects away.

D Some insects, for example, ants, termites, and some bees, are social insects. They must live in groups in order to survive. Members of the group work together to gather food, maintain the nest, and care for the offspring. Often with social insects, just one female, called a queen, produces and lays eggs.

Develop Critical Thinking

COMPARE AND CONTRAST To help students understand the structure of the insect, ask: How are the segments of the insect's body different from those of an earthworm? *The insect has three distinct segments that are different from each other. The earthworm has segments that look the same.*

Teacher Demo

To demonstrate the importance of segmented joints, invite a volunteer to the front of the room. Place a rolled cardboard tube around the student's arm up to the shoulder. Challenge the student to write on the board, lift an object, or eat a cracker. Reinforce the concept that jointed appendages are an adaptation that enabled arthropods to live on land.

EXPLORE the BIG idea

Revisit "Insects and You" on p. 121. Ask students if their idea of insects has changed based on their observations and what they have learned from the text.

Ongoing Assessment

CHECK YOUR READING *Answer: Adult insects have three body segments, six jointed legs.*

DIFFERENTIATE INSTRUCTION

More Reading Support

C Through which structures do insects obtain oxygen? *spiracles*

D What term describes insects that live in groups? *social insects*

Advanced Invite interested students to choose one group of social insects, such as ants, termites, or bees. The dance of bees might be of particular interest, a dance of communication that informs other honeybees as to the type, quality, and location of nectar. Ask students to display their findings using illustrations, captions, and other graphics.

R Challenge and Extension, p. 261

INVESTIGATE Insect Metamorphosis

PURPOSE To observe the development of mealworms to discover how often they molt

TIPS *20 min.* Make holes in lids of jars prior to activity so mealworms will have enough air. They will not escape. The oats are the mealworms' food; the potato provides moisture they need. Purchase mealworms at pet stores or biological supply companies.

WHAT DO YOU THINK? *Eggs will be visible with a magnifying glass. Larva hatches from egg in one to two weeks. Larval stage lasts an average of ten weeks and includes several molts. Pupal stage lasts two to three weeks. Newly hatched adults have pale soft exoskeletons, which later harden and darken.*

CHALLENGE *There are approximately four to six molts during the larval stage, so students should find about four to six times the number of shed exoskeletons as living mealworms.*

R Datasheet, Insect Metamorphosis, p. 262

Technology Resources

Customize this student lab as needed or look for an alternative. Print rubrics to assess student lab reports.

Lab Generator CD-ROM

Teach Difficult Concepts

Pupae, chrysalis, and eggs are among the stages of a life cycle. In each stage the organism may not be moving, but its cells continue to grow, feed, respond, and maintain internal conditions. The cells have not died and reanimated. They emerge from living cells that were their parents.

Ongoing Assessment

CHECK YOUR READING *Answer: During metamorphosis, an insect's body systems change dramatically in three stages. Larva: the insect eats; pupa: it develops in a protective casing. It emerges as an adult, capable of reproducing.*

INVESTIGATE Insect Metamorphosis

How often do mealworms molt?

PROCEDURE

1. Prepare the jar for the mealworms. Fill it halfway with oat bran for food. Place a slice of potato and a piece of carrot on top for moisture.
2. In your notebook, note how many mealworms you have. Carefully place the mealworms inside the jar and close the lid. Wash your hands.
3. Without opening the jar, look for signs of activity every day. Once a week, open the container and pour some of the contents into a tray. Examine this sample for molted exoskeletons. Then return it to the jar. Replace the vegetables and add new oats as needed. Wash your hands.

WHAT DO YOU THINK?

- What changes did you observe in the mealworms?
- Did you see any sign of other stages of development?

CHALLENGE Use tweezers and a petri dish to collect the molted exoskeletons. How do the number of molts compare to the number of worms? Estimate how often the worms molt.

SKILL FOCUS
Observing

MATERIALS
- glass jar
- lid with air holes
- oat bran
- potato
- carrot
- mealworms
- *for Challenge:* petri dish tweezers

TIME
20 minutes

All insects reproduce sexually. Females lay eggs, often a large number of eggs. The queen honey bee can lay over a million eggs in her lifetime. Many insect eggs have a hard outer covering. This adaptation protects the egg from drying out and can allow hatching to be delayed until conditions are right.

? E

During their life cycle, insects undergo a process in which their appearance and body systems may change dramatically. This process is called **metamorphosis.** There are three stages to a complete metamorphosis. The first stage is the larva, which spends its time eating. The second stage is the pupa. During this stage, the insect body develops within a protective casing. The final stage is the adult, which is capable of going on to produce a new generation.

READING TIP The word *metamorphosis* means "many changes."

CHECK YOUR READING What happens to an insect during metamorphosis?

? F

Not all insects go through complete metamorphosis. Some insects, such as grasshoppers, have a simple metamorphosis. When a young grasshopper hatches from an egg, its form is similar to an adult's, just smaller. A grasshopper grows and molts several times before reaching adult size.

DIFFERENTIATE INSTRUCTION

More Reading Support

E What term describes the changes in an insect's life? *metamorphosis*

F Give an example of an insect that undergoes simple metamorphosis. *Sample answer: grasshopper*

English Learners Allow English learners to draw their observations in the investigation, p. 146, or to narrate their observations to a partner or group.

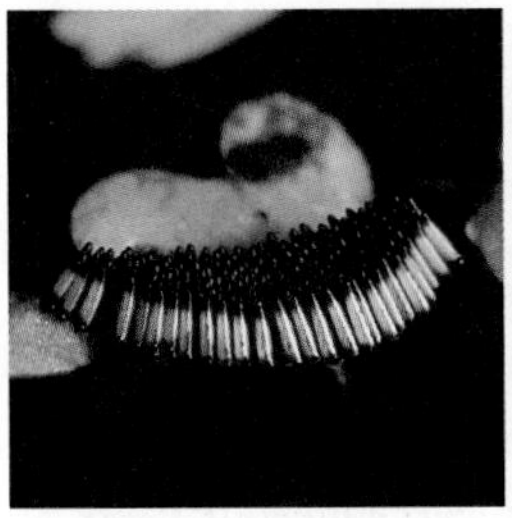

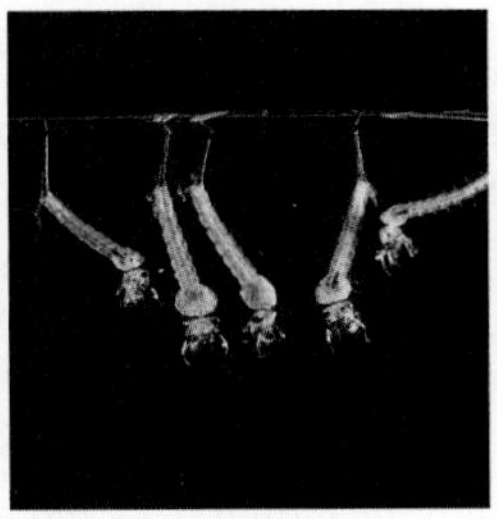

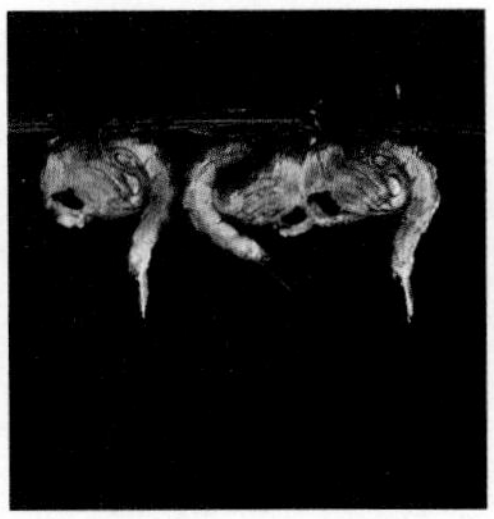

1 A female mosquito lays a mass of eggs on the surface of the water.

2 Each egg develops into a larva that swims head down, feeding on algae.

3 The larva develops into a pupa. Inside, the body of the insect matures.

4 At the adult stage, the mosquito leaves the water and flies away.

You have probably seen many insects in their larval form. A caterpillar is a larva, so is an inchworm. Often the larval form of an insect lives in a way very different from its adult form. A mosquito, for example, begins its life in the water. The larva swims about, feeding on algae. The pupa forms at the water's surface. The developing mosquito is encased in a protective covering. The adult form of the mosquito, the flying insect, leaves the water. It is a parasite that feeds off the blood of other animals.

Crustaceans live in water and on land.

Most crustacean species live in the water. Several of these, including the Atlantic lobster and the Dungeness crab, are used by people as a source of food. Crustaceans are important to the ocean food web. Tiny crustaceans such as krill and copepods are a food source for many other animals, including other invertebrates, fish, and whales. Some species of crustaceans live in freshwater and a few, such as pill bugs, live on land.

CHECK YOUR READING Where do most crustaceans live?

? G

Crustaceans have three or more pairs of legs and two pairs of sensory antennae. Many of the larger, water-living crustaceans, such as crabs, have gills. Most crustaceans, like other arthropods, have a circulatory system that includes a heart but no blood vessels. Crustaceans reproduce sexually. Their young hatch from eggs.

The eating habits of crustaceans vary. Lobsters and shrimp eat plants and small animals. Many crustaceans are scavengers, feeding off the remains of other organisms. Some, such as barnacles, are filter feeders. The larval form of a barnacle is free swimming. However, as an adult this arthropod attaches itself to a rock or another hard surface, such as a mollusk's shell or the hull of a ship. It uses its tentacles to capture food from the surrounding water.

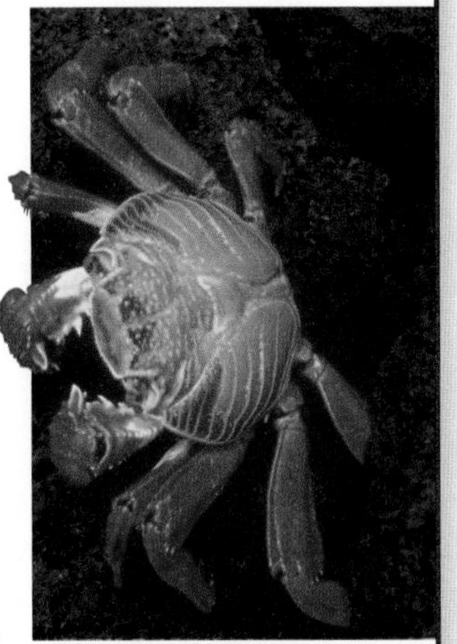

IDENTIFY How many pairs of legs does this crab have?

DIFFERENTIATE INSTRUCTION

? More Reading Support

G How many pairs of antennae do crustaceans have? *two pairs*

Below Level Have each student write a one-page story entitled "A Day in the Life of Insect X." Each story should include scientific information such as life cycle stages, how the insect moves, what the insect eats, and how the insect's body plan suits its environment. The story could end with the question: "Who am I?" and students can try to guess the identity of each other's subjects as they read them in a class book or display.

Teach from Visuals

To help students understand the photographs of mosquito metamorphosis, ask:

- What are the four stages of insect development? *egg, larva, pupa, adult*
- When is metamorphosis complete? *when the pupa becomes an adult*

History of Science

Some diseases, such as West Nile virus and malaria, are transmitted to humans by the bites of mosquitoes. Malaria has been documented as far back as 400 B.C. by Hippocrates. It most likely made its way to the Americas by import through Spanish explorers. It was not until 1880 that French army surgeon Alphonse Laveran became the first to recognize the malarial parasite. West Nile virus has been studied for several decades and has become an increasing public health concern in North America. It made its way to the United States in 1999 and spread into Canada and the Caribbean.

Mathematics Connection

To date, scientists recognize about 800,000 different species of insects. This represents about three-fourths of all known species on Earth. To help students visualize the vast numbers of insects that inhabit Earth, draw a circle graph on the board. Shade in three fourths of the circle, pointing out that this portion of the circle represents the number of insect species on Earth compared to all other animal species.

Ongoing Assessment

Describe the process of metamorphosis.

Ask students to compare the metamorphosis of a mosquito to that of a butterfly. *Each metamorphosis takes the insect from larval stage to a stage within a protective case (pupa) to a fully developed adult.*

CHECK YOUR READING *Answer: Most crustaceans live in water.*

PHOTO CAPTION *Answer: It has ten jointed legs covered by an exoskeleton.*

Real World Example

Commercial silk is produced by the cocoons of moth species known as silkworms. Silk weaving dates back to the 3rd millennium B.C. in China, where its production was a highly guarded secret. To produce silk, the larvae inside the cocoons of the silkworms are killed by steam or hot air. The strands of silk that make up the cocoons are unwrapped and twisted into yarn.

Language Arts Connection

The name *Arachnida* comes from the Greek mythological character Arachne, a peasant girl who challenged the weaving talent of the goddess Athena. Arachne was able to match Athena's talent in a competition, but succumbed in response to Athena's anger. Athena turned Arachne into a spider, allowing her to keep her weaving talent.

Teaching with Technology

Invite students to compile an electronic database of "Fascinating Facts" about arthropods. For example, some species of centipedes and millipedes can grow up to 8 inches long. Students can print their databases or post them on a school Web site.

Ongoing Assessment

CHECK YOUR READING *Answer: Some spiders weave webs, which are unusual because they are traps for prey.*

Arachnids are eight-legged arthropods.

This mite is an arachnid that lives in dust. This micrograph shows it magnified 150×. H

Spiders, mites, ticks, and scorpions belong to a group called the arachnids. Like all arthropods, arachnids have an exoskeleton, jointed limbs, and segmented bodies. But the bodies of arachnids have some characteristics that distinguish them from other arthropods. Arachnids always have four pairs of legs and only two body segments. Arachnids do not have antennae.

Some arachnids, including ticks and chigger mites, are parasites. Other arachnids, such as spiders and scorpions, are predators. Recall that predators get their food by capturing and consuming other animals. Predatory arachnids kill their prey by stinging them, biting them, or injecting them with venom.

The spiders are the largest group of arachnids. Many spiders have a unique adaptation for capturing their prey. They produce an extremely strong material, called silk, inside their bodies and use the silk to make webs for capturing food. The spider spins strands of silk out from tubes called spinnerets at the rear of its abdomen. It weaves the strands into a nearly invisible web. The web serves as a net for catching insects and other small organisms that the spider eats. This adaptation allows web-building spiders to wait for their prey to come to them. Other invertebrates, such as silkworms, produce silk, but they do not weave webs.

CHECK YOUR READING How is the way some spiders capture prey unusual?

Some arachnids obtain oxygen through spiracles, as insects do. However, certain species of spiders have a unique type of respiratory organ referred to as book lungs. Book lungs are like moist pockets with folds. They are located inside the animal's abdomen.

This spider has wrapped its prey in silk.

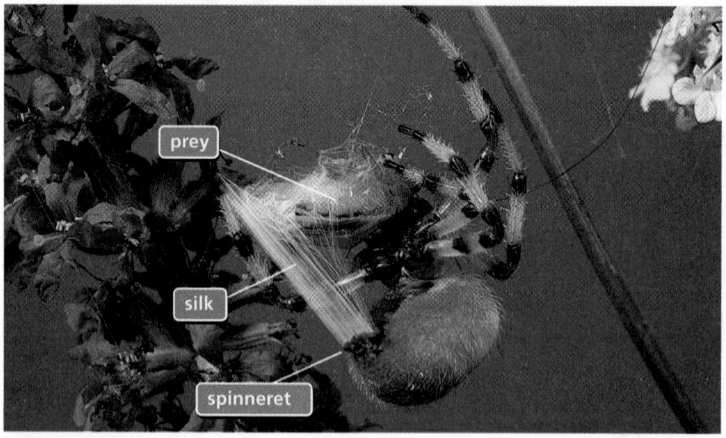

DIFFERENTIATE INSTRUCTION

More Reading Support

H What is the largest group of arachnids? *spiders*

Inclusion To help students learn the differences among insects, crustaceans, and arachnids, help them to build models using clay or pipe cleaners. Allow time for students to display and discuss their models with other students.

C 148 Unit: Diversity of Living Things

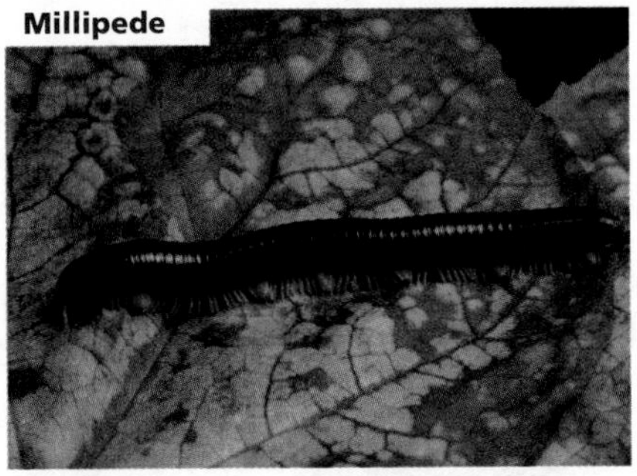

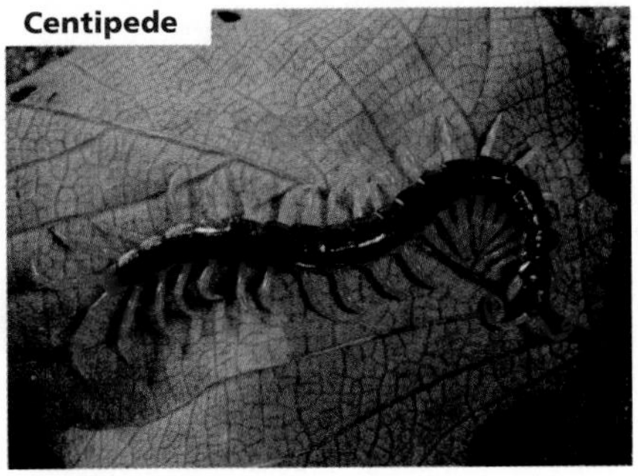

READING VISUALS COMPARE AND CONTRAST With their long segmented bodies, a millipede and a centipede look very similar. How are they different?

Millipedes and centipedes are arthropods.

At first glance, the members of two other arthropod groups look similar. Both centipedes and millipedes have long, segmented bodies and many legs. However, animals from these groups differ in their body features and their behavior.

Millipedes are arthropods with two pairs of walking legs on each body segment. Millipedes move rather slowly and eat decaying leaves and plant matter. When disturbed, many millipedes emit a foul odor that can be harmful to predators.

Centipedes can move more quickly. They have one pair of walking legs per body segment. They have antennae and jawlike mouthparts. Many centipedes also have pincers on their rearmost segment. Centipedes are predators. They can use their jaws, and pincers to paralyze prey and protect themselves from predators.

4.4 Review

KEY CONCEPTS

1. Describe the characteristics of insects, crustaceans, and arachnids.
2. What is molting and how does it relate to an exoskeleton?
3. Name three arthropods and the adaptations they have for feeding.

CRITICAL THINKING

4. **Analyze** How does the form of an exoskeleton relate to its function?
5. **Connect** Mosquitoes can spread disease, such as the West Nile virus. People are advised not to leave open containers of water in the yard. How does standing water contribute to an increase in the number of mosquitoes?

CHALLENGE

6. **Evaluate** Many plant-eating insects live less than a year. An adult will lay eggs in the fall and then die as winter comes. The eggs hatch the next spring. How does the life cycle of the insect fit in with the life cycle of plants? What role does the egg play in the survival of the insect species in this case?

ANSWERS

1. Insects have three segments, six legs, and one pair of antennae. Crustaceans have two pairs of antennae and specialized legs. Arachnids have two body segments, eight legs, and no antennae.

2. Molting is the shedding of an exoskeleton in order to grow a larger one.

3. Sample answer: Butterflies have tubelike mouths, barnacles have tentacles, and spiders make webs.

4. Exoskeletons are hard in order to protect the animal, and have joints that allow the animal to move.

5. Standing water gives the mosquitoes a place to breed.

6. Insect life cycles may be timed so they mature at same time plant is ready to be pollinated. Egg protects offspring during winter.

EXPLORE the BIG idea

Revisit "Internet Activity: Invertebrate Diversity" on p. 121. Now that students have studied the variety of invertebrates, ask them if their understandings about invertebrates have changed.

Ongoing Assessment

READING VISUALS *Answer: The centipede has fewer legs than the millipede, and appears to be longer and thicker. The centipede has curved structures at one end of its body. A millipede has more body segments than the centipede, and more legs per segment.*

Reinforce the BIG idea

Have students relate the section to the Big Idea.

R Reinforcing Key Concepts, p. 263

4.4 ASSESS & RETEACH

Assess

A Section 4.4 Quiz, p. 66

Reteach

Ask students to name a type of arthropod. Then have students tell whether it is an insect, crustacean, or arachnid. Ask for volunteers to describe some body characteristics about the organism such as number of body segments, presence of antennae, and the like. Students can then describe one fact about the organism, such as it lives in water, spins silk, or goes through metamorphosis.

Technology Resources

Have students visit **ClassZone.com** for reteaching of Key Concepts.

CONTENT REVIEW

CONTENT REVIEW CD-ROM

BACK TO

the BIG idea

Have students name and describe two body plans and two adaptations of invertebrate animals. *Bilaterally symmetrical bodies have halves that are mirror images. Radially symmetric bodies are organized around a central point. Adaptations may include jointed appendages and an external skeleton.*

KEY CONCEPTS SUMMARY

SECTION 4.1
Ask students to describe how specialized cells in sponges enable them to feed on and digest food. *Specialized cells filter out food particles as water moves through the sponge. Other specialized cells digest food.*

SECTION 4.2
Ask students how the body plans of cnidarians and worms differ. *All cnidarians are radially symmetrical and have tentacles that surround a mouth cavity. Worms are bilaterally symmetrical and have organs that are organized into body systems.*

SECTION 4.3
Ask students to name two adaptations that distinguish mollusks as a group. *a mantle and a muscular foot* Then ask students to explain the purpose of the water vascular system in echinoderms. *Water-filled tubes end in a series of tube feet. As muscles close off the top of the tubes, suction is produced at the feet, enabling the animal to move and feed on prey.*

SECTION 4.4
Have students describe why the exoskeletons and jointed appendages are adaptations that allow arthropods to live on land. *They provide support, protection, and movement, allowing arthropods to live in many different environments.*

Review Concepts

- Big Idea Flow Chart, p. T25
- Chapter Outline, pp. T31–T32

4 Chapter Review

the BIG idea

Invertebrate animals have a variety of body plans and adaptations.

CONTENT REVIEW
CLASSZONE.COM

KEY CONCEPTS SUMMARY

4.1 Most animals are invertebrates.

Invertebrates are a diverse group of animals. Species of invertebrates live in almost every environment.

Sponges are simple invertebrates that have several types of specialized cells.

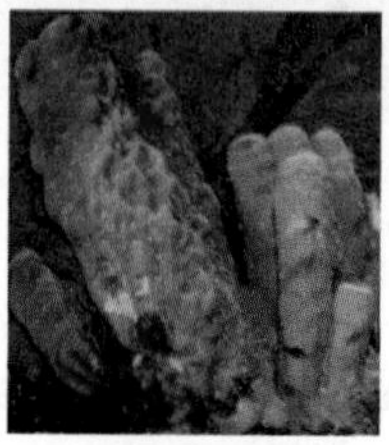

VOCABULARY
invertebrate p. 123
sponge p. 125
sessile p. 125
larva p. 126

4.2 Cnidarians and worms have different body plans.

Cnidarians have simple bodies with specialized cells and tissues.

Most **worms** have organs and complex body systems.

VOCABULARY
cnidarian p.128
tentacle p. 128
mobile p. 130

4.3 Most mollusks have shells, and echinoderms have spiny skeletons.

Mollusks include bivalves, gastropods, and cephalopods.

Echinoderms have a water vascular system and tube feet.

VOCABULARY
mollusk p. 136
gill p. 137
lung p. 137
echinoderm p. 139

4.4 Arthropods have exoskeletons and joints.

Arthropods, which include insects, crustaceans, and arachnids, are the most abundant and diverse group of animals.

VOCABULARY
arthropod p. 142
exoskeleton p. 143
molting p. 143
insect p. 145
metamorphosis p. 146

Technology Resources

Have students visit **ClassZone.com** or use the CD-ROM for a cumulative review of concepts.

CONTENT REVIEW

CONTENT REVIEW CD-ROM

Engage students in a whole-class interactive review of Key Concepts. Edit content as you wish.

POWER PRESENTATIONS

C 150 Unit: Diversity of Living Things

Reviewing Vocabulary

Copy and complete the chart below.

Word	Definition	Example
1. mollusk		clam, snail, squid
2. arthropod	invertebrate with jointed legs, segmented body, and an exoskeleton	
3.	ocean-dwelling animal with spiny skeleton	sea star
4. sessile		sponge
5. larva		caterpillar
6. metamorphosis		caterpillar changing into a butterfly
7. molting	process by which an arthropod sheds its exoskeleton	
8.	arthropod with three body segments, one pair of antennae, and six legs	grasshopper, mosquito, beetle

Reviewing Key Concepts

Multiple Choice *Choose the letter of the best answer.*

9. Which of the following groups of animals is the most abundant?
a. worms
b. mollusks
c. echinoderms
d. arthropods

10. In what way are all invertebrates alike?
a. They do not have backbones.
b. They live in the ocean.
c. They are predators.
d. They have a closed circulatory system.

11. Sponges bring food into their bodies through a
a. system of pores
b. water vascular system
c. mouth
d. digestive tract

12. Which group of invertebrates has a mantle?
a. echinoderms
b. crustaceans
c. cnidarians
d. mollusks

13. Bivalves, cephalopods, and gastropods are all types of
a. echinoderms
b. mollusks
c. crustaceans
d. cnidarians

14. As they grow, arthropods shed their exoskeleton in a process called
a. metamorphosis
b. symmetry
c. molting
d. siphoning

15. Which invertebrate animals always have three body segments: a head, a thorax, and an abdomen?
a. segmented worms
b. adult insects
c. arachnids
d. echinoderms

16. Which group of invertebrates have a water vascular system and tube feet?
a. echinoderms
b. crustaceans
c. cnidarians
d. mollusks

Short Answer *Write a short answer to each question.*

17. Describe the stages in the life cycle of an insect that has complete metamorphosis.

18. Explain one advantage and one disadvantage an exoskeleton has for an organism.

19. Is a spider an insect? Explain.

ASSESSMENT RESOURCES

UNIT ASSESSMENT BOOK
- Chapter Test, Level A, pp. 67–70
- Chapter Test, Level B, pp. 71–74
- Chapter Test, Level C, pp. 75–78
- Alternative Assessment, pp. 79–80

SPANISH ASSESSMENT BOOK
Spanish Chapter Test, pp. 53–56

Technology Resources

Edit test items and answer choices.

Test Generator CD-ROM

Visit **ClassZone.com** to extend test practice.

Test Practice

Reviewing Vocabulary

1. *invertebrate with a strong muscular foot and a mantle*
2. *sample answers: lobster, insect, spider*
3. *echinoderm*
4. *lives and grows while remaining in one place*
5. *an immature, or early, stage of life for an organism, in which it may look different from the adult form*
6. *process of physical transformation as an organism matures from one stage of life to the next*
7. *sample answers: mealworm, larva, softshell crabs, cicada*
8. *insect*

Reviewing Key Concepts

9. d
10. a
11. a
12. d
13. b
14. c
15. b
16. a

17. *The egg hatches into a larva. The larva eats, grows, and becomes a pupa. During the pupa stage, the animal may not move or eat. The final stage is the adult stage.*
18. *allows for attachment of muscles and provides animal with protection; can be heavy, impair movement, or cause periods of vulnerability when some animals move or grow and molt*
19. *A spider is an arachnid. Insects, arachnids, and crustaceans are examples of a larger group of invertebrates called arthropods.*

(Answers to items that appear on p. 152)

Thinking Critically

20. *Plantlike: A sponge is sessile, does not have a mouth, eyes, or legs, nor a symmetrical body plan. Animal: It consumes and digests food.*
21. *Sample answers: legs, wings, antennae for sensing, eyes for seeing*
22. *Food materials can be taken in at one tube end, be broken down and absorbed as they are moved through the tube, wastes excreted out the far end.*

(Answers for items 20–22 appear on p. 151.)

Thinking Critically

23. *Both start as larva. Jellyfish larvae develop into polyps, mosquito larvae into pupae. A jellyfish medusa is mobile; the polyp is not. Mosquito larvae swim; adults fly.*

24. *86%*

25. *14%*

26. *Shade the red slice to show what percentage of it are insects, crustaceans, and arachnids.*

27. *sea star: mouth in center, spreads its arms around prey, bringing it to the middle of its body; spider: mouth at front end, prey is brought in front*

28. *Insects, crustaceans, and arachnids all have exoskeletons, segmented bodies, paired legs, and distinct life stages. Crustaceans usually have 2 sets of antennae, insects 1, and arachnids none. Arachnids have 8 legs; insects 6.*

29. *A sponge's body has stiff spicules that provide support. The bivalve is surrounded by a shell; the body is inside. The insect has an exoskeleton that completely covers its body.*

30. *If insects became extinct, many plants and animals would die for lack of food. We would need a new way to pollinate plants. It would likely lead to extinction of most life.*

the BIG idea

31. *Jellyfish has a nerve net and tentacles that allow it to sense and respond. Some have tentacles and venom. It does not select what it stings. Its stinging cells are released as it brushes an object. The diver may be in danger of being stung.*

32. *Answers should relate body plan to environment and feeding method.*

UNIT PROJECTS

Evaluate designs and drafts for students' models, visuals, or text. Encourage students to cite research sources.

Unit Projects, pp. 5–10

Thinking Critically

20. **CLASSIFY** What characteristics does a sponge have that make it seem like a plant? What characteristic makes a sponge an animal?

21. **PROVIDE EXAMPLES** Arthropods are the most diverse and abundant group of animals on Earth. Give three examples of arthropod features that enable them to be active in their environment.

22. **INFER** Worms have a tube-shaped body with openings at either end. How does this body plan relate to the way a worm obtains its food and processes it?

23. **COMPARE** Jellyfish go through a life cycle that involves different stages of development. Insects also go through different stages of development in a process called metamorphosis. What are some similarities between metamorphosis in a mosquito, for example, and a jellyfish life cycle? Use the terms in the table below in your answer.

larva	polyp	medusa
pupa	adult	mobile
sessile		

Refer to the chart below as you answer the next three questions.

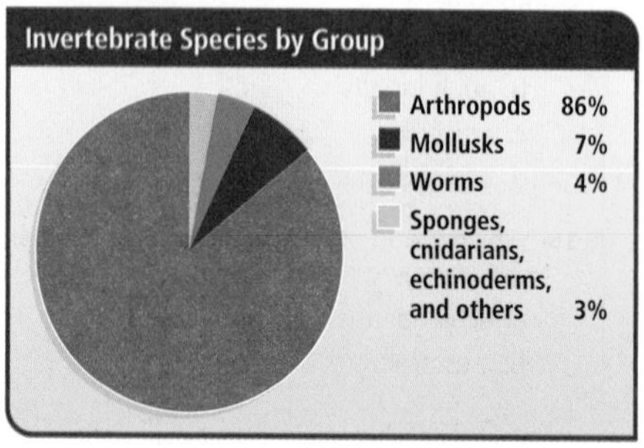

24. **APPLY** What percentage of invertebrate species are arthropods?

25. **CALCULATE** What is the combined percentage of all other invertebrate species, not including arthropods?

26. **APPLY** How could you modify this pie chart to show that insects, crustaceans, and arachnids are types of arthropods?

27. **APPLY** A sea star has a radial body plan. A spider has a bilateral body plan. How does the shape of these animal's bodies affect how they capture their food?

28. **COMPARE AND CONTRAST** How are the three main groups of arthropods similar? How are they different?

29. **SYNTHESIZE** Animal bodies need support as well as protection. What structures do the bodies of a sponge, a bivalve, and an insect have for protection and support? Use the terms in the table below in your answer.

spicule	shell	exoskeleton

30. **PREDICT** Many people think of insects as pests, but some species of insects are important pollinators for many flowering plants. Also, many animals eat insects. What problems would humans face if Earth's insect species became extinct?

the BIG idea

31. **INFER** How does a cnidarian such as the jellyfish in the photograph on pages 120–121 respond to its environment? Is the diver in the photograph in danger? Explain your answer.

32. **COMPARE AND CONTRAST** Make a chart showing the key features of the body plans of three different invertebrate animal groups. For each group, identify one characteristic that is an adaptation.

UNIT PROJECTS

If you need to create graphs or other visuals for your project, be sure you have grid paper, poster board, markers, or other supplies.

MONITOR AND RETEACH

If students are having trouble applying the concepts in Chapter Review item 23, suggest that they review the illustration of Mosquito Metamorphosis on p. 147. Then have students create a flow chart to show the process.

Students may benefit from summarizing one or more sections of the chapter.

Summarizing the Chapter, pp. 289–290

Standardized Test Practice

For practice on your state test, go to . . .
TEST PRACTICE
CLASSZONE.COM

Analyzing Data

Reef-building corals are invertebrates that live in clear, warm ocean water. As a coral grows, it produces a hard external skeleton. If many generations of corals grow near each other over a long period of time, their accumulated skeletons form a structure called a reef. Many ocean life-forms live in and around coral reefs. This table shows the maximum growth rates of five species of reef-building corals.

Species	Rate of Growth (mm per year)	Number of Years to Grow a 1400-m Reef
A	143	9,790
B	99	14,100
C	120	11,700
D	100	14,000
E	226	6,190

Choose the letter of the best response

1. Which has the fastest rate of growth?
- **a.** Species C
- **b.** Species B
- **c.** Species D
- **d.** Species E

2. What is the growth rate for Species B ?
- **a.** 99 mm per year
- **b.** 120 mm per year
- **c.** 143 mm per year
- **d.** 14,100 mm per year

3. Which takes the shortest amount of time to grow to 1400 meters?
- **a.** Species B
- **b.** Species C
- **c.** Species D
- **d.** Species E

4. Which have about the same rate of growth?
- **a.** Species A and C
- **b.** Species B and D
- **c.** Species C and E
- **d.** Species D and E

5. How many years does it take Species A to grow into a 1400-m reef?
- **a.** 143
- **b.** 226
- **c.** 9790
- **d.** 14,000

6. Based on the information in the table, which statement is true?
- **a.** Coral species with the fastest growth rates take the greatest amount of time to grow.
- **b.** Coral species with the slowest growth rates take the least amount of time to grow.
- **c.** Coral species have different rates of growth that affect how long it takes them to grow.
- **d.** Coral species that grow more than 100 mm per year take the longest to grow.

Extended Response

7. Corals are cnidarians. They are sessile animals that live attached to one place. Other ocean-dwelling animals are mobile and can move about their environment. Crustaceans like the lobster are mobile, so are mollusks like the octopus. How does being sessile or mobile affect the feeding behaviors of an animal? Do you think the bodies and systems of sessile animals are going to be different from those of mobile animals? Use some of the terms in the word box in your answer.

digestive system	sessile	mobile
nervous system	filter	mouth
muscle tissue	food	sensory organs

Analyzing Data

1. d   *4. b*
2. a   *5. c*
3. d   *6. c*

Extended Response

7. RUBRIC

4 points for a response that correctly answers the questions and includes at least five of the following terms:

- digestive system
- nervous system
- muscle tissue
- sessile
- filter
- food
- mobile
- mouth
- sensory organs

Sample: Because they are not mobile, sessile animals cannot chase prey. They need to have specialized mouths and sensory organs to allow them to filter food from the water or grab it as it passes by. Their nervous systems could be simpler because they do not need complex behaviors for finding and catching prey or escaping predators. Mobile animals need to have more developed muscle tissue for movement. The digestive systems of mobile and sessile animals differ because of their different eating habits.

3 points correctly answers the question and includes at least four of the terms shown

2 points correctly answers the question and includes at least three of the terms shown

1 point correctly answers the question and includes at least two terms shown

METACOGNITIVE ACTIVITY

Have students answer the following questions in their **Science Notebook:**

1. What seems most puzzling to you about invertebrate animals?
2. Did you learn anything about invertebrate animals that surprised you? Explain.
3. What strategies are you using to keep work on your Unit Project organized?

CHAPTER 5

Vertebrate Animals

Life Science
UNIFYING PRINCIPLES

PRINCIPLE 1
All living things share common characteristics.

PRINCIPLE 2
All living things share common needs.

PRINCIPLE 3
Living things meet their needs through interactions with the environment.

PRINCIPLE 4
The types and numbers of living things change over time.

Unit: Diversity of Living Things
BIG IDEAS

CHAPTER 1
Single-Celled Organisms and Viruses
Bacteria and protists have the characteristics of living things, while viruses are not alive.

CHAPTER 2
Introduction to Multicellular Organisms
Multicellular organisms live in and get energy from a variety of environments.

CHAPTER 3
Plants
Plants are a diverse group of organisms that live in many land environments.

CHAPTER 4
Invertebrate Animals
Invertebrate animals have a variety of body plans and adaptations.

CHAPTER 5
Vertebrate Animals
Vertebrate animals live in most of Earth's environments.

CHAPTER 5
KEY CONCEPTS

SECTION 5.1

Vertebrates are animals with endoskeletons.
1. Vertebrate animals have backbones.
2. Most vertebrates are fish.
3. Fish can be classified in three groups.
4. Most young fish develop inside an egg.

SECTION 5.2

Amphibians and reptiles are adapted for life on land.
1. Vertebrates adapted to live on land.
2. Amphibians have moist skin and lay eggs without shells.
3. Reptiles have dry, scaly skin and lay eggs with shells.
4. The body temperatures of amphibians and reptiles change with the environment.

SECTION 5.3

Birds meet their needs on land, in water, and in the air.
1. Bird species live in most environments.
2. Birds can maintain body temperature.
3. Most birds can fly.
4. Birds lay eggs with hard shells.
5. Most birds take care of their offspring.

SECTION 5.4

Mammals live in many environments.
1. Mammals are a diverse group.
2. Mammals are endotherms.
3. Mammals have adapted to many environments.
4. Mammals have reproductive adaptations.

The Big Idea Flow Chart is available on p. T33 in the **UNIT TRANSPARENCY BOOK.**

Previewing Content

SECTION

5.1 Vertebrates are animals with endoskeletons. pp. 157–163

1. Vertebrate animals have backbones.
About five percent of all animals are **vertebrates.** They are distinguished from other animals by an **endoskeleton.** Specialized bones called vertebrae form the backbone of these animals.

2. Most vertebrates are fish.
- Fish are the most diverse group of vertebrates. They are adapted to life in water.
- Fish breathe through gills. Most fish have a swim bladder that allows the fish to adjust the depth at which it floats. A lateral line is a sensory organ that allows the fish to sense objects and other organisms that are nearby.

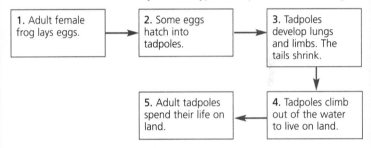

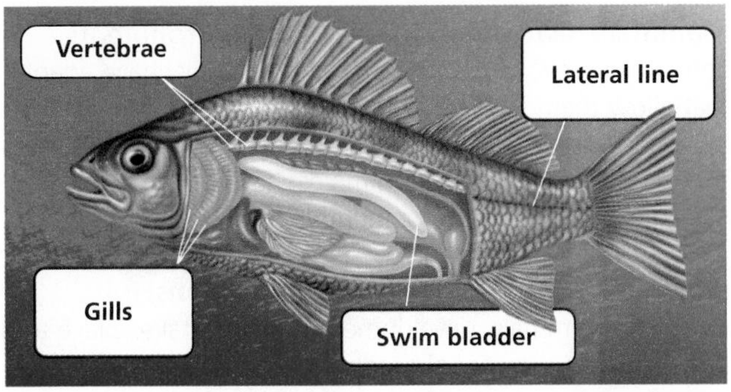

3. Fish can be classified in three groups.
- Jawless fish include lampreys and hagfish. This group is distinguished by tube-shaped bodies and jawless mouths.
- Cartilaginous fish include sharks, rays, and skates. They are distinguished by skeletons that are made of cartilage.
- Bony fish make up the largest group of fish. Most bony fish are covered with **scales.**

4. Most young fish develop inside an egg.
Female fish release eggs into the water where they are fertilized by sperm. Fish eggs get water and oxygen directly from the water. The eggs have a yolk that provides nutrients.

SECTION

5.2 Amphibians and reptiles are adapted for life on land. pp. 164–172

1. Vertebrates adapted to live on land.
The first vertebrates to live on land were **amphibians,** followed by **reptiles.** Some of the organisms in Earth's ancient seas developed limbs and lungs and made the move to land.

2. Amphibians have moist skin and lay eggs without shells.
- All amphibians share certain characteristics. Some of these are: they have two pairs of legs; they lay their eggs in water; and they obtain oxygen through their skin and as adults through lungs.
- The amphibian life cycle includes a larval stage. The flow chart shows the life cycle of a typical amphibian wood frog.

1. Adult female frog lays eggs. →
2. Some eggs hatch into tadpoles. →
3. Tadpoles develop lungs and limbs. The tails shrink. →
4. Tadpoles climb out of the water to live on land. →
5. Adult tadpoles spend their life on land.

3. Reptiles have dry, scaly skin and lay eggs with shells.
- Most reptiles have two pairs of legs, breathe through lungs, and lay eggs with shells on land. Dry, scaly skin protects the reptile from predators and the environment.
- The reptilian egg is an adaptation that allows vertebrates to live completely on land. The egg has a shell to protect the embryo and a yolk to provide it with nutrients.

4. The body temperatures of amphibians and reptiles change with the environment.
Amphibians and reptiles are **ectotherms**—their body temperatures change with the temperature of the environment in which they live.

Common Misconceptions

MISCONCEPTION DATABASE
CLASSZONE.COM Background on student misconceptions

REPTILES AND AMPHIBIANS Many studies have shown that students at all levels correctly identify mammals and birds as vertebrates. Students often fail to classify amphibians and reptiles as vertebrates or even as animals. The misconceptions arise through students' preconceptions of external features of these organisms such as segmentation, body covering, and appendages. Some students classify vertebrates such as snakes and fish as invertebrates because they lack external segmentation and limbs. Certain invertebrates, such as grasshoppers, are misclassified as vertebrates because they have segmentation and appendages. Students also frequently classify amphibians as reptiles and vice versa.

TE This misconception is addressed on pp. 165 and 169.

Previewing Content

SECTION

5.3 Birds meet their needs on land, in water, and in the air. pp. 173–181

1. Bird species live in most environments.
Bird species inhabit all environments. Birds are characterized by feathers and a beak, and a pair of legs and a pair of wings.

2. Birds can maintain body temperature.
Birds are **endotherms**—animals that maintain a constant body temperature. Down feathers insulate bird bodies against cold and contour feathers are water-resistant.

3. Most birds can fly.
- A bird's body has external and internal structures that are adapted for flight.
- Some bones of the endoskeleton are fused, or connected without joints. This makes the body strong and light. The bones of the legs and wings are hollow. Contour feathers are adapted for flight.
- The respiratory system contains air sacs, which allow the bird to take in more oxygen. Instead of teeth, birds have a beak and an internal organ called a gizzard to grind food.
- Many birds migrate to warmer climates in cold winter seasons.

Wings and feathers are structural adaptations for flight.

4. Birds lay eggs with hard shells.
Fertilization occurs internally. Hard shells allow the parent bird to **incubate** the eggs by sitting on top of them.

5. Most birds take care of their offspring.
In some bird species, both parents provide the young hatchlings with food, warmth, and protection.

SECTION

5.4 Mammals live in many environments. pp. 182–187

1. Mammals are a diverse group.
Mammals have characteristics that distinguish them, including:
- All mammals have hair for some part of their lives.
- Mammals have specialized teeth.
- Mammals produce milk to feed their young.

2. Mammals are endotherms.
- Hair allows mammals to keep warm or to protect them from extreme heat. Some hair is specialized, such as whiskers for sensing and quills for protection.
- Mammals have a layer of body fat for maintaining body temperature. The body fat also stores energy for future use.

3. Mammals have adapted to many environments.
Some mammals, such as whales and dolphins, have adaptations that allow them to live in water. These adaptations include blowholes, flippers, tail flukes, and blubber. Other mammals, such as otters, monkeys, and moles, have adapted to completely different environments.

4. Mammals have reproductive adaptations.
- In most mammals, development of young takes place using the **placenta,** a special organ that transports nutrients, water, and oxygen from the mother's blood to the growing embryo.
- Some mammals, such as the duck-billed platypus, lay eggs. Marsupials are pouched mammals in which the young compete their **gestation** inside the mother's external pouch.

Mammary glands in females produce milk, allowing mammals to feed their young.

Common Misconceptions

MISCONCEPTION DATABASE
CLASSZONE.COM Background on student misconceptions

VERTEBRATES VS. MAMMALS Students assume that animal is synonymous with vertebrate, more specifically mammal. Mammals are often the prototypical animal followed by birds, reptiles, and fish in descending order of usage. Students also think that any animal in the water is a fish, including whales and other mammals.

TE This misconception is addressed on p. 183.

Previewing Labs

Lab Generator CD-ROM
Edit these Pupil Edition labs and generate alternative labs.

EXPLORE the BIG idea

What Animals Live Near You? p. 155 Students consider the diversity of animals by listing organisms that live in their area and classifying them into groups.	**TIME** 10 minutes **MATERIALS** notebook, pencil
How Is a Bird Like a Frog? p. 155 Students compare vertebrate internal structures to learn about vertebrates.	**TIME** 10 minutes **MATERIALS** notebook, pencil
Internet Activity: Where in the World? p. 155 Students learn about U.S. wildlife by observing where different animals are found in North America.	**TIME** 20 minutes **MATERIALS** computer with Internet access

SECTION 5.1

EXPLORE Streamlined Shapes, p. 157 Students use a tub of water to explore how the shape of a fish helps it move.	**TIME** 10 minutes **MATERIALS** tub of water

SECTION 5.2

EXPLORE Moving on Land, p. 164 Students use a meter stick to calculate jumping distance.	**TIME** 15 minutes **MATERIALS** meter stick
INVESTIGATE Eggs, p. 170 Students examine a hard-boiled egg to determine the characteristics of eggs.	**TIME** 20 minutes **MATERIALS** hard-boiled egg, plastic knife

SECTION 5.3

EXPLORE Feathers, p. 173 Students explore bird feathers to learn how they differ.	**TIME** 15 minutes **MATERIALS** assorted bird feathers, including down and contour feathers
CHAPTER INVESTIGATION **Bird Beak Adaptations,** pp. 180–181 Students determine how different bird beaks are adapted for gathering and eating different types of food in order to learn how structure relates to function.	**TIME** 40 minutes **MATERIALS** tweezers, eyedropper, slotted spoon, pliers, test tubes in rack, water, dried pasta, millet seeds, jar of rubber bands, empty containers, stopwatch

SECTION 5.4

INVESTIGATE How Body Fat Insulates, p. 184 Students make a "blubber glove" to model how fat keeps a mammal warm.	**TIME** 15 minutes **MATERIALS** 2 zip-lip plastic bags, 1/2 gallon size; can of vegetable shortening; bowl of ice water

R **Additional INVESTIGATION,** Adaptations of a Frog's Tongue, A, B, & C, pp. 349–357; Teacher Instructions, pp. 360–361

Previewing Chapter Resources

	INTEGRATED TECHNOLOGY	LABS AND ACTIVITIES
CHAPTER 5 **Vertebrate Animals**	CLASSZONE.COM • eEdition Plus • EasyPlanner Plus • Misconception Database • Content Review • Test Practice • Resource Centers • Internet Activity: Where In the World? • Math Tutorial SCILINKS.ORG SCI LINKS CD-ROMS • eEdition • EasyPlanner Plus • Power Presentations • Content Review • Lab Generator • Test Generator AUDIO CDS • Audio Readings • Audio Readings in Spanish	PE EXPLORE the Big Idea, p. 155 • What Animals Live Near You? • How is a Frog Like a Bird? • Internet Activity: Where in the World? R **UNIT RESOURCE BOOK** Unit Projects, pp. 5–10 **Lab Generator CD-ROM** Generate customized labs.
SECTION 5.1 **Vertebrates are animals with endoskeletons.** pp. 157–163 Time: 2 periods (1 block) R Lesson Plan, pp. 291–292	• **RESOURCE CENTER,** Fish • **MATH TUTORIAL** T **UNIT TRANSPARENCY BOOK** • Big Idea Flow Chart, p. T33 • Daily Vocabulary Scaffolding, p. T34 • Note-Taking Model, p. T35 • 3-Minute Warm-Up, p. T36	PE • EXPLORE Streamlined Shapes, p. 157 • Math in Science, p. 163 R **UNIT RESOURCE BOOK** • Math Support, p. 338 • Math Practice, p. 339
SECTION 5.2 **Amphibians and reptiles are adapted for life on land.** pp. 164–172 Time: 2 periods (1 block) R Lesson Plan, pp. 301–302	• **RESOURCE CENTERS,** Amphibians, Reptiles T **UNIT TRANSPARENCY BOOK** • Daily Vocabulary Scaffolding, p. T34 • 3-Minute Warm-Up, p. T36	PE • EXPLORE Moving on Land, p. 164 • INVESTIGATE Eggs, p. 170 • Connecting Sciences, p. 172 R **UNIT RESOURCE BOOK** • Datasheet, Eggs, p. 310 • Additional INVESTIGATION, Adaptations of a Frog's Tongue, Levels A, B, & C, pp. 349–357
SECTION 5.3 **Birds meet their needs on land, in water, and in the air.** pp. 173–181 Time: 3 periods (1.5 blocks) R Lesson Plan, pp. 312–313	T **UNIT TRANSPARENCY BOOK** • Daily Vocabulary Scaffolding, p. T34 • 3-Minute Warm-Up, p. T37 • "Adaptations for Flight" Visual, p. T38	PE • EXPLORE Feathers, p. 173 • CHAPTER INVESTIGATION, Bird Beak Adaptations, pp. 180–181 R **UNIT RESOURCE BOOK** CHAPTER INVESTIGATION Bird Beak Adaptations, Levels A, B, & C, pp. 340–348
SECTION 5.4 **Mammals live in many environments.** pp. 182–187 Time: 3 periods (1.5 blocks) R Lesson Plan, pp. 322–333	**RESOURCE CENTER,** Mammals T **UNIT TRANSPARENCY BOOK** • Big Idea Flow Chart, p. T33 • Daily Vocabulary Scaffolding, p. T34 • 3-Minute Warm-Up, p. T37 • Chapter Outline, pp. T39–T40	PE INVESTIGATE How Body Fat Insulates, p. 184 R **UNIT RESOURCE BOOK** Datasheet, How Body Fat Insulates, p. 331

KEY TO ICONS CD/CD-ROM | Teacher Edition | UNIT TRANSPARENCY BOOK | SPANISH ASSESSMENT BOOK | INTERNET | Pupil Edition | UNIT RESOURCE BOOK | UNIT ASSESSMENT BOOK | SCIENCE TOOLKIT

READING AND REINFORCEMENT	ASSESSMENT	STANDARDS
• Magnet Words, B24–25 • Choose Your Own Strategy, C35–44 • Daily Vocabulary Scaffolding, H1–8 **UNIT RESOURCE BOOK** • Vocabulary Practice, pp. 335–336 • Decoding Support, p. 337 • Summarizing the Chapter, pp. 358–359 **Audio Readings CD** Listen to Pupil Edition. **Audio Readings in Spanish CD** Listen to Pupil Edition in Spanish.	• Chapter Review, pp.189–190 • Standardized Test Practice, p. 191 **UNIT ASSESSMENT BOOK** • Diagnostic Test, pp. 81–82 • Chapter Test, A, B, & C, pp. 87–98 • Alternative Assessment, pp. 99–100 • Unit Test, A, B, & C, pp. 101–112 • Spanish Chapter Test, pp. 57–60 • Spanish Unit Test, pp. 61–64 **Test Generator CD-ROM** Generate customized tests. **Lab Generator CD-ROM** Rubrics for Labs	National Standards A.1–8, A.9.a–c, A.9.e–g, C.3.b, C.5.b, G.1.b See p. 154 for the standards.
UNIT RESOURCE BOOK • Reading Study Guide, A & B, pp. 293–296 • Spanish Reading Study Guide, pp. 297–298 • Challenge and Extension, p. 299 • Reinforcing Key Concepts, p. 300	Ongoing Assessment, pp. 157–162 Section 5.1 Review, p. 162 **UNIT ASSESSMENT BOOK** Section 5.1 Quiz, p. 83	National Standards A.2.8, A.9.c, C.3.b, C.5.b, G.1.b
UNIT RESOURCE BOOK • Reading Study Guide, A & B, pp. 303–306 • Spanish Reading Study Guide, pp. 307–308 • Challenge and Extension, p. 309 • Reinforcing Key Concepts, p. 311	Ongoing Assessment, pp. 165–169, 171 Section 5.2 Review, p. 171 **UNIT ASSESSMENT BOOK** Section 5.2 Quiz, p. 84	National Standards A.2–7, A.9.a–b, A.9.e–f, C.3.b, C.5.b, G.1.b
UNIT RESOURCE BOOK • Reading Study Guide, A & B, pp. 314–317 • Spanish Reading Study Guide, pp. 318–319 • Challenge and Extension, p. 320 • Reinforcing Key Concepts, p. 321	Ongoing Assessment, pp. 174, 176–178 Section 5.3 Review, p. 179 **UNIT ASSESSMENT BOOK** Section 5.3 Quiz, p. 85	National Standards A.1–7, A.9.a–b, A.9.e–g, C.3.b, C.5.b, G.1.b
UNIT RESOURCE BOOK • Reading Study Guide, A & B, pp. 324–327 • Spanish Reading Study Guide, pp. 328–329 • Challenge and Extension, p. 330 • Reinforcing Key Concepts, p. 332 • Challenge Reading, pp. 333–334	Ongoing Assessment, pp. 183–184, 186–187 Section 5.4 Review, p. 187 **UNIT ASSESSMENT BOOK** Section 5.4 Quiz, p. 86	National Standards A.2–7, A.9.a–b, A.9.e–f, C.3.b, C.5.b, G.1.b

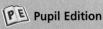

Previewing Resources for Differentiated Instruction

CHAPTER INVESTIGATION

Leveled resources present the same concepts for different abilities.

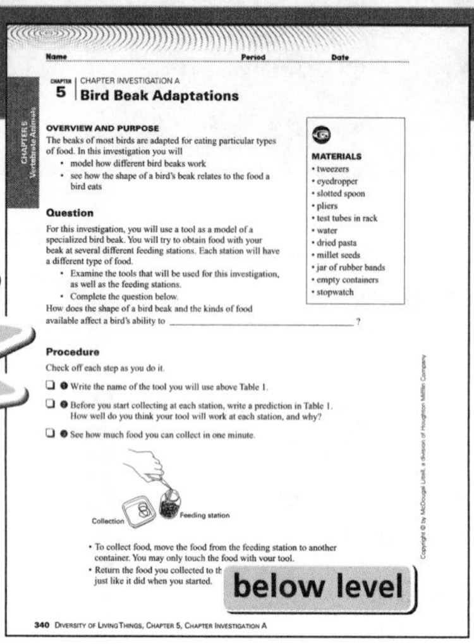
CHAPTER 5 | CHAPTER INVESTIGATION A
Bird Beak Adaptations

below level

340 DIVERSITY OF LIVING THINGS, CHAPTER 5, CHAPTER INVESTIGATION A

UNIT RESOURCE BOOK, pp. 340–343

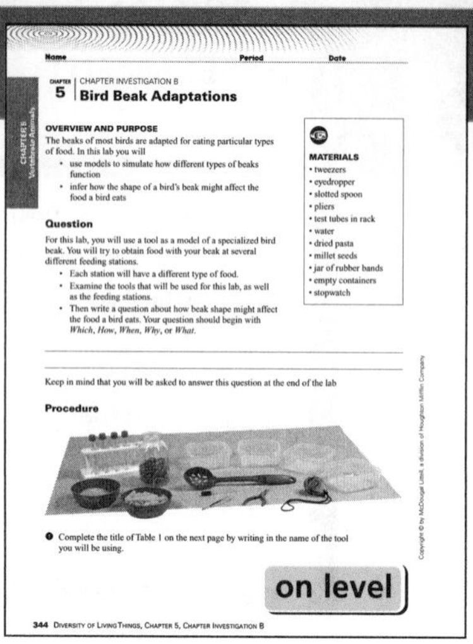
CHAPTER 5 | CHAPTER INVESTIGATION B
Bird Beak Adaptations

on level

344 DIVERSITY OF LIVING THINGS, CHAPTER 5, CHAPTER INVESTIGATION B

pp. 344–347

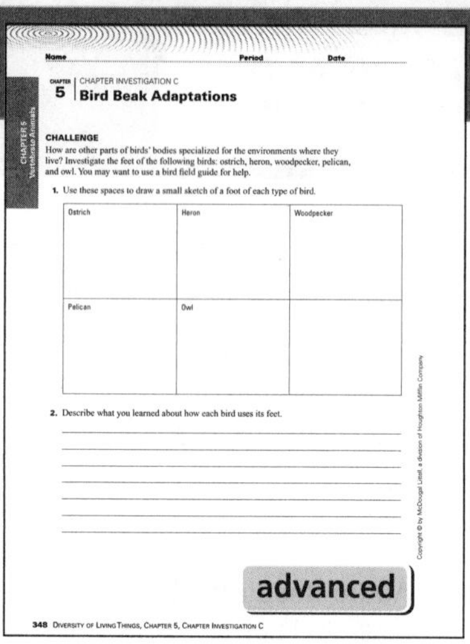
CHAPTER 5 | CHAPTER INVESTIGATION C
Bird Beak Adaptations

advanced

348 DIVERSITY OF LIVING THINGS, CHAPTER 5, CHAPTER INVESTIGATION C

pp. 344–348

READING STUDY GUIDE

Reading Study Guide is also in Spanish.

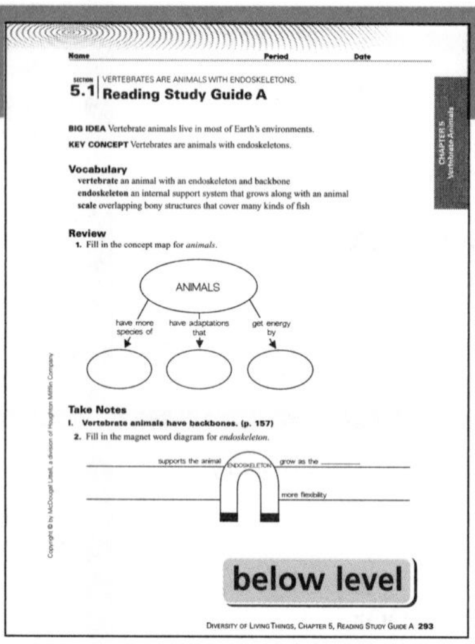
SECTION 5.1 | VERTEBRATES ARE ANIMALS WITH ENDOSKELETONS.
Reading Study Guide A

below level

DIVERSITY OF LIVING THINGS, CHAPTER 5, READING STUDY GUIDE A 293

UNIT RESOURCE BOOK, pp. 293–294

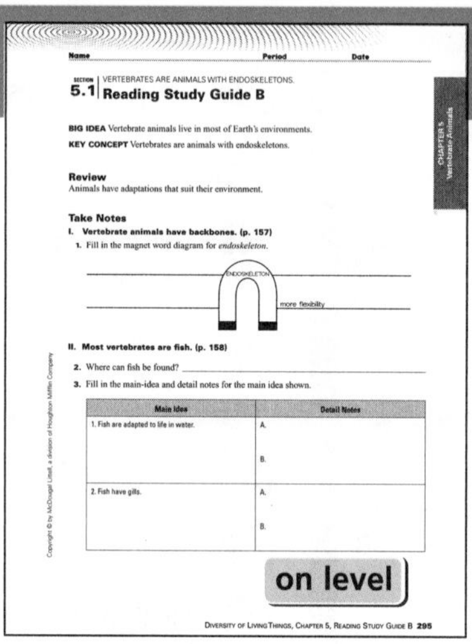
SECTION 5.1 | VERTEBRATES ARE ANIMALS WITH ENDOSKELETONS.
Reading Study Guide B

on level

DIVERSITY OF LIVING THINGS, CHAPTER 5, READING STUDY GUIDE B 295

pp. 295–296

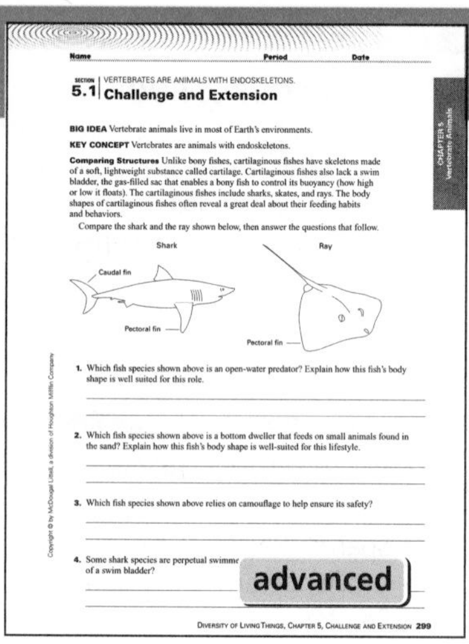
SECTION 5.1 | VERTEBRATES ARE ANIMALS WITH ENDOSKELETONS.
Challenge and Extension

advanced

DIVERSITY OF LIVING THINGS, CHAPTER 5, CHALLENGE AND EXTENSION 299

p. 299

CHAPTER TEST

Chapter Test is also in Spanish.

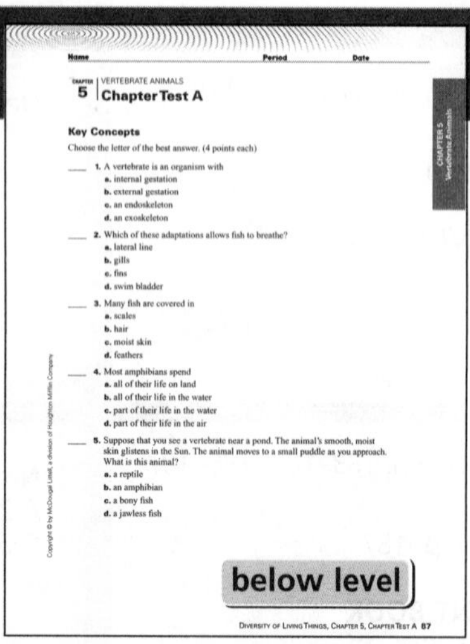
CHAPTER 5 | VERTEBRATE ANIMALS
Chapter Test A

below level

DIVERSITY OF LIVING THINGS, CHAPTER 5, CHAPTER TEST A 87

UNIT ASSESSMENT BOOK, pp. 87–90

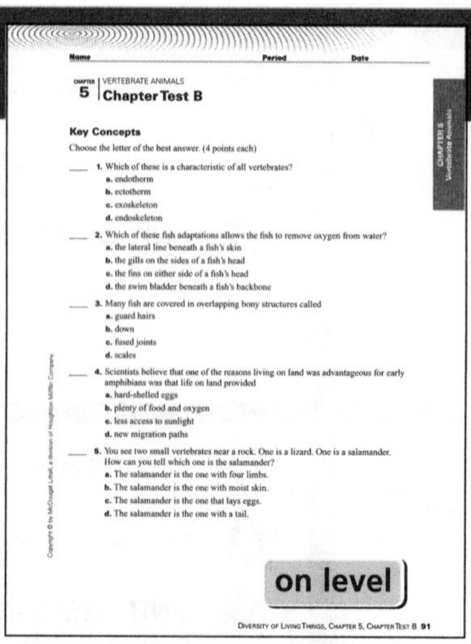
CHAPTER 5 | VERTEBRATE ANIMALS
Chapter Test B

on level

DIVERSITY OF LIVING THINGS, CHAPTER 5, CHAPTER TEST B 91

pp. 91–94

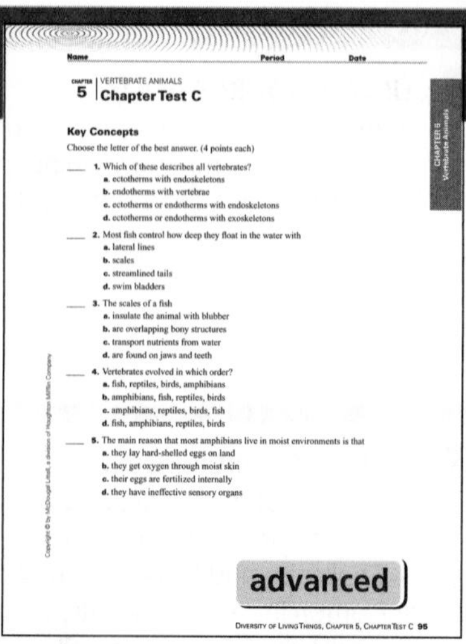
CHAPTER 5 | VERTEBRATE ANIMALS
Chapter Test C

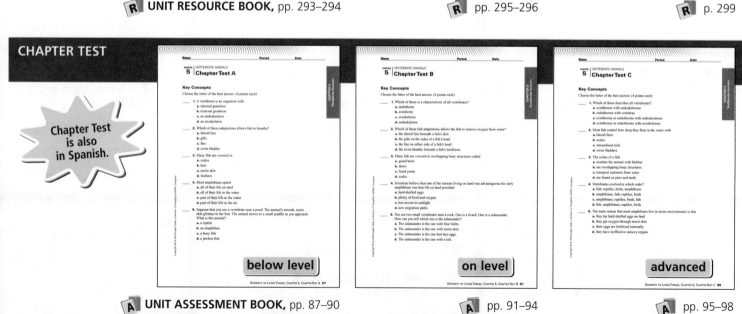

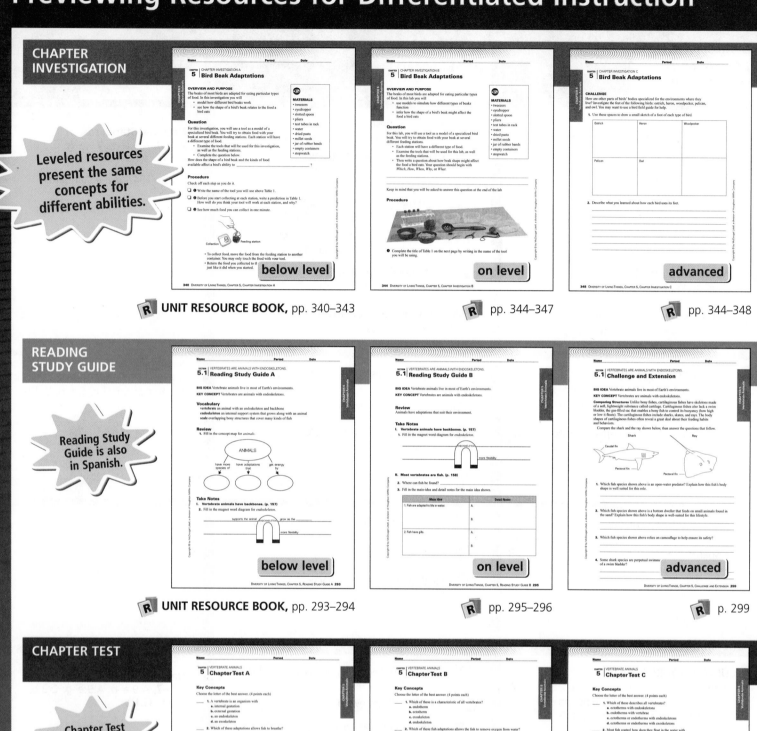

advanced

DIVERSITY OF LIVING THINGS, CHAPTER 5, CHAPTER TEST C 95

pp. 95–98

TECHNOLOGY

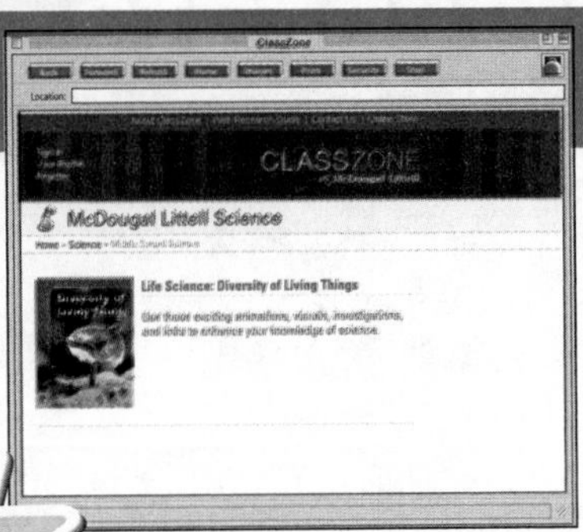

There are four Resource Centers for this chapter.

CLASSZONE.COM

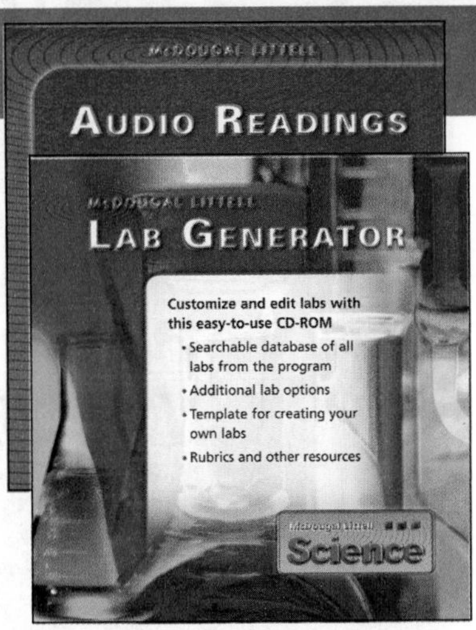

CD/CD-ROMS

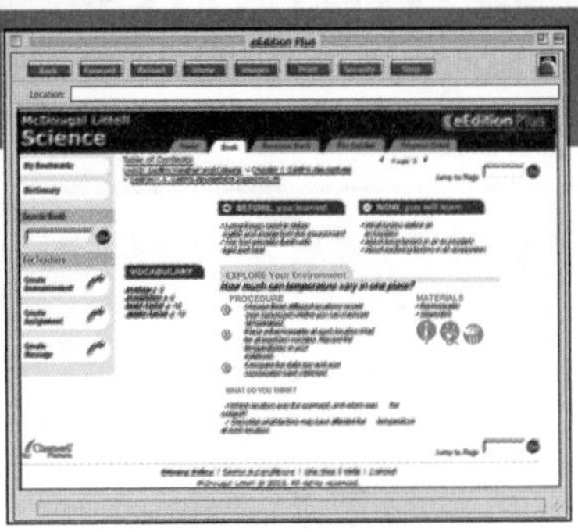

CLASSZONE.COM

VISUAL CONTENT

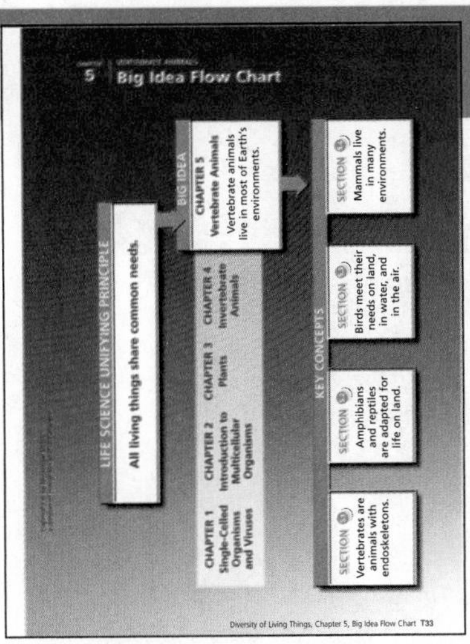

UNIT TRANSPARENCY BOOK, p. T33

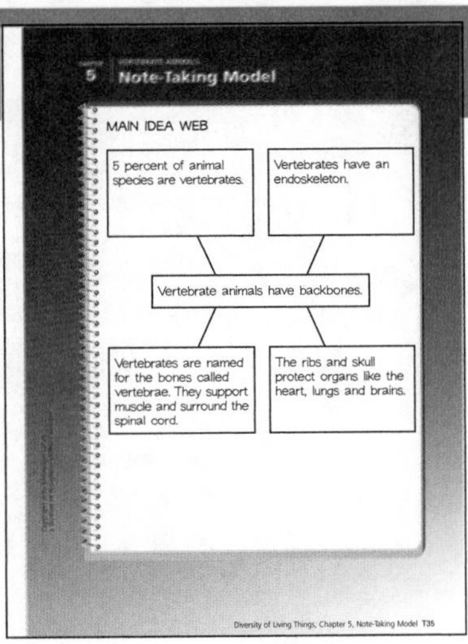

p. T35

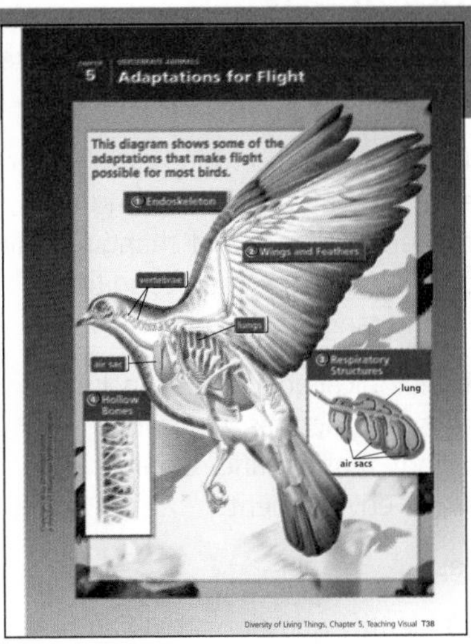

p. T38

MORE SUPPORT

Reinforcing Key Concepts for each section

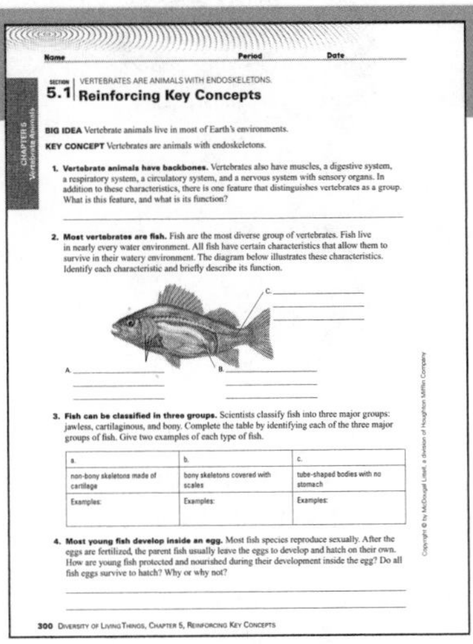
Name Period Date

SECTION 5.1 VERTEBRATES ARE ANIMALS WITH ENDOSKELETONS

Reinforcing Key Concepts

BIG IDEA Vertebrate animals live in most of Earth's environments.

KEY CONCEPT Vertebrates are animals with endoskeletons.

1. **Vertebrate animals have backbones.** Vertebrates also have muscles, a digestive system, a respiratory system, a circulatory system, and a nervous system with sensory organs. In addition to these characteristics, there is one feature that distinguishes vertebrates as a group. What is this feature, and what is its function?

2. **Most vertebrates are fish.** Fish are the most diverse group of vertebrates. Fish live in nearly every water environment. All fish have certain characteristics that allow them to survive in their watery environment. The diagram below illustrates these characteristics. Identify each characteristic and briefly describe its function.

3. **Fish can be classified in three groups.** Scientists classify fish into three major groups: jawless, cartilaginous, and bony. Complete the table by identifying each of the three major groups of fish. Give two examples of each type of fish.

a.	b.	c.
non-bony skeletons made of cartilage	bony skeletons covered with scales	tube-shaped bodies with no stomach
Examples:	Examples:	Examples:

4. **Most young fish develop inside an egg.** Most fish species reproduce sexually. After the eggs are fertilized, the parent fish usually leave the eggs to develop and hatch on their own. How are young fish protected and nourished during their development inside the egg? Do all fish eggs survive to hatch? Why or why not?

300 Diversity of Living Things, Chapter 5, Reinforcing Key Concepts

UNIT RESOURCE BOOK, p. 300

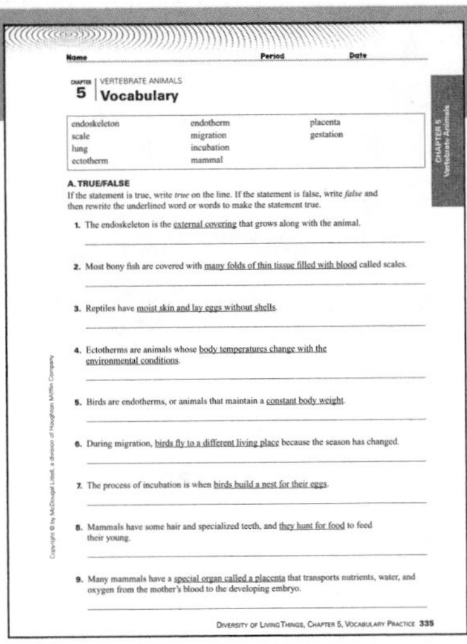
Name Period Date

CHAPTER 5 VERTEBRATE ANIMALS

Vocabulary

endoskeleton	endotherm	placenta
scale	migration	gestation
lung	incubation	
ectotherm	mammal	

A. TRUE/FALSE
If the statement is true, write *true* on the line. If the statement is false, write *false* and then rewrite the underlined word or words to make the statement true.

1. The endoskeleton is the external covering that grows along with the animal.
2. Most bony fish are covered with many folds of thin tissue filled with blood called scales.
3. Reptiles have moist skin and lay eggs without shells.
4. Ectotherms are animals whose body temperatures change with the environmental conditions.
5. Birds are endotherms, or animals that maintain a constant body weight.
6. During migration, birds fly to a different living place because the season has changed.
7. The process of incubation is when birds build a nest for their eggs.
8. Mammals have some hair and specialized teeth, and they hunt for food to feed their young.
9. Many mammals have a special organ called a placenta that transports nutrients, water, and oxygen from the mother's blood to the developing embryo.

Diversity of Living Things, Chapter 5, Vocabulary Practice 335

pp. 335–336

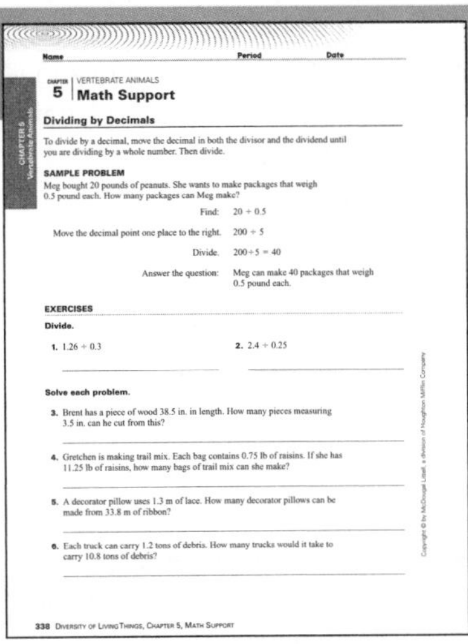
Name Period Date

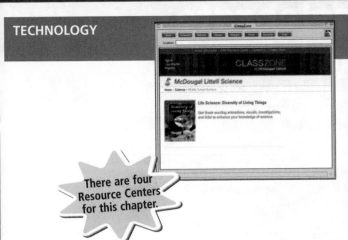

CHAPTER 5 VERTEBRATE ANIMALS

Math Support

Dividing by Decimals

To divide by a decimal, move the decimal in both the divisor and the dividend until you are dividing by a whole number. Then divide.

SAMPLE PROBLEM
Meg bought 20 pounds of peanuts. She wants to make packages that weigh 0.5 pound each. How many packages can Meg make?

Find: 20 ÷ 0.5
Move the decimal point one place to the right. 200 ÷ 5
Divide. 200 ÷ 5 = 40
Answer the question. Meg can make 40 packages that weigh 0.5 pound each.

EXERCISES
Divide.

1. 1.26 ÷ 0.3
2. 2.4 ÷ 0.25

Solve each problem.

3. Brent has a piece of wood 38.5 in. in length. How many pieces measuring 3.5 in. can he cut from this?
4. Gretchen is making trail mix. Each bag contains 0.75 lb of raisins. If she has 11.25 lb of raisins, how many bags of trail mix can she make?
5. A decorator pillow uses 1.3 m of lace. How many decorator pillows can be made from 33.8 m of ribbon?
6. Each truck can carry 1.2 tons of debris. How many trucks would it take to carry 10.8 tons of debris?

338 Diversity of Living Things, Chapter 5, Math Support

p. 338

CHAPTER 5

Vertebrate Animals

INTRODUCE

the BIG idea

Have students look at the photograph of penguins and a sea lion. Discuss the question in the box:

- Ask: In what type of environment do the penguin and the seal live?
- Ask: How are their bodies adapted to the cold and to the water?
- Brainstorm a list of environments in which vertebrate animals live.

National Science Education Standards

Content

C.3.b Regulation of an organism's internal environment involves sensing the internal environment and changing physiological activities to keep conditions within the range required to survive.

C.5.b Biological adaptations include changes in structures, behaviors, or physiology that enhance survival in a particular environment.

Process

A.1–8 Identify questions that can be answered by scientific investigations; design and conduct an investigation; use tools to gather and interpret data; use evidence to describe, predict, explain, model; think critically to make relationships between evidence and explanation; recognize different explanations and predictions; communicate scientific procedures and explanations; use mathematics.

A.9.a–c, A.9.e–g Understand scientific inquiry by using different investigations, methods, mathematics, and explanations based on logic, evidence, and skepticism. Data often results in new investigations.

E.1–5 Identify a problem; design, implement, and evaluate a solution or product; communicate technological design.

G.1.b Science requires different abilities.

CHAPTER 5

Vertebrate Animals

the BIG idea

Vertebrate animals live in most of Earth's environments.

What do these penguins have in common with this seal?

Key Concepts

SECTION 5.1
Vertebrates are animals with endoskeletons.
Learn how most of the vertebrates on Earth are fish.

SECTION 5.2
Amphibians and reptiles are adapted for life on land.
Learn how most amphibians hatch in water and most reptiles hatch on land.

SECTION 5.3
Birds meet their needs on land, in water, and in the air.
Learn how adaptations for flight affect how birds meet their needs.

SECTION 5.4
Mammals live in many environments.
Learn about mammals' many adaptations.

Internet Preview

CLASSZONE.COM
Chapter 5 online resources: Content Review, two Visualizations, four Resource Centers, Math Tutorial, Test Practice.

C 154 Unit: Diversity of Living Things

INTERNET PREVIEW

CLASSZONE.COM For student use with the following pages:

Review and Practice
- Content Review, pp. 157, 188
- Math Tutorial: Dividing by Decimals, p. 163
- Test Practice, p. 191

Activities and Resources
- Internet Activity: Where in the World?, p. 155
- Resource Centers: Fish, p. 160; Amphibians, p. 167; Reptiles, p. 169; Mammals, p. 183

NSTA scilinks.org SCILINKS
Bird Characteristics
Code: MDL043

EXPLORE the BIG idea

What Animals Live Near You?

Make a list of animals you think live in your neighborhood. Remember that some animals are small! Organize the animals on your list into groups.

Observe and Think Where do you think you would see the most animals? What about the widest variety?

How Is a Bird Like a Frog?

Fish, frogs, snakes, birds, dogs, and humans are all vertebrate animals. Choose two vertebrates and quickly sketch their body plans, including their skeletons.

Observe and Think Where do you think each animal's brain, heart, and stomach are located?

Internet Activity: Where in the World?

Go to **ClassZone.com** to learn more about where different types of vertebrate animals are found in North America.

Observe and Think What sorts of adaptations would an animal need to live near the North Pole? What about in a desert?

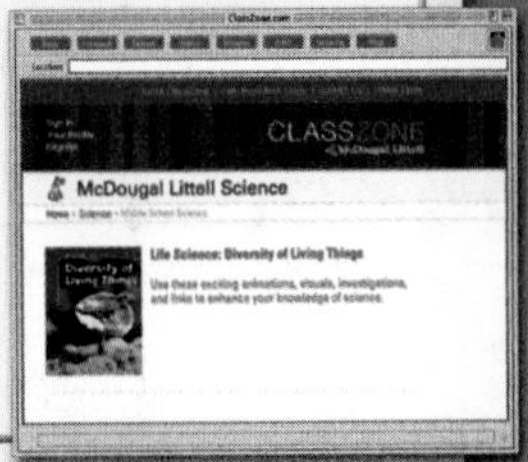

NSTA scilinks.org SCILINKS
Bird Characteristics Code: MDL043

TEACHING WITH TECHNOLOGY

Electronic Encyclopedias Students can use electronic encyclopedias to investigate the body structures of vertebrate groups, beginning with the study of fish on p. 158. Have students print pictures and create a display.

Computer Dissections Students can use computer dissections to investigate the internal structures of vertebrates beginning with the internal structure of fish on p. 159.

EXPLORE the BIG idea

These inquiry-based activities are appropriate for use at home or as a supplement to classroom instruction.

What Animals Live Near You?

PURPOSE To list local animals and classify them into groups.

TIPS *10 min.* Have students work in groups. Compile their groupings of animals on a chart on the board.

Answers depend on the area in which you live and organisms that students listed. The most and widest variety of animals should be where there are the most resources.

REVISIT after p. 187.

How Is a Bird Like a Frog?

PURPOSE To compare vertebrate internal structures.

TIP *10 min.* Guide students to choose animals that belong in different classes.

Check student diagrams for realism. Help students research information of which they are unsure.

REVISIT after p. 158.

Internet Activity: Where in the World?

PURPOSE To observe where different animals are found in North America.

TIP *20 min.* Challenge students to learn to recognize vertebrates, especially local species.

Sample answers: North Pole: ways of keeping warm and getting water in freezing temperatures; Desert: ways of staying cool and getting water in an environment that is hot and dry during daylight hours.

REVISIT after p. 185.

PREPARE

CONCEPT REVIEW

Activate Prior Knowledge

- Ask students to name some of the common characteristics that all living organisms share. *organization, growth, reproduction, and response*
- Discuss how organisms meet their needs by obtaining food, water, and oxygen.
- Have students brainstorm a list of body structures that characterize vertebrate animals. *digestive systems, skeletal systems, circulatory systems, reproductive systems*

TAKING NOTES

Choose Your Own Strategy

Students can rely on the blue and red headings in the text as a basis for main ideas. Specific information from text and visuals can serve as details in the Main Idea Web, Main Idea and Details Notes, and the Mind Map. Encourage students to compare their choices in strategies.

Vocabulary Strategy

Students can include as many details as they wish in a magnet diagram. Words and phrases, descriptions, and definitions that are associated with the topic are useful study devices.

Vocabulary and Note-Taking Resources

R
- Vocabulary Practice, pp. 335–336
- Decoding Support, p. 337

T
- Daily Vocabulary Scaffolding, p. T34
- Note-Taking Model, p. T35

- Magnet Words, B24–25
- Choose Your Own Strategy, C35–44
- Daily Vocabulary Scaffolding, H1–8

CHAPTER 5

Getting Ready to Learn

CONCEPTS REVIEW

- All living things have common needs.
- Plants and some invertebrates have adaptations for life on land.
- Most multicellular organisms can reproduce sexually.

VOCABULARY REVIEW

migration p. 64
embryo p. 98
gill p. 137
lung p. 137
exoskeleton p. 143

CONTENT REVIEW
CLASSZONE.COM
Review concepts and vocabulary.

TAKING NOTES

CHOOSE YOUR OWN STRATEGY

Take notes using one or more of the strategies from earlier chapters – **main idea webs, main idea and details, mind maps,** or **combination notes.** You can also use other note-taking strategies that you may already know.

VOCABULARY STRATEGY

Think about a vocabulary term as a **magnet word** diagram. Write other terms and ideas related to that term around it.

See the Note-Taking Handbook on pages R45–R51.

SCIENCE NOTEBOOK

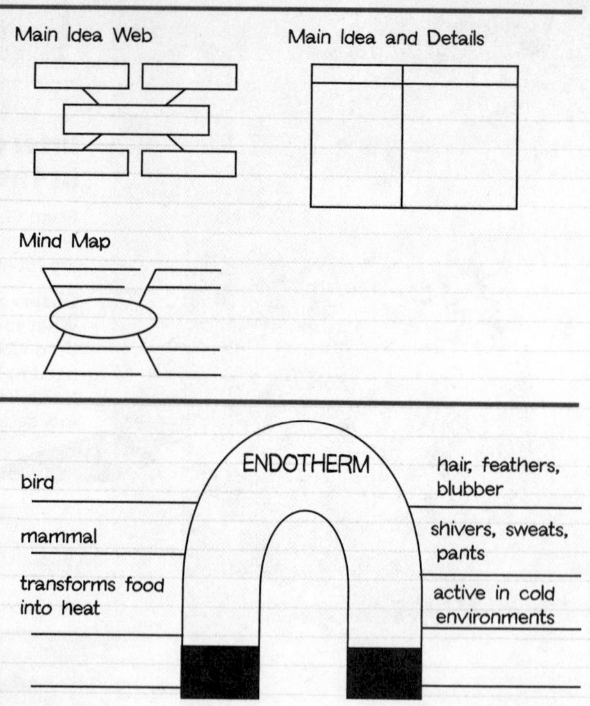

CHECK READINESS

Administer the Diagnostic Test to determine students' readiness for new science content and their mastery of requisite math skills.

A Diagnostic Test, pp. 81–82

Technology Resources

Students needing content and math skills should visit **ClassZone.com**.

- **CONTENT REVIEW**
- **MATH TUTORIAL**

CONTENT REVIEW CD-ROM

KEY CONCEPT

Vertebrates are animals with endoskeletons.

BEFORE, you learned
- Most animals are invertebrates
- Animals have adaptations that suit their environment
- Animals get energy by consuming food

NOW, you will learn
- About the skeletons of vertebrate animals
- About the characteristics of fish
- About three groups of fish

VOCABULARY
vertebrate p. 157
endoskeleton p. 157
scale p. 161

EXPLORE Streamlined Shapes

How does a fish's shape help it move?

PROCEDURE

1. Place your hand straight up and down in a tub of water. Keep your fingers together and your palm flat.
2. Move your hand from one side of the tub to the other, using your palm to push the water.
3. Move your hand across the tub again, this time using the edge of your hand as if you were cutting the water.

MATERIALS
tub of water

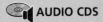

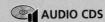

WHAT DO YOU THINK?
- In which position was the shape of your hand most like the shape of a fish's body?
- How might the shape of a fish's body affect its ability to move through water?

Vertebrate animals have backbones.

If you asked someone to name an animal, he or she would probably name a vertebrate. Fish, frogs, snakes, birds, dogs, and humans are all **vertebrates,** or animals with backbones. Even though only about 5 percent of animal species are vertebrates, they are among the most familiar and thoroughly studied organisms on Earth.

VOCABULARY
Add magnet word diagrams for *vertebrate* and *endoskeleton* to your notebook.

Vertebrate animals have muscles, a digestive system, a respiratory system, a circulatory system, and a nervous system with sensory organs. The characteristic that distinguishes vertebrates from other animals is the **endoskeleton,** an internal support system that grows along with the animal. Endoskeletons allow more flexibility and ways of moving than exoskeletons do.

RESOURCES FOR DIFFERENTIATED INSTRUCTION

Below Level
UNIT RESOURCE BOOK
- Reading Study Guide A, pp. 293–294
- Decoding Support, p. 337

AUDIO CDS

Advanced
UNIT RESOURCE BOOK
- Challenge and Extension, p. 299

English Learners
UNIT RESOURCE BOOK
Spanish Reading Study Guide, pp. 297–298

AUDIO CDS
- Audio Readings in Spanish
- Audio Readings (English)

5.1 FOCUS

Set Learning Goals

Students will
- Explain that vertebrate animals have internal skeletons with backbones.
- Discuss how fish are adapted for life in water.
- Characterize the three main groups of fish.
- Explore how the shape of a fish helps it move.

3-Minute Warm-Up

Display Transparency 36 or copy this exercise on the board:

Decide which of these statements are true. If not true, correct them.

1. All invertebrate animals have external skeletons. *Arthropods have external skeletons.*
2. Sponges are the simplest invertebrates. *true*
3. Cephalopods are mollusks with two shells that are held together at a hinge. *Bivalves are mollusks that have two shells.*

3-Minute Warm-Up, p. T36

5.1 MOTIVATE

EXPLORE Streamlined Shapes

PURPOSE To explore how the shape of a fish helps it move

TIPS *10 min.* Note that students' hands appear larger when submerged in water. Students will be more comfortable if the water is warm.

WHAT DO YOU THINK? *The second hand position is most like the shape of a fish's body. A more streamlined body will move through water more efficiently.*

Ongoing Assessment

Explain that vertebrate animals have internal skeletons with backbones.

Ask: What is an endoskeleton? *an internal support system found in vertebrates that grows as the animal grows*

5.1 INSTRUCT

Teach from Visuals

To help students interpret the diagram of the cheetah's skeleton, ask:

- What structures aid in movement? *muscles attached to bones*
- What is the function of the cheetah's ribs? *protects organs like heart and lungs*

Teaching with Technology

If students have access to electronic encyclopedias, have them find and print pictures of the various vertebrate groups as they read through the chapter.

EXPLORE the BIG idea

Revisit "How Is a Bird Like a Frog?" on p. 155. Have students take another look at the sketches they made and be sure that backbones are included.

Ongoing Assessment

READING VISUALS *Answer: The cheetah's endoskeleton creates the shape of the cheetah.*

CHECK YOUR READING *Answer: The endoskeleton supports muscles and surrounds and protects the spinal cord.*

Vertebrate Skeleton and Muscles

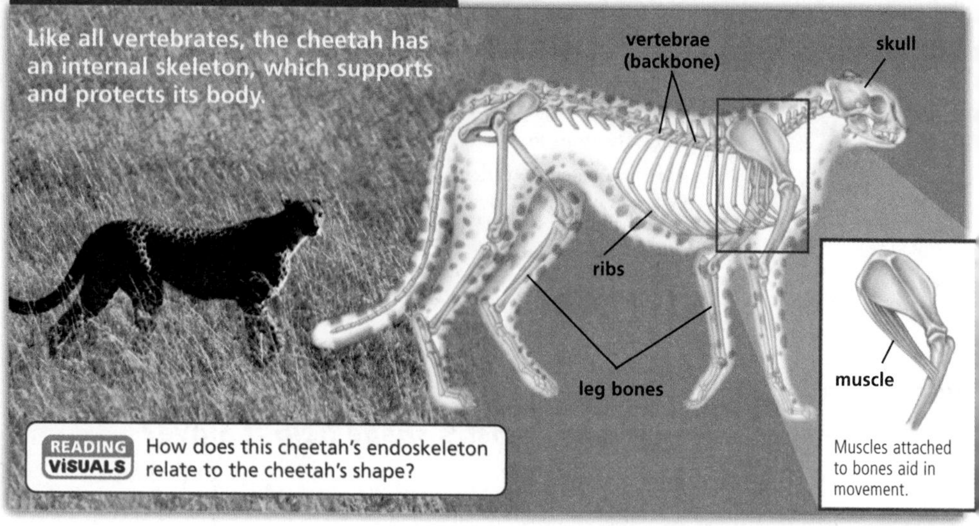

Like all vertebrates, the cheetah has an internal skeleton, which supports and protects its body.
vertebrae (backbone)
skull
ribs
leg bones
muscle
Muscles attached to bones aid in movement.

READING VISUALS How does this cheetah's endoskeleton relate to the cheetah's shape?

Vertebrates are named for specialized bones called vertebrae. These bones are located in the middle of each vertebrate animal's central body segment. Together, the vertebrae are sometimes called a backbone. The vertebrae support muscles and surround the spinal cord, which connects the animal's brain to its nerves. Other bones, such as the ribs and skull, protect organs like the heart, lungs, and brain.

CHECK YOUR READING What is one function of the endoskeleton for vertebrate animals?

Most vertebrates are fish.

NOTETAKING STRATEGY Choose a strategy from an earlier chapter to take notes on the idea that most vertebrates are fish. Be sure to include information on adaptations to water.

Fish are the most diverse group of vertebrate animals. There are more than 20,000 species of fish, ranging in size from tiny minnows to huge whale sharks. Fish live in nearly every aquatic environment, from freshwater lakes to the bottom of the sea. Some fish even are able to survive below the ice in the Antarctic!

Fish are adapted for life in water. Like all living things, fish need to get materials from their environment. For example, fish must be able to get oxygen from water. Fish must also be able to move through water in order to find food. Fish that live in water where sunlight does not penetrate need special organs to help them find food.

Most fish move by using muscles and fins to push their streamlined bodies through water. These muscles allow fish to move more

DIFFERENTIATE INSTRUCTION

More Reading Support

A Together, what are vertebrae called? *a backbone*

B Which organ does the skull protect? *brain*

English Learners When introducing new vocabulary, encourage students to use all the information that a dictionary offers. For example, when learning *endoskeleton* and *exoskeleton*, have students look up the meanings of the prefixes *endo-* and *exo-*. Have students look up the term *vertebrate* in the dictionary. Ask them what the Latin meaning of the term—"having joints"—has to do with the definition.

quickly than most other invertebrates. Most fish also have an organ called a swim bladder, which allows them to control the depth at which they float.

Fish have sensory organs for taste, odor, and sound. Most fish species have eyes that allow them to see well underwater. Most fish also have a sensory system unlike other vertebrates. This system includes an organ called a lateral line, which allows fish to sense vibrations from objects nearby without touching or seeing them.

Fish, like some invertebrates, remove oxygen from water with specialized respiratory organs called gills. You can locate most fishs' gills by looking for the openings, called gill slits, on the sides of their head. You can see what gills look like in the diagram of a fish below.

VISUALIZATION CLASSZONE.COM

Explore how fish breathe.

CHECK YOUR READING How are gills similar to lungs?

Fish gills are made up of many folds of tissue and are filled with blood. When a fish swims, it takes water in through its mouth and then pushes the water back over its gills. In the gills, oxygen dissolved in the water moves into the fish's blood. Carbon dioxide, a waste product of respiration, moves from the blood into the water. Then the water is forced out of the fish's body through its gill slits. The oxygen is transported to the fish's cells. It is a necessary material for releasing energy.

Inside a Fish

Fish are vertebrates that live in water.

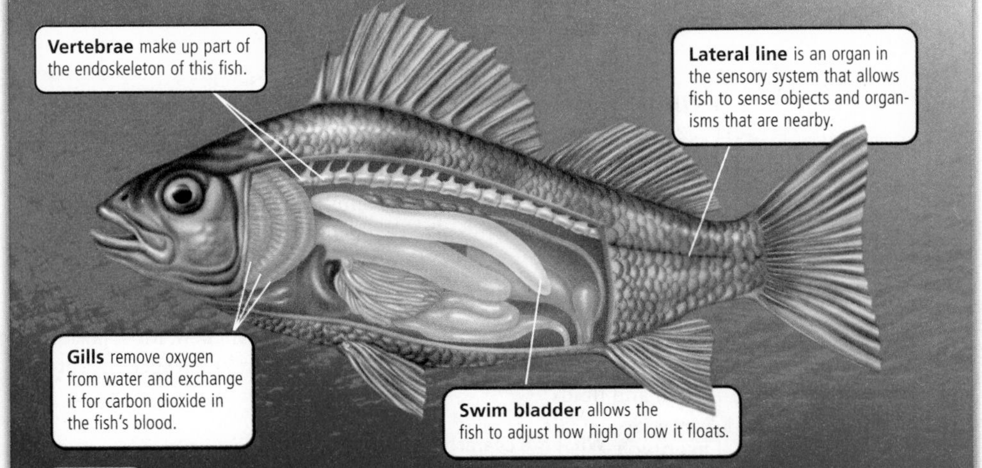

Vertebrae make up part of the endoskeleton of this fish.
Lateral line is an organ in the sensory system that allows fish to sense objects and organisms that are nearby.
Gills remove oxygen from water and exchange it for carbon dioxide in the fish's blood.
Swim bladder allows the fish to adjust how high or low it floats.

READING VISUALS What is the function of the vertebrae on the fish's backbone?

Teaching with Technology

If computers are available, have students use them to perform computer dissections to study the internal organs of vertebrates.

Teacher Demo

Model the swim bladder of a fish with a partially inflated balloon and a weight in a tank of water. The weight should be attached to the balloon with string and the combination should be suspended motionless in the water.

Teach from Visuals

To help students interpret the diagram of the internal structure of the fish, ask:

- What structures enable the fish to breathe? *gills*
- How might the lateral line help protect the fish from possible predators? *The lateral line allows the fish to sense other fish that might be nearby.*

Ongoing Assessment

Explain how fish are adapted for life in water.

Ask: What are three adaptations that enable fish to live in water? Explain each one. *Gills enable fish to obtain oxygen from water. Swim bladders allow fish to control the depth at which they swim, and lateral lines enables fish to sense nearby organisms and objects.*

CHECK YOUR READING *Answer: Both gills and lungs transfer oxygen into blood.*

READING VISUALS *Answer: make up backbone*

DIFFERENTIATE INSTRUCTION

More Reading Support

C Where does the exchange of oxygen and carbon dioxide in fish take place? *gills*

Below Level Have students write mystery vertebrate riddles that mirror the following example: I have a backbone, moist skin, and I lay eggs. What am I? Answer: amphibian. Have students write two mystery questions for each type of vertebrate.

Teach from Visuals

To help students interpret the photographs of the fish, ask:

- Which fish is the least complex? Explain. *The jawless fish is the least complex because it lacks a jawbone.*
- What is the difference between cartilaginous fish and bony fish? *Cartilaginous fish have skeletons made of cartilage. Bony fish have skeletons made of bone and are covered with scales.*

Develop Critical Thinking

COMPARE AND CONTRAST Point out that vertebrate bones are made of bone tissue, a substance that is not the same as a mollusk's shell or an arthropod's exoskeleton, even though these structures have similar functions in supporting the animals' bodies. Have students complete this chart to help them compare and contrast the three kinds of fish.

Kind	Traits	Examples
Jawless	*tubelike shape, digestive system with no stomach*	*lampreys, hagfish*
Cartilaginous	*skeletons made of cartilage, feed on small animals*	*sharks, rays, skates*
Bony	*skeletons made of hard bone, covered with scales*	*tuna, flounder, goldfish, eels*

Ongoing Assessment

READING VISUALS *Answer: The fish all have mouths at the front (or head) end of their bodies. They have eyes and gill slits.*

CHECK YOUR READING *Answer: They have tube-shaped bodies.*

Three Groups of Fish

Scientists usually classify fish in three groups.

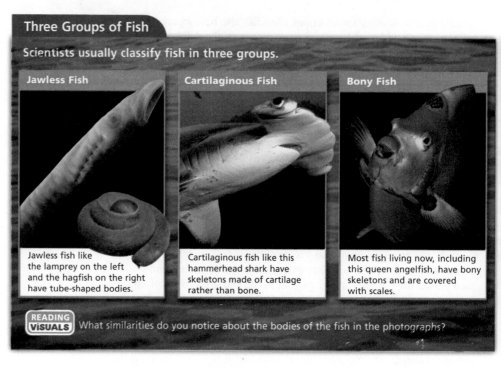

Jawless Fish — Jawless fish like the lamprey on the left and the hagfish on the right have tube-shaped bodies.

Cartilaginous Fish — Cartilaginous fish like this hammerhead shark have skeletons made of cartilage rather than bone.

Bony Fish — Most fish living now, including this queen angelfish, have bony skeletons and are covered with scales.

READING VISUALS What similarities do you notice about the bodies of the fish in the photographs?

Fish can be classified in three groups.

Scientists classify fish into three major groups: jawless fish, cartilaginous fish, and bony fish. Each group is characterized by body features. As you read on and learn about each group of fish, look at the photographs above.

Jawless Fish

D Scientists think that fish in this group, which includes lampreys and hagfish, are the living animals most similar to the first fish that lived on Earth. Jawless fish have simpler bodies than the other fish. They have a slender tubelike shape and a digestive system without a stomach.

As the name of the group implies, these fish do not have jaw bones. Although they do have teeth, they cannot chew. Most jawless fish eat by biting into another animal's body and then sucking out flesh and fluids. E

RESOURCE CENTER CLASSZONE.COM
Learn more about fish.

CHECK YOUR READING What is a characteristic of jawless fish?

DIFFERENTIATE INSTRUCTION

More Reading Support

D Which group of fish includes lampreys and hagfish? *jawless fish*

E How do jawless fish eat? *They bite into animals' bodies and suck out flesh and fluids.*

Advanced Offer the following challenge: Choose three animals: a fish, a mammal, and a reptile. Research what their skeletons look like. Draw a picture of each skeleton making sure to highlight the backbone. Explain how each species' structure affects how it moves. Include an explanation of how each structure is different from the other structures.

R Challenge and Extension, p. 299

Cartilaginous Fish

This group includes sharks, rays, and skates. Their skeletons are not made of hard bone, but of a flexible tissue called cartilage (KAHR-tuhl-ihj). Some species of sharks are dangerous to humans, but most cartilaginous fish feed primarily on small animals such as mollusks and crustaceans. Whale sharks and basking sharks, which are the largest fish on Earth, feed by filtering small organisms from the water as they swim.

Rays are flat-bodied cartilaginous fish that live most of their lives on the ocean floor. Their mouths are on the underside of their bodies. Most rays eat by pulling small animals out of the sand. A ray's flat body has fins that extend on either side of its vertebrae like wings. When rays swim, these fins wave so it looks as if the fish is flying through the water.

CHECK YOUR READING Describe three ways cartilaginous fish species obtain food.

Bony Fish

Most fish species, including tuna, flounder, goldfish, and eels, are classified in this large, diverse group. Of the nearly 20,000 fish species, about 96 percent are bony fish. Bony fish have skeletons made of hard bone, much like the skeleton in your body. Most bony fish are covered with overlapping bony structures called **scales.** They have jaws and teeth and several pairs of fins.

The range of body shapes and behavior in bony fish show how living things are adapted to their environments. Think of the bright colors and patterns of tropical fish in an aquarium. These eye-catching features are probably adaptations for survival in the fishes' natural environment. In a coral reef, for example, bright stripes and spots might provide camouflage or might advertise the fish's presence to other animals, including potential mates.

Most young fish develop inside an egg.

Most fish species reproduce sexually. The female produces eggs, and the male produces sperm. In many fish species, individual animals select a mate. For example, a female fish might release eggs into the water at a time and place where a male can fertilize them. After the eggs are fertilized, the parent fish usually leave the eggs to develop and hatch on their own. Most fish reproduce this way, but there are many exceptions.

REMINDER
In sexual reproduction the genetic material from two parents is combined in their offspring.

Integrate the Sciences

The first fish, which were the first vertebrates, appeared during the Cambrian Period about 510 million years ago. They were jawless fish that lived mainly in fresh water. Lampreys and hagfish are believed to have evolved from this early group of jawless fish. The first fish with jaws appeared in the late Silurian Period, about 410 million years ago. Modern bony fish did not appear until the late Silurian or early Devonian Period, about 395 million years ago.

Develop Critical Thinking

COMPARE AND CONTRAST Ask students why they think that marine animals such as starfish, crayfish, and sea anemones are not classified as fish. *Because fish are classified as vertebrates (animals with backbones) and these animals lack backbones.*

Ongoing Assessment

Characterize the three main groups of fish.

Ask: What differentiates, or defines, the three main types of fish? *jawless—have tube-shaped bodies; cartilaginous—have skeletons made of cartilage; bony—have bony skeletons and scales*

CHECK YOUR READING *Answer: biting with their jaws, filtering from water, pulling from sand*

DIFFERENTIATE INSTRUCTION

More Reading Support

F Which group of fish includes sharks, rays, and skates? *cartilaginous fish*

G What term describes the overlapping structures that cover the bodies of bony fish? *scales*

Below Level There is great diversity in how fish species reproduce, tend their eggs, and care for their young. There are fish that bury their eggs, fish that make nests, and a group called *mouthbrooders* that incubate eggs and protect offspring by keeping them in their mouths. Invite interested students to find out more about how fish reproduce and tend to their eggs. Have them create a bulletin board display of their findings.

Ongoing Assessment

CHECK YOUR READING *Answer: Fish eggs have a soft case. Water and other materials pass into the egg directly from the environment.*

Reinforce the BIG idea

Have students relate the section to the Big Idea.

R Reinforcing Key Concepts, p. 300

5.1 ASSESS & RETEACH

Assess

A Section 5.1 Quiz, p. 83

Reteach

Have students create trading cards for the different types of fish by observable traits. On one side they can draw a picture of the skeleton of the fish and on the other side they can write facts about the species including adaptations to the environment. They should divide the deck of trading cards into three "suits" for the three major types of fish. As a starting point, they might trace or sketch the fish pictured on p. 159, and list the adaptations explained in the captions of the photographs.

Technology Resources

Have students visit **ClassZone.com** for reteaching of Key Concepts.

CONTENT REVIEW

CONTENT REVIEW CD-ROM

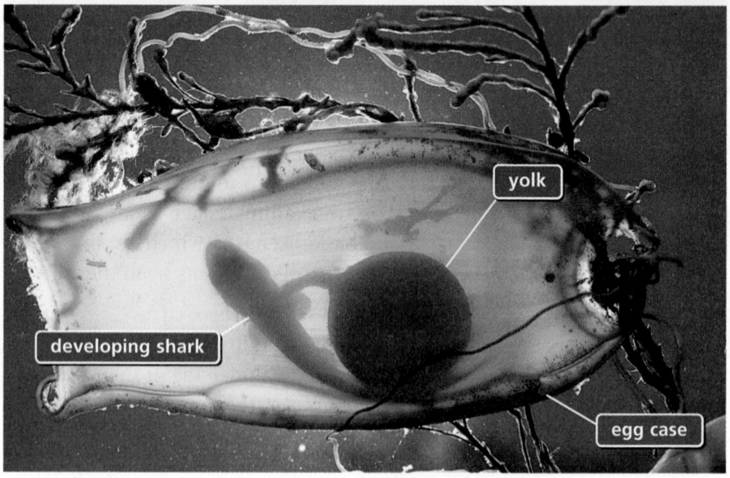

This young shark is developing from an egg that is covered by an egg case. In this photograph, light shining through the egg case allows you to see the shark and the yolk inside.

All organisms that can reproduce sexually produce egg cells and sperm cells. However, the structure and size of eggs varies among species. In Chapter 3, you learned that the eggs of flowering plants are found inside the seeds in fruit. In Chapter 4, you read about the eggs of many different types of invertebrates. You learned that some eggs have a food supply and a protective covering. For animals, the food supply is called yolk, and the covering is called an egg case.

REMINDER: An embryo is the immature form of an organism that has the potential to grow to maturity.

Most fish eggs are surrounded by a soft egg case that water can pass through. Since fish lay eggs in the water, this means that a fish embryo inside an egg gets the water and oxygen it needs directly from its surroundings. The egg's yolk provides the developing fish with food. Such eggs can develop on their own, without needing care from adults. However, many animals eat fish eggs. Fish often lay and fertilize many eggs, but few of them survive to maturity.

CHECK YOUR READING How are fish eggs different from invertebrate eggs?

5.1 Review

KEY CONCEPTS

1. Why are fish classified as vertebrate animals?
2. What are three adaptations that suit fish for life in water?
3. Name a feature for fish from each of these groups: jawless, cartilaginous, and bony.

CRITICAL THINKING

4. **Apply** If you wanted to choose tropical fish that could live comfortably in the same tank, what body features or behaviors might you look for?
5. **Infer** Some fish do not lay eggs. Their eggs develop inside the female fish. How might the offspring of one of these fish differ from those of an egg-laying fish?

CHALLENGE

6. **Synthesize** Fossils indicate that species of fish with bodies very similar to today's sharks have lived in aquatic environments for hundreds of millions of years. What can you infer about the adaptations of sharks from this?

ANSWERS

1. because they have an endoskeleton that includes vertebrae, or a backbone

2. Sample answers: gills, a swim bladder, a lateral line, fins, streamlined body

3. jawless: no jaw, tube-shaped body; cartilaginous: skeleton made of cartilage; bony: skeleton made of bone, scales

4. Sample answers: similar coloration, body shapes, and similarities in behavior

5. A fish that retained the eggs inside its body would produce fewer eggs than one that laid its eggs in water. However, eggs protected inside the parent's body would be less vulnerable to predators.

6. Sharks are very well adapted to their environment and have been able to meet their needs without changing.

MATH in SCIENCE

SKILL: DIVIDING BY DECIMALS

Great Growth

The leatherback sea turtle is one of the largest reptiles alive. Full-grown, adult leatherbacks can weigh 880 kilograms. This huge turtle starts out life weighing just 44 grams.

MATH TUTORIAL

Click on Math Tutorial for more help dividing by decimals.

Example

How many times heavier is the 880 kg adult than the 44 g baby?

(1) Convert the units so they are all in kilograms.
44 g × 0.001 kg/g = 0.044 kg

(2) Divide 880 by 0.044 to get the answer.

(3) To divide by a decimal, multiply the divisor and the dividend by a multiple of 10. Since the decimal number is in thousandths, multiply by 1000.

$0.044\overline{)880000}$ $\qquad$ $\begin{array}{r} 20{,}000 \\ 44\overline{)880000} \end{array}$

ANSWER The adult leatherback is 20,000 times heavier than the baby leatherback hatchling.

Answer the following questions.

1. An adult leatherback has been measured as 1.5 m from nose to tail. The same animal measured just 6 cm as a baby hatchling. How many times longer is the adult?
2. A typical box turtle grows to 12.5 cm long. How many times longer is the adult leatherback than the adult box turtle?
3. Suppose the box turtle hatched with a length of 2.5 cm. By how many times has its length grown at adulthood?
4. How many times longer is the leatherback hatchling than the box turtle hatchling?

CHALLENGE What fraction of its adult weight is the leatherback hatchling in the example?

ANSWERS

1. 150 ÷ 6 = 25 times

2. 150 ÷ 12.5 = 1500 ÷ 125 = 12 times

3. 12.5 ÷ 2.5 = 125 ÷ 25 = 5 times

4. 6 ÷ 2.5 = 60 ÷ 25 = 2.4 times

CHALLENGE $\frac{.044}{880} = \frac{1}{20{,}000}$

MATH IN SCIENCE

Math Skills Practice for Science

Set Learning Goal

To compare animal measurements by dividing by decimals

Present the Science

The leatherback turtle is the largest living species of turtle. The leatherback lives in the open sea, but goes ashore to lay its eggs. It is a threatened species world-wide because of over-hunting of its eggs, fishing, pollution, and a reduction of nesting sites.

Develop Measurement Skills

Review metric equivalents with students. Remind them that there are 1000 grams in one kilogram. So, 1 g = 1/1000 kg. To convert from grams to kilograms, multiply by 1/1000 or 0.001. Also remind students that 1 meter = 100 centimeters.

DIFFERENTIATION TIP Prepare a chart of metric conversions for students to use as a reference.

Close

Ask students why knowing how to compute with decimals is helpful to scientists. *Many scientific measurements involve the metric system. In order to convert measurements within the system, division with decimals is used.*

R
- Math Support, p. 338
- Math Practice, p. 339

Technology Resources

Students can visit **ClassZone.com** to practice measurement skills.

MATH TUTORIAL

5.2 FOCUS

Set Learning Goals

Students will

- Discuss the characteristics of amphibians, vertebrates that can live on land for part of their lives.
- Recognize the characteristics of reptiles, vertebrates that can live on land for their entire lives.
- Explain that amphibians and reptiles are *ectotherms.*
- Observe and describe characteristics of eggs.

3-Minute Warm-Up

Display Transparency 36 or copy this exercise on the board:

Complete the graphic organizer.

Main Idea: Fish are adapted to live in water.	
Adaptation	Function
Gills	*Allow fish to breathe*
Lateral line	*Allows fish to sense objects*
Swim bladder	*Adjusts how high or low fish float*

3-Minute Warm-Up, p. T36

KEY CONCEPT

5.2 Amphibians and reptiles are adapted for life on land.

BEFORE, you learned

- Fish are vertebrates that live in water
- Fish gills remove oxygen from water
- Most young fish develop inside eggs laid in the water

NOW, you will learn

- About amphibians, vertebrates that can live on land for part of their lives
- About reptiles, vertebrates that can live on land for their whole lives
- About the body temperature of amphibians and reptiles

VOCABULARY

amphibian p. 167
reptile p. 168
ectotherm p. 170

EXPLORE Moving on Land

What good are legs?

PROCEDURE

1. Measure and record your height in meters.
2. Jump as far as you can, and have your partner record the distance.
3. Divide the distance you jumped by your height.
4. Some frogs can jump a distance that's equal to 10 times their body length.Calculate the distance you would be able to jump if you were a frog.

MATERIALS
meter stick

WHAT DO YOU THINK?
How might the ability to jump help a frog survive on land?

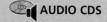

Vertebrates adapted to live on land.

Most of the groups of invertebrates and all of the vertebrates you have read about so far live in water. Organisms such as plants and insects became very diverse after adapting to live on land. Some vertebrate animals adapted to live on land as well. In this section you will learn about the first vertebrates to live on land, a group called the amphibians, and the group that came next, the reptiles.

Amphibians living today include frogs, toads, and salamanders. Reptiles include turtles, snakes, lizards, and crocodiles. Some people find it hard to tell animals from these two groups apart, but there are some important characteristics that distinguish them.

More than 350 million years ago, Earth was already inhabited by many species of vertebrate animals. All of them were fish. They

5.2 MOTIVATE

EXPLORE Moving on Land

PURPOSE To calculate jumping distance

TIP *15 min.* Have students convert their measurements to centimeters. (1 m = 100 cm)

WHAT DO YOU THINK? *Jumping ability might enable a frog to escape predators and to catch food whereas the human body is adapted to walk upright and use hands with thumbs.*

RESOURCES FOR DIFFERENTIATED INSTRUCTION

Below Level

UNIT RESOURCE BOOK
- Reading Study Guide A, pp. 303–304
- Decoding Support, p. 337

AUDIO CDS

Additional INVESTIGATION,
Adaptations of a Frog's Tongue, A, B, & C, pp. 349–357; Teacher Instructions, pp. 360–361

Advanced

UNIT RESOURCE BOOK
Challenge and Extension, p. 309

English Learners

UNIT RESOURCE BOOK
Spanish Reading Study Guide, pp. 307–308

AUDIO CDS
- Audio Readings in Spanish
- Audio Readings (English)

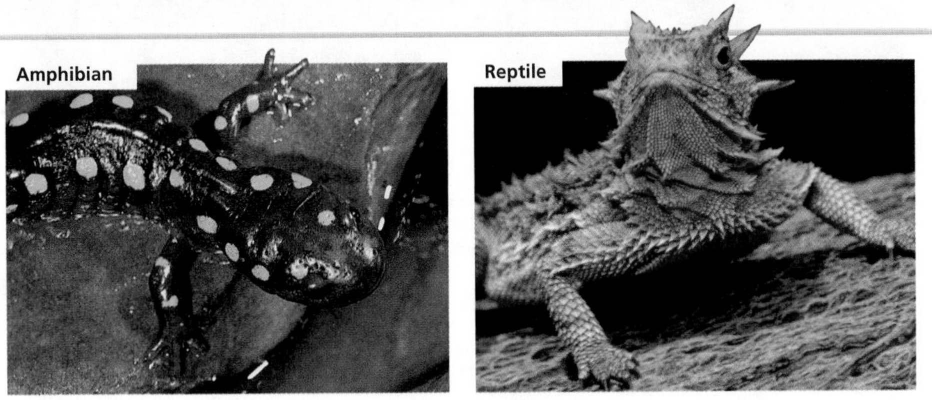

READING VISUALS COMPARE AND CONTRAST Just by looking at these two animals, what physical differences can you see? What similarities do you see?

lived in salt water and fresh water, consumed other organisms as food, and obtained oxygen using specialized organs called gills.

Recall the pond you imagined in Chapter 3, when you learned that plants adapted to land. Now imagine the same pond a hundred million years later. The pond is crowded with invertebrates and fish, all competing for oxygen and food.

Suppose a period of dry weather makes the the pond start to dry up. Many animals die, and food and oxygen become scarce. On the banks of the pond it might be less crowded. Invertebrates living there are sources of food. Air on land contains more oxygen than water does. Fish that could survive on land would be better off than the fish in the pond in this situation.

However, the gills of fish work only when they are wet. Fins can function to make a fish move through water, but they are are not good for moving on land. Water provides more support for the body than air. Plus, fish sensory organs are specialized for detecting sounds and smells in water, not in air.

A It took millions of years and many generations before different adaptations occurred and amphibians became a distinct group. These early amphibians were able to survive on land. Today there are fish that can breathe air and fish that can walk for short distances on land. There are also some modern amphibian species that have adapted to life only in water.

CHECK YOUR READING How are amphibians different from fish?

NOTETAKING STRATEGY
Choose a strategy from an earlier chapter or use one of your own to take notes on how vertebrates adapted to life on land.

5.2 INSTRUCT

Address Misconceptions

IDENTIFY Ask students to make a list of some vertebrate animals. If students list snakes and turtles, they may hold the misconception that if an organism lacks obvious external segmentation and limbs, such as a snake, or if an organism contains an external shell, as a turtle does, that the organism must be an invertebrate.

CORRECT Ask students to create a comparison poster depicting two organisms as examples, and stating the main difference between invertebrates and vertebrates. *Vertebrates have backbones.*

REASSESS Ask students to explain why an amphibian is classified as a vertebrate. *Amphibians have backbones.*

Technology Resources

Visit **ClassZone.com** for background on common student misconceptions.

MISCONCEPTION DATABASE

Ongoing Assessment

READING VISUALS *Answer: The skin texture is different: the amphibian appears to have smooth skin, while the reptile appears to have bumpy- or rough-looking skin. Both animals have heads with eyes, mouths, jaws, and front limbs with toes.*

CHECK YOUR READING *Answer: Amphibians can live on land.*

DIFFERENTIATE INSTRUCTION

More Reading Support

A Which group of vertebrates evolved after fish? *amphibians*

English Learners Have students create a picture dictionary of amphibians and reptiles. Students can print or sketch pictures from electronic encyclopedias and label the pictures with the animal names in English and other languages.

Additional Investigation To reinforce Section 5.2 learning goals, use the following full-period investigation:

Additional INVESTIGATION, Adaptations of a Frog's Tongue, A, B, & C, pp. 349–357, 360–361 (Advanced students should complete Levels B and C.)

Teach from Visuals

To help students interpret the wood frog life cycle, ask:

- Where do wood frogs begin their lives? *in water*
- How are tadpoles similar to fish? *They breathe with gills and swim and live in the water.*
- What physical developments occur in tadpoles that allow them to move onto land? *They develop legs and lungs.*

Real World Example

An adult wood frog is about two inches long. Each female lays a mass of 2,000 to 3,000 eggs. The transformation from tadpole to frog is completed quickly in wood frogs, which typically lay their eggs in fish-free ponds, puddles, or vernal pools. These habitats form as a result of snow melt and spring rain. They disappear quickly and so the young amphibians that hatch in them must complete their development and move to land before their ponds dry out.

Ongoing Assessment

READING VISUALS *Answer: The most obvious changes are the growth of legs (back pair first) and the shrinking tail. The shape of the frog's head also changes. Its eyes grow toward the top of the head, and its round mouth widens and develops an upper and lower jaw.*

Wood Frog Life Cycle

Wood frogs live in moist forest environments and breed once a year, in early spring.

A female adult wood frog deposits a mass of eggs in a pool of fresh water.

Some of the eggs hatch and become tadpoles. Tadpoles swim and breathe like fish.

The tadpole's legs develop, and its tail shrinks. Many changes occur inside the tadpole's body as well.

The young wood frog climbs out of the water. From now on, it will live on land and breathe with lungs.

The wood frog will grow until fall and slow its activity in winter. In spring females lay eggs and the cycle repeats.

READING VISUALS What visible changes occur in a wood frog tadpole's body as it transforms into an adult wood frog?

C 166 Unit: Diversity of Living Things

DIFFERENTIATE INSTRUCTION

Alternative Assessment Have students work independently to create a flow chart to illustrate the life cycle of the wood frog. Students should include these terms in their charts: *eggs, tadpole, gills,* and *lungs.*

Advanced Have pairs of students work together to create a game about amphibians and reptiles. The game should include Q-and-A or pitfalls and opportunities that address the habitat, adaptations, and any threats to the animals' survival. As part of the game design, students might research the worldwide decline of frogs and amphibians and include suggestions about what can be done to address this problem.

Amphibians have moist skin and lay eggs without shells.

As adults, most **amphibians** have these characteristics:

- They have two pairs of legs, or a total of four limbs.
- They lay their eggs in water.
- They obtain oxygen through their smooth, moist skin. Many also have respiratory organs called lungs.
- Their sensory organs are adapted for sensing on land.

Most amphibians live in moist environments. Their skin is a respiratory organ that functions only when it is wet. Most species of amphibians live close to water or in damp places. Some are most active at night, when the ground is wet with dew. Others live mostly underground, beneath wet leaves, or under decaying trees.

RESOURCE CENTER
CLASSZONE.COM
Learn more about amphibians.

Amphibians reproduce sexually. In most amphibian species, a female lays eggs in water, a male fertilizes them with sperm, and then the offspring develop and hatch on their own. Yolk inside the eggs provides developing embryos with nutrients. Like fish eggs, amphibian eggs do not have hard shells. This means developing amphibians can get water and oxygen directly from their surroundings.

CHECK YOUR READING How is the way most amphibians reproduce similar to the way most fish reproduce?

Amphibian Life Cycle

The diagram shows the life cycle of one amphibian, the wood frog. When a young amphibian hatches, it is a larva. In Chapter 4 you learned that a larva is an early stage that is very different from the animal's adult form. For example, the larvae of frogs and toads are called tadpoles. Tadpoles look and behave like small fish. They breathe with gills, eat mostly algae, and move by pushing against the water with their tails.

READING TIP
As you read about the amphibian life cycle in these paragraphs, look at the diagram on page 166 to see what a wood frog looks like at each stage.

After a few weeks, a tadpole's body begins to change. Inside, the lungs develop and parts of the digestive system transform. The tadpole begins to have some of the external features of a frog. It develops legs, its tail shrinks, and its head changes shape.

As a young wood frog's body changes, its gills stop functioning, and it begins breathing air with its lungs. The frog starts using its tongue to capture and eat small animals. It leaves the water and begins using its legs to move around on land. Some amphibians, such as sirens and bullfrogs, remain in or near water for all of their lives. Others, like wood frogs, most toads, and some salamanders, live in moist land environments as adults.

DIFFERENTIATE INSTRUCTION

More Reading Support

B What are two ways adult amphibians get oxygen? *skin and lungs*

C How do tadpoles move? *They push against water with tails.*

Advanced The three categories of amphibians are salamanders, frogs and toads, and caecilians. Caecilians live primarily in South America; there are no living species in the United States. There is considerable variety among amphibians. Invite students to research and produce single pages for a field guide on "Amazing Amphibians." They can include pictures, facts, and scientific names.

Challenge and Extension, p. 309

Develop Critical Thinking

ANALYZE Have students study and interpret the list of amphibian characteristics. Ask:

- Why do most adult amphibians stay near the water throughout their lives? *They need to stay near water to keep their skin moist. They need to lay their eggs in a moist environment for the eggs to survive.*
- Would a tadpole be able to live on land? Explain. *No, a tadpole does not have legs to move around or lungs to breathe.*
- What invertebrate animal also breathes through its skin? *earthworms*

Metacognitive Strategy

Ask students to recall the stages of insect metamorphosis they studied in Chapter 4. Ask: How are the stages of a tadpole's metamorphosis similar to those of a butterfly? *Some insects, such as the butterfly begin as eggs and hatch into larva. The larva stages of both the butterfly and the tadpole look very different physically from the adult forms.*

Ongoing Assessment

Describe the characteristics of amphibians.

Ask: What adaptations do amphibians have that suit them to life in both water and land environments? *They lay eggs in water, have lungs as well as the ability to breathe through skin, have legs and webbed feet.*

CHECK YOUR READING *Answer: Similarities include soft eggs that are laid in water, eggs that are fertilized externally, and offspring that develop and hatch with little or no care from adults.*

Teach Difficult Concepts

Some students might think that the hard, outer coverings of some reptiles, such as turtles, are made of bone tissue. Point out that the shells are not made of bone, but are fused with the vertebrae so their backs are not flexible. Explain that the bone on the inside, including the vertebrae, is made up of bone tissue.

After reviewing the list of reptilian characteristics, point out that not all of the animals classified as reptiles have all of the characteristics listed. Snakes, for example, do not have visible legs, nor do legless lizards. Many reptilian species are aquatic and have adaptations for sensing and responding in water that complement or may even be better than their senses on land.

Ongoing Assessment

CHECK YOUR READING *Answer: Differences include tough, dry skin covered with scales; sensory organs adapted for sensing on land; eggs that have shells and are laid on land; only lungs for respiration.*

Reptiles have dry, scaly skin and lay eggs with shells.

VOCABULARY
Add a magnet word diagram for *reptile* to your notebook.

Reptiles evolved soon after amphibians and are closely related to them. However, animals in the reptile group have adaptations that allow them to survive in hotter, drier places than amphibians. For many millions of years they were the largest and most diverse vertebrate animal group living on land. Most of the animals classified as reptiles have these characteristics:

- They have two pairs of legs, for a total of four limbs.
- They have tough, dry skin covered by scales.
- D They obtain oxygen from air with respiratory organs called lungs.
- Their sensory organs are adapted for sensing on land.
- E They lay their eggs, which have shells, on land.

CHECK YOUR READING What characteristics of reptiles are different from the characteristics of amphibians listed on page 167?

Lungs

Reptiles do not get oxygen through their skin the way amphibians do. They are born with lungs that provide their bodies with all the oxygen they need. Lungs, like gills, are internal organs made up of many folds of thin tissue filled with blood. When an animal with lungs inhales, it takes air in through its nostrils or mouth and moves the air into its lungs. There, oxygen is transported across the tissues and into the blood, and carbon dioxide is moved from the blood to the lungs and exhaled.

Reptiles like these garter snakes are covered with scales and breathe through their nostrils.

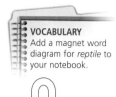

DIFFERENTIATE INSTRUCTION

More Reading Support

D How do reptiles obtain oxygen? *through lungs*

E Where do reptiles lay their eggs? *on land*

Below Level To help students understand the difference between reptiles and amphibians, have them complete this compare/contrast graphic organizer:

Trait	Amphibians	Reptiles
Number of limbs	*four*	*four*
Outer skin	*wet and smooth*	*dry with scales*
Way of getting oxygen	*moist skin*	*lungs*
Eggs	*in water*	*on land*

CONTRAST How does the egg this gecko hatched from differ from the wood frog eggs shown on page 166?

Dry, Scaly Skin

Reptile skin is hard, dry, and covered with scales made of keratin, a substance much like your fingernails. The thick, waterproof skin of reptiles protects them from the environment and from predators. However, this means that reptiles cannot obtain water through their skin.

Eggs with Shells

The reptile egg is an important adaptation that allows vertebrate animals to survive in hot, dry environments. The eggs of reptiles contain everything an embryo needs: water, nutrients, a system for gas exchange, and a place to store waste. Membranes separate the internal parts of the egg, which is covered by a protective shell.

Reptiles reproduce sexually. The egg cell of the female joins with the sperm cell of the male in the process of fertilization. After fertilization, a protective case, or shell, forms around each egg while it is still inside the female's body. The female selects a place to lay the eggs on land. Many species of reptiles build or dig nests. Some female reptiles, including alligators, guard their nests and care for their offspring after they hatch. Most reptiles, however, leave soon after the eggs are laid. As you can see in the photograph above, when young reptiles hatch, they look like small adults.

RESOURCE CENTER
CLASSZONE.COM
Find out more about reptiles.

DIFFERENTIATE INSTRUCTION

More Reading Support

F What are the scales of reptile skin made of? *keratin*

G Why is the reptile egg an important adaptation? *Reptiles may survive in hot, dry environments.*

Advanced Have students create a timeline showing the geological periods in which fish, amphibians, and reptiles made their appearances on Earth. They will need to do research in paleontology to complete the chart. Students can add to the timeline as they study birds and mammals later in Chapter 5. Encourage students to include pictures in their timelines of extinct and present-day species of animals.

Address Misconceptions

IDENTIFY Ask students to give some examples of reptiles and amphibians. Many students may hold the misconception that amphibians and reptiles are not vertebrates. Students also will frequently confuse the two classes or lump them together.

CORRECT Have students create a Venn diagram of the characteristics of amphibians and reptiles. In the intersecting section of the diagram they should note how the organisms are similar. For example, both have backbones, are vertebrates, and so on.

REASSESS Ask students to explain why a turtle or a snake is classified as a reptile. *Turtles lay eggs, have four legs, and their legs and heads are covered with dry, scaly skin. Snakes have more characteristics of reptiles than of other vertebrates, only lacking limbs.*

Technology Resources

Visit **ClassZone.com** for background on common student misconceptions.

MISCONCEPTION DATABASE

Real World Example

Aquatic reptiles, such as many species of sea turtles, have lungs and must come to the surface to breathe. These turtles are adapted to sea life with forelimbs that are modified into flippers for swimming. Most sea turtles only leave the sea during the breeding season when the females lay and bury their eggs on land. The hatchlings then make their journey from the nest to the sea.

Ongoing Assessment

Describe the characteristics of reptiles.

Ask: What characteristics make reptiles suited to life in a land environment? *eggs with shells; protective, thick skin; lungs and respiratory organs; legs; sensory organs*

PHOTO CAPTION Answer: The gecko egg is on land and has a shell that is hard and protective; the wood frog egg is in water and is soft.

INVESTIGATE Eggs

PURPOSE Students observe a hard-boiled egg to determine the characteristics of eggs

TIPS *20 min.* Make sure students handle their eggs with care to avoid dropping and breaking the eggs. Put down sheets of newspaper or plastic bags over work surfaces. Have students peel eggs directly over their work surfaces.

WHAT DO YOU THINK? *Eggs with shells have a shell, a protective membrane, an egg white, and a yolk. The shell protects the egg, the membrane regulates and holds the embryo's environment, the yolk provides nutrients.*

CHALLENGE *As the embryo develops, it consumes the yolk and grows. The membrane contains the embryo, and the egg white contains structures that help maintain the egg's internal environment.*

R Datasheet, Eggs, p. 310

Technology Resources

Customize this student lab as needed or look for an alternative. Print rubrics to assess student lab reports.

Lab Generator CD-ROM

Develop Critical Thinking

ANALYZE Point out that ectothermic animals are sometimes referred to as "cold-blooded." Ask: Why do you think that this term is inaccurate? *An ectothermic animal that is in a hot environment would have a high body temperature.*

Real World Example

Most fish are considered ectotherms, but the general temperature in water environments does not vary as dramatically or change as rapidly as it does on land. Some fish, such as tuna, generate body heat and are considered endothermic. Other fish have anti-freeze proteins in the blood that allow them to remain active in extremely cold water.

INVESTIGATE Eggs

SKILL FOCUS
Observing

MATERIALS
- hard-boiled egg
- plastic knife

TIME:
20 minutes

What are some of the characteristics of eggs?

PROCEDURE

1. Carefully examine the outside of the hard-boiled egg. Try to notice as many details as you can. Write your observations in your notebook.
2. Gently crack the eggshell and remove it. Try to keep the shell in large pieces and the egg whole. Set the egg aside, and examine the pieces of shell. Look for details you could not see before. Write your observations in your notebook.
3. Examine the outside of the egg. Make notes about what you see. Include a sketch.
4. Use the knife to cut the egg in half. Take one half apart carefully, trying to notice as many parts as you can. Use the other half for comparison. Write up your observations.

WHAT DO YOU THINK?

- Reptiles, like birds, have eggs with hard shells. What structures does an egg with a shell contain?
- What might the function of each structure be?

CHALLENGE How might the egg's structures support a developing embryo?

The body temperatures of amphibians and reptiles change with the environment.

? H Amphibians and reptiles are **ectotherms,** animals whose body temperatures change with environmental conditions. You are not an ectotherm. Whether the air temperature of your environment is −4°C (25°F) or 43°C (110°F), your body temperature remains around 37°C (99°F). A tortoise's body temperature changes with the temperature of the air or water surrounding it. On a cool day, a tortoise's body will be cooler than it is on a hot one.

Many ectothermic animals can move and respond more quickly when their bodies are warm. Many ectotherms warm themselves in the Sun. You may have seen turtles or snakes sunning themselves. Ectothermic animals transform most of the food they consume directly into energy. Some ectotherms, even large ones such as alligators, or the Galápagos tortoise in the photograph, can survive for a long time without consuming much food.

DIFFERENTIATE INSTRUCTION

? More Reading Support

H What term describes an animal whose body temperature changes with environmental conditions? *ectotherm*

Below Level Have students research what a herpatologist does. Ask them to find out about the different areas of studies, what kind of training is needed, and what kind of jobs are available.

Students might also find out about the needs of amphibians and reptiles that are kept as pets. They can write interview questions for a local pet store owner and then conduct the interview. They can then share their findings with the class.

This sand-diving lizard can reduce the amount of heat that transfers from the sand into its body by standing on two feet.

Although amphibians and reptiles do not have a constant body temperature, their bodies stop functioning well if they become too hot or too cold. Most amphibians and reptiles live in environments where the temperature of the surrounding air or water does not change too much. Others, like wood frogs and painted turtles, have adaptations that allow them to slow their body processes during the winter.

Amphibians and reptiles also have behaviors that allow them to adjust their body temperature in less extreme ways. The sand-diving lizard in the photograph above is able to control how much heat enters its body through the sand by standing on just two of its four feet. Many amphibians and reptiles live near water and use it to cool off their bodies.

5.2 Review

KEY CONCEPTS

1. What are three adaptations that allowed the first amphibians to survive on land?
2. What are two adaptations reptiles have that allow them to live their whole live on land?
3. A crocodile has been lying in the sun for hours. When it slides into the cool river, how will its body temperature change? Why?

CRITICAL THINKING

4. **Compare and Contrast** Make a diagram to show how amphibians and reptiles are different and how they are similar.
5. **Infer** Some reptiles, like sea turtles, live almost their whole lives in water. What differences would you expect to see between the bodies of a sea turtle and a land turtle?

CHALLENGE

6. **Hypothesize** For many millions of years, reptiles were the most diverse and successful vertebrate animals on land. Now many of these ancient reptiles are extinct. Give some reasons that might explain the extinction of these reptiles.

ANSWERS

1. limbs, lungs, sensory organs adapted for land

2. born with lungs, eggs have shells

3. Its body temperature should go down. The water will cool its body.

4. Diagrams should include adaptations, methods of reproduction, and how animals obtain oxygen.

5. Sample answer: a more streamlined body in sea turtle (for moving through water), webbing on feet of sea turtle, better underwater eyesight and other senses, and more lung capacity in sea turtle.

6. Answers should show understanding of how environmental factors affect survival rates. Sample answer: If climate changed and became significantly cooler, reptiles might be adversely affected.

Ongoing Assessment

Explain that reptiles and amphibians are ectotherms.

Ask: How might the body temperature of an ectotherm become hot? *The animal could warm itself in the Sun.*

Reinforce the BIG idea

Have students relate the section to the Big Idea.

R Reinforcing Key Concepts, p. 310

5.2 ASSESS & RETEACH

Assess

A Section 5.2 Quiz, p. 84

Reteach

Have students create a compare and contrast chart about amphibians and reptiles. Charts should include:

- body structures suited to environments
- how oxygen is obtained
- how the animal group reproduces

Technology Resources

Have students visit **ClassZone.com** for reteaching of Key Concepts.

CONTENT REVIEW

CONTENT REVIEW CD-ROM

CONNECTING SCIENCES
Integration of Sciences

Set Learning Goal

To understand how physics principles, such as the bonding of atoms, apply to a trait of certain reptiles

Present the Science

Geckos are nocturnal insect-eating lizards that are found in warm climates. There are about 700 species of geckos. Most are about four to six inches (10 cm to 15 cm) in length. The largest can grow to a length of 14 inches (35.5 cm). Feet of geckos vary. Some toes have claws, others, extensive webbing and climbing pads.

The explanation for why a gecko's spatulae help it stick to walls relies on physical science. When two atoms from different molecules get very close together, the movement of the electrons within each atom are forced to change. This produces electrical charges that fluctuate and move along the surface of each atom. The net result is a force that attracts the two atoms towards each other. This force, called the van der Waals force, can bond spatulae atoms to a wall's atoms.

Discussion Questions

Ask: How are the hairs in a gecko's foot arranged? *The hairs branch into hundreds of tiny hairs called* spatulae.

Ask: Why do you think the gecko's ability to stick to walls is called *dry adhesion*? *Adhesion occurs between two dry surfaces.*

Close

Ask: How does the gecko's ability to stick to surfaces rely on the physical activity of atoms? *The interaction between the atoms of the gecko's feet and the surface is based on one property of atoms that is the attraction of electrons to one another.*

CONNECTING SCIENCES

LIFE SCIENCE AND PHYSICAL SCIENCE

Sticky Feet

Imagine having the ability to walk up a polished glass window, across ceiling tiles, and down a mirror—without using suction or glue. Small tropical lizards called geckos have this unique ability, and scientists are beginning to understand what makes these animals stick.

A gecko's foot is covered in billions of tiny hairs. Each hair branches into hundreds of smaller hairs called spatulae, and the atoms in these spatulae are attracted to atoms in the surfaces the geckos walk on.

All atoms contain particles called electrons, which constantly move around. When two atoms get very close, the motion of their electrons can change, attracting the two atoms to one another.

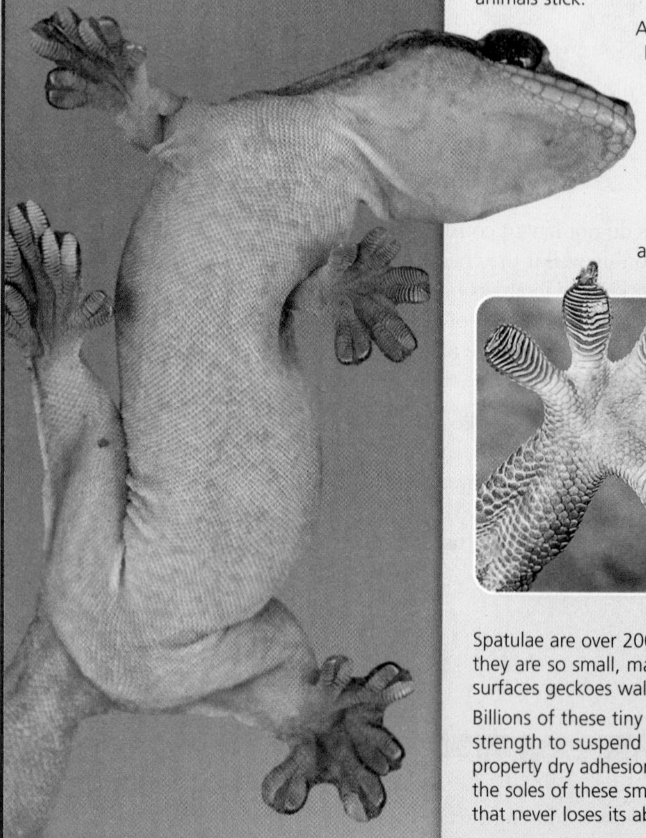

Shown life-size, the gecko is a reptile with tiny sticky hairs on its feet. Why are chemists and engineers interested in this adaptation?

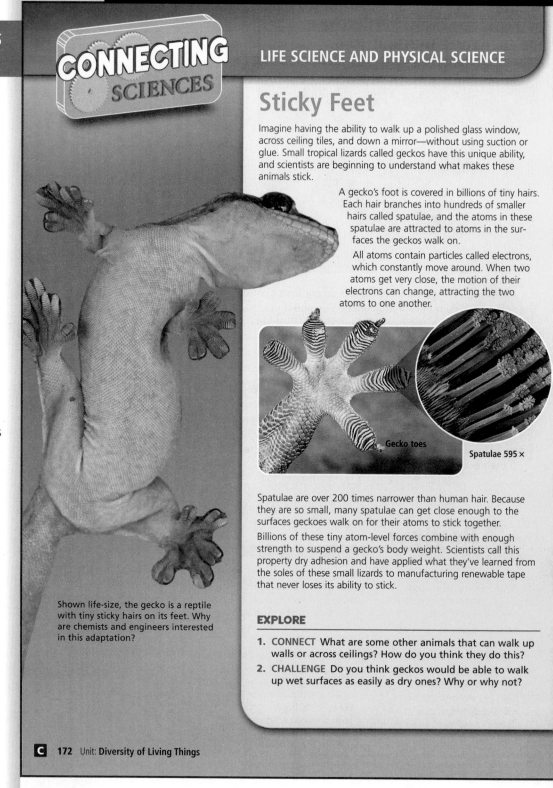

Spatulae are over 200 times narrower than human hair. Because they are so small, many spatulae can get close enough to the surfaces geckoes walk on for their atoms to stick together.

Billions of these tiny atom-level forces combine with enough strength to suspend a gecko's body weight. Scientists call this property dry adhesion and have applied what they've learned from the soles of these small lizards to manufacturing renewable tape that never loses its ability to stick.

EXPLORE

1. **CONNECT** What are some other animals that can walk up walls or across ceilings? How do you think they do this?
2. **CHALLENGE** Do you think geckos would be able to walk up wet surfaces as easily as dry ones? Why or why not?

EXPLORE

1. ***CONNECT*** *Sample answers: insects, some reptiles such as lizards. Accept all reasonable responses.*
2. ***CHALLENGE*** *If the surface is wet, the spatulae will not be able to get as much suction on the surface. Therefore, the gecko will have more trouble walking up wet surfaces.*

C 172 Unit: Diversity of Living Things

KEY CONCEPT

5.3 Birds meet their needs on land, in water, and in the air.

BEFORE, you learned
- Vertebrate animals have endoskeletons with backbones
- Amphibians and reptiles have adaptations for life on land
- Ectotherms do not maintain a constant body temperature

NOW, you will learn
- About birds as endotherms
- How the adaptations of birds allow them to live in many environments
- About adaptations for flight

VOCABULARY

endotherm p. 174
incubation p. 179

EXPLORE Feathers

How do feathers differ?

PROCEDURE
1. Examine several feathers.
2. Make a list of ways some of your feathers differ from others.
3. Make a diagram of each feather, showing the characteristics you listed.
4. Compare your list and your diagram with those of your classmates.

MATERIALS
assorted feathers

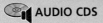

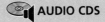

WHAT DO YOU THINK?
- Of all the characteristics of feathers that you and your classmates listed, which do you think are the most important? Why?
- On the basis of your observations, what are some functions of feathers for birds?

Bird species live in most environments.

NOTETAKING STRATEGY
Be sure to take notes on the main idea, *Bird species live in most environments.* Choose an earlier strategy or one of your own.

Penguins live in Antarctica, and parrots inhabit the tropics. Pelicans scoop their food from the water, while cardinals crack open seeds and eat the insides. Swallows skim insects from above the surface of a pond. A soaring hawk swoops down, and a smaller animal becomes its prey. There are nearly 10,000 species of the vertebrate animals called birds. Their adaptations allow them to live all over the world.

Some bird species, such as pigeons, are adapted to live in a wide range of environments, while others have adaptations that limit them to living in one place. Many birds travel long distances during their lives. Some migrate as the seasons change, and others cover long distances while searching for food.

RESOURCES FOR DIFFERENTIATED INSTRUCTION

Below Level
UNIT RESOURCE BOOK
- Reading Study Guide A, pp. 314–315
- Decoding Support, p. 337

AUDIO CDS

Advanced
UNIT RESOURCE BOOK
Challenge and Extension, p. 320

English Learners
UNIT RESOURCE BOOK
Spanish Reading Study Guide, pp. 318–319

AUDIO CDS
- Audio Readings in Spanish
- Audio Readings (English)

5.3 FOCUS

Set Learning Goals

Students will
- Recognize that birds are endotherms.
- Describe how birds' adaptations allow them to live in many environments.
- Explain the adaptations that enable most birds to fly.

3-Minute Warm-Up

Display Transparency 37 or copy this exercise on the board:

Write *reptile* or *amphibian* for each description.

Descriptions
1. begins life in water breathing through gills *amphibian*
2. lays eggs on land *reptile*
3. has moist skin for obtaining oxygen *amphibian*
4. has scales that are made up of keratin *reptile*

3-Minute Warm-Up, p. T37

5.3 MOTIVATE

EXPLORE Feathers

PURPOSE To inquire about the structures and functions of bird feathers

TIPS *15 min.* Bags of feathers can be purchased at art supply or craft stores. The best sample will include naturally colored feathers of varying shapes and sizes that are all from the same species of bird. Examining feathers from different species will also work.

WHAT DO YOU THINK? *Students should note that birds have contour feathers of varying sizes and shapes, as well as down feathers. Color, shape, form, and function of the feathers vary depending on where the feather is located on the body. Most feathers have a central shaft. Smaller shafts project from the central shaft. Some feathers (not down) have barbs that knit the feathers together.*

5.3 INSTRUCT

Develop Critical Thinking

ANALYZE Explain that bird beaks and feathers are made of keratin as are the scales on their legs. Then ask:

- What other animal has body parts that are made of keratin? *Reptiles have scales made of keratin.*
- How are feathers like the scales of a reptile? *Both provide a covering for the body.*
- How do these features indicate that birds share common ancestors with reptiles? *Bird feathers are modified scales. Feathers may have evolved from reptilian scales showing that birds and reptiles have common ancestors.*

Ongoing Assessment

Recognize that birds are endotherms.

What are endotherms? *animals that maintain a constant body temperature*

CHECK YOUR READING *Answer: Endotherms generate more body heat than ectotherms. Endotherms have adaptations and behavior for conserving body heat so they can maintain a constant body temperature.*

It probably seems easy for you to recognize which animals are birds. Birds are distinguished by these characteristics:

A
- They have feathers and a beak.
- They have four limbs: a pair of scaly legs and a pair of wings.
- Their eggs have hard shells.

Birds can maintain body temperature.

In the last section, you learned that the body temperature of ectotherms, such as amphibians and reptiles, changes with their environment. Birds are **endotherms,** or animals that maintain a constant body temperature. Maintaining temperature allows endotherms to live in some places where frogs, turtles, and alligators cannot.

This chickadee is an endotherm. This means its body remains warm, even in very cold weather.

When an ectothermic animal's body is cool, its systems slow down and it becomes less active. A less active animal consumes little or no food and is unlikely to reproduce. It moves slowly or not at all and breathes less often. Its nervous system becomes less responsive, and its heart pumps more slowly. An ectothermic animal that stays cool for too long will die. Even if it has enough food, its body lacks the energy needed to digest the food.

All animals are affected by the air temperature of their environment and will die if they become too cold. However, birds and other endotherms can stay active in colder climates than ectotherms. This is because endothermic animals have adaptations for generating more body heat and keeping it near their bodies.

CHECK YOUR READING How can endotherms stay alive in colder climates than ectotherms?

Generating Heat

The energy birds produce as body heat comes from food. This means that birds and other endotherms need to eat a lot. An ectotherm such as a frog might be able to survive for days on the energy it gets from just one worm, while a bird on the same diet might starve. Also, the amount of food an endotherm needs is affected by climate. House sparrows and other birds that do not migrate need to eat more food and produce more energy to survive in winter than they do during warmer seasons.

C 174 Unit: Diversity of Living Things

DIFFERENTIATE INSTRUCTION

More Reading Support

A How many limbs do birds have? What are they? *two legs, two wings*

B What happens to an endotherm in cold weather? *It uses activity to generate body heat.*

English Learners Students new to English may add these words to their Science Notebooks: *endotherm, temperature, endoskeleton, migration,* and *incubation.* Compare the prefixes *endo-* in *endotherm* and *endoskeleton.* Tell students that *endo-* means "inner." Point out that *migration* and *incubation* are forms of the verbs *migrate* and *incubate.*

Insulating with Down

Just as warm air trapped between down feathers helps keep geese warm, feathers in a jacket keep a student warm.

Controlling Body Temperature

Birds have soft feathers, called down, that keep warm air close to their bodies. If you have ever slept with a down comforter or worn a down jacket, you know that these feathers are good insulation, even though they are not very heavy. Other feathers, called contour feathers, cover the down on birds. In most species, contour feathers are water-resistant and protect birds from getting wet. C

Birds shiver when they are cold, and this muscular movement generates heat. They also have ways of cooling their bodies down when the weather is hot. Birds do not sweat, but they can fluff their feathers out to release heat. D Birds, like other animals, have behaviors for maintaining body temperature, such as resting in a shady place during the hottest part of a summer day.

Most birds can fly.

Of all the animals on Earth today, only three groups have evolved adaptations for flight: insects, bats, and birds. Fossil evidence suggests that the first birds appeared on Earth about 150 million years ago and that they were reptiles with adaptations for flight. Scientists think that all birds are descended from these flying ancestors, even modern species such as ostriches and penguins, which cannot fly.

Teach from Visuals

To help students interpret the photograph, have them recall what they observed about down feathers in the explore activity on p. 173. Then ask: Why do you think down feathers are able to trap air? *Down feathers are soft, fluffy, and flexible. They have air pockets within the feathers, where the warm air is trapped in pockets.*

Integrate the Sciences

Materials that do not conduct heat well are insulators. Substances such as wool, wood, straw, cork, paper, and gases such as air are good insulators of heat. Materials such as metals are good conductors of heat. The heat travels quickly through the material. That is why a metal spoon will heat up and be hot to the touch when placed in hot water and a wooden spoon will still be cool to the touch. Insulators such as down feathers slow the transfer of heat away from a bird's body.

Teacher Demo

Demonstrate how feathers make good insulation. Display two plastic cups filled with warm water. Wrap aluminum foil around one cup. Fold a paper towel in half and fill it with down feathers. Wrap the other cup with the paper towel holding the feathers. Secure each wrapping with a rubber band. Place a thermometer in each cup. Have students note which cup retains its temperature over a period of time.

DIFFERENTIATE INSTRUCTION

More Reading Support

C Which bird feathers are water-resistant? *contour feathers*

D How do birds release heat from their bodies? *They fluff up their feathers.*

Below Level Encourage groups of interested students to find out what birds are in your school yard by creating a school bird feeder. A bird feeder can be as simple as dried bread spread with honey and seeds. Students can do a simple tally count of which birds visit the feeder. Provide field guides or an Internet field guide for them to identify the birds. Then have them prepare a bar graph or pictograph showing, by species, the birds who visited.

Develop Critical Thinking

ANALYZE Discuss the difference between sustained flight and what flying squirrels or flying foxes do, which is glide. Point out that many birds, such as ostriches and penguins do not have the ability to fly. Ask:

- Why do you think that these birds are unable to fly? *They are large and heavy in size and their wings would have to be much larger to compensate for their weight.*
- Do you think humans should be added to the list of animals that fly? *Accept all reasonable responses.*

Integrate the Sciences

Bernoulli's principle of lift helps explain why birds are able to fly. Bernoulli's principle states that the pressure exerted by a moving stream of air is less than pressure of the air surrounding it. As air moves over a bird's wings, the air pressure below the wings is less than the pressure above the wings. Because a bird's wings are curved, the air above the wing moves at a faster rate. This difference in pressure creates the upward lift, enabling the bird to fly.

Ongoing Assessment

Describe how bird adaptations allow them to live in many environments.

Ask: Why do you think birds may have been the first vertebrates adapted to live on islands? *Birds have wings that allow them to cross bodies of water as well as to live in high or low places.*

CHECK YOUR READING *Answer: Birds do not have teeth or strong jaws, so they cannot chew. Some parts of their skeletons are fused, making the trunk parts of their bodies less flexible.*

Adaptations for Flight

To lift its body into the air and fly, an animal's body has to be very strong, but also light. Many adaptations and many millions of years were needed before birds' body plans and systems became capable of flight. With these adaptations, birds lost the ability to do some things that other vertebrates can. As you read about birds' adaptations for flight, match the numbers with the diagram on the next page.

READING TIP As you read the numbered text on this page, find the matching number on page 177.

1. **Endoskeleton** Some of the bones in a bird's body are fused, or connected without joints. This makes those parts of a bird's skeleton light and strong, but not as flexible. A specialized bone supports the bird's powerful flight muscles.
2. **Wings and Feathers** Birds do not have hands or paws on their wings. Contour feathers along the wing are called flight feathers, and are specialized for lifting and gliding. Feathers are a strong and adjustable surface for pushing against air.
3. **Specialized Respiratory System** Flying takes a lot of energy, so birds need a lot of oxygen. They breathe using a system of air sacs and lungs. Air follows a path through this system that allows oxygen to move constantly through a bird's body.
4. **Hollow Bones** Many of the bones in a bird's skeleton are hollow. Inside, crisscrossing structures provide strength without adding much weight.

E Other body systems within birds are more suited for flight than systems in other invertebrates. Instead of heavy jaws and teeth, an internal organ called a gizzard grinds up food. This adaptation makes a bird lighter in weight and makes flight easier. Birds also have highly developed senses of hearing and vision, senses which are important for flight. Their senses of taste and smell are not as well developed.

CHECK YOUR READING Give two examples of things that birds cannot do that relate to their adaptations for flight.

Benefits of Flight

Flight allows animals to get food from places where animals living on land or in water cannot. For example, some species of birds spend most of their lives flying over the ocean, hunting for fish. Also, a flying bird can search a large area for food more effectively than it could if it walked, ran, or swam.

F For many species of birds, flight makes migration possible. In Chapter 2, you learned that some animals migrate to different living places in different seasons. Most migratory birds have two living places, one for the summer and one for the winter.

DIFFERENTIATE INSTRUCTION

More Reading Support

E What is a gizzard? *an internal organ in birds that grinds food*

F What do birds having two living places do in different seasons? *They migrate.*

English Learners Have English learners draw their own pictures of a bird. Then, have them label the *endoskeleton, wings and feathers, hollow bones,* and *specialized respiratory system.* Students can compare their drawings and labels to the diagram on p. 177.

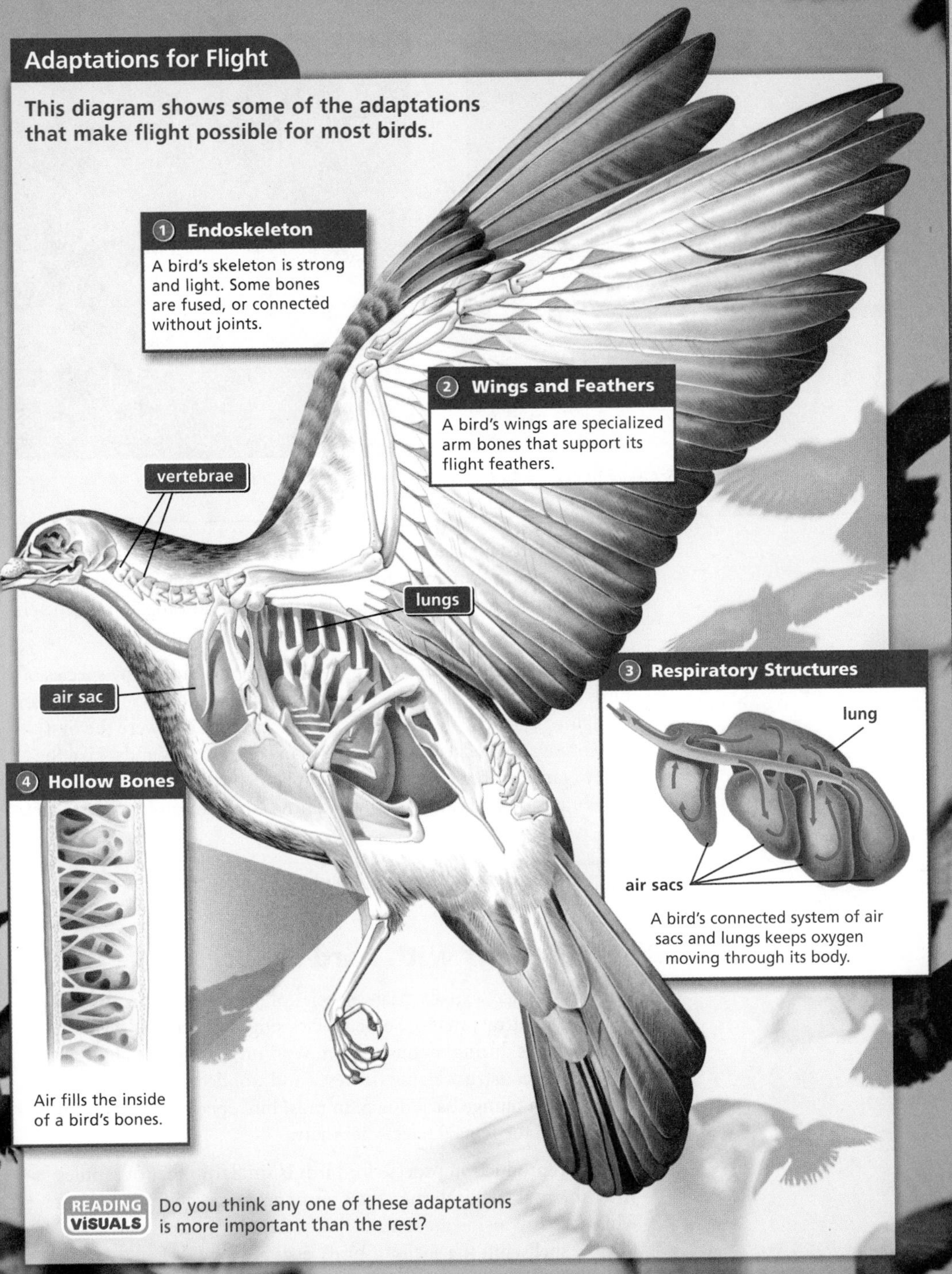
Adaptations for Flight
This diagram shows some of the adaptations that make flight possible for most birds.
① Endoskeleton
A bird's skeleton is strong and light. Some bones are fused, or connected without joints.
② Wings and Feathers
A bird's wings are specialized arm bones that support its flight feathers.
vertebrae
lungs
air sac
③ Respiratory Structures
lung
air sacs
A bird's connected system of air sacs and lungs keeps oxygen moving through its body.
④ Hollow Bones
Air fills the inside of a bird's bones.
READING VISUALS Do you think any one of these adaptations is more important than the rest?

DIFFERENTIATE INSTRUCTION

Below Level Have students create models of wings and birds by using paper and other objects such as drinking straws and brass fasteners. Point out that a bird moves its wings in a circular motion when in flight, not up and down.

Teach from Visuals

To help students interpret the diagram illustrating adaptations for flight, ask:

- How might a flexible neck be beneficial to a bird? *A flexible neck allows a bird to catch food, clean itself, and increases the field of vision.*
- Why would a bird need powerful chest muscles? *The chest muscles are needed to enable the wings to fly for long periods of time.*
- How are the wings specialized for flight? *They are hollow; they have flight feathers attached to them.*
- What are two structures in the bird body that allow air inside? How is this beneficial to the bird? *Air sacs and hollow bones are filled with air. This makes the bird lighter and enables it to fly.*

T This visual is also available as T38 in the Unit Transparency Book.

Ongoing Assessment

Explain the adaptations that enable most birds to fly.

Ask: How are hollow bones an adaptation for flight? *Hollow bones are lightweight.*

READING VISUALS *Sample answer: Adaptations to the skeleton are especially important. Many of the features described in the diagram are especially important.*

Teach from Visuals

To help students interpret the photographs showing flight, ask:

How is the osprey's body adapted for capturing prey? *The feet have strong talons, or claws, that enable the bird to capture its prey.*

History of Science

Many recent bird extinctions are due to the adverse effect of humans on the natural habitats of birds. One of the most notable extinctions was that of the dodo bird, once found on the island of Mauritius in the Indian Ocean. First encountered by Portuguese sailors in the early 1500s, the dodo was a flightless bird that did not have any natural predators. By 1681, the dodo was extinct. Sailors had hunted the dodo for food and had introduced dogs and swine, which ate dodo eggs and the young.

Ongoing Assessment

CHECK YOUR READING *Answer: Flight gives birds access to food in a larger area, access to food and living places that are not reachable on foot or by swimming. Flight allows birds to escape from predators and other dangers, and find safe places to raise their young.*

Escaping Predators

Flight allows these birds to escape from predators such as this wolf.

Catching Food

Flight allows this osprey to hunt for fish in bodies of water.

Some birds migrate very long distances. Ruby-throated hummingbirds, for example, migrate from Canada and the United States to Mexico and Central America each winter, flying nonstop 700 kilometers across the Gulf of Mexico.

By flying, birds can escape danger on the ground. Many species of birds lay their eggs and raise their young in places that are difficult for predators to reach. Fossil evidence suggests that birds were the first vertebrate animal species to live on many of Earth's islands. Like the first organisms that adapted to live on land, birds that were the first vertebrates on an island usually had little competition for food, water, and living space.

CHECK YOUR READING How does flight benefit birds? Give three examples.

Birds lay eggs with hard shells.

Birds reproduce sexually. Many species of birds have distinctive ways of trying to attract mates. Some species sing, and others develop colorful feathers during mating season. Wild turkeys fight each other, bowerbirds construct elaborate nests, and woodcocks fly high in the air and then plunge back down. In most bird species, the male animals display and the female selects her mate.

The reproduction process for birds is similar to that of reptiles. After internal fertilization, a shell forms around each fertilized egg while it is still inside the female's body. Reptiles' eggs usually have flexible shells, but the shells of birds' eggs are hard.

G

DIFFERENTIATE INSTRUCTION

More Reading Support

G How do the shells of bird eggs differ from reptile's eggs? *Bird eggs are hard; reptile eggs are flexible.*

Advanced Invite students to investigate the different orders of birds and provide examples of the special adaptations that allow them to live in certain environments. Bird groups to investigate include perching birds, birds of prey, cavity-nesting birds, shore birds, diving birds, game birds, song birds, and ostriches and their relatives.

R Challenge and Extension, p. 320

The female bird chooses a place to lay the eggs. Often this is a nest. In some bird species either the male or the female builds the nest, while in others, mated birds build a nest together. Bird eggs have to be kept at a constant, warm temperature or they will not develop. Most birds use their body heat to keep the eggs warm. They sit on the eggs, which can support the adult birds' weight because the shells are hard. This process is called **incubation**. When the eggs hatch, the young birds are not yet able to fly. They must be cared for until they can meet their own needs.

These 3- to 8-day-old tanagers are being fed by both of their parents.

Most birds take care of their offspring.

In some bird species, male and female mates care for their offspring together. This is often the case for birds, such as the tanagers in the top photograph, whose young hatch before their eyes open or their feathers grow. It takes two adult birds to provide them with enough food, warmth, and protection.

The offspring of some species, like the ducks in the bottom photograph, hatch at a later state of development, with open eyes. They are already covered with down feathers and able to walk. In such species it's common for just one adult, usually the female, to incubate the eggs and care for the young.

These 7-day-old ducklings are able to find their own food while their mother watches over them.

5.3 Review

KEY CONCEPTS

1. What does an endothermic animal need to do in order to generate body heat?
2. What are two types of feathers, and what are their functions?
3. Describe three adaptations that make flight possible for birds.

CRITICAL THINKING

4. **Compare and Contrast** How is reproduction in birds similar to reproduction in reptiles? How does it different?
5. **Synthesize** Most flightless birds live on islands or in remote places where there are few predators. Explain why, in such an environment, flying birds' bodies might have adapted and become flightless.

CHALLENGE

6. **Hypothesize** Many species of birds begin migration while food is still plentiful in their current environment. Understanding what triggers migration in birds is an active area of scientific research. Develop a hypothesis about migration triggers in one species of bird. Describe how to test your hypothesis.

ANSWERS

1. Endotherms need to consume food to produce body heat.

2. Down feathers provide insulation. Contour feathers protect the body.

3. Sample answers: feathers; fused bones and hollow bones; air sacs, crop and gizzard; and keen sight and hearing.

4. Both birds and reptiles lay eggs with shells. Bird eggs have hard shells while the shells of reptiles' eggs are flexible. Most reptiles do not incubate their eggs.

5. Flight requires a great deal of energy. If food was plentiful and there was little threat of predation, the birds may not need to fly.

6. Accept all reasonable hypotheses and tests.

Teacher Demo

To demonstrate the strength of eggs, bring in half of an egg carton and place four raw eggs in the corners. Ask students to predict what might happen if you place a heavy book on top of the eggs. The eggs should be strong enough to support a heavy dictionary.

Reinforce the BIG idea

Have students relate the section to the Big Idea.

R Reinforcing Key Concepts, p. 321

5.3 ASSESS & RETEACH

Assess

A Section 5.3 Quiz, p. 85

Reteach

Have students describe how each of these structures is an adaptation for flight.

- hollow bones
- air sacs
- contour features

Have students describe how each of these structures is an adaptation for a bird to live in different environments.

- down feathers
- wings
- hard-shelled eggs

Technology Resources

Have students visit **ClassZone.com** for reteaching of Key Concepts.

CONTENT REVIEW

CONTENT REVIEW CD-ROM

CHAPTER INVESTIGATION

Focus

PURPOSE To determine how different bird beaks are adapted for gathering and eating different types of food

OVERVIEW Students will model bird beaks using everyday tools. They will:

- simulate how the different types of beaks are adapted for gathering food.
- make inferences about factors that influence how a bird meets its needs.

Lab Preparation

- Include tools that model beaks suited for significantly different foods, and "food" that can be collected with varying degrees of success.
- Prior to the investigation have students read through the investigation and prepare their data tables. Or you may wish to copy and distribute datasheets and rubrics.

UNIT RESOURCE BOOK, pp. 340–348

SCIENCE TOOLKIT, F15

Lab Management

Students can work in groups and rotate from station to station. Set up the feeding stations as follows:

- A rack of test tubes containing water simulates nectar in flowers.
- Dried pasta represents the diet of an omnivorous bird. Add "non-food" material such as rice that will filter through the slotted spoon.
- Millet seeds, spread in a thin layer over a flat surface, represent small seeds or insects consumed by a small bird. To make this task more challenging, bury the seeds under a layer of sand or dirt.
- Fill the jar halfway with rubber bands to simulate food for robins that pluck worms from beneath the ground.

INCLUSION Some students may have difficulty manipulating tools. Have them record observations while a partner performs the activities. Alternatively, station them at an area that is manageable.

CHAPTER INVESTIGATION

Bird Beak Adaptations

OVERVIEW AND PURPOSE The beaks of most birds are adapted for eating particular types of food. In this lab you will

- use models to simulate how different types of beaks function
- infer how the shape of a bird's beak might affect the food a bird eats

Question

Write It Up

For this lab, you will use a tool as a model of a specialized bird beak. You will try to obtain food with your beak at several different feeding stations. Each station will have a different type of food. Examine the tools that will be used for this lab, as well as the feeding stations. Then write a question about how beak shape might affect the food a bird eats. Your question should begin with *Which, How, When, Why,* or *What*. Keep in mind that you will be asked to answer this question at the end of the lab.

MATERIALS
- tweezers
- eyedropper
- slotted spoon
- pliers
- test tubes in rack
- water
- dried pasta
- millet seeds
- jar of rubber bands
- empty containers
- stopwatch

Procedure

1. Make a data table like the one shown on page 181.
2. Complete the title of your data table by writing in the name of the tool you will be using.
3. Before you start collecting food at one of the feeding stations, write a prediction in your data table. Predict how well you think the tool will function and why.

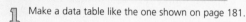

INVESTIGATION RESOURCES

CHAPTER INVESTIGATION, Bird Beak Adaptations
- Level A, pp. 340–343
- Level B, pp. 344–347
- Level C, p. 348

Advanced students should complete Levels B & C.

Writing a Lab Report, D12–13

Technology Resources

Customize this student lab as needed or look for an alternative. Print rubrics to assess student lab reports.

Lab Generator CD-ROM

4 See how much food you can collect in one minute. To collect food, you must move it from the feeding station into a different container, and you may only touch the food with the tool.

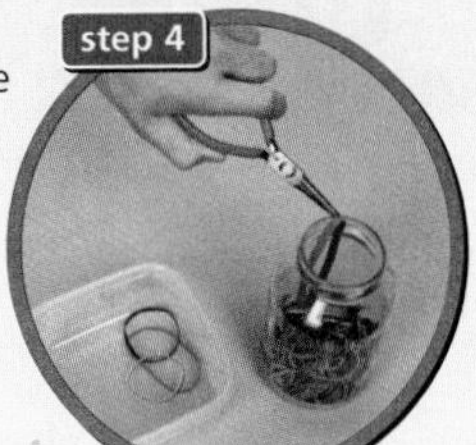

5 Describe your results in your data table.

6 Return the food you collected to the feeding station. Try to make it look just like it did when you started.

7 Repeat steps 3–6 at each of the other feeding stations.

Observe and Analyze (Write It Up)

1. **INTERPRET DATA** At which feeding station did you have the best results? Why? Explain your answer.
2. **EVALUATE** How accurate were the predictions you made?
3. **APPLY** If you could visit each feeding station again, how would you change the way you used the tool to collect food? Explain your answer.

Conclude (Write It Up)

1. **INFER** What answers do you have for the question you wrote at the beginning of this lab?
2. **INFER** In what ways do you think the experience you had during this lab is similar to the ways real birds obtain food?
3. **IDENTIFY LIMITATIONS** What unexpected factors or problems might have affected your results?
4. **SYNTHESIZE** What environmental and physical factors can a bird actually control when it is getting food?
5. **APPLY** Examine the beaks of the birds in the photographs below. Write a brief description of the shape of each bird's beak. Then, for each bird, name one type of food its beak might be suited for and one type of food each bird would probably not be able to eat.

INVESTIGATE Further

CHALLENGE How are other parts of birds' bodies specialized for the environments where they live? Investigate the feet of the following birds: ostrich, heron, woodpecker, pelican, and owl.

Bird Beak Adaptations

Question:

Table 1. Collecting Food with________

Station	Prediction	Results
Water in test tubes		
Dried pasta		
Millet seeds		
Rubber bands in jar		

Observe and Analyze (Write It Up)

1. *The eyedropper will be the best tool for gathering water from the test tubes. The forceps, spoon, and tweezers can collect this food from the pasta bowls in varying amounts. Tweezers will be most successful at gathering millet seed. The only tool that should easily reach the rubber bands is the forceps.*
2. *Accuracy will depend on original predictions. Answers should cite results as evidence of accuracy.*
3. *Accept all reasonable improvements upon feeding techniques.*

Conclude (Write It Up)

1. *Answers depend on students' original questions. Check that they correlate.*
2. *Real birds have beaks adapted to allow them to get certain types of food successfully. A bird with a small, pointy beak cannot get much nourishment from an acorn.*
3. *Students might suggest different tools, designs for different feeding stations, or varying sizes and metabolic rates of birds.*
4. *A bird can choose the territory in which it searches for and obtains food. It can decide when to eat, and when to do other things. It can probably learn how to be a more effective food-obtainer by improving upon its "technique" for opening seeds, locating worms or insects.*
5. *Pictured birds are* (left to right) *hummingbird, parrot, and robin. Sample answer: hummingbird: needle shape, nectar; parrot: hook shape, hard seeds; robin: dart shape, worms, insects, small seeds*

INVESTIGATE Further

CHALLENGE Students should find that birds' feet are different sizes and shapes according to the environments in which they live and the needs of the birds. They may be adapted for running, wading in water, grabbing onto the sides of trees, and grabbing prey.

Post-Lab Discussion

- On the board, keep a tally of results. Students can infer why one tool worked better than others at the various stations.
- Ask students to describe the features of each "beak." Have them describe how each model bird beak compares with the real beak. *Sample answer: The eyedropper represents a beak that is long and narrow. This type of beak would feed from nectar in long, slender flowers.*
- Discuss any questions raised by the lab. How might students investigate or observe to find answers?

5.4 FOCUS

Set Learning Goals

Students will

- Recognize that mammals are endotherms.
- Describe the diversity of adaptations that are found in mammals.
- Explain that mammals produce milk, which is food for their young.
- Make models to show how fat keeps a mammal warm.

3-Minute Warm-Up

Display Transparency 37 or copy this exercise on the board:

Complete the Venn diagram to show how reptiles and birds are similar in the intersecting circles. In the outer circles, describe the features of each that differ.

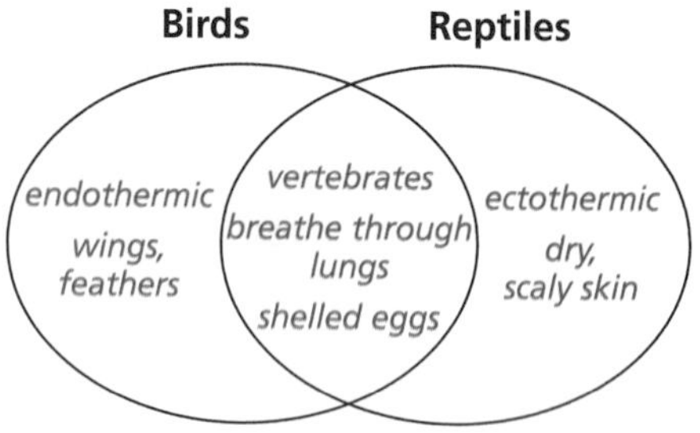

3-Minute Warm-Up, p. T37

5.4 MOTIVATE

THINK ABOUT

PURPOSE To examine the diversity of mammals

DISCUSS Ask students what characteristics a tiny shrew, an elephant, and a polar bear have that make them suitable for different environments. *Shrews are very small and can live in small cracks and crevasses and escape the notice of predators; elephants have a trunk that can be used to drink large amounts of water and to grip things; polar bears have thick fur to keep them warm.*

KEY CONCEPT

5.4 Mammals live in many environments.

BEFORE, you learned

- Endotherms can stay active in cold environments
- Many bird adaptations are related to flight
- Birds lay hard-shelled eggs and usually take care of their young

NOW, you will learn

- About mammals as endotherms
- About the diversity of adaptations in mammals
- That mammals produce milk, which is food for their young

VOCABULARY

mammal p. 183
placenta p. 186
gestation p. 186

THINK ABOUT

How diverse are mammals?

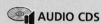

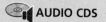

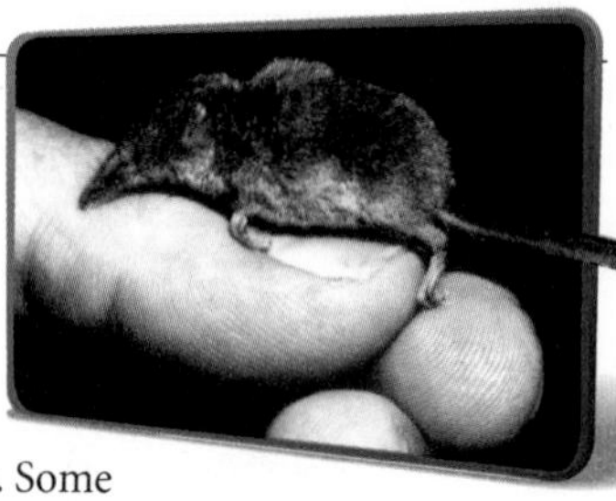

Mammals have adapted to survive in many environments and come in many shapes and sizes. Whales live in the ocean, and goats may live near mountain peaks. Some monkeys live in tropical forests, and polar bears survive in frozen areas. An elephant might not fit in your classroom, but the tiny shrew shown here could fit on your finger. As you read this chapter, think about the characteristics of mammals that help them to survive in such a variety of ways.

NOTETAKING STRATEGY
Using a strategy of your choice, take notes on the idea that mammals are a diverse group. Be sure to include mammal characteristics.

Mammals are a diverse group.

The group of vertebrates called mammals includes many familiar animals. Mice are mammals, and so are cows, elephants, and chimpanzees. You are a mammal, too. Bats are mammals that can fly. Some mammals, including whales, live in water.

Some mammal species, such as raccoons and skunks, have adapted to live in many sorts of environments, including cities. Others, such as cheetahs and polar bears, have adaptations for meeting their needs in just a few environments.

Mammals are a diverse animal group. Although there are less than 5000 species of mammals on Earth, mammal species come in many shapes and sizes, and have many different ways of moving, finding

C 182 Unit: Diversity of Living Things

RESOURCES FOR DIFFERENTIATED INSTRUCTION

Below Level

UNIT RESOURCE BOOK
- Reading Study Guide A, pp. 324–325
- Decoding Support, p. 337

AUDIO CDS

Advanced

UNIT RESOURCE BOOK
- Challenge and Extension, p. 330
- Challenge Reading, pp. 333–334

English Learners

UNIT RESOURCE BOOK
Spanish Reading Study Guide, pp. 328–329

AUDIO CDS
- Audio Readings in Spanish
- Audio Readings (English)

food, and eating. These are some of the characteristics that distinguish **mammals** from other animals:

- All mammals have hair during some part of their lives.
- Most mammal species have teeth specialized for consuming particular kinds of food.
- All mammal species produce milk, with which they feed their young.

RESOURCE CENTER
CLASSZONE.COM
Learn more about mammals.

Mammals are endotherms.

You have learned that endothermic animals are able to stay active in cold environments. This is because endotherms maintain a constant body temperature. Mammals are endotherms. This means that they use some of the food they consume to generate body heat. Mammals also have adaptations for controlling body temperature.

Hair

Many species of mammals have bodies covered with hair. Like birds' feathers, hair is an adaptation that allows mammals to have some control over the warmth or coldness of their bodies. Mammals that live in cold regions, like polar bears, have hair that keeps them warm. Desert mammals, such as camels, have hair that protects them from extreme heat.

Most mammals have at least two types of hair. Soft, fluffy underhairs keep heat close to their bodies, like the down feathers of birds. Water-resistant guard hairs cover the underhairs and give the animal's fur its color.

Some species of mammals also have specialized hairs. A specialized structure is one that performs a particular function. For example, whiskers are sensory hairs that are part of an animal's sense of touch. Porcupines' quills are hairs that function in self defense.

CHECK YOUR READING What are three functions of hair?

Colored guard hairs give this tiger its stripes. Its whiskers are specialized sensory hairs. It also has underhairs, which you cannot see.

This porcupine's quills are specialized guard hairs.

DIFFERENTIATE INSTRUCTION

More Reading Support

A How do all mammals feed their young? *They produce milk.*

B How are some mammals' hairs like down feathers? *They keep heat close to the body.*

English Learners Place the terms *mammal, placenta,* and *gestation* on a classroom Science Word Wall with brief definitions of each. Use the chapter summary on p. 188 to help students focus on key concepts and vocabulary.

5.4 INSTRUCT

Address Misconceptions

IDENTIFY Ask students what characteristics they use to identify a vertebrate. Students might hold the misconception that animal is synonymous with vertebrate, more specifically mammal. Mammals are often the prototypical animal followed by birds, reptiles, and fish in descending order of usage.

CORRECT Construct a bulletin board with the main heading *Vertebrates* and subheadings *Backbones, Ectothermic, Endothermic.* Have students put the names of vertebrate species on scraps of paper and tack them under the appropriate headings.

REASSESS Based on the bulletin board activity above, ask students to re-evaluate their criteria for characteristics of vertebrates.

Technology Resources

Visit **ClassZone.com** for background on common student misconceptions.

MISCONCEPTION DATABASE

Teach from Visuals

To help students interpret the photographs showing animal hair, ask:

What are the specialized hairs found on these mammals? *whiskers; quills*

What are the hairs' functions? *sense of touch; self defense*

Ongoing Assessment

Recognize that mammals are endotherms.

Ask: What adaptation enables mammals to survive in cold climates? *Their bodies are insulated and they are endothermic.*

CHECK YOUR READING *Sample answers: Hair controls heat loss (keeps heat in or lets it out); hair can be water-resistant; it provides color for camouflage; hairs can be sensory structures such as whiskers, or used for defense such as quills.*

Teach Difficult Concepts

Many students think that any animal living in water is a fish, including whales, dolphins, and other marine mammals. Ask students to describe the characteristics of mammals and compare them to those of fish. Then ask them to make a two-column chart of marine animals, with headings "Breathe with Gills" and "Breathe with Lungs." (The "Breathe with Lungs" list may include reptiles, amphibians, and mammals.) Remind students that animals are classified according to body characteristics, DNA evidence, and fossil histories, not where the animals live.

INVESTIGATE How Body Fat Insulates

PURPOSE To make a model of how well fat keeps a mammal warm

TIPS *15 min.* The "blubber gloves" can be made (following procedure step 1) prior to the lesson and reused. You may want to spread newspaper on work surfaces to control spills.

WHAT DO YOU THINK? *Students should find that the hand without the glove gets cold more quickly than the hand in the glove.*

CHALLENGE *A mammal that lives in a hot environment would not be as likely to have a thick layer of body fat. The animal would most likely have very little body fat.*

R Datasheet, How Body Fat Insulates, p. 331

Technology Resources

Customize this student lab as needed or look for an alternative. Print rubrics to assess student lab reports.

Lab Generator CD-ROM

Ongoing Assessment

CHECK YOUR READING *Answer: Body fat insulates mammals and stores energy obtained from food.*

Body Fat

? C

Some mammal species that live in water, such as dolphins, have very little hair. These mammals have a layer of fat, called blubber, that plays an important role in maintaining body temperature. The blubber is located between the animal's skin and muscles, and provides its organs with insulation from heat and cold.

Body fat can also be a storage place for energy. When a mammal consumes more food than it needs, the extra energy may be stored in its fat cells. Later, if the animal needs energy but cannot find food, it can use the energy stored in the fat.

For example, animals that hibernate, such as woodchucks, may eat a lot more than they need to survive at times when plenty of food is available. This makes them fat. Then, while they are hibernating, they do not have to eat, because their body fat provides them with the energy they need.

CHECK YOUR READING What are two ways that body fat functions in mammals?

INVESTIGATE How Body Fat Insulates

How well does fat keep a mammal warm?

A layer of body fat between the muscles and the skin allows some mammals to survive in very cold places. In this investigation, you will experience how well that adaptation works by making a blubber glove model.

PROCEDURE

1. Half fill one large, zip-lip plastic bag with vegetable shortening. Turn another bag inside out and place it inside the bag with the shortening. Zip the edges of the two bags together so that the shortening is sealed between them.
2. Place one hand inside this "blubber glove." Then submerge both your gloved hand and your free hand in a tub of ice water. Which hand stays warmer? For how long? Record your observations.

WHAT DO YOU THINK?

How well did the layer of fat in your blubber glove insulate your hand?

CHALLENGE Would you expect an animal that lives in a very hot environment to have a thick layer of body fat?

SKILL FOCUS
Making models

MATERIALS
- 2 zip-lip plastic bags, 1/2-gallon size
- can of vegetable shortening
- bowl of ice water

TIME
15 minutes

DIFFERENTIATE INSTRUCTION

More Reading Support

C Where is a mammal's blubber located? *between skin and muscle*

Below Level Invite interested students to find out more about marine mammals. Students should make a chart listing the characteristics that help to classify these animals as mammals. Alternatively, they might draw the animals and label the characteristics, as in the diagram of the dolphin, p. 185.

C 184 Unit: Diversity of Living Things

Mammal Adaptations for a Water Environment

A dolphin's blowhole, flippers, tail flukes, and blubber are features that allow it to meet its needs in water.

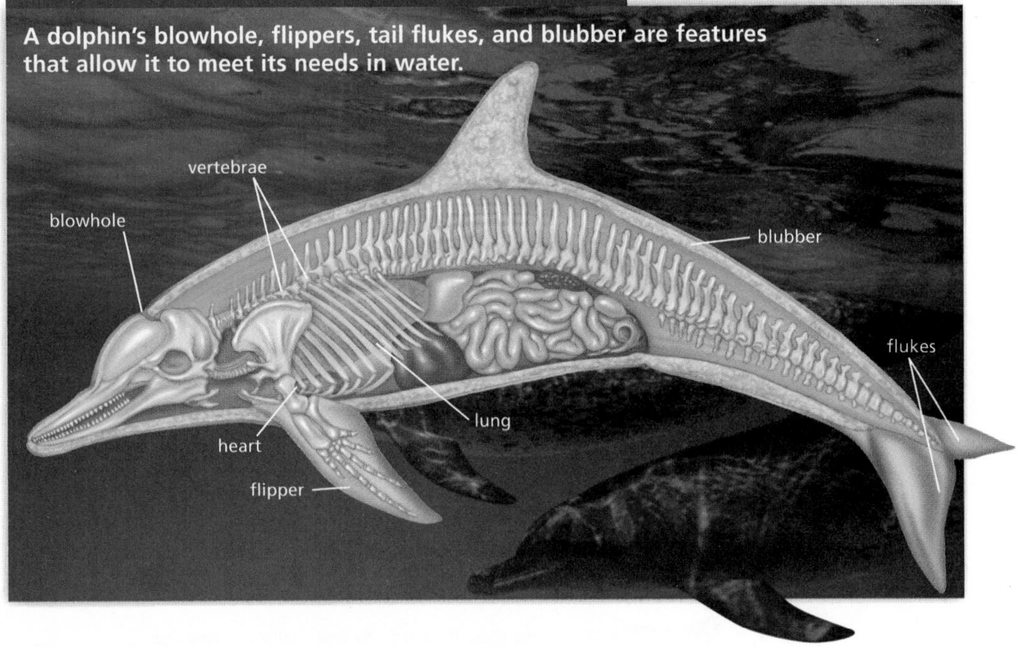

Mammals have adapted to many environments.

Scientists think mammals appeared on Earth about 200 million years ago. Fossil evidence suggests that the first mammals were small land vertebrates with four limbs, a tail, and specialized teeth. They probably had fur and were most active at night.

Over millions of years, those early mammals adapted to live in many different environments and became the diverse group of species they are today. Moles, for example, live almost entirely underground. They have strong limbs for digging and organs specialized for sensing invertebrate prey in the dark. Spider monkeys, on the other hand, live mostly in trees. A spider monkey can use all four of its long limbs and its flexible, grasping tail to move through a forest without touching the ground.

The first mammals lived on land, but over time some species adapted to live in watery environments. Some of them, such as otters and walruses, live mostly in water but can also be found on land. Others, such as dolphins and whales, have bodies so completely adapted for life in the water that they no longer have a way of moving on land. If you look carefully at the diagram above, however, you will see that a dolphin's body plan differs in some important ways from that of a fish.

Teach from Visuals

To help students interpret the illustration of mammal adaptations for water, ask:

- Which features of marine mammals enable them to live in water? *blowhole, flippers, flukes, blubber*
- What is the purpose of the blowhole? *It allows the mammal to come to the surface of the water and breathe.*
- What do the bones inside the flipper resemble in other mammals? *forelimbs and fingers*

Integrate the Sciences

The first mammals appeared during the Triassic Period more than 190 million years ago. These early mammals evolved from mammal-like reptiles that dominated Earth about 280 million years ago during the Permian Period before dinosaurs appeared on Earth. The early mammals were as small as mice and most likely fed on insects. Mammals persisted during the reign of dinosaurs most likely because they were nocturnal, small, and inconspicuous. When dinosaurs became extinct about 65 million years ago, mammals flourished, due to lack of competition from the dinosaurs and the many adaptations that they had evolved over time. The Cenozoic Era, from 65 million years ago to the present is considered the Age of Mammals.

EXPLORE (the BIG idea)

Revisit "Internet Activity: Where in the World?" on p. 155. Ask students to compare the list of adaptations from the activity with those they have read about concerning mammals.

DIFFERENTIATE INSTRUCTION

More Reading Support

D When did mammals appear on Earth? *about 200 million years ago*

E Which land mammals spend most of their time in the water? *otters and walruses*

Advanced Have students change one feature of the skeleton of a mammal to design a fictional mammal species that is well suited to live in a real or fictional environment. Students should create a poster showing both the original and altered skeletons. On the altered skeleton, students should extrapolate the animal's behaviors or capabilities in relation to its environment.

R
- Challenge and Extension, p. 330
- Challenge Reading, pp. 333–334

Develop Critical Thinking

INFER Point out that the major difference among the three groups of mammals is the way in which they reproduce and give birth. Ask: What advantage do mammals with young that develop with help from placentas have over mammals that lay eggs and have young develop in pouches? *Eggs are less likely to be eaten by predators. Marsupial young might not make it into their mother's pouch right after they are born. Placental mammals are more developed when born.*

Teach from Visuals

To help students interpret the graph showing gestation, point out that this is a bar graph that is used to compare sets of data. Ask:

Which animal has the longest period of gestation? *The elephant—84–88 weeks.* How do you know? *The bar for elephant is the longest on the graph.*

History of Science

Domestication of mammals occurred gradually, perhaps before the development of agriculture. Domestication of goats and sheep occurred around 7000 B.C. Cattle and pigs followed around 4000 B.C. to 3000 B.C. The dog was most likely the first pet, dating back almost 10,000 years ago. Cats, revered in ancient Egypt, most likely appeared as pets 5000 years ago.

Ongoing Assessment

CHECK YOUR READING *Answer: inside the mother's body*

Mammals have reproductive adaptations.

Mammals reproduce sexually. Before a mammal can produce offspring, it finds a mate. Some mammals, such as lions, live in groups that include both males and females. However, most mammals live alone most of the time. They find a mate when they are ready to reproduce. Some mammal species breed only at certain times of the year. Others can reproduce throughout the year.

Development Before Birth

VOCABULARY Remember to make a word magnet for the terms *placenta* and *gestation.*

F Fertilization occurs internally in mammals. In almost every species of mammal, the offspring develop inside the female's body. Many mammals have a special organ called a **placenta** that transports nutrients, water, and oxygen from the mother's blood to the developing embryo. The embryo's waste materials leave through the placenta and are transported out of the mother's body along with her waste.

G The time when a mammal is developing inside its mother is called **gestation.** As you can see in the diagram, the length of the gestation period is different for different species. Gestation ends when the young animal's body has grown and developed enough for it to survive outside the mother. Then it is born.

CHECK YOUR READING Where do the offspring of most mammals develop before they are born?

Length of Gestation

The length of gestation varies greatly in placental mammals.

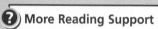

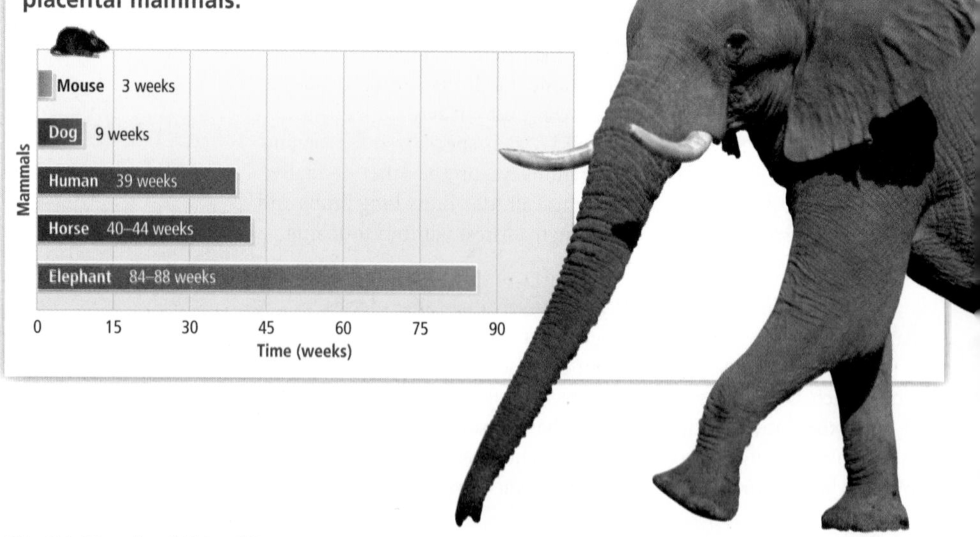

DIFFERENTIATE INSTRUCTION

More Reading Support

F What is the organ that provides a mammalian embryo with nutrients? *the placenta*

G When does gestation end? *when the mammal is born*

Below Level Invite interested students to find out about the gestation periods of other mammals and create a bar graph displaying their findings.

C 186 Unit: Diversity of Living Things

Not all mammals fully develop inside their mothers. The duck-billed platypus and two species of spiny anteaters have young that hatch from eggs. Mammals in a group called the marsupials (mahr-SOO-pee-uhlz), which includes kangaroos, develop inside the mother at first but are born when they are still extremely small. Right after birth, a young marsupial climbs across its mother to a special pouch on the outside of her body. It completes its development there. Only one marsupial, the opossum, lives in North America.

Raising Young

Milk, a high-energy liquid food full of proteins, fats, sugars, and other nutrients, is the first food young mammals consume. Each species' milk has a different combination of these ingredients.

Mammals' bodies have special glands for producing milk, called mammary glands. In almost all mammal species, only females' mammary glands function. This means that only female mammals can feed the young with milk. In most mammal species, females gestate and care for the offspring alone. However, in some mammal species, the male helps raise the young.

This dog is feeding her puppies with milk.

Different species of mammals are born at varying stages of development. Most mice are helpless, blind, and naked at birth, while giraffes can walk soon after they are born. The length of time a young mammal needs care from an adult varies. Some seals nurse their young for less than a month before leaving them to survive on their own. Some whales live alongside their young for a much longer time. However, humans may be the mammal species that takes care of their offspring for the longest time of all.

5.4 Review

KEY CONCEPTS

1. Why is a bat classified as a mammal instead of a bird? Why are whales classified as mammals and not fish?
2. Name two adaptations that allow mammals to have control over body temperature.
3. How does the way that mammals feed their young differ from the ways other animals feed their young?

CRITICAL THINKING

4. **Apply** One day on your way home from school, you see an animal that you have never seen before. What clues would you look for to tell you what type of animal it is?
5. **Synthesize** Make a Venn diagram that shows how the vertebrate animal groups you learned about in this chapter are similar and different.

CHALLENGE

6. **Evaluate** This section begins with the statement "mammals are a diverse group." Explain what this means and why it is a true statement. How does the diversity of mammals compare to the diversity of other living things, such as bacteria, plants, arthropods, or fish?

ANSWERS

1. Bats and whales are classified as mammals because they have body hair, body fat, and mammary glands.

2. hair and body fat

3. Young mammals consume milk that is produced by female mammals' bodies.

4. Student answers should reflect knowledge that physical features and behavior can be used to identify animals.

5. The best diagrams will show overlaps between the groups clearly and accurately.

6. Mammals have larger variety in size, shape, behavior, and environment than any other type of organism. There are more species of arthropods, fish, etc., than mammals; however, they are limited to certain environments and body types.

Ongoing Assessment

Explain that mammals produce milk which is food for their young.

Ask: How does producing milk help mammals to care for their young? *Milk is a high-energy food which is easily taken in and digested by newborns. It provides young mammals with the nutrients they need.*

EXPLORE the BIG idea

Revisit "What Animals Live Near You?" on p. 155. Ask students if they would like to add to or revise their lists.

Reinforce the BIG idea

Have students relate the section to the Big Idea.

R Reinforcing Key Concepts, p. 332

5.4 ASSESS & RETEACH

Assess

A Section 5.4 Quiz, p. 86

Reteach

Ask students to describe each of the following characteristics of mammals and explain how each allows mammals to live in different environments.

- hair
- blubber
- specialized teeth
- placenta

Technology Resources

Have students visit **ClassZone.com** for reteaching of Key Concepts.

CONTENT REVIEW

CONTENT REVIEW CD-ROM

BACK TO

the BIG idea

Have students choose one of the vertebrate groups in the chapter and describe how the animals in this group are adapted to live in a variety of environments. *Students should give specific examples of body traits and characteristics.*

KEY CONCEPTS SUMMARY

SECTION 5.1
Ask students to explain why the endoskeleton is an important adaptation for vertebrate animals. *The endoskeleton provides support and grows as the animal grows, allowing vertebrates to become larger in size than invertebrates.*

SECTION 5.2
Which structures of amphibians and reptiles allow them to live on land? Explain. *They have lungs for breathing, limbs for movement; reptiles have eggs with shells and yolks that allow the enclosed embryos to develop on land.*

SECTION 5.3
How are bird bodies adapted for flight? *They have wings, contour feathers, hollow bones, and air sacs to increase the amount of oxygen in the body.*

SECTION 5.4
What are three ways in which mammals bear their young? *Monotremes lay eggs; marsupials give birth to live young that develop in pouches; the young of placental mammals develop fully inside placentas until they are born.*

Review Concepts

- Big Idea Flow Chart, p. T33
- Chapter Outline, pp. T39–T40

5 Chapter Review

the BIG idea

Vertebrate animals live in most of Earth's environments.

CONTENT REVIEW
CLASSZONE.COM

KEY CONCEPTS SUMMARY

5.1 Vertebrates are animals with endoskeletons.

All **vertebrate** animals have an **endoskeleton** which includes vertebrae, or backbones.

Most vertebrates are fish. Fish are adapted for life in a water environment.

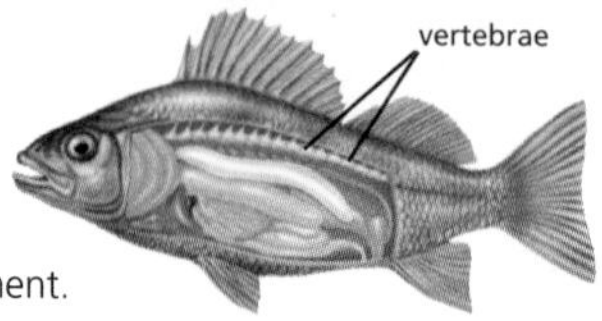

VOCABULARY
vertebrate p. 157
endoskeleton p. 157
scale p. 161

5.2 Amphibians and reptiles are adapted for life on land.

Amphibians and **reptiles** have adaptations for moving, getting food, and breathing on land.

Most amphibians live in moist places.

Many reptiles live in hot, dry places.

VOCABULARY
ectotherm p. 170
amphibian p. 167
reptile p. 168

5.3 Birds meet their needs on land, in water, and in the air.

Birds have adaptations that allow them to survive in many environments. Many of the features and behaviors of birds relate to flight.

VOCABULARY
endotherm p. 174
incubation p. 179

Mammals live in many environments.

Mammal species live in many environments. Mammals have adaptations that allow them to survive in cold places. They also have distinctive reproductive adaptations.

VOCABULARY
mammal p. 183
placenta p. 186
gestation p. 186

Technology Resources

Have students visit **ClassZone.com** or use the CD-ROM for a cumulative review of concepts.

CONTENT REVIEW

CONTENT REVIEW CD-ROM

Engage students in a whole-class interactive review of Key Concepts. Edit content as you wish.

POWER PRESENTATIONS

C 188 Unit: Diversity of Living Things

Reviewing Vocabulary

Place each vocabulary term listed below at the center of a description wheel diagram. Write some words describing it on the spokes.

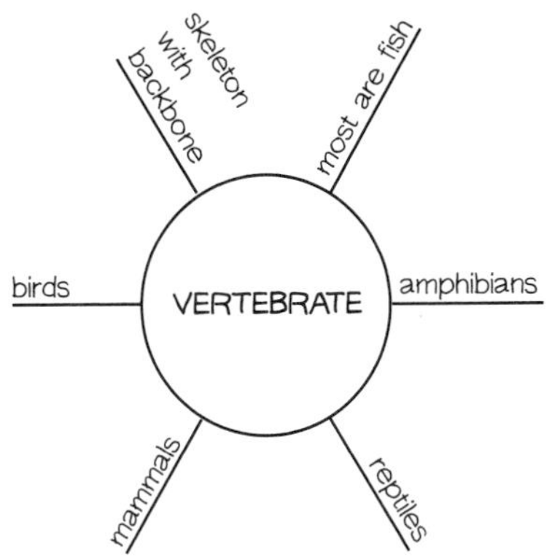

1. endotherm
2. migration
3. ectotherm
4. scale
5. incubation
6. gestation

Reviewing Key Concepts

Multiple Choice *Choose the letter of the best response.*

7. The vertebrate endoskeleton differs from an arthopod's exoskeleton because it
- **a.** supports muscles
- **b.** protects organs
- **c.** is inside the animal's body
- **d.** has joints

8. Most vertebrate species are
- **a.** fish
- **b.** amphibians
- **c.** birds
- **d.** mammals

9. Fish obtain oxygen from water with structures called
- **a.** vertebrae
- **b.** fins
- **c.** lungs
- **d.** gills

10. Down feathers provide insulation for a bird's body because they
- **a.** trap warm air between them
- **b.** are waterproof
- **c.** overlap each other
- **d.** are slightly oily

11. Which organ do both fish and amphibians have?
- **a.** lateral line
- **b.** lungs
- **c.** gills
- **d.** scaly skin

12. Mammals differ from other vertebrates because mammals can
- **a.** lay eggs
- **b.** produce milk
- **c.** are endotherms
- **d.** are able to swim

13. The two types of vertebrates that lay their eggs in water are
- **a.** amphibians and birds
- **b.** reptiles and fish
- **c.** fish and amphibians
- **d.** mollusks and fish

14. Which statement is true?
- **a.** All birds are ectotherms.
- **b.** All birds can fly.
- **c.** All birds have two wings and two legs.
- **d.** All birds have teeth and beaks.

15. The organ that transports materials between a female mammal and the offspring developing inside her body is called a
- **a.** yolk
- **b.** placenta
- **c.** blubber
- **d.** vertebrae

Short Answer *Write a short answer to the questions below.*

16. Describe how you would determine if an animal was a salamander or a lizard.

17. Why do bird eggs have to be incubated?

Reviewing Vocabulary

1. Sample answer: bird, mammal, survives cold, constant temperature, internal source of heat, needs lots of energy

2. Sample answer: birds, Monarch butterflies, seasonal, winter home, summer home, warmer weather

3. Sample answer: reptiles, amphibians, fish, changing temperature, can't survive cold, needs less energy

4. Sample answer: overlapping, bird's feet, reptile's skin, fish, bony fish, keratin

5. Sample answer: birds, monotremes, hard shells, marsupials, keeps warm, endotherms

6. Sample answer: mammals, live young, inside mother, mice: 3 weeks, elephants: 88 weeks, humans: 39 weeks

Reviewing Key Concepts

7. c

8. a

9. d

10. a

11. c

12. b

13. c

14. c

15. b

16. A salamander has smooth, moist, skin. A lizard has dry, scaly skin.

17. Bird eggs must be incubated because their body temperature must be maintained and kept warm by the parent bird.

ASSESSMENT RESOURCES

UNIT ASSESSMENT BOOK
- Chapter Test, Level A, pp. 87–90
- Chapter Test, Level B, pp. 91–94
- Chapter Test, Level C, pp. 95–98
- Alternative Assessment, pp. 99–100
- Unit Test, A, B, & C, pp. 101–112

SPANISH ASSESSMENT BOOK
- Spanish Chapter Test, pp. 57–60
- Spanish Unit Test, pp. 61–64

Technology Resources

Edit test items and answer choices.

Test Generator CD-ROM

Visit **ClassZone.com** to extend test practice.

Test Practice

Thinking Critically

18. The person in Mexico, because reptiles are ectotherms and more common in warmer, drier climates.

19. Polar bears should have more body fat because fat provides insulation from the cold.

20. The bird's skeleton provides support, a means of movement, and grows as the birds' body grows.

21. digestive system—has a gizzard that helps in digestion, circulatory system—heart

22. Answers may vary. Sample answers include the hypothesis that birds that do not migrate would have a more varied diet than migratory birds.

23. The salamander will be extinct by the year 2019. Divide 825 by 50 to get 16.5. Round to up 17, then add 2002 + 17. The sum is 2019.

24. a. reptile; b. amphibians; c. mammal; d. birds

the BIG idea

25. Penguins and seals are both vertebrates and endotherms. They have body coverings (feathers for penguin and hair for the seal) that help keep heat in. They burn food to create body heat, and they store excess food energy as fat.

26. Answers will vary depending on the animal that students chose.

UNIT PROJECTS

Have students present their projects. Use appropriate rubrics from the Unit Resource Book to evaluate their work.

R Unit Projects, pp. 5–10

Thinking Critically

18. PREDICT Imagine that you live in Mexico and you have a pen pal who lives in Iceland. Both of you want to know about animals in the other person's country. Which of you will be more likely to have seen wild reptiles? Why?

19. APPLY Polar bears live in an arctic environment. Jaguars live in a rain forest environment. Which animal would you expect to have more body fat? Explain.

Refer to diagram below as you answer parts of the next two questions.

20. COMPARE Birds, like all vertebrates, have internal skeletons. What functions does a bird's skeleton have in common with all other vertebrates?

21. INFER This diagram shows only some of a bird's body systems. Name two systems that are not shown here, including organs that belong to them and their functions.

22. HYPOTHESIZE Not all birds migrate. Some birds, such as pigeons and house sparrows, stay in one living place through the winter. What do you think you would find if you compared the diets of birds that migrate with those of birds that do not?

23. MATH AND SCIENCE Scientists studying an endangered species of rainforest salamanders estimated that only 875 of these animals were still living in 2002. If this salamander population decreases by 50 animals per year, in what year will it become extinct? Explain how you found your answer.

24. CLASSIFY Imagine that you move to a new place and spend a year watching nearby animals. Read each description and identify each animal described below as a fish, amphibian, reptile, bird, or mammal.

- **a.** A scaly animal warms itself on your sidewalk when the sun is out. It has dry skin, four legs, and a tail.
- **b.** Small animals swim in a pond near your home. They have no legs, but they do have tails. As they grow older, their tails shrink and they develop four limbs. Then they disappear from the pond.
- **c.** A furry animal chews a hole under your porch and seems to be living there. You later see it with smaller animals that appear to be its young.
- **d.** A pair of flying, feathered animals collect objects and carry them into an opening under the gutter of your neighbor's house. At first you see them carrying twigs and grass, but later it looks as if they are carrying worms.

the BIG idea

25. DRAW CONCLUSIONS Look again at the picture on pages 154–155. What do penguins and seals have in common? What adaptations do some species of birds and mammals have that allow them to survive in cold environments like Antarctica?

26. PROVIDE EXAMPLES Think of an example of a species from each of the vertebrate animal groups you've learned about, and describe an adaptation that suits each to the environment where it lives. Explain your answers.

UNIT PROJECTS

Evaluate all the data, results, and information from your project folder. Prepare to present your project. Be ready to answer questions posed by your classmates about your results.

MONITOR AND RETEACH

If students are having trouble applying the concepts in Chapter Review items 20 and 21, suggest that they review the diagram of the bird on p. 177. Then have students create concept maps for the following:

- structures in the body that are adapted for flight
- structures in the skin that are characteristic of all birds

Students may benefit from summarizing one or more sections of the chapter.

R Summarizing the Chapter, pp. 358–359

Standardized Test Practice

For practice on your state test, go to . . . TEST PRACTICE CLASSZONE.COM

Analyzing Diagrams

Read the text and study the diagram, and then choose the best response for the questions that follow.

Vertebrates, such as birds, fish, and mammals, have endoskeletons. This internal skeleton is made up of a system of bones that extends throughout the body. Muscles can attach directly to the bones, around joints—the place where two bones meet. As shown in the generalized diagram below, at least two muscles are needed to produce movement. One of the muscles contracts, or shortens, pulling on the bone, while the other muscle extends, or is stretched.

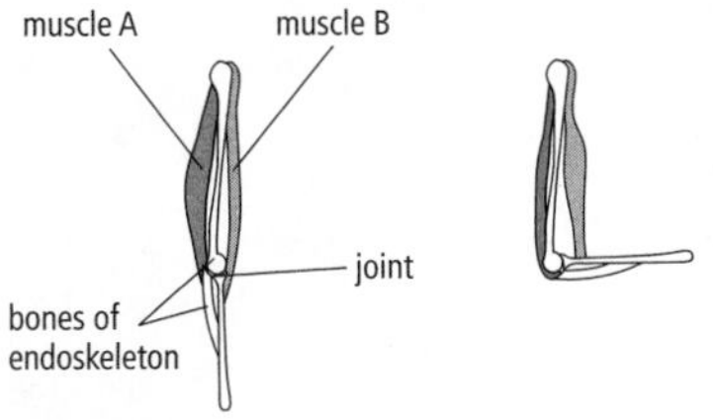

1. Endoskeletons are made up of
   - **a.** contracting muscles
   - **b.** extending muscles
   - **c.** internal bones
   - **d.** external bones
2. A muscle produces movement by
   - **a.** pulling
   - **b.** relaxing
   - **c.** extending
   - **d.** rotating
3. A joint is where
   - **a.** one muscle connects to another
   - **b.** one bone connects to another
   - **c.** a muscle connects to a bone
   - **d.** the bone ends
4. The leg in the diagram bends when
   - **a.** muscle A contracts
   - **b.** muscle A relaxes
   - **c.** muscle B contracts
   - **d.** muscles A and B contract
5. What is the main point of the diagram and the text above?
   - **a.** Two muscles are needed to produce movement in vertebrates.
   - **b.** Vertebrates have internal skeletons.
   - **c.** When one muscle shortens, another muscle gets longer.
   - **d.** Movement in vertebrates is the result of the interaction of muscles and bones.

Extended Response

Use terms from the word box to answer the next question. Underline each term you use in your answer.

6. Many fish go forward by moving their tails from side to side. Describe the action of the muscles on each side of the fish as the tail moves from one side to the other.

contract	pull
extend	muscle

7. Using what you know about joints and muscles and the diagram, describe what happens to muscles and bones as you bend your leg at the knee.

METACOGNITIVE ACTIVITY

Have students answer the following questions in their **Science Notebook:**

1. What are some strategies you used to learn about the similarities and differences among the various groups of vertebrates?
2. What questions do you still have about the characteristics of fish, amphibians, reptiles, birds, or mammals?
3. How did your unit project help you gain a better understanding of the vertebrates studied in this chapter?

Analyzing Diagrams

1. c *3. b* *5. d*
2. a *4. c*

Extended Response

6. RUBRIC

4 points for a response that correctly answers the question and uses the following terms accurately:

- contract
- pull
- extend
- muscle

Sample answer: The fish will have <u>muscles</u> on both sides of its body that attach to its tail. When muscles on one side <u>contract</u> they <u>pull</u> the tail to that side. When this happens the muscles on the other side are <u>extended</u>. For the tail to move to the other side, the extended muscles must contract, pulling the tail back to the other side. The muscles that were contracted are now relaxed.

3 points correctly answers the question and uses three terms accurately
2 points correctly answers the question and uses two terms accurately
1 point correctly answers the question and uses one term accurately

7. RUBRIC

4 points for a response that correctly answers the question and meets the following criteria:

- reflects that the knee is a joint
- describes accurately how the bones move
- understanding of how muscles contract to pull bones
- understanding how a pair of muscles works together

Sample answer: When the knee bends, the two bones that meet at the knee change from forming a straight line to forming an angle. The center of the angle is the joint where the two leg bones meet. One muscle contracts and forces one bone to swivel around the joint. At the same time, a second muscle relaxes to allow this movement. The second muscle must contract for the bones to move back to their original place.

3 points includes three of the above criteria
2 points includes two of the above criteria
1 point includes one of the above criteria

Student Resource Handbooks

Scientific Thinking Handbook

Making Observations

An **observation** is an act of noting and recording an event, characteristic, behavior, or anything else detected with an instrument or with the senses.

Observations allow you to make informed hypotheses and to gather data for experiments. Careful observations often lead to ideas for new experiments. There are two categories of observations:

- **Quantitative observations** can be expressed in numbers and include records of time, temperature, mass, distance, and volume.
- **Qualitative observations** include descriptions of sights, sounds, smells, and textures.

EXAMPLE

A student dissolved 30 grams of Epsom salts in water, poured the solution into a dish, and let the dish sit out uncovered overnight. The next day, she made the following observations of the Epsom salt crystals that grew in the dish.

To determine the mass, the student found the mass of the dish before and after growing the crystals and then used subtraction to find the difference.

The student measured several crystals and calculated the mean length. (To learn how to calculate the mean of a data set, see page R36.)

Table 1. Observations of Epsom Salt Crystals

Quantitative Observations	Qualitative Observations
• mass = 30 g • mean crystal length = 0.5 cm • longest crystal length = 2 cm	• Crystals are clear. • Crystals are long, thin, and rectangular. • White crust has formed around edge of dish.

Photographs or sketches are useful for recording qualitative observations.

Epsom salt crystals

MORE ABOUT OBSERVING

- Make quantitative observations whenever possible. That way, others will know exactly what you observed and be able to compare their results with yours.
- It is always a good idea to make qualitative observations too. You never know when you might observe something unexpected.

Predicting and Hypothesizing

A **prediction** is an expectation of what will be observed or what will happen. A **hypothesis** is a tentative explanation for an observation or scientific problem that can be tested by further investigation.

EXAMPLE

Suppose you have made two paper airplanes and you wonder why one of them tends to glide farther than the other one.

1. Start by asking a question.
2. Make an educated guess. After examination, you notice that the wings of the airplane that flies farther are slightly larger than the wings of the other airplane.
3. Write a prediction based upon your educated guess, in the form of an "If . . . , then . . ." statement. Write the independent variable after the word *if,* and the dependent variable after the word *then.*
4. To make a hypothesis, explain why you think what you predicted will occur. Write the explanation after the word *because.*

1. Why does one of the paper airplanes glide farther than the other?
2. The size of an airplane's wings may affect how far the airplane will glide.
3. Prediction: If I make a paper airplane with larger wings, then the airplane will glide farther.
4. Hypothesis: If I make a paper airplane with larger wings, then the airplane will glide farther, because the additional surface area of the wing will produce more lift.

To read about independent and dependent variables, see page R30.

Notice that the part of the hypothesis after *because* adds an explanation of why the airplane will glide farther.

MORE ABOUT HYPOTHESES

- The results of an experiment cannot prove that a hypothesis is correct. Rather, the results either support or do not support the hypothesis.
- Valuable information is gained even when your hypothesis is not supported by your results. For example, it would be an important discovery to find that wing size is not related to how far an airplane glides.
- In science, a hypothesis is supported only after many scientists have conducted many experiments and produced consistent results.

Inferring

An **inference** is a logical conclusion drawn from the available evidence and prior knowledge. Inferences are often made from observations.

EXAMPLE

A student observing a set of acorns noticed something unexpected about one of them. He noticed a white, soft-bodied insect eating its way out of the acorn.

The student recorded these observations.

Observations

- There is a hole in the acorn, about 0.5 cm in diameter, where the insect crawled out.
- There is a second hole, which is about the size of a pinhole, on the other side of the acorn.
- The inside of the acorn is hollow.

Here are some inferences that can be made on the basis of the observations.

Inferences

- The insect formed from the material inside the acorn, grew to its present size, and ate its way out of the acorn.
- The insect crawled through the smaller hole, ate the inside of the acorn, grew to its present size, and ate its way out of the acorn.
- An egg was laid in the acorn through the smaller hole. The egg hatched into a larva that ate the inside of the acorn, grew to its present size, and ate its way out of the acorn.

When you make inferences, be sure to look at all of the evidence available and combine it with what you already know.

MORE ABOUT INFERENCES

Inferences depend both on observations and on the knowledge of the people making the inferences. Ancient people who did not know that organisms are produced only by similar organisms might have made an inference like the first one. A student today might look at the same observations and make the second inference. A third student might have knowledge about this particular insect and know that it is never small enough to fit through the smaller hole, leading her to the third inference.

Identifying Cause and Effect

In a **cause-and-effect relationship,** one event or characteristic is the result of another. Usually an effect follows its cause in time.

There are many examples of cause-and-effect relationships in everyday life.

Cause	Effect
Turn off a light.	Room gets dark.
Drop a glass.	Glass breaks.
Blow a whistle.	Sound is heard.

Scientists must be careful not to infer a cause-and-effect relationship just because one event happens after another event. When one event occurs after another, you cannot infer a cause-and-effect relationship on the basis of that information alone. You also cannot conclude that one event caused another if there are alternative ways to explain the second event. A scientist must demonstrate through experimentation or continued observation that an event was truly caused by another event.

EXAMPLE

Make an Observation

Suppose you have a few plants growing outside. When the weather starts getting colder, you bring one of the plants indoors. You notice that the plant you brought indoors is growing faster than the others are growing. You cannot conclude from your observation that the change in temperature was the cause of the increased plant growth, because there are alternative explanations for the observation. Some possible explanations are given below.

- The humidity indoors caused the plant to grow faster.
- The level of sunlight indoors caused the plant to grow faster.
- The indoor plant's being noticed more often and watered more often than the outdoor plants caused it to grow faster.
- The plant that was brought indoors was healthier than the other plants to begin with.

To determine which of these factors, if any, caused the indoor plant to grow faster than the outdoor plants, you would need to design and conduct an experiment.

See pages R28–R35 for information about designing experiments.

SCIENTIFIC THINKING HANDBOOK

Recognizing Bias

Television, newspapers, and the Internet are full of experts claiming to have scientific evidence to back up their claims. How do you know whether the claims are really backed up by good science?

Bias is a slanted point of view, or personal prejudice. The goal of scientists is to be as objective as possible and to base their findings on facts instead of opinions. However, bias often affects the conclusions of researchers, and it is important to learn to recognize bias.

When scientific results are reported, you should consider the source of the information as well as the information itself. It is important to critically analyze the information that you see and read.

SOURCES OF BIAS

There are several ways in which a report of scientific information may be biased. Here are some questions that you can ask yourself:

1. **Who is sponsoring the research?**

   Sometimes, the results of an investigation are biased because an organization paying for the research is looking for a specific answer. This type of bias can affect how data are gathered and interpreted.

2. **Is the research sample large enough?**

   Sometimes research does not include enough data. The larger the sample size, the more likely that the results are accurate, assuming a truly random sample.

3. **In a survey, who is answering the questions?**

   The results of a survey or poll can be biased. The people taking part in the survey may have been specifically chosen because of how they would answer. They may have the same ideas or lifestyles. A survey or poll should make use of a random sample of people.

4. **Are the people who take part in a survey biased?**

   People who take part in surveys sometimes try to answer the questions the way they think the researcher wants them to answer. Also, in surveys or polls that ask for personal information, people may be unwilling to answer questions truthfully.

SCIENTIFIC BIAS

It is also important to realize that scientists have their own biases because of the types of research they do and because of their scientific viewpoints. Two scientists may look at the same set of data and come to completely different conclusions because of these biases. However, such disagreements are not necessarily bad. In fact, a critical analysis of disagreements is often responsible for moving science forward.

R6 Student Resources

Identifying Faulty Reasoning

Faulty reasoning is wrong or incorrect thinking. It leads to mistakes and to wrong conclusions. Scientists are careful not to draw unreasonable conclusions from experimental data. Without such caution, the results of scientific investigations may be misleading.

EXAMPLE

Scientists try to make generalizations based on their data to explain as much about nature as possible. If only a small sample of data is looked at, however, a conclusion may be faulty. Suppose a scientist has studied the effects of the El Niño and La Niña weather patterns on flood damage in California from 1989 to 1995. The scientist organized the data in the bar graph below.

The scientist drew the following conclusions:

1. The La Niña weather pattern has no effect on flooding in California.
2. When neither weather pattern occurs, there is almost no flood damage.
3. A weak or moderate El Niño produces a small or moderate amount of flooding.
4. A strong El Niño produces a lot of flooding.

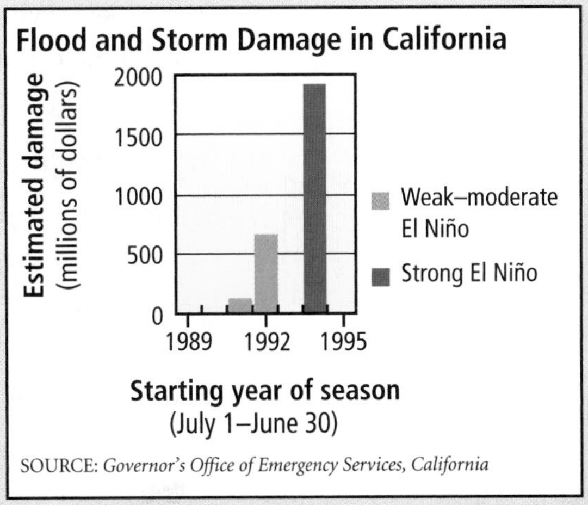

For the six-year period of the scientist's investigation, these conclusions may seem to be reasonable. However, a six-year study of weather patterns may be too small of a sample for the conclusions to be supported. Consider the following graph, which shows information that was gathered from 1949 to 1997.

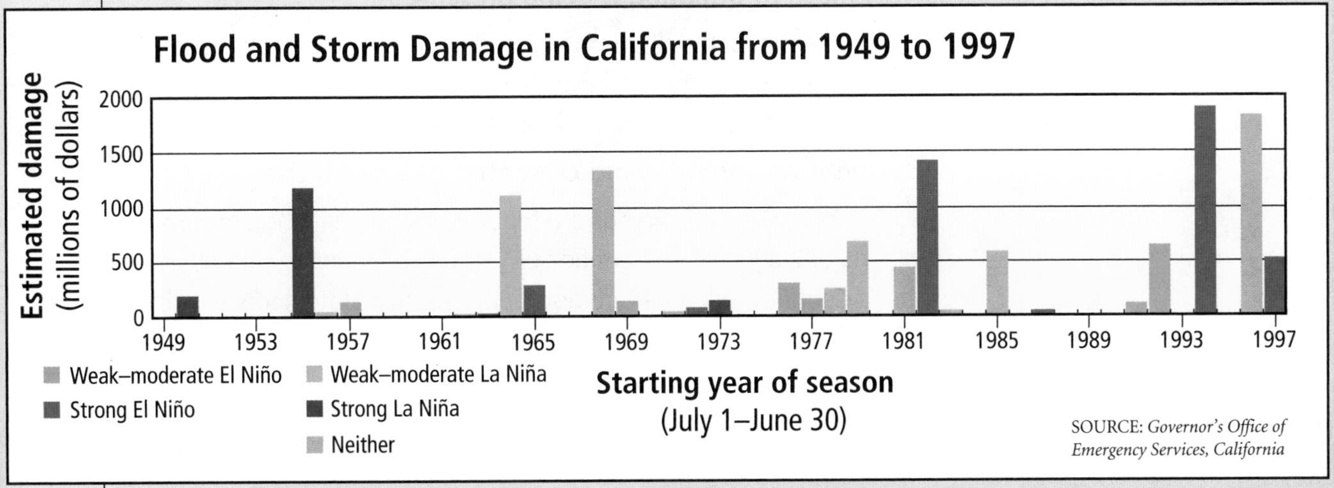

The only one of the conclusions that all of this information supports is number 3: a weak or moderate El Niño produces a small or moderate amount of flooding. By collecting more data, scientists can be more certain of their conclusions and can avoid faulty reasoning.

Analyzing Statements

To **analyze** a statement is to examine its parts carefully. Scientific findings are often reported through media such as television or the Internet. A report that is made public often focuses on only a small part of research. As a result, it is important to question the sources of information.

Evaluate Media Claims

To **evaluate** a statement is to judge it on the basis of criteria you've established. Sometimes evaluating means deciding whether a statement is true.

Reports of scientific research and findings in the media may be misleading or incomplete. When you are exposed to this information, you should ask yourself some questions so that you can make informed judgments about the information.

1. **Does the information come from a credible source?**

   Suppose you learn about a new product and it is stated that scientific evidence proves that the product works. A report from a respected news source may be more believable than an advertisement paid for by the product's manufacturer.

2. **How much evidence supports the claim?**

   Often, it may seem that there is new evidence every day of something in the world that either causes or cures an illness. However, information that is the result of several years of work by several different scientists is more credible than an advertisement that does not even cite the subjects of the experiment.

3. **How much information is being presented?**

   Science cannot solve all questions, and scientific experiments often have flaws. A report that discusses problems in a scientific study may be more believable than a report that addresses only positive experimental findings.

4. **Is scientific evidence being presented by a specific source?**

   Sometimes scientific findings are reported by people who are called experts or leaders in a scientific field. But if their names are not given or their scientific credentials are not reported, their statements may be less credible than those of recognized experts.

R8

Differentiate Between Fact and Opinion

Sometimes information is presented as a fact when it may be an opinion. When scientific conclusions are reported, it is important to recognize whether they are based on solid evidence. Again, you may find it helpful to ask yourself some questions.

1. **What is the difference between a fact and an opinion?**

   A **fact** is a piece of information that can be strictly defined and proved true. An **opinion** is a statement that expresses a belief, value, or feeling. An opinion cannot be proved true or false. For example, a person's age is a fact, but if someone is asked how old they feel, it is impossible to prove the person's answer to be true or false.

2. **Can opinions be measured?**

   Yes, opinions can be measured. In fact, surveys often ask for people's opinions on a topic. But there is no way to know whether or not an opinion is the truth.

HOW TO DIFFERENTIATE FACT FROM OPINION

Opinions
Notice words or phrases that express beliefs or feelings. The words *unfortunately* and *careless* show that opinions are being expressed.

Opinion
Look for statements that speculate about events. These statements are opinions, because they cannot be proved.

Human Activities and the Environment

Unfortunately, human use of fossil fuels is one of the most significant developments of the past few centuries. Humans rely on fossil fuels, a nonrenewable energy resource, for more than 90 percent of their energy needs.

This careless misuse of our planet's resources has resulted in pollution, global warming, and the destruction of fragile ecosystems. For example, oil pipelines carry more than one million barrels of oil each day across tundra regions. Transporting oil across such areas can only result in oil spills that poison the land for decades.

Facts
Statements that contain statistics tend to be facts. Writers often use facts to support their opinions.

Safety Rules

Before you work in the laboratory, read these safety rules twice. Ask your teacher to explain any rules that you do not completely understand. Refer to these rules later on if you have questions about safety in the science classroom.

Directions

- Read all directions and make sure that you understand them before starting an investigation or lab activity. If you do not understand how to do a procedure or how to use a piece of equipment, ask your teacher.
- Do not begin any investigation or touch any equipment until your teacher has told you to start.
- Never experiment on your own. If you want to try a procedure that the directions do not call for, ask your teacher for permission first.
- If you are hurt or injured in any way, tell your teacher immediately.

Dress Code

goggles

apron

gloves

- Wear goggles when
  — using glassware, sharp objects, or chemicals
  — heating an object
  — working with anything that can easily fly up into the air and hurt someone's eye
- Tie back long hair or hair that hangs in front of your eyes.
- Remove any article of clothing—such as a loose sweater or a scarf—that hangs down and may touch a flame, chemical, or piece of equipment.
- Observe all safety icons calling for the wearing of eye protection, gloves, and aprons.

Heating and Fire Safety

fire safety

heating safety

- Keep your work area neat, clean, and free of extra materials.
- Never reach over a flame or heat source.
- Point objects being heated away from you and others.
- Never heat a substance or an object in a closed container.
- Never touch an object that has been heated. If you are unsure whether something is hot, treat it as though it is. Use oven mitts, clamps, tongs, or a test-tube holder.
- Know where the fire extinguisher and fire blanket are kept in your classroom.
- Do not throw hot substances into the trash. Wait for them to cool or use the container your teacher puts out for disposal.

Electrical Safety

electrical safety

- Never use lamps or other electrical equipment with frayed cords.
- Make sure no cord is lying on the floor where someone can trip over it.
- Do not let a cord hang over the side of a counter or table so that the equipment can easily be pulled or knocked to the floor.
- Never let cords hang into sinks or other places where water can be found.
- Never try to fix electrical problems. Inform your teacher of any problems immediately.
- Unplug an electrical cord by pulling on the plug, not the cord.

Chemical Safety

chemical safety

poison

fumes

- If you spill a chemical or get one on your skin or in your eyes, tell your teacher right away.
- Never touch, taste, or sniff any chemicals in the lab. If you need to determine odor, waft. Wafting consists of holding the chemical in its container 15 centimeters (6 in.) away from your nose, and using your fingers to bring fumes from the container to your nose.
- Keep lids on all chemicals you are not using.
- Never put unused chemicals back into the original containers. Throw away extra chemicals where your teacher tells you to.
- Pour chemicals over a sink or your work area, not over the floor.
- If you get a chemical in your eye, use the eyewash right away.
- Always wash your hands after handling chemicals, plants, or soil.

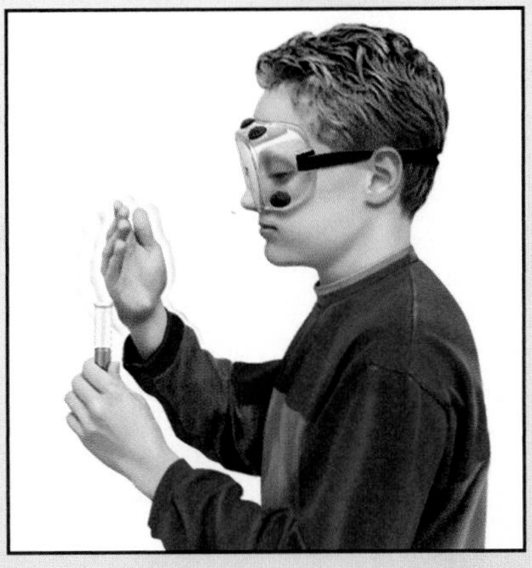
Wafting

Glassware and Sharp-Object Safety

sharp objects

- If you break glassware, tell your teacher right away.
- Do not use broken or chipped glassware. Give these to your teacher.
- Use knives and other cutting instruments carefully. Always wear eye protection and cut away from you.

Animal Safety

- Never hurt an animal.
- Touch animals only when necessary. Follow your teacher's instructions for handling animals.
- Always wash your hands after working with animals.

Cleanup

disposal

- Follow your teacher's instructions for throwing away or putting away supplies.
- Clean your work area and pick up anything that has dropped to the floor.
- Wash your hands.

R11

Using Lab Equipment

Different experiments require different types of equipment. But even though experiments differ, the ways in which the equipment is used are the same.

Beakers

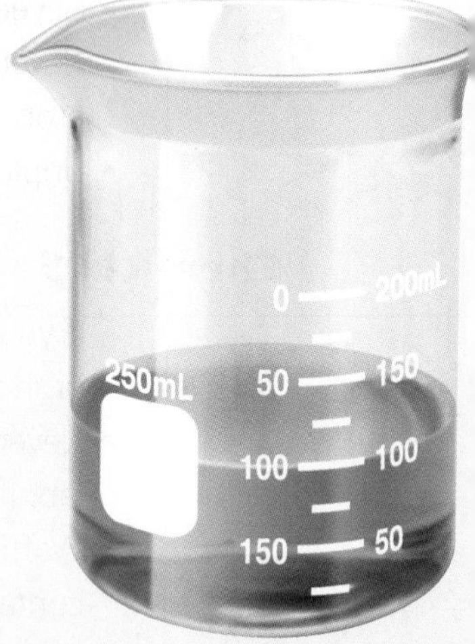

- Use beakers for holding and pouring liquids.
- Do not use a beaker to measure the volume of a liquid. Use a graduated cylinder instead. (See page R16.)
- Use a beaker that holds about twice as much liquid as you need. For example, if you need 100 milliliters of water, you should use a 200- or 250-milliliter beaker.

Test Tubes

- Use test tubes to hold small amounts of substances.
- Do not use a test tube to measure the volume of a liquid.
- Use a test tube when heating a substance over a flame. Aim the mouth of the tube away from yourself and other people.
- Liquids easily spill or splash from test tubes, so it is important to use only small amounts of liquids.

Test-Tube Holder

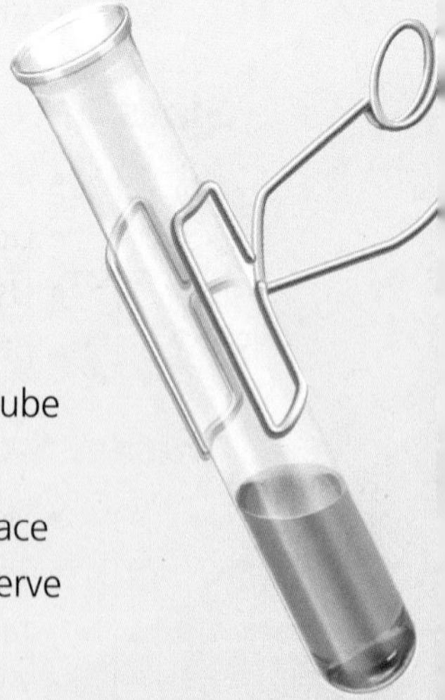

- Use a test-tube holder when heating a substance in a test tube.
- Use a test-tube holder if the substance in a test tube is dangerous to touch.
- Make sure the test-tube holder tightly grips the test tube so that the test tube will not slide out of the holder.
- Make sure that the test-tube holder is above the surface of the substance in the test tube so that you can observe the substance.

LAB HANDBOOK

Test-Tube Rack

- Use a test-tube rack to organize test tubes before, during, and after an experiment.
- Use a test-tube rack to keep test tubes upright so that they do not fall over and spill their contents.
- Use a test-tube rack that is the correct size for the test tubes that you are using. If the rack is too small, a test tube may become stuck. If the rack is too large, a test tube may lean over, and some of its contents may spill or splash.

Forceps

- Use forceps when you need to pick up or hold a very small object that should not be touched with your hands.
- Do not use forceps to hold anything over a flame, because forceps are not long enough to keep your hand safely away from the flame. Plastic forceps will melt, and metal forceps will conduct heat and burn your hand.

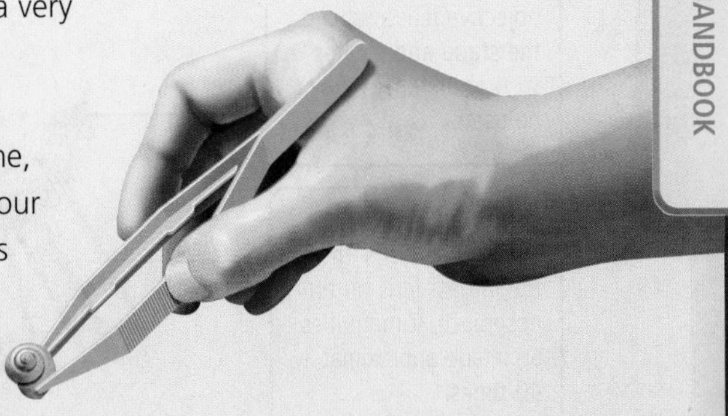

Hot Plate

- Use a hot plate when a substance needs to be kept warmer than room temperature for a long period of time.
- Use a hot plate instead of a Bunsen burner or a candle when you need to carefully control temperature.
- Do not use a hot plate when a substance needs to be burned in an experiment.
- Always use "hot hands" safety mitts or oven mitts when handling anything that has been heated on a hot plate.

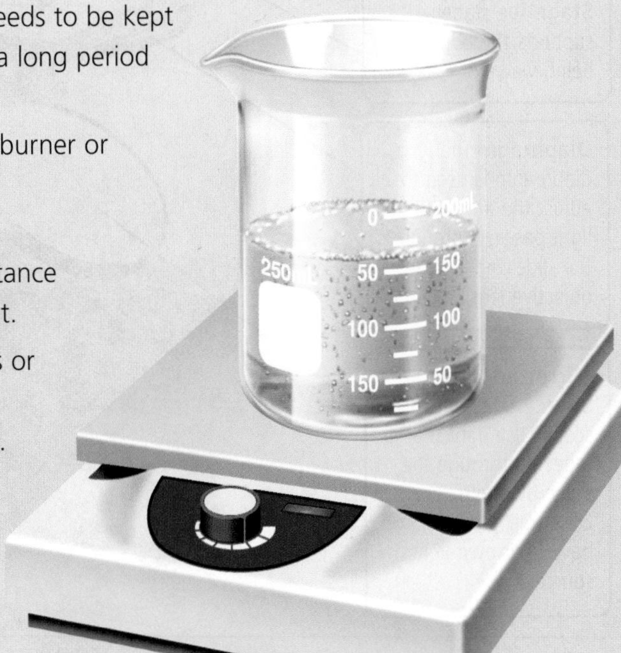

Microscope

Scientists use microscopes to see very small objects that cannot easily be seen with the eye alone. A microscope magnifies the image of an object so that small details may be observed. A microscope that you may use can magnify an object 400 times—the object will appear 400 times larger than its actual size.

Body The body separates the lens in the eyepiece from the objective lenses below.

Nosepiece The nosepiece holds the objective lenses above the stage and rotates so that all lenses may be used.

High-Power Objective Lens This is the largest lens on the nosepiece. It magnifies an image approximately 40 times.

Stage The stage supports the object being viewed.

Diaphragm The diaphragm is used to adjust the amount of light passing through the slide and into an objective lens.

Mirror or Light Source Some microscopes use light that is reflected through the stage by a mirror. Other microscopes have their own light sources.

Eyepiece Objects are viewed through the eyepiece. The eyepiece contains a lens that commonly magnifies an image 10 times.

Coarse Adjustment This knob is used to focus the image of an object when it is viewed through the low-power lens.

Fine Adjustment This knob is used to focus the image of an object when it is viewed through the high-power lens.

Low-Power Objective Lens This is the smallest lens on the nosepiece. It magnifies an image approximately 10 times.

Arm The arm supports the body above the stage. Always carry a microscope by the arm and base.

Stage Clip The stage clip holds a slide in place on the stage.

Base The base supports the microscope.

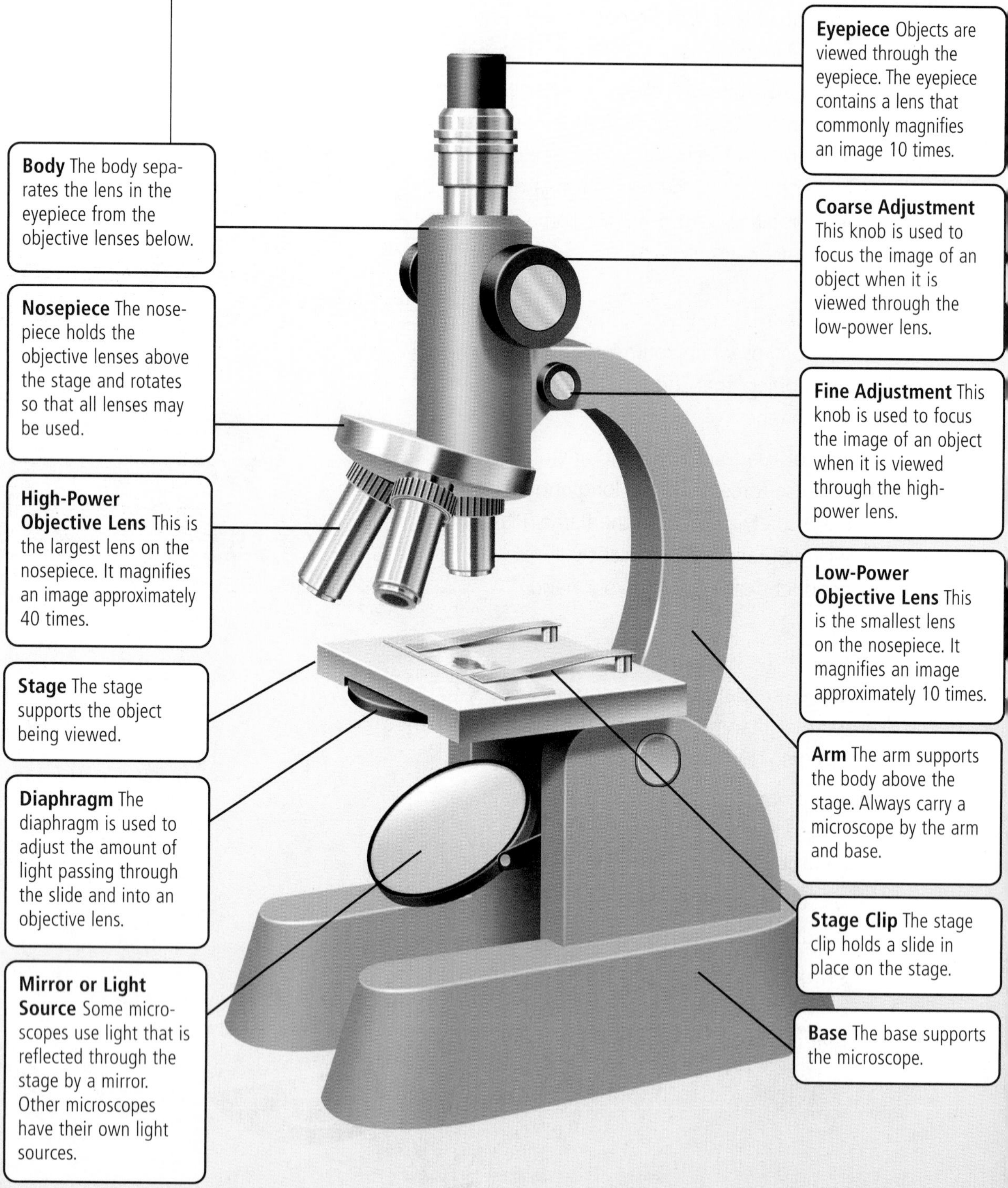

VIEWING AN OBJECT

1. Use the coarse adjustment knob to raise the body tube.
2. Adjust the diaphragm so that you can see a bright circle of light through the eyepiece.
3. Place the object or slide on the stage. Be sure that it is centered over the hole in the stage.
4. Turn the nosepiece to click the low-power lens into place.
5. Using the coarse adjustment knob, slowly lower the lens and focus on the specimen being viewed. Be sure not to touch the slide or object with the lens.
6. When switching from the low-power lens to the high-power lens, first raise the body tube with the coarse adjustment knob so that the high-power lens will not hit the slide.
7. Turn the nosepiece to click the high-power lens into place.
8. Use the fine adjustment knob to focus on the specimen being viewed. Again, be sure not to touch the slide or object with the lens.

MAKING A SLIDE, OR WET MOUNT

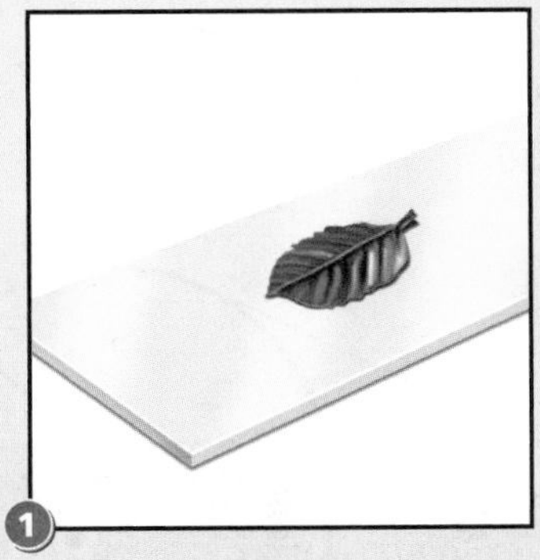

1 Place the specimen in the center of a clean slide.

2 Place a drop of water on the specimen.

3 Place a cover slip on the slide. Put one edge of the cover slip into the drop of water and slowly lower it over the specimen.

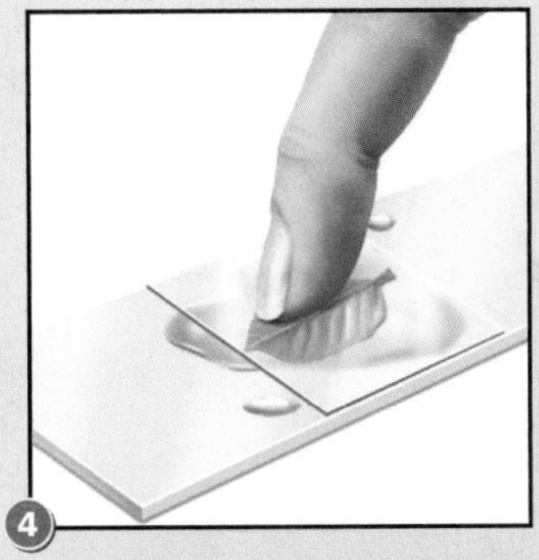

4 Remove any air bubbles from under the cover slip by gently tapping the cover slip.

5 Dry any excess water before placing the slide on the microscope stage for viewing.

Spring Scale (Force Meter)

- Use a spring scale to measure a force pulling on the scale.
- Use a spring scale to measure the force of gravity exerted on an object by Earth.
- To measure a force accurately, a spring scale must be zeroed before it is used. The scale is zeroed when no weight is attached and the indicator is positioned at zero.
- Do not attach a weight that is either too heavy or too light to a spring scale. A weight that is too heavy could break the scale or exert too great a force for the scale to measure. A weight that is too light may not exert enough force to be measured accurately.

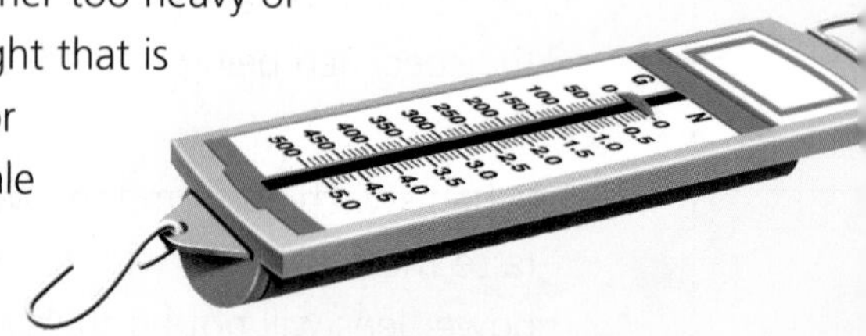

Graduated Cylinder

- Use a graduated cylinder to measure the volume of a liquid.
- Be sure that the graduated cylinder is on a flat surface so that your measurement will be accurate.
- When reading the scale on a graduated cylinder, be sure to have your eyes at the level of the surface of the liquid.
- The surface of the liquid will be curved in the graduated cylinder. Read the volume of the liquid at the bottom of the curve, or meniscus (muh-NIHS-kuhs).
- You can use a graduated cylinder to find the volume of a solid object by measuring the increase in a liquid's level after you add the object to the cylinder.

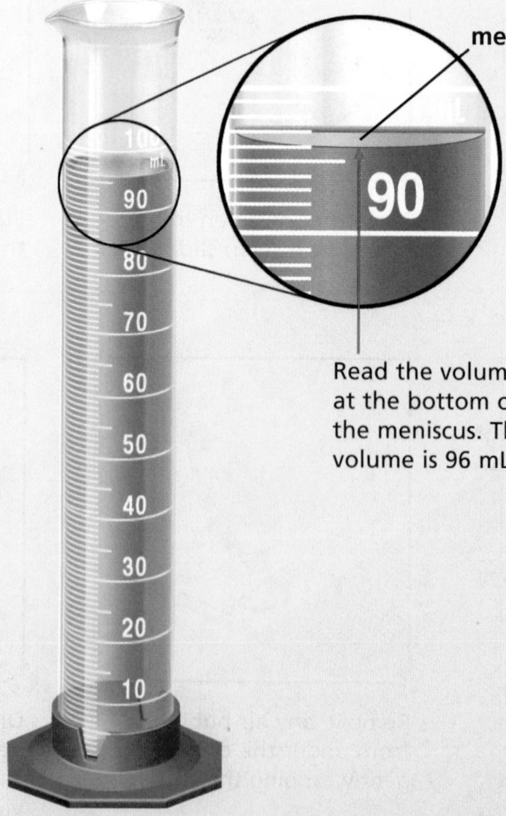

Read the volume at the bottom of the meniscus. The volume is 96 mL.

Metric Rulers

- Use metric rulers or meter sticks to measure objects' lengths.
- Do not measure an object from the end of a metric ruler or meter stick, because the end is often imperfect. Instead, measure from the 1-centimeter mark, but remember to subtract a centimeter from the apparent measurement.
- Estimate any lengths that extend between marked units. For example, if a meter stick shows centimeters but not millimeters, you can estimate the length that an object extends between centimeter marks to measure it to the nearest millimeter.
- **Controlling Variables** If you are taking repeated measurements, always measure from the same point each time. For example, if you're measuring how high two different balls bounce when dropped from the same height, measure both bounces at the same point on the balls—either the top or the bottom. Do not measure at the top of one ball and the bottom of the other.

EXAMPLE

How to Measure a Leaf

1. Lay a ruler flat on top of the leaf so that the 1-centimeter mark lines up with one end. Make sure the ruler and the leaf do not move between the time you line them up and the time you take the measurement.
2. Look straight down on the ruler so that you can see exactly how the marks line up with the other end of the leaf.
3. Estimate the length by which the leaf extends beyond a marking. For example, the leaf below extends about halfway between the 4.2-centimeter and 4.3-centimeter marks, so the apparent measurement is about 4.25 centimeters.
4. Remember to subtract 1 centimeter from your apparent measurement, since you started at the 1-centimeter mark on the ruler and not at the end. The leaf is about 3.25 centimeters long (4.25 cm – 1 cm = 3.25 cm).

R17

Triple-Beam Balance

This balance has a pan and three beams with sliding masses, called riders. At one end of the beams is a pointer that indicates whether the mass on the pan is equal to the masses shown on the beams.

1. Make sure the balance is zeroed before measuring the mass of an object. The balance is zeroed if the pointer is at zero when nothing is on the pan and the riders are at their zero points. Use the adjustment knob at the base of the balance to zero it.
2. Place the object to be measured on the pan.
3. Move the riders one notch at a time away from the pan. Begin with the largest rider. If moving the largest rider one notch brings the pointer below zero, begin measuring the mass of the object with the next smaller rider.
4. Change the positions of the riders until they balance the mass on the pan and the pointer is at zero. Then add the readings from the three beams to determine the mass of the object.

300 g	position of largest rider
90 g	position of middle rider
+ 3 g	position of smallest rider
393 g	mass of beaker

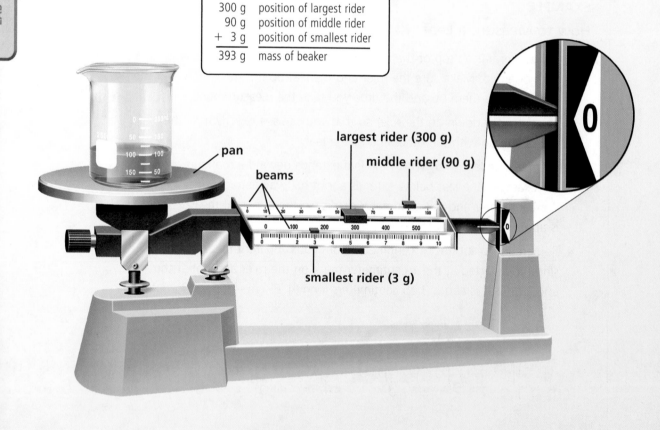

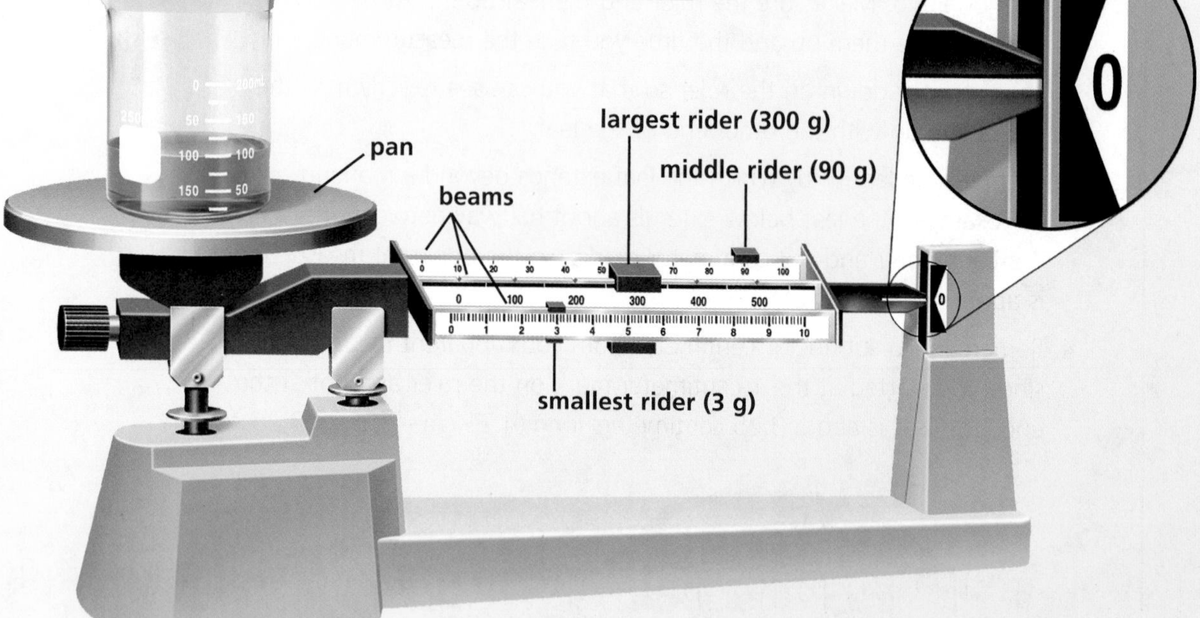

Double-Pan Balance

This type of balance has two pans. Between the pans is a pointer that indicates whether the masses on the pans are equal.

1. Make sure the balance is zeroed before measuring the mass of an object. The balance is zeroed if the pointer is at zero when there is nothing on either of the pans. Many double-pan balances have sliding knobs that can be used to zero them.
2. Place the object to be measured on one of the pans.
3. Begin adding standard masses to the other pan. Begin with the largest standard mass. If this adds too much mass to the balance, begin measuring the mass of the object with the next smaller standard mass.
4. Add standard masses until the masses on both pans are balanced and the pointer is at zero. Then add the standard masses together to determine the mass of the object being measured.

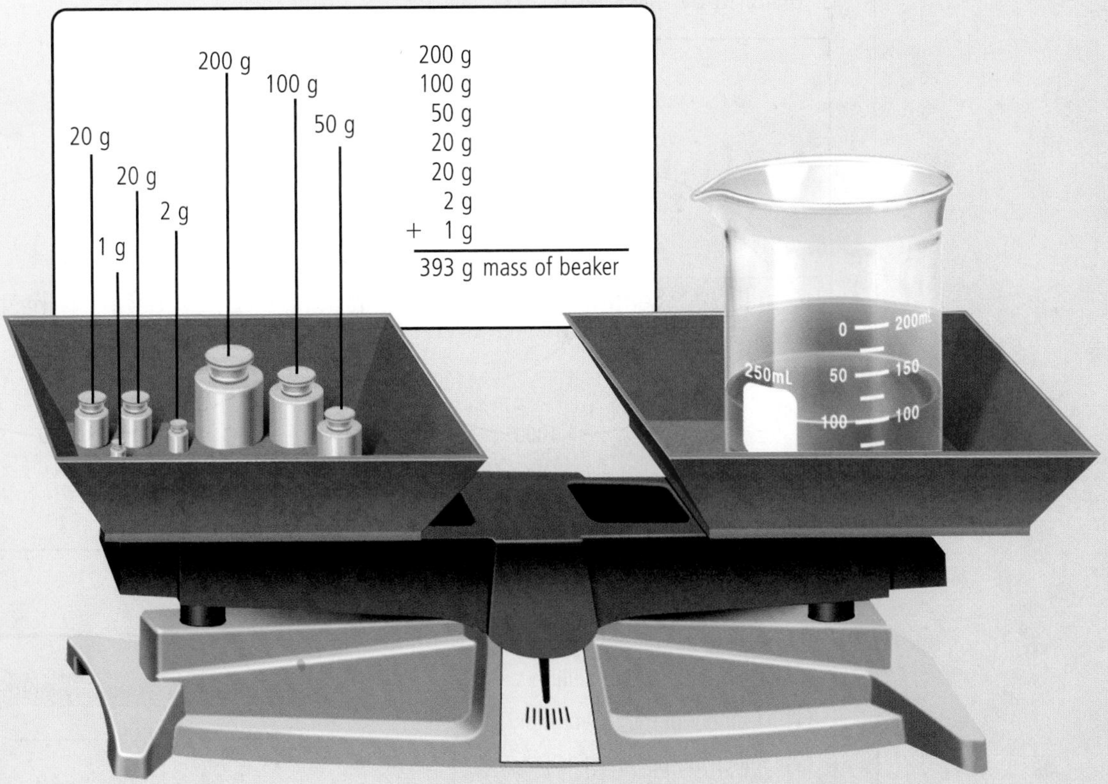

Never place chemicals or liquids directly on a pan. Instead, use the following procedure:

1. Determine the mass of an empty container, such as a beaker.
2. Pour the substance into the container, and measure the total mass of the substance and the container.
3. Subtract the mass of the empty container from the total mass to find the mass of the substance.

LAB HANDBOOK

The Metric System and SI Units

Scientists use International System (SI) units for measurements of distance, volume, mass, and temperature. The International System is based on multiples of ten and the metric system of measurement.

Basic SI Units

Property	Name	Symbol
length	meter	m
volume	liter	L
mass	kilogram	kg
temperature	kelvin	K

SI Prefixes

Prefix	Symbol	Multiple of 10
kilo-	k	1000
hecto-	h	100
deca-	da	10
deci-	d	0.1 $(\frac{1}{10})$
centi-	c	0.01 $(\frac{1}{100})$
milli-	m	0.001 $(\frac{1}{1000})$

Changing Metric Units

You can change from one unit to another in the metric system by multiplying or dividing by a power of 10.

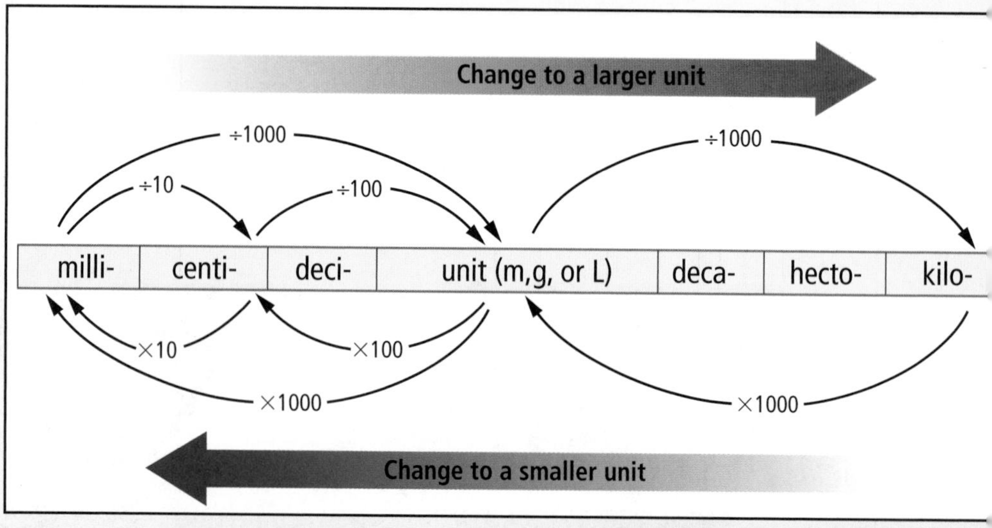

Example

Change 0.64 liters to milliliters.

(1) Decide whether to multiply or divide.

(2) Select the power of 10.

ANSWER 0.64 L = 640 mL

Change to a smaller unit by multiplying.

mL ← × 1000 — **L**

0.64 × 1000 = **640.**

Example

Change 23.6 grams to kilograms.

(1) Decide whether to multiply or divide.

(2) Select the power of 10.

ANSWER 23.6 g = 0.0236 kg

Change to a larger unit by dividing.

g — ÷ 1000 → **kg**

23.6 ÷ 1000 = **0.0236**

Temperature Conversions

Even though the kelvin is the SI base unit of temperature, the degree Celsius will be the unit you use most often in your science studies. The formulas below show the relationships between temperatures in degrees Fahrenheit (°F), degrees Celsius (°C), and kelvins (K).

$$°C = \frac{5}{9}(°F - 32)$$

$$°F = \frac{9}{5}°C + 32$$

$$K = °C + 273$$

See page R42 for help with using formulas.

Examples of Temperature Conversions

Condition	Degrees Celsius	Degrees Fahrenheit
Freezing point of water	0	32
Cool day	10	50
Mild day	20	68
Warm day	30	86
Normal body temperature	37	98.6
Very hot day	40	104
Boiling point of water	100	212

Converting Between SI and U.S. Customary Units

Use the chart below when you need to convert between SI units and U.S. customary units.

SI Unit	From SI to U.S. Customary			From U.S. Customary to SI		
Length	**When you know**	**multiply by**	**to find**	**When you know**	**multiply by**	**to find**
kilometer (km) = 1000 m	kilometers	0.62	miles	miles	1.61	kilometers
meter (m) = 100 cm	meters	3.28	feet	feet	0.3048	meters
centimeter (cm) = 10 mm	centimeters	0.39	inches	inches	2.54	centimeters
millimeter (mm) = 0.1 cm	millimeters	0.04	inches	inches	25.4	millimeters
Area	**When you know**	**multiply by**	**to find**	**When you know**	**multiply by**	**to find**
square kilometer (km^2)	square kilometers	0.39	square miles	square miles	2.59	square kilometers
square meter (m^2)	square meters	1.2	square yards	square yards	0.84	square meters
square centimeter (cm^2)	square centimeters	0.155	square inches	square inches	6.45	square centimeters
Volume	**When you know**	**multiply by**	**to find**	**When you know**	**multiply by**	**to find**
liter (L) = 1000 mL	liters	1.06	quarts	quarts	0.95	liters
	liters	0.26	gallons	gallons	3.79	liters
	liters	4.23	cups	cups	0.24	liters
	liters	2.12	pints	pints	0.47	liters
milliliter (mL) = 0.001 L	milliliters	0.20	teaspoons	teaspoons	4.93	milliliters
	milliliters	0.07	tablespoons	tablespoons	14.79	milliliters
	milliliters	0.03	fluid ounces	fluid ounces	29.57	milliliters
Mass	**When you know**	**multiply by**	**to find**	**When you know**	**multiply by**	**to find**
kilogram (kg) = 1000 g	kilograms	2.2	pounds	pounds	0.45	kilograms
gram (g) = 1000 mg	grams	0.035	ounces	ounces	28.35	grams

Precision and Accuracy

When you do an experiment, it is important that your methods, observations, and data be both precise and accurate.

low precision

precision, but not accuracy

precision and accuracy

Precision

In science, **precision** is the exactness and consistency of measurements. For example, measurements made with a ruler that has both centimeter and millimeter markings would be more precise than measurements made with a ruler that has only centimeter markings. Another indicator of precision is the care taken to make sure that methods and observations are as exact and consistent as possible. Every time a particular experiment is done, the same procedure should be used. Precision is necessary because experiments are repeated several times and if the procedure changes, the results will change.

EXAMPLE

Suppose you are measuring temperatures over a two-week period. Your precision will be greater if you measure each temperature at the same place, at the same time of day, and with the same thermometer than if you change any of these factors from one day to the next.

Accuracy

In science, it is possible to be precise but not accurate. **Accuracy** depends on the difference between a measurement and an actual value. The smaller the difference, the more accurate the measurement.

EXAMPLE

Suppose you look at a stream and estimate that it is about 1 meter wide at a particular place. You decide to check your estimate by measuring the stream with a meter stick, and you determine that the stream is 1.32 meters wide. However, because it is hard to measure the width of a stream with a meter stick, it turns out that you didn't do a very good job. The stream is actually 1.14 meters wide. Therefore, even though your estimate was less precise than your measurement, your estimate was actually more accurate.

Making Data Tables and Graphs

Data tables and graphs are useful tools for both recording and communicating scientific data.

Making Data Tables

You can use a **data table** to organize and record the measurements that you make. Some examples of information that might be recorded in data tables are frequencies, times, and amounts.

EXAMPLE

Suppose you are investigating photosynthesis in two elodea plants. One sits in direct sunlight, and the other sits in a dimly lit room. You measure the rate of photosynthesis by counting the number of bubbles in the jar every ten minutes.

1. Title and number your data table.
2. Decide how you will organize the table into columns and rows.
3. Any units, such as seconds or degrees, should be included in column headings, not in the individual cells.

Table 1. Number of Bubbles from Elodea

Always number and title data tables.

Time (min)	Sunlight	Dim Light
0	0	0
10	15	5
20	25	8
30	32	7
40	41	10
50	47	9
60	42	9

The data in the table above could also be organized in a different way.

Table 1. Number of Bubbles from Elodea

Put units in column heading.

Light Condition	Time (min)						
	0	10	20	30	40	50	60
Sunlight	0	15	25	32	41	47	42
Dim light	0	5	8	7	10	9	9

LAB HANDBOOK

Making Line Graphs

You can use a **line graph** to show a relationship between variables. Line graphs are particularly useful for showing changes in variables over time.

EXAMPLE

Suppose you are interested in graphing temperature data that you collected over the course of a day.

Table 1. Outside Temperature During the Day on March 7

	Time of Day						
	7:00 A.M.	9:00 A.M.	11:00 A.M.	1:00 P.M.	3:00 P.M.	5:00 P.M.	7:00 P.M.
Temp (°C)	8	9	11	14	12	10	6

1. Use the vertical axis of your line graph for the variable that you are measuring—temperature.
2. Choose scales for both the horizontal axis and the vertical axis of the graph. You should have two points more than you need on the vertical axis, and the horizontal axis should be long enough for all of the data points to fit.
3. Draw and label each axis.
4. Graph each value. First find the appropriate point on the scale of the horizontal axis. Imagine a line that rises vertically from that place on the scale. Then find the corresponding value on the vertical axis, and imagine a line that moves horizontally from that value. The point where these two imaginary lines intersect is where the value should be plotted.
5. Connect the points with straight lines.

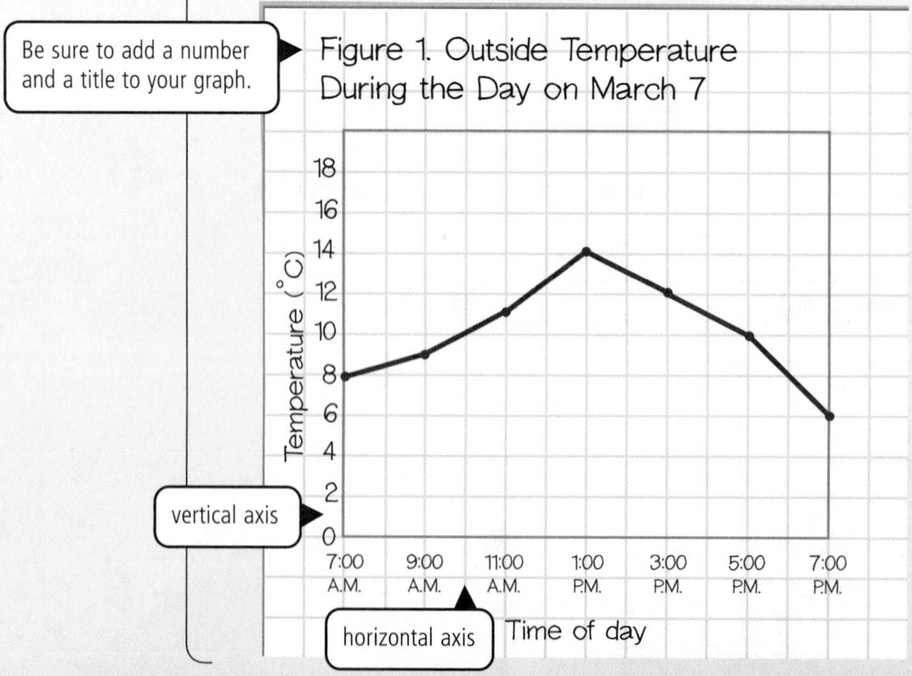

Figure 1. Outside Temperature During the Day on March 7

Making Circle Graphs

You can use a **circle graph,** sometimes called a pie chart, to represent data as parts of a circle. Circle graphs are used only when the data can be expressed as percentages of a whole. The entire circle shown in a circle graph is equal to 100 percent of the data.

EXAMPLE

Suppose you identified the species of each mature tree growing in a small wooded area. You organized your data in a table, but you also want to show the data in a circle graph.

1. To begin, find the total number of mature trees.

   $56 + 34 + 22 + 10 + 28 = 150$

2. To find the degree measure for each sector of the circle, write a fraction comparing the number of each tree species with the total number of trees. Then multiply the fraction by 360°.

   Oak: $\frac{56}{150} \times 360° = 134.4°$

3. Draw a circle. Use a protractor to draw the angle for each sector of the graph.
4. Color and label each sector of the graph.
5. Give the graph a number and title.

Table 1. Tree Species in Wooded Area

Species	Number of Specimens
Oak	56
Maple	34
Birch	22
Willow	10
Pine	28

Figure 1. Tree Species in Wooded Area

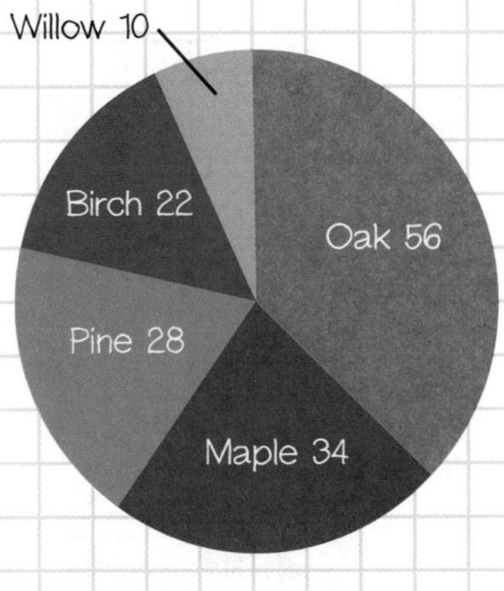

Instead of labeling each sector, you could make a color key.

Bar Graph

A **bar graph** is a type of graph in which the lengths of the bars are used to represent and compare data. A numerical scale is used to determine the lengths of the bars.

EXAMPLE

To determine the effect of water on seed sprouting, three cups were filled with sand, and ten seeds were planted in each. Different amounts of water were added to each cup over a three-day period.

Table 1. Effect of Water on Seed Sprouting

Daily Amount of Water (mL)	Number of Seeds That Sprouted After 3 Days in Sand
0	1
10	4
20	8

1. Choose a numerical scale. The greatest value is 8, so the end of the scale should have a value greater than 8, such as 10. Use equal increments along the scale, such as increments of 2.
2. Draw and label the axes. Mark intervals on the vertical axis according to the scale you chose.
3. Draw a bar for each data value. Use the scale to decide how long to make each bar.

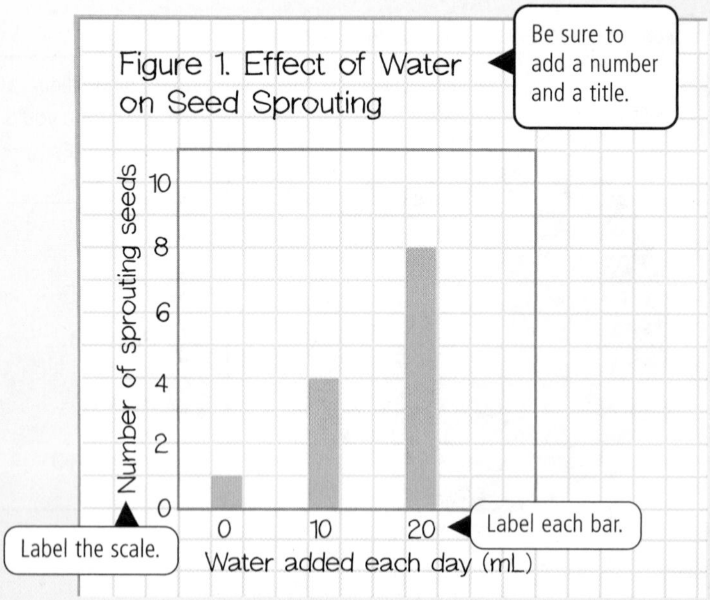

R26

Double Bar Graph

A **double bar graph** is a bar graph that shows two sets of data. The two bars for each measurement are drawn next to each other.

EXAMPLE

The same seed-sprouting experiment was repeated with potting soil. The data for sand and potting soil can be plotted on one graph.

1. Draw one set of bars, using the data for sand, as shown below.
2. Draw bars for the potting-soil data next to the bars for the sand data. Shade them a different color. Add a key.

Table 2. Effect of Water and Soil on Seed Sprouting

Daily Amount of Water (mL)	Number of Seeds That Sprouted After 3 Days in Sand	Number of Seeds That Sprouted After 3 Days in Potting Soil
0	1	2
10	4	5
20	8	9

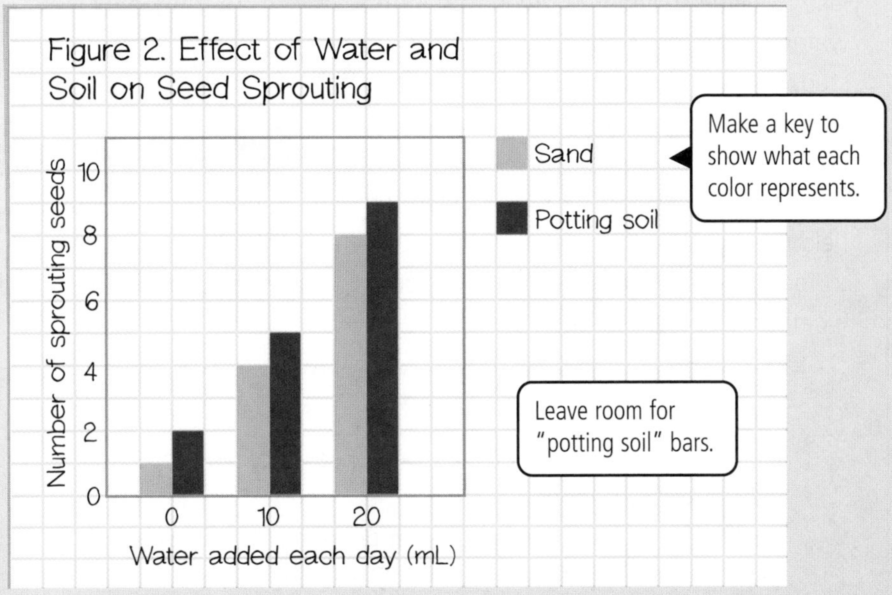

R27

LAB HANDBOOK

Designing an Experiment

Use this section when designing or conducting an experiment.

Determining a Purpose

You can find a purpose for an experiment by doing research, by examining the results of a previous experiment, or by observing the world around you. An **experiment** is an organized procedure to study something under controlled conditions.

1. Write the purpose of your experiment as a question or problem that you want to investigate.
2. Write down research questions and begin searching for information that will help you design an experiment. Consult the library, the Internet, and other people as you conduct your research.

Don't forget to learn as much as possible about your topic before you begin.

EXAMPLE

Middle school students observed an odor near the lake by their school. They also noticed that the water on the side of the lake near the school was greener than the water on the other side of the lake. The students did some research to learn more about their observations. They discovered that the odor and green color in the lake came from algae. They also discovered that a new fertilizer was being used on a field nearby. The students inferred that the use of the fertilizer might be related to the presence of the algae and designed a controlled experiment to find out whether they were right.

Problem

How does fertilizer affect the presence of algae in a lake?

Research Questions

- Have other experiments been done on this problem? If so, what did those experiments show?
- What kind of fertilizer is used on the field? How much?
- How do algae grow?
- How do people measure algae?
- Can fertilizer and algae be used safely in a lab? How?

Research
As you research, you may find a topic that is more interesting to you than your original topic, or learn that a procedure you wanted to use is not practical or safe. It is OK to change your purpose as you research.

Writing a Hypothesis

A **hypothesis** is a tentative explanation for an observation or scientific problem that can be tested by further investigation. You can write your hypothesis in the form of an "If . . . , then . . . , because . . ." statement.

Hypothesis

If the amount of fertilizer in lake water is increased, then the amount of algae will also increase, because fertilizers provide nutrients that algae need to grow.

Hypotheses
For help with hypotheses, refer to page R3.

Determining Materials

Make a list of all the materials you will need to do your experiment. Be specific, especially if someone else is helping you obtain the materials. Try to think of everything you will need.

Materials

- 1 large jar or container
- 4 identical smaller containers
- rubber gloves that also cover the arms
- sample of fertilizer-and-water solution
- eyedropper
- clear plastic wrap
- scissors
- masking tape
- marker
- ruler

Determining Variables and Constants

EXPERIMENTAL GROUP AND CONTROL GROUP

An experiment to determine how two factors are related always has two groups—a control group and an experimental group.

1. Design an experimental group. Include as many trials as possible in the experimental group in order to obtain reliable results.
2. Design a control group that is the same as the experimental group in every way possible, except for the factor you wish to test.

Experimental Group: two containers of lake water with one drop of fertilizer solution added to each
Control Group: two containers of lake water with no fertilizer solution added

Go back to your materials list and make sure you have enough items listed to cover both your experimental group and your control group.

VARIABLES AND CONSTANTS

Identify the variables and constants in your experiment. In a controlled experiment, a **variable** is any factor that can change. **Constants** are all of the factors that are the same in both the experimental group and the control group.

1. Read your hypothesis. The **independent variable** is the factor that you wish to test and that is manipulated or changed so that it can be tested. The independent variable is expressed in your hypothesis after the word *if*. Identify the independent variable in your laboratory report.
2. The **dependent variable** is the factor that you measure to gather results. It is expressed in your hypothesis after the word *then*. Identify the dependent variable in your laboratory report.

Hypothesis
If the amount of fertilizer in lake water is increased, then the amount of algae will also increase, because fertilizers provide nutrients that algae need to grow.

Table 1. Variables and Constants in Algae Experiment

Independent Variable	Dependent Variable	Constants
Amount of fertilizer in lake water	Amount of algae that grow	• Where the lake water is obtained • Type of container used • Light and temperature conditions where water will be stored

Set up your experiment so that you will test only one variable.

LAB HANDBOOK

MEASURING THE DEPENDENT VARIABLE

Before starting your experiment, you need to define how you will measure the dependent variable. An **operational definition** is a description of the one particular way in which you will measure the dependent variable.

Your operational definition is important for several reasons. First, in any experiment there are several ways in which a dependent variable can be measured. Second, the procedure of the experiment depends on how you decide to measure the dependent variable. Third, your operational definition makes it possible for other people to evaluate and build on your experiment.

EXAMPLE 1

An operational definition of a dependent variable can be qualitative. That is, your measurement of the dependent variable can simply be an observation of whether a change occurs as a result of a change in the independent variable. This type of operational definition can be thought of as a "yes or no" measurement.

Table 2. Qualitative Operational Definition of Algae Growth

Independent Variable	Dependent Variable	Operational Definition
Amount of fertilizer in lake water	Amount of algae that grow	Algae grow in lake water

A qualitative measurement of a dependent variable is often easy to make and record. However, this type of information does not provide a great deal of detail in your experimental results.

EXAMPLE 2

An operational definition of a dependent variable can be quantitative. That is, your measurement of the dependent variable can be a number that shows how much change occurs as a result of a change in the independent variable.

Table 3. Quantitative Operational Definition of Algae Growth

Independent Variable	Dependent Variable	Operational Definition
Amount of fertilizer in lake water	Amount of algae that grow	Diameter of largest algal growth (in mm)

A quantitative measurement of a dependent variable can be more difficult to make and analyze than a qualitative measurement. However, this type of data provides much more information about your experiment and is often more useful.

Writing a Procedure

Write each step of your procedure. Start each step with a verb, or action word, and keep the steps short. Your procedure should be clear enough for someone else to use as instructions for repeating your experiment.

If necessary, go back to your materials list and add any materials that you left out.

Procedure

1. Put on your gloves. Use the large container to obtain a sample of lake water.
2. Divide the sample of lake water equally among the four smaller containers.
3. Use the eyedropper to add one drop of fertilizer solution to two of the containers.
4. Use the masking tape and the marker to label the containers with your initials, the date, and the identifiers "Jar 1 with Fertilizer," "Jar 2 with Fertilizer," "Jar 1 without Fertilizer," and "Jar 2 without Fertilizer."
5. Cover the containers with clear plastic wrap. Use the scissors to punch ten holes in each of the covers.
6. Place all four containers on a window ledge. Make sure that they all receive the same amount of light.
7. Observe the containers every day for one week.
8. Use the ruler to measure the diameter of the largest clump of algae in each container, and record your measurements daily.

Controlling Variables The same amount of fertilizer solution must be added to two of the four containers.

Controlling Variables All four containers must receive the same amount of light.

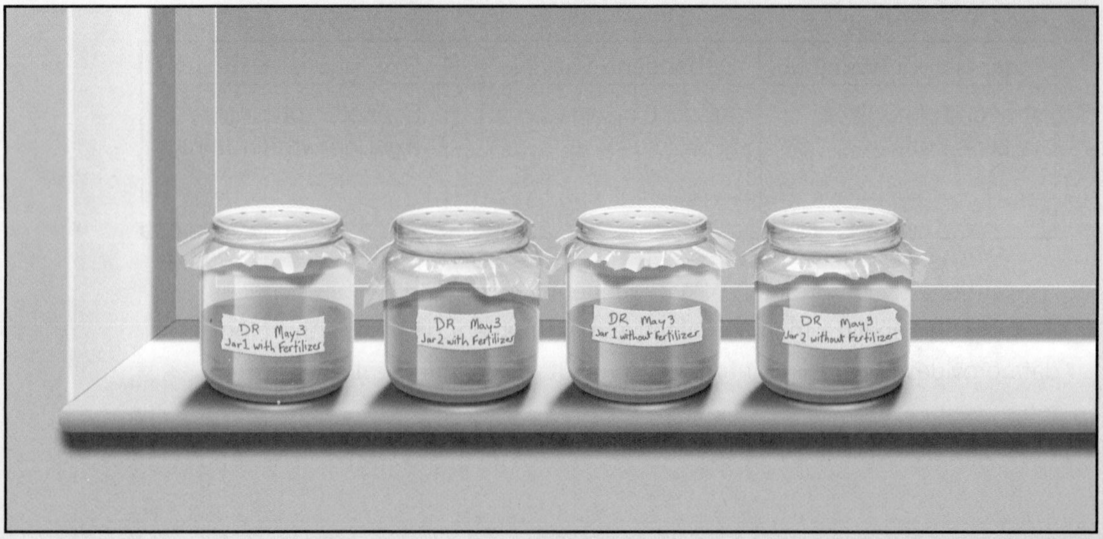

Recording Observations

Once you have obtained all of your materials and your procedure has been approved, you can begin making experimental observations. Gather both quantitative and qualitative data. If something goes wrong during your procedure, make sure you record that too.

Observations
For help with making qualitative and quantitative observations, refer to page R2.

For more examples of data tables, see page R23.

Table 4. Fertilizer and Algae Growth

	Experimental Group		Control Group		
Date and Time	Jar 1 with Fertilizer (diameter of algae in mm)	Jar 2 with Fertilizer (diameter of algae in mm)	Jar 1 without Fertilizer (diameter of algae in mm)	Jar 2 without Fertilizer (diameter of algae in mm)	Observations
5/3 4:00 P.M.	0	0	0	0	condensation in all containers
5/4 4:00 P.M.	0	3	0	0	tiny green blobs in jar 2 with fertilizer
5/5 4:15 P.M.	4	5	0	3	green blobs in jars 1 and 2 with fertilizer and jar 2 without fertilizer
5/6 4:00 P.M.	5	6	0	4	water light green in jar 2 with fertilizer
5/7 4:00 P.M.	8	10	0	6	water light green in jars 1 and 2 with fertilizer and in jar 2 without fertilizer
5/8 3:30 P.M.	10	18	0	6	cover off jar 2 with fertilizer
5/9 3:30 P.M.	14	23	0	8	drew sketches of each container

Notice that on the sixth day, the observer found that the cover was off one of the containers. It is important to record observations of unintended factors because they might affect the results of the experiment.

Drawings of Samples Viewed Under Microscope on 5/9 at 100x

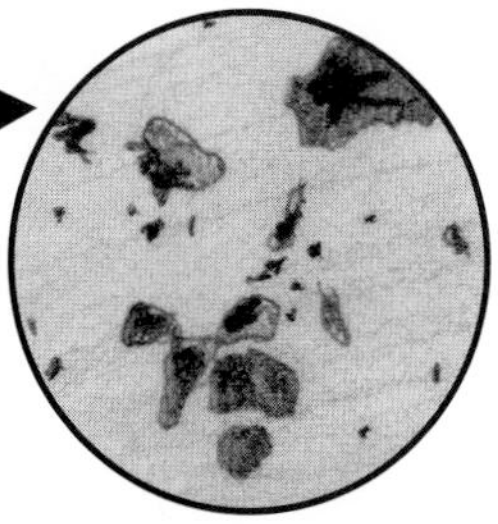

Jar 1 with Fertilizer

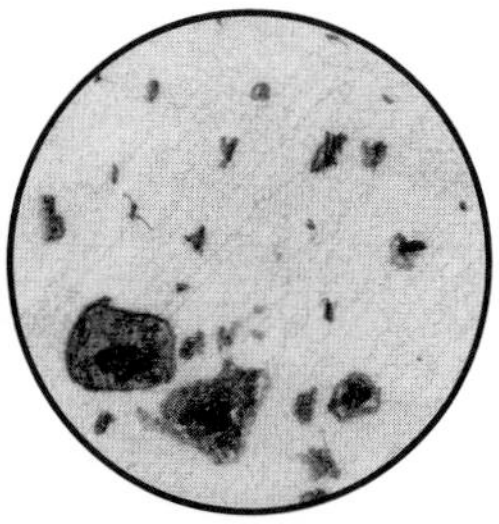

Jar 2 with Fertilizer

Jar 1 without Fertilizer

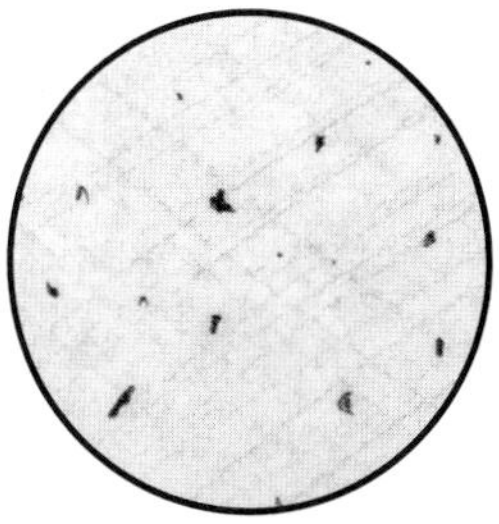

Jar 2 without Fertilizer

Use technology, such as a microscope, to help you make observations when possible.

LAB HANDBOOK

Summarizing Results

To summarize your data, look at all of your observations together. Look for meaningful ways to present your observations. For example, you might average your data or make a graph to look for patterns. When possible, use spreadsheet software to help you analyze and present your data. The two graphs below show the same data.

EXAMPLE 1

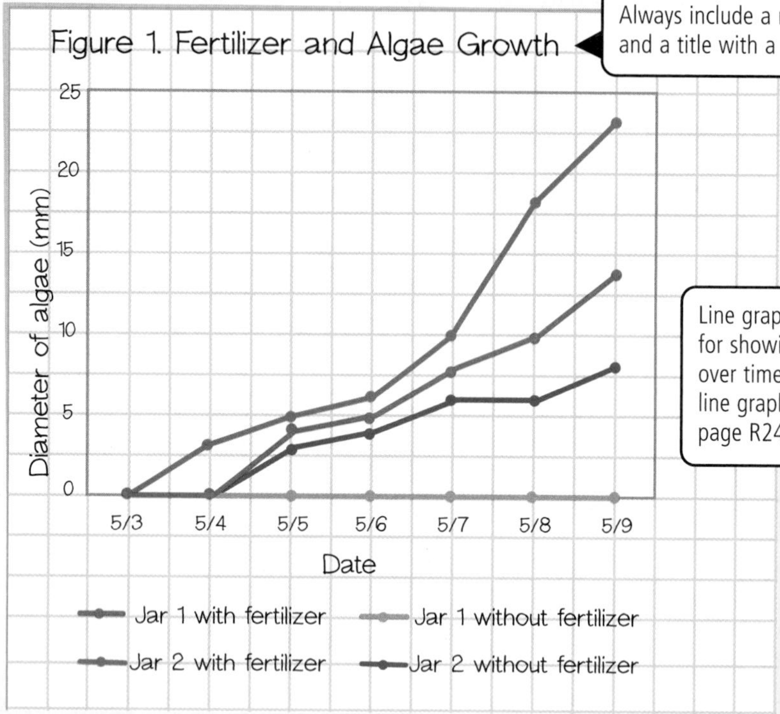

Always include a number and a title with a graph.

Line graphs are useful for showing changes over time. For help with line graphs, refer to page R24.

EXAMPLE 2

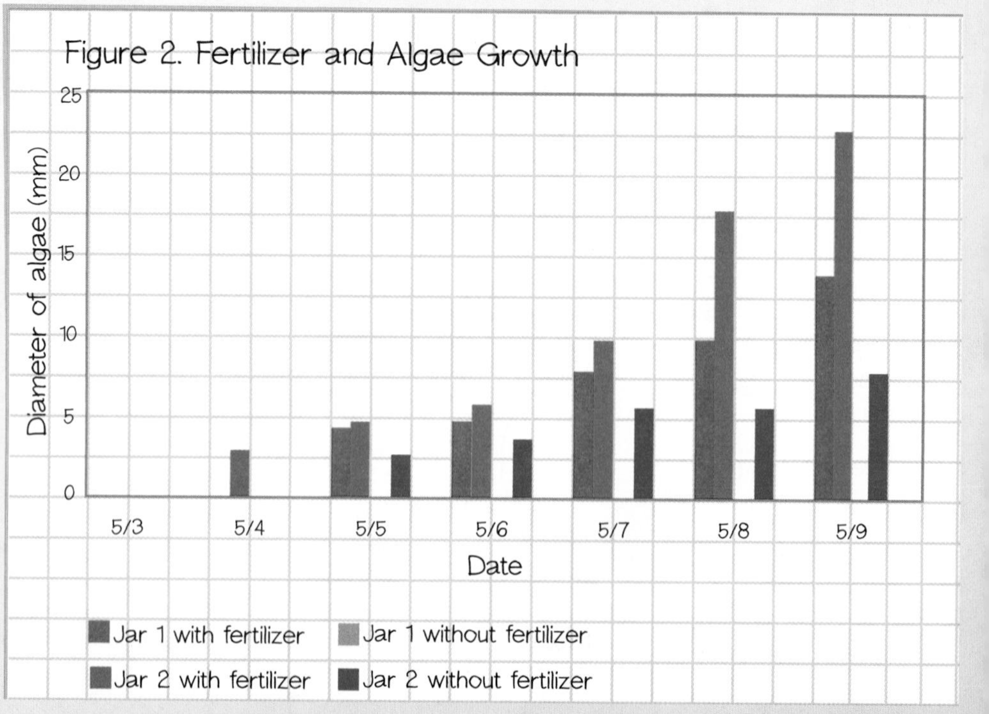

Bar graphs are useful for comparing different data sets. This bar graph has four bars for each day. Another way to present the data would be to calculate averages for the tests and the controls, and to show one test bar and one control bar for each day.

LAB HANDBOOK

Drawing Conclusions

RESULTS AND INFERENCES

To draw conclusions from your experiment, first write your results. Then compare your results with your hypothesis. Do your results support your hypothesis? Be careful not to make inferences about factors that you did not test.

For help with making inferences, see page R4.

Results and Inferences

The results of my experiment show that more algae grew in lake water to which fertilizer had been added than in lake water to which no fertilizer had been added. My hypothesis was supported. I infer that it is possible that the growth of algae in the lake was caused by the fertilizer used on the field.

Notice that you cannot conclude from this experiment that the presence of algae in the lake was due only to the fertilizer.

QUESTIONS FOR FURTHER RESEARCH

Write a list of questions for further research and investigation. Your ideas may lead you to new experiments and discoveries.

Questions for Further Research

- What is the connection between the amount of fertilizer and algae growth?
- How do different brands of fertilizer affect algae growth?
- How would algae growth in the lake be affected if no fertilizer were used on the field?
- How do algae affect the lake and the other life in and around it?
- How does fertilizer affect the lake and the life in and around it?
- If fertilizer is getting into the lake, how is it getting there?

Describing a Set of Data

Means, medians, modes, and ranges are important math tools for describing data sets such as the following widths of fossilized clamshells.

13 mm 25 mm 14 mm 21 mm 16 mm 23 mm 14 mm

Mean

The **mean** of a data set is the sum of the values divided by the number of values.

Example

To find the mean of the clamshell data, add the values and then divide the sum by the number of values.

$$\frac{13\text{ mm} + 25\text{ mm} + 14\text{ mm} + 21\text{ mm} + 16\text{ mm} + 23\text{ mm} + 14\text{ mm}}{7} = \frac{126\text{ mm}}{7} = 18\text{ mm}$$

ANSWER The mean is 18 mm.

Median

The **median** of a data set is the middle value when the values are written in numerical order. If a data set has an even number of values, the median is the mean of the two middle values.

Example

To find the median of the clamshell data, arrange the values in order from least to greatest. The median is the middle value.

13 mm 14 mm 14 mm 16 mm 21 mm 23 mm 25 mm

ANSWER The median is 16 mm.

MATH HANDBOOK

Mode

The **mode** of a data set is the value that occurs most often.

Example

To find the mode of the clamshell data, arrange the values in order from least to greatest and determine the value that occurs most often.

13 mm 14 mm 14 mm 16 mm 21 mm 23 mm 25 mm

ANSWER The mode is 14 mm.

A data set can have more than one mode or no mode. For example, the following data set has modes of 2 mm and 4 mm:

2 mm 2 mm 3 mm 4 mm 4 mm

The data set below has no mode, because no value occurs more often than any other.

2 mm 3 mm 4 mm 5 mm

Range

The **range** of a data set is the difference between the greatest value and the least value.

Example

To find the range of the clamshell data, arrange the values in order from least to greatest.

13 mm 14 mm 14 mm 16 mm 21 mm 23 mm 25 mm

Subtract the least value from the greatest value.

13 mm is the least value.
25 mm is the greatest value.

25 mm – 13 mm = 12 mm

ANSWER The range is 12 mm.

MATH HANDBOOK

Using Ratios, Rates, and Proportions

You can use ratios and rates to compare values in data sets. You can use proportions to find unknown values.

Ratios

A **ratio** uses division to compare two values. The ratio of a value a to a nonzero value b can be written as $\frac{a}{b}$.

Example

The height of one plant is 8 centimeters. The height of another plant is 6 centimeters. To find the ratio of the height of the first plant to the height of the second plant, write a fraction and simplify it.

$$\frac{8\text{ cm}}{6\text{ cm}} = \frac{4 \times \overset{1}{\cancel{2}}}{3 \times \underset{1}{\cancel{2}}} = \frac{4}{3}$$

ANSWER The ratio of the plant heights is $\frac{4}{3}$.

You can also write the ratio $\frac{a}{b}$ as "a to b" or as $a:b$. For example, you can write the ratio of the plant heights as "4 to 3" or as 4:3.

Rates

A **rate** is a ratio of two values expressed in different units. A unit rate is a rate with a denominator of 1 unit.

Example

A plant grew 6 centimeters in 2 days. The plant's rate of growth was $\frac{6\text{ cm}}{2\text{ days}}$. To describe the plant's growth in centimeters per day, write a unit rate.

Divide numerator and denominator by 2: $\frac{6\text{ cm}}{2\text{ days}} = \frac{6\text{ cm} \div 2}{2\text{ days} \div 2}$

Simplify: $= \frac{3\text{ cm}}{1\text{ day}}$

You divide 2 days by 2 to get 1 day, so divide 6 cm by 2 also.

ANSWER The plant's rate of growth is 3 centimeters per day.

MATH HANDBOOK

Proportions

A **proportion** is an equation stating that two ratios are equivalent. To solve for an unknown value in a proportion, you can use cross products.

Example

If a plant grew 6 centimeters in 2 days, how many centimeters would it grow in 3 days (if its rate of growth is constant)?

Write a proportion: $\frac{6 \text{ cm}}{2 \text{ days}} = \frac{x \text{ cm}}{3 \text{ days}}$

Set cross products: $6 \cdot 3 = 2x$

Multiply 6 and 3: $18 = 2x$

Divide each side by 2: $\frac{18}{2} = \frac{2x}{2}$

Simplify: $9 = x$

ANSWER The plant would grow 9 centimeters in 3 days.

Using Decimals, Fractions, and Percents

Decimals, fractions, and percentages are all ways of recording and representing data.

Decimals

A **decimal** is a number that is written in the base-ten place value system, in which a decimal point separates the ones and tenths digits. The values of each place is ten times that of the place to its right.

Example

A caterpillar traveled from point *A* to point *C* along the path shown.

ADDING DECIMALS To find the total distance traveled by the caterpillar, add the distance from *A* to *B* and the distance from *B* to *C*. Begin by lining up the decimal points. Then add the figures as you would whole numbers and bring down the decimal point.

$$\begin{array}{r} 36.9 \text{ cm} \\ +\ 52.4 \text{ cm} \\ \hline 89.3 \text{ cm} \end{array}$$

ANSWER The caterpillar traveled a total distance of 89.3 centimeters.

Example *continued*

SUBTRACTING DECIMALS To find how much farther the caterpillar traveled on the second leg of the journey, subtract the distance from *A* to *B* from the distance from *B* to *C*.

```
  52.4 cm
– 36.9 cm
---------
  15.5 cm
```

ANSWER The caterpillar traveled 15.5 centimeters farther on the second leg of the journey.

Example

A caterpillar is traveling from point *D* to point *F* along the path shown. The caterpillar travels at a speed of 9.6 centimeters per minute.

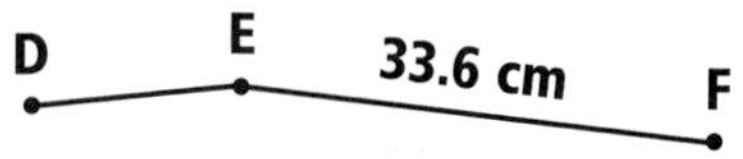

MULTIPLYING DECIMALS You can multiply decimals as you would whole numbers. The number of decimal places in the product is equal to the sum of the number of decimal places in the factors.

For instance, suppose it takes the caterpillar 1.5 minutes to go from *D* to *E*. To find the distance from *D* to *E*, multiply the caterpillar's speed by the time it took.

```
   9.6      1   decimal place
 × 1.5    + 1   decimal place
 -----    ---
   480
   96
 -----
 14.40      2   decimal places
```

Align as shown.

ANSWER The distance from *D* to *E* is 14.4 centimeters.

DIVIDING DECIMALS When you divide by a decimal, move the decimal points the same number of places in the divisor and the dividend to make the divisor a whole number.

For instance, to find the time it will take the caterpillar to travel from *E* to *F*, divide the distance from *E* to *F* by the caterpillar's speed.

```
9.6 )33.6

      3.5
96 )336.
    288
    ---
     480
     480
     ---
       0
```

Move each decimal point one place to the right.

Line up decimal points.

ANSWER The caterpillar will travel from *E* to *F* in 3.5 minutes.

Fractions

A **fraction** is a number in the form $\frac{a}{b}$, where b is not equal to 0. A fraction is in **simplest form** if its numerator and denominator have a greatest common factor (GCF) of 1. To simplify a fraction, divide its numerator and denominator by their GCF.

Example

A caterpillar is 40 millimeters long. The head of the caterpillar is 6 millimeters long. To compare the length of the caterpillar's head with the caterpillar's total length, you can write and simplify a fraction that expresses the ratio of the two lengths.

Write the ratio of the two lengths: $\frac{\text{Length of head}}{\text{Total length}} = \frac{6 \text{ mm}}{40 \text{ mm}}$

Write numerator and denominator as products of numbers and the GCF: $= \frac{3 \times 2}{20 \times 2}$

Divide numerator and denominator by the GCF: $= \frac{3 \times \overset{1}{\cancel{2}}}{20 \times \underset{1}{\cancel{2}}}$

Simplify: $= \frac{3}{20}$

ANSWER In simplest form, the ratio of the lengths is $\frac{3}{20}$.

Percents

A **percent** is a ratio that compares a number to 100. The word *percent* means "per hundred" or "out of 100." The symbol for *percent* is %.

For instance, suppose 43 out of 100 caterpillars are female. You can represent this ratio as a percent, a decimal, or a fraction.

Percent	Decimal	Fraction
43%	0.43	$\frac{43}{100}$

Example

In the preceding example, the ratio of the length of the caterpillar's head to the caterpillar's total length is $\frac{3}{20}$. To write this ratio as a percent, write an equivalent fraction that has a denominator of 100.

Multiply numerator and denominator by 5: $\frac{3}{20} = \frac{3 \times 5}{20 \times 5}$

$= \frac{15}{100}$

Write as a percent: $= 15\%$

ANSWER The caterpillar's head represents 15 percent of its total length.

MATH HANDBOOK

Using Formulas

A mathematical **formula** is a statement of a fact, rule, or principle. It is usually expressed as an equation.

The term *variable* is also used in science to refer to a factor that can change during an experiment.

In science, a formula often has a word form and a symbolic form. The formula below expresses Ohm's law.

Word Form	**Symbolic Form**
$\text{Current} = \frac{\text{voltage}}{\text{resistance}}$	$I = \frac{V}{R}$

In this formula, *I*, *V*, and *R* are variables. A mathematical **variable** is a symbol or letter that is used to represent one or more numbers.

Example

Suppose that you measure a voltage of 1.5 volts and a resistance of 15 ohms. You can use the formula for Ohm's law to find the current in amperes.

Write the formula for Ohm's law: $I = \frac{V}{R}$

Substitute 1.5 volts for V and 15 ohms for R: $I = \frac{1.5 \text{ volts}}{15 \text{ ohms}}$

Simplify: $I = 0.1 \text{ amp}$

ANSWER The current is 0.1 ampere.

If you know the values of all variables but one in a formula, you can solve for the value of the unknown variable. For instance, Ohm's law can be used to find a voltage if you know the current and the resistance.

Example

Suppose that you know that a current is 0.2 amperes and the resistance is 18 ohms. Use the formula for Ohm's law to find the voltage in volts.

Write the formula for Ohm's law: $I = \frac{V}{R}$

Substitute 0.2 amp for I and 18 ohms for R: $0.2 \text{ amp} = \frac{V}{18 \text{ ohms}}$

Multiply both sides by 18 ohms: $0.2 \text{ amp} \cdot 18 \text{ ohms} = V$

Simplify: $3.6 \text{ volts} = V$

ANSWER The voltage is 3.6 volts.

MATH HANDBOOK

Finding Areas

The area of a figure is the amount of surface the figure covers.

Area is measured in square units, such as square meters (m^2) or square centimeters (cm^2). Formulas for the areas of three common geometric figures are shown below.

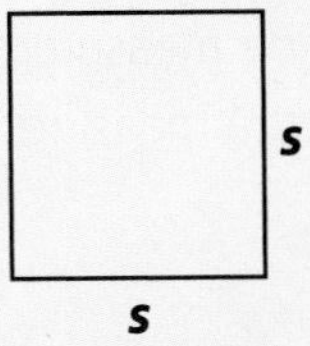

Area = (side length)^2
$A = s^2$

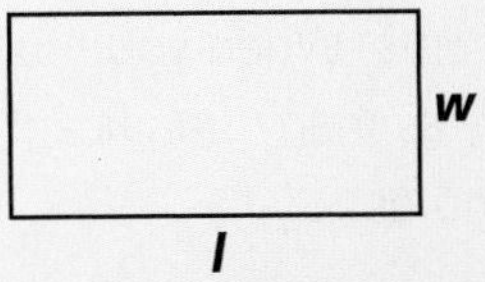

Area = length × width
$A = lw$

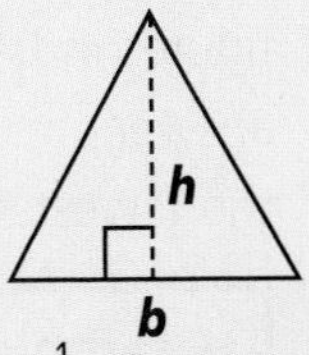

Area = $\frac{1}{2}$ × base × height
$A = \frac{1}{2}bh$

Example

Each face of a halite crystal is a square like the one shown. You can find the area of the square by using the steps below.

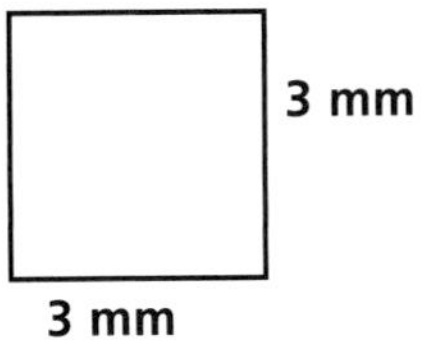

Write the formula for the area of a square:	$A = s^2$
Substitute 3 mm for s*:*	$= (3 \text{ mm})^2$
Simplify:	$= 9 \text{ mm}^2$

ANSWER The area of the square is 9 square millimeters.

Finding Volumes

The volume of a solid is the amount of space contained by the solid.

Volume is measured in cubic units, such as cubic meters (m^3) or cubic centimeters (cm^3). The volume of a rectangular prism is given by the formula shown below.

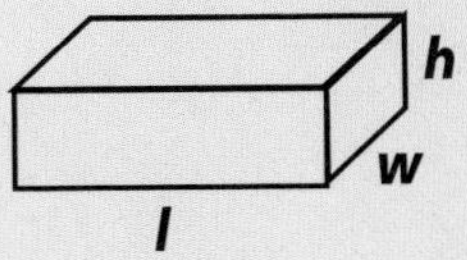

Volume = length × width × height
$V = lwh$

Example

A topaz crystal is a rectangular prism like the one shown. You can find the volume of the prism by using the steps below.

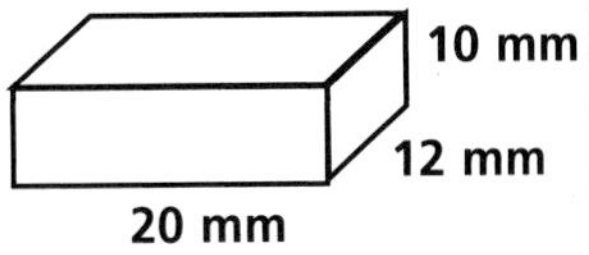

Write the formula for the volume of a rectangular prism:	$V = lwh$
Substitute dimensions:	$= 20 \text{ mm} \times 12 \text{ mm} \times 10 \text{ mm}$
Simplify:	$= 2400 \text{ mm}^3$

ANSWER The volume of the rectangular prism is 2400 cubic millimeters.

Using Significant Figures

The significant figures in a decimal are the digits that are warranted by the accuracy of a measuring device.

When you perform a calculation with measurements, the number of significant figures to include in the result depends in part on the number of significant figures in the measurements. When you multiply or divide measurements, your answer should have only as many significant figures as the measurement with the fewest significant figures.

Example

Using a balance and a graduated cylinder filled with water, you determined that a marble has a mass of 8.0 grams and a volume of 3.5 cubic centimeters. To calculate the density of the marble, divide the mass by the volume.

Write the formula for density: $\text{Density} = \frac{\text{mass}}{\text{Volume}}$

Substitute measurements: $= \frac{8.0\text{ g}}{3.5\text{ cm}^3}$

Use a calculator to divide: $\approx 2.285714286\text{ g/cm}^3$

ANSWER Because the mass and the volume have two significant figures each, give the density to two significant figures. The marble has a density of 2.3 grams per cubic centimeter.

Using Scientific Notation

Scientific notation is a shorthand way to write very large or very small numbers. For example, 73,500,000,000,000,000,000,000 kg is the mass of the Moon. In scientific notation, it is 7.35×10^{22} kg.

Example

You can convert from standard form to scientific notation.

Standard Form	Scientific Notation
720,000 5 decimal places left	7.2×10^5 Exponent is 5.
0.000291 4 decimal places right	2.91×10^{-4} Exponent is –4.

You can convert from scientific notation to standard form.

Scientific Notation	Standard Form
4.63×10^7 Exponent is 7.	46,300,000 7 decimal places right
1.08×10^{-6} Exponent is –6.	0.00000108 6 decimal places left

MATH HANDBOOK

Note-Taking Handbook

Note-Taking Strategies

Taking notes as you read helps you understand the information. The notes you take can also be used as a study guide for later review. This handbook presents several ways to organize your notes.

Content Frame

1. Make a chart in which each column represents a category.
2. Give each column a heading.
3. Write details under the headings.

NAME	GROUP	CHARACTERISTICS	DRAWING
snail	mollusks	mantle, shell	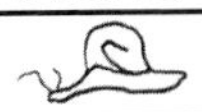
ant	arthropods	six legs, exoskeleton	
earthworm	segmented worms	segmented body, circulatory and digestive systems	
heartworm	roundworms	digestive system	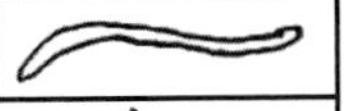
sea star	echinoderms	spiny skin, tube feet	
jellyfish	cnidarians	stinging cells	

categories

details

Combination Notes

1. For each new idea or concept, write an informal outline of the information.
2. Make a sketch to illustrate the concept, and label it.

NOTES

Types of forces
- contact force
- gravity
- friction

informal outline

forces on a box being pushed

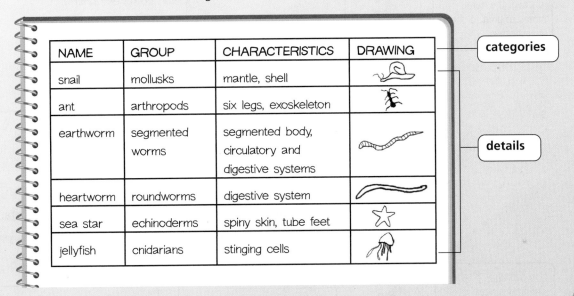

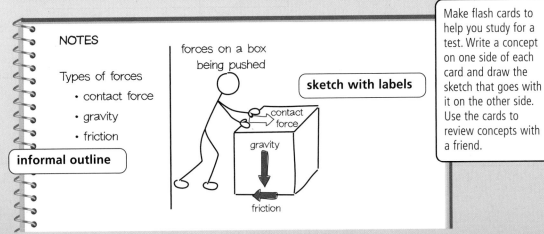

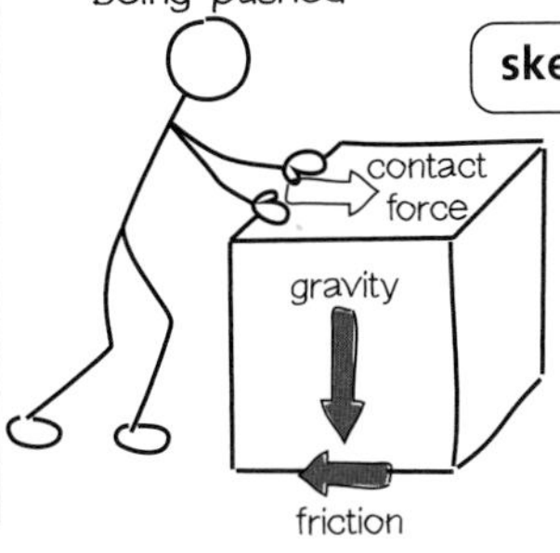

sketch with labels

Make flash cards to help you study for a test. Write a concept on one side of each card and draw the sketch that goes with it on the other side. Use the cards to review concepts with a friend.

NOTE-TAKING HANDBOOK

Main Idea and Detail Notes

1. In the left-hand column of a two-column chart, list main ideas. The blue headings express main ideas throughout this textbook.
2. In the right-hand column, write details that expand on each main idea.

You can shorten the headings in your chart. Be sure to use the most important words.

When studying for tests, cover up the detail notes column with a sheet of paper. Then use each main idea to form a question—such as "How does latitude affect climate?" Answer the question, and then uncover the detail notes column to check your answer.

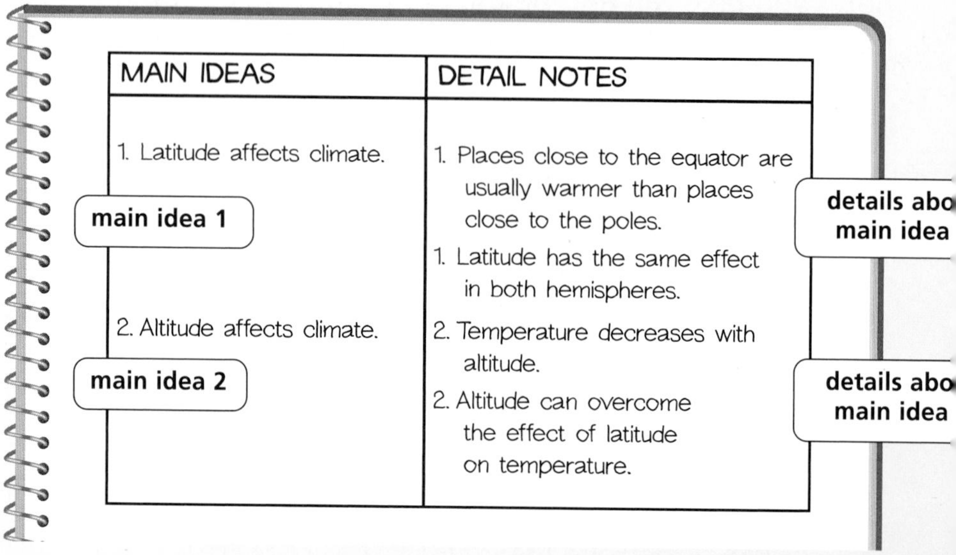
MAIN IDEAS
DETAIL NOTES
1. Latitude affects climate.
main idea 1
2. Altitude affects climate.
main idea 2
1. Places close to the equator are usually warmer than places close to the poles.
1. Latitude has the same effect in both hemispheres.
2. Temperature decreases with altitude.
2. Altitude can overcome the effect of latitude on temperature.
details abo
main idea
details abo
main idea

Main Idea Web

1. Write a main idea in a box.
2. Add boxes around it with related vocabulary terms and important details.

You can find definitions near highlighted terms.

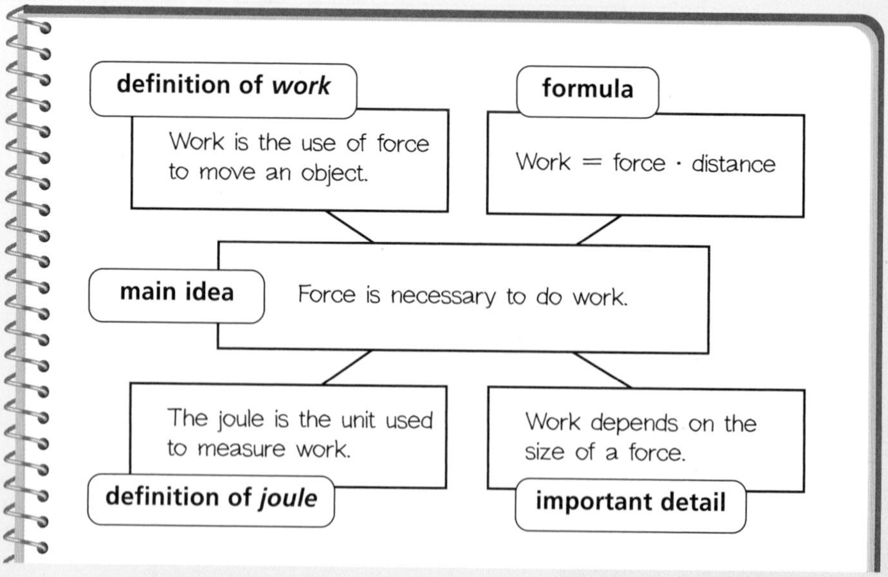
definition of *work*
Work is the use of force to move an object.
formula
Work = force · distance
main idea
Force is necessary to do work.
The joule is the unit used to measure work.
definition of *joule*
Work depends on the size of a force.
important detail

Mind Map

1. Write a main idea in the center.
2. Add details that relate to one another and to the main idea.

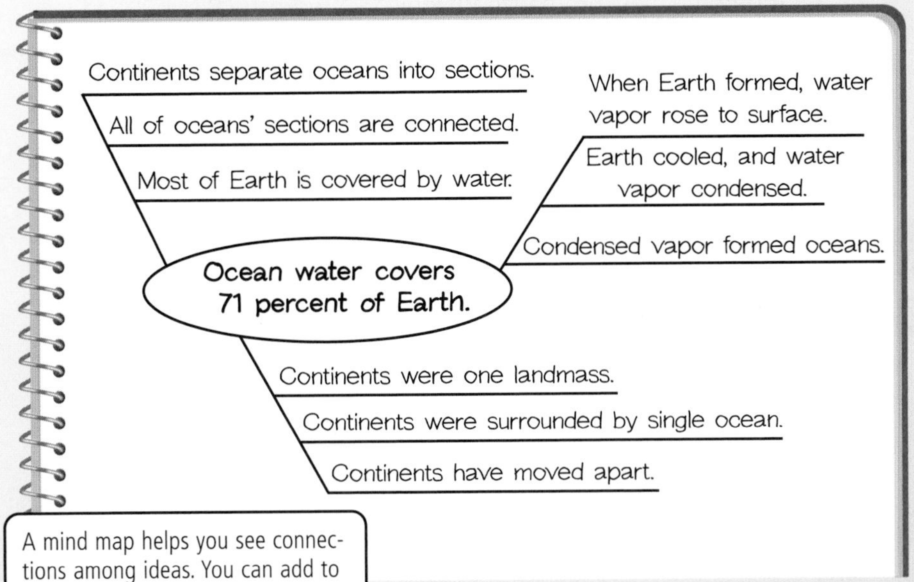

Continents separate oceans into sections.
All of oceans' sections are connected.
Most of Earth is covered by water.
When Earth formed, water vapor rose to surface.
Earth cooled, and water vapor condensed.
Condensed vapor formed oceans.
Ocean water covers 71 percent of Earth.
Continents were one landmass.
Continents were surrounded by single ocean.
Continents have moved apart.

A mind map helps you see connections among ideas. You can add to the map as you continue reading.

Supporting Main Ideas

1. Write a main idea in a box.
2. Add boxes underneath with information—such as reasons, explanations, and examples—that supports the main idea.

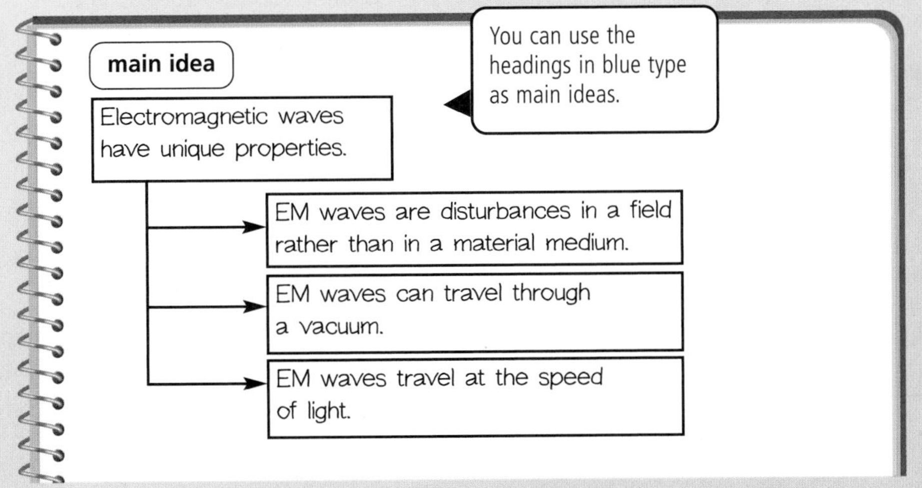

main idea
Electromagnetic waves have unique properties.
EM waves are disturbances in a field rather than in a material medium.
EM waves can travel through a vacuum.
EM waves travel at the speed of light.
You can use the headings in blue type as main ideas.

NOTE-TAKING HANDBOOK

Outline

1. Copy the chapter title and headings from the book in the form of an outline.
2. Add notes that summarize in your own words what you read.

1st key idea
1st subpoint of I
2nd subpoint of I
1st detail about B
2nd detail about B

Cell Processes

I. Cells capture and release energy.
   A. All cells need energy.
   B. Some cells capture light energy.
      1. Process of photosynthesis
      2. Chloroplasts (site of photosynthesis)
      3. Carbon dioxide and water as raw materials
      4. Glucose and oxygen as products
   C. All cells release energy.
      1. Process of cellular respiration
      2. Fermentation of sugar to carbon dioxide
      3. Bacteria that carry out fermentation
II. Cells transport materials through membranes.
   A. Some materials move by diffusion.
      1. Particle movement from higher to lower concentrations
      2. Movement of water through membrane (osmosis)
   B. Some transport requires energy.
      1. Active transport
      2. Examples of active transport

Correct Outline Form

Include a title.

Arrange key ideas, subpoints, and details as shown.

Indent the divisions of the outline as shown.

Use the same grammatical form for items of the same rank. For example, if A is a sentence, B must also be a sentence.

You must have at least two main ideas or subpoints. That is, every A must be followed by a B, and every 1 must be followed by a 2.

Concept Map

1. Write an important concept in a large oval.
2. Add details related to the concept in smaller ovals.
3. Write linking words on arrows that connect the ovals.

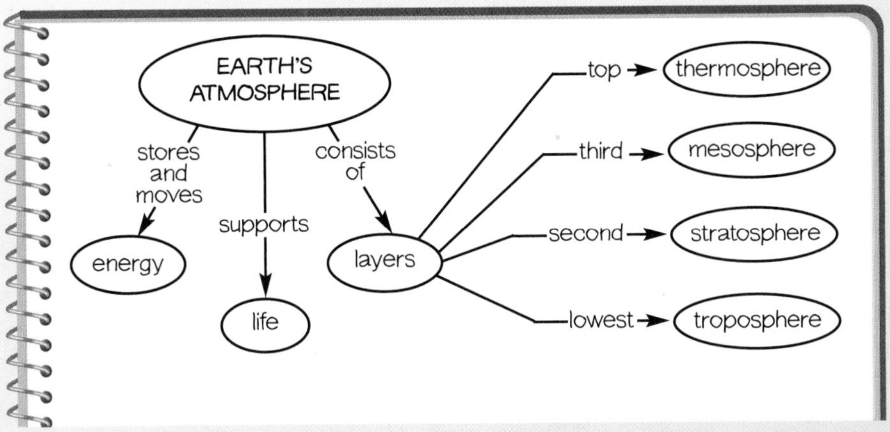

The main ideas or concepts can often be found in the blue headings. An example is "The atmosphere stores and moves energy." Use nouns from these concepts in the ovals, and use the verb or verbs on the lines.

Venn Diagram

1. Draw two overlapping circles, one for each item that you are comparing.
2. In the overlapping section, list the characteristics that are shared by both items.
3. In the outer sections, list the characteristics that are peculiar to each item.
4. Write a summary that describes the information in the Venn diagram.

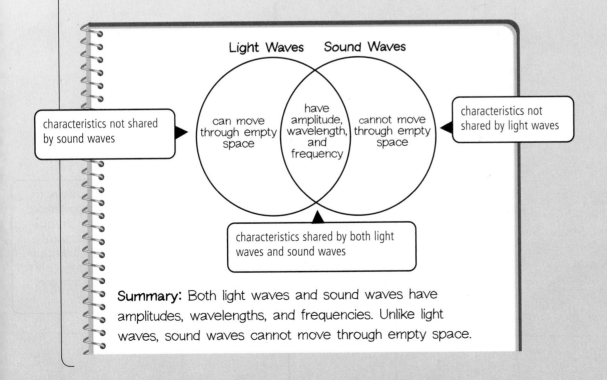

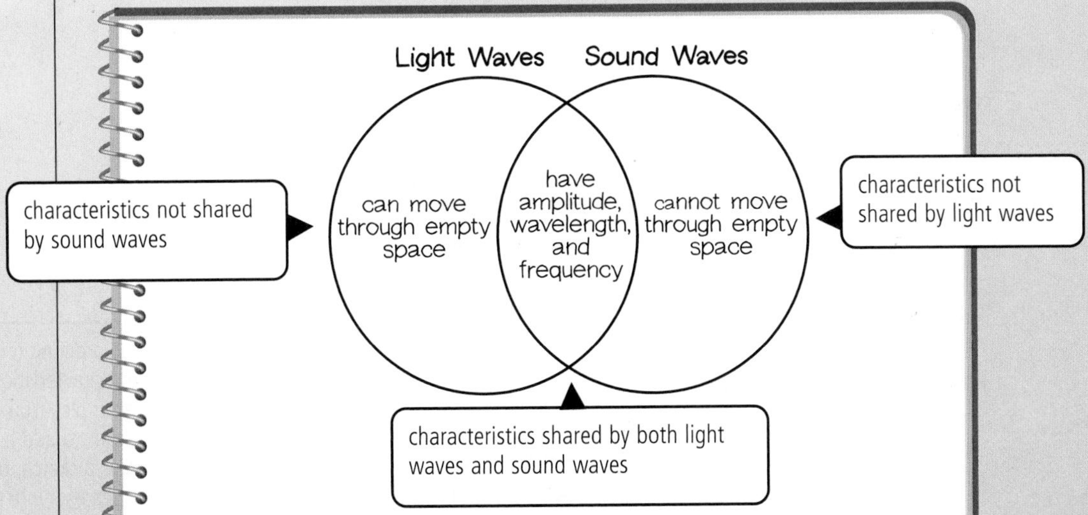

Summary: Both light waves and sound waves have amplitudes, wavelengths, and frequencies. Unlike light waves, sound waves cannot move through empty space.

Vocabulary Strategies

Important terms are highlighted in this book. A definition of each term can be found in the sentence or paragraph where the term appears. You can also find definitions in the Glossary. Taking notes about vocabulary terms helps you understand and remember what you read.

Description Wheel

1. Write a term inside a circle.
2. Write words that describe the term on "spokes" attached to the circle.

When studying for a test with a friend, read the phrases on the spokes one at a time until your friend identifies the correct term.

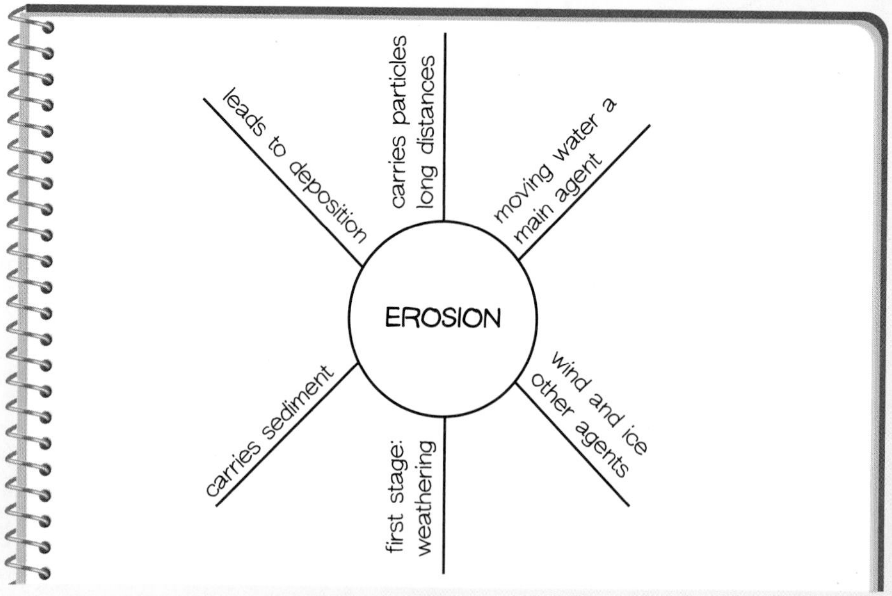

Four Square

1. Write a term in the center.
2. Write details in the four areas around the term.

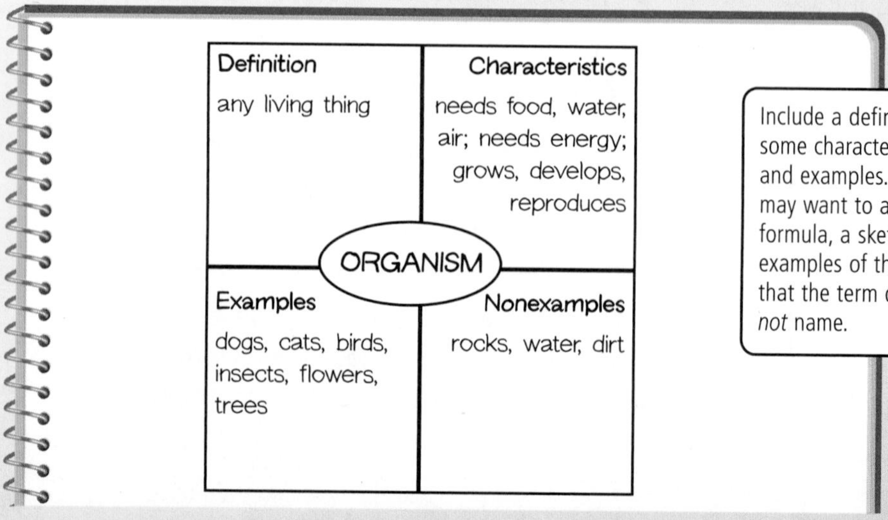

Include a definition, some characteristics, and examples. You may want to add a formula, a sketch, or examples of things that the term does *not* name.

Frame Game

1. Write a term in the center.
2. Frame the term with details.

Include examples, descriptions, sketches, or sentences that use the term in context. Change the frame to fit each new term.

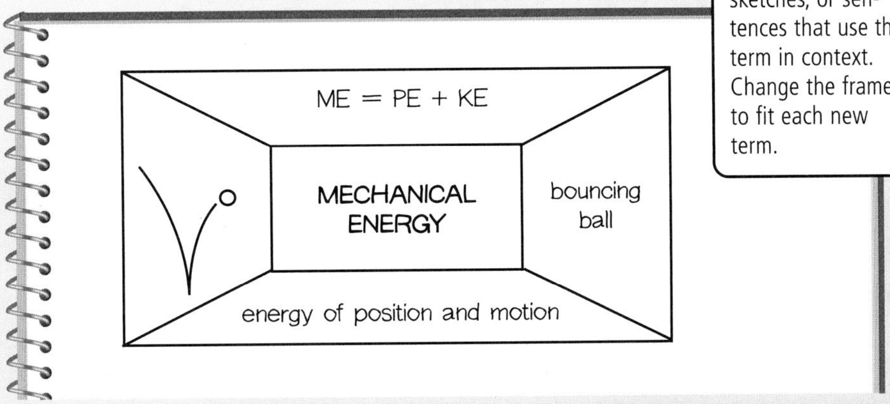

Magnet Word

1. Write a term on the magnet.
2. On the lines, add details related to the term.

You can also use phrases or sentences on the lines.

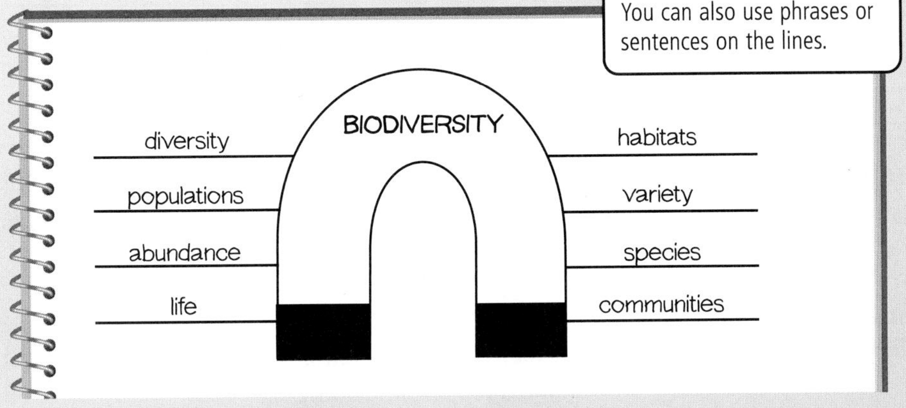

Word Triangle

1. Write a term and its definition in the bottom section.
2. In the middle section, write a sentence in which the term is used correctly.
3. In the top section, draw a small picture to illustrate the term.

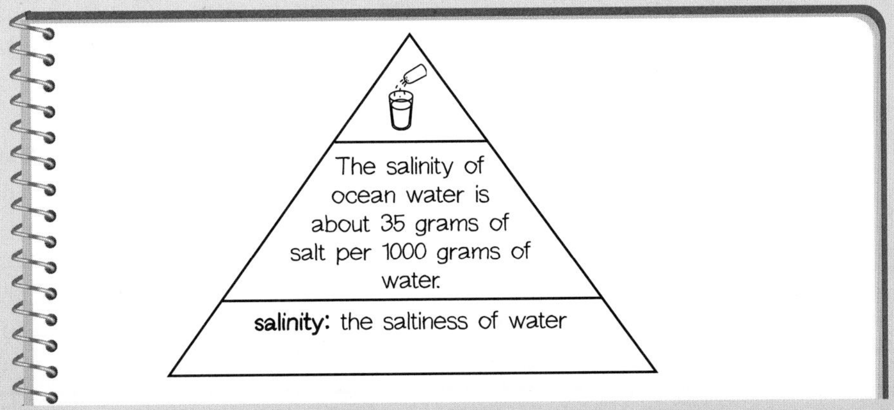

R51

Appendix

Classification of Living Things

Living things are classified into three domains. These domains are further divided into kingdoms, and then phyla. Major phyla are described in the table below, along with important features that are used to distinguish each group.

Classification of Living Things

Domain	Kingdom	Phylum	Common Name and Description
Archaea	**Archaea**		Single-celled, with no nucleus. Live in some of Earth's most extreme environments, including salty, hot, and acid environments, and the deep ocean.
Bacteria	**Bacteria**		Single-celled, with no nucleus, but chemically different from Archaea. Live in all types of environments, including the human body; reproduce by dividing from one cell into two. Includes blue-green bacteria (cyanobacteria), *Streptococcus,* and *Bacillus.*
Eukarya			Cells are larger than archaea or bacteria and are eukaryotic (have a nucleus containing DNA). Single-celled or multicellular.
	Protista		Usually single-celled, but sometimes multicellular. DNA contained in a nucleus. Many phyla resemble plants, fungi, or animals but are usually smaller or simpler in structure.
	Animal-like protists	Ciliophora	Ciliates; have many short, hairlike extensions called cilia, which they use for feeding and movement. Includes paramecium.
		Zoomastigina	Zooflagellates; have usually one or two long, hairlike extensions called flagella.
		Sporozoa	Cause diseases in animals such as birds, fish, and humans. Includes *Plasmodium,* which causes malaria.
		Sarcodina	Use footlike extensions to move and feed. Includes foraminifers and amoebas. Sometimes called Rhizopoda.
	Plantlike protists	Euglenozoa	Single-celled, with one flagellum. Some have chloroplasts that carry out photosynthesis. Includes euglenas and *Trypanosoma,* which causes African sleeping sickness.
		Dinoflagellata	Dinoflagellates; usually single-celled; usually have chloroplasts and flagellum. In great numbers, some species can cause red tides along coastlines.

APPENDIX

Classification of Living Things (cont.)

Domain	Kingdom	Phylum	Common Name and Description
		Chrysophyta	Yellow algae, golden-brown algae, and diatoms; single-celled; named for the yellow pigments in their chloroplasts (*chrysophyte,* in Greek, means "golden plant").
		Chlorophyceae	Green algae; have chloroplasts and are chemically similar to land plants. Unicellular or forms simple colonies of cells. Includes *Chlamydomonas, Ulva* (sea lettuce), and *Volvox.*
		Phaeophyta	Brown seaweed; contain a special brown pigment that gives these organisms their color. Multicellular, live mainly in salt water; includes kelp.
		Rhodophyta	Red algae; contain a red pigment that makes these organisms red, purple, or reddish-black. Multicellular, live in salt water.
	Funguslike protists	Acrasiomycota	Cellular slime molds; live partly as free-living single-celled organisms, then fuse together to form a many-celled mass. Live in damp, nutrient-rich environments; decomposers.
		Myxomycota	Acellular slime molds; form large, slimy masses made of many nuclei but technically a single cell.
		Oomycota	Water molds and downy mildews; produce thin, cottonlike extensions called hyphae. Feed off of dead or decaying material, often in water.
	Fungi		Usually multicellular; eukaryotic; cells have a thick cell wall. Obtain nutrients through absorption; often function as decomposers.
		Chytridiomycota	Oldest and simplest fungi; usually aquatic (fresh water or brackish water); single-celled or multicellular.
		Basidiomycota	Multicellular; reproduce with a club-shaped structure that is commonly seen on forest floors. Includes mushrooms, puffballs, rusts, and smuts.
		Zygomycota	Mostly disease-causing molds; often parasitic.
		Ascomycota	Includes single-celled yeasts and multicellular sac fungi. Includes *Penicillium.*

Classification of Living Things (cont.)

Domain	Kingdom	Phylum	Common Name and Description
	Plantae		Multicellular and eukaryotic; make sugars using energy from sunlight. Cells have a thick cell wall of cellulose.
		Bryophyta	Mosses; small, grasslike plants that live in moist, cool environments. Includes sphagnum (peat) moss. Seedless, nonvascular plants.
		Hepatophyta	Liverworts; named for the liver-shaped structure of one part of the plant's life cycle. Live in moist environments. Seedless, nonvascular plants.
		Anthoceratophyta	Hornworts; named for the visible hornlike structures with which they reproduce. Live on forest floors and other moist, cool environments. Seedless, nonvascular plants.
		Psilotophyta	Simple plant, just two types. Includes whisk ferns found in tropical areas, a common greenhouse weed. Seedless, vascular plants.
		Lycophyta	Club mosses and quillworts; look like miniature pine trees; live in moist, wooded environments. Includes *Lycopodium* (ground pine). Seedless vascular plants.
		Sphenophyta	Plants with simple leaves, stems, and roots. Grow about a meter tall, usually in moist areas. Includes *Equisetum* (scouring rush). Seedless, vascular plants.
		Pterophyta	Ferns; fringed-leaf plants that grow in cool, wooded environments. Includes many species. Seedless, vascular plants.
		Cycadophyta	Cycads; slow-growing palmlike plants that grow in tropical environments. Reproduce with seeds.
		Ginkgophyta	Includes only one species: *Ginkgo biloba,* a tree that is often planted in urban environments. Reproduce with seeds in cones.
		Gnetophyta	Small group includes desert-dwelling and tropical species. Includes *Ephedra* (Mormon tea) and *Welwitschia,* which grows in African deserts. Reproduce with seeds.
		Coniferophyta	Conifers, including pines, spruces, firs, sequoias. Usually evergreen trees; tend to grow in cold, dry environments; reproduce with seeds produced in cones.

Classification of Living Things (cont.)

Domain	Kingdom	Phylum	Common Name and Description
		Anthophyta	Flowering plants; includes grasses and flowering trees and shrubs. Reproduce with seeds produced in flowers, becoming fruit.
	Animalia		Multicellular and eukaryotic; obtain energy by consuming food. Usually able to move around.
		Porifera	Sponges; spend most of their lives fixed to the ocean floor. Feed by filtering water (containing nutrients and small organisms) through their body.
		Cnidaria	Aquatic animals with a radial (spokelike) body shape; named for their stinging cells (cnidocytes). Includes jellyfish, hydras, sea anemones, and corals.
		Ctenophora	Comb jellies; named for the comblike rows of cilia (hairlike extensions) that are used for movement.
		Platyhelminthes	Flatworms; thin, flattened worms with simple tissues and sensory organs. Includes planaria and tapeworms, which cause diseases in humans and other hosts.
		Nematoda	Roundworms; small, round worms; many species are parasites, causing diseases in humans, such as trichinosis and elephantiasis.
		Annelida	Segmented worms; body is made of many similar segments. Includes earthworms, leeches, and many marine worms.
		Mollusca	Soft-bodied, aquatic animals that usually have an outer shell. Includes snails, mussels, clams, octopus, and squid.
		Arthropoda	Animals with an outer skeleton (exoskeleton) and jointed appendages (for example, legs or wings). Very large group that includes insects, spiders and ticks, centipedes, millipedes, and crustaceans.
		Echinodermata	Marine animals with a radial (spokelike) body shape. Includes feather stars, sea stars (starfish), sea urchins, sand dollars, and sea cucumbers.
		Chordata	Mostly vertebrates (animals with backbones) that share important stages of early development. Includes tunicates (sea squirts), fish, sharks, amphibians, reptiles, birds, and mammals.

Plant and Animal Cells

Plants and animals are eukaryotes, that is, their cells contain a nucleus and other membrane-bound structures called organelles. The diagrams on page R57 show the different structures that can be found in plant and animal cells. The table below lists the functions of the structures.

Cell Structures and Their Functions	Plant Cell	Animal Cell
Nucleus	✔	✔
stores genetic material that enables a cell to function and divide		
Cell Membrane	✔	✔
controls what comes into and goes out of a cell		
Cell wall	✔	
tough outer covering provides support		
Ribosome	✔	✔
uses genetic material to assemble materials needed to make proteins		
Endoplasmic reticulum	✔	✔
manufactures proteins and other materials a cell needs to function		
Golgi apparatus	✔	✔
finishes processing proteins and transports them		
Vesicle	✔	✔
stores and transports materials and wastes		
Mitochondrion	✔	✔
releases chemical energy stored in sugars		
Chloroplast	✔	
uses energy from sunlight to make sugars		
Lysosome		✔
breaks down food particles and wastes		

APPENDIX

Plant Cell

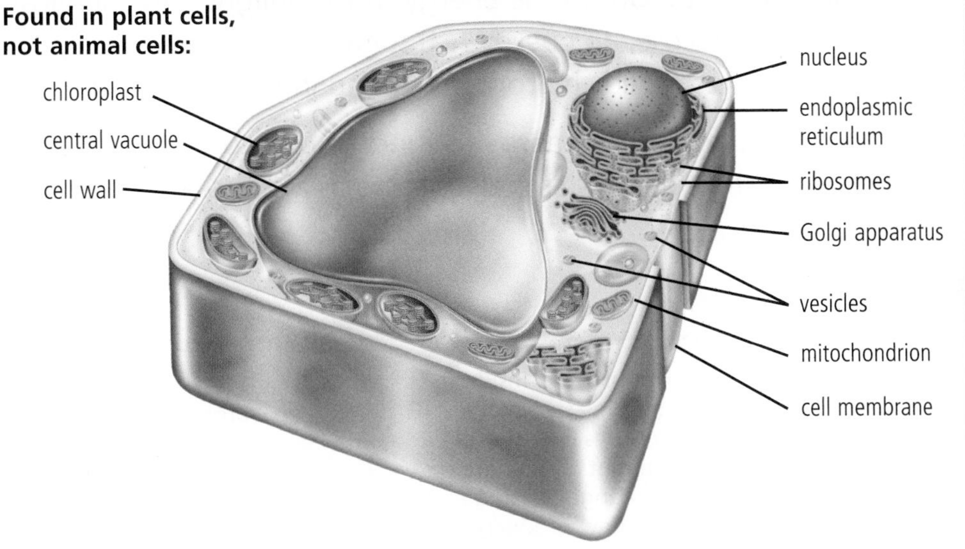

Found in plant cells, not animal cells:
chloroplast
central vacuole
cell wall
nucleus
endoplasmic reticulum
ribosomes
Golgi apparatus
vesicles
mitochondrion
cell membrane

Animal Cell

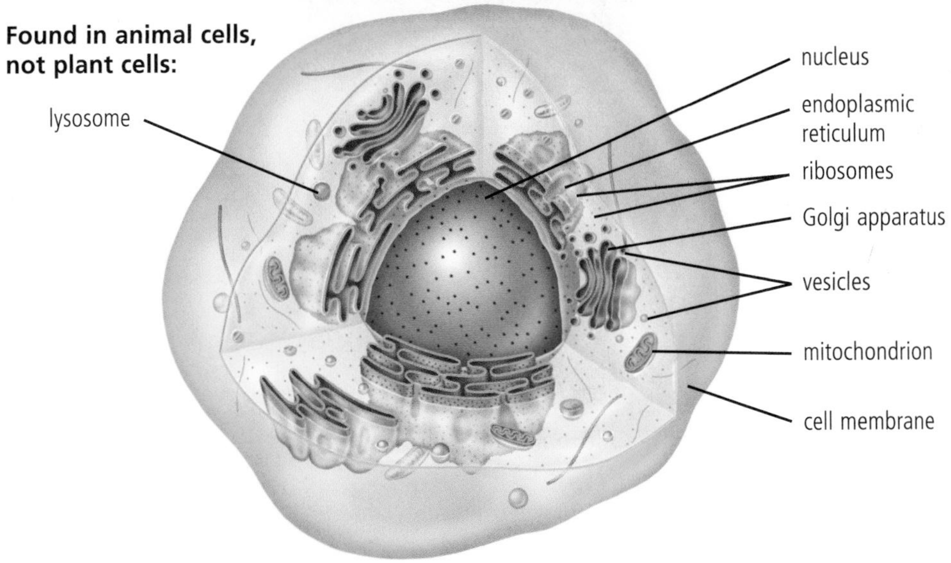

Found in animal cells, not plant cells:
lysosome
nucleus
endoplasmic reticulum
ribosomes
Golgi apparatus
vesicles
mitochondrion
cell membrane

APPENDIX

Photosynthesis and Cellular Respiration

The source of energy for almost all organisms is the Sun. Plants and other photosynthetic organisms such as algae change the energy in sunlight into a form of energy that cells can use—chemical energy. Photosynthesis is the process that changes the energy from sunlight into chemical energy and produces glucose, an energy-rich sugar. The process takes place in cellular structures called chloroplasts, found in the cytoplasm.

All cells must have energy to function. Glucose and other sugars and starches store energy, as well as serve as a source of material for cells.

Photosynthesis

1. **The starting materials** Carbon dioxide from the air and water from the soil enter the chloroplasts.
2. **The process** Inside the chloroplasts, chlorophyll captures energy from sunlight. This energy is used to change starting materials into new products.
3. **The products** Glucose provides energy and materials for the plant; most oxygen is released into the air.

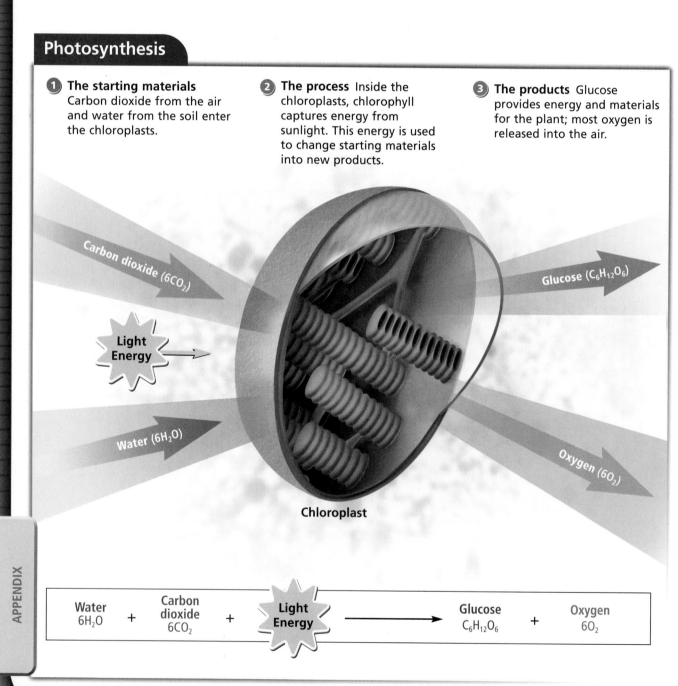

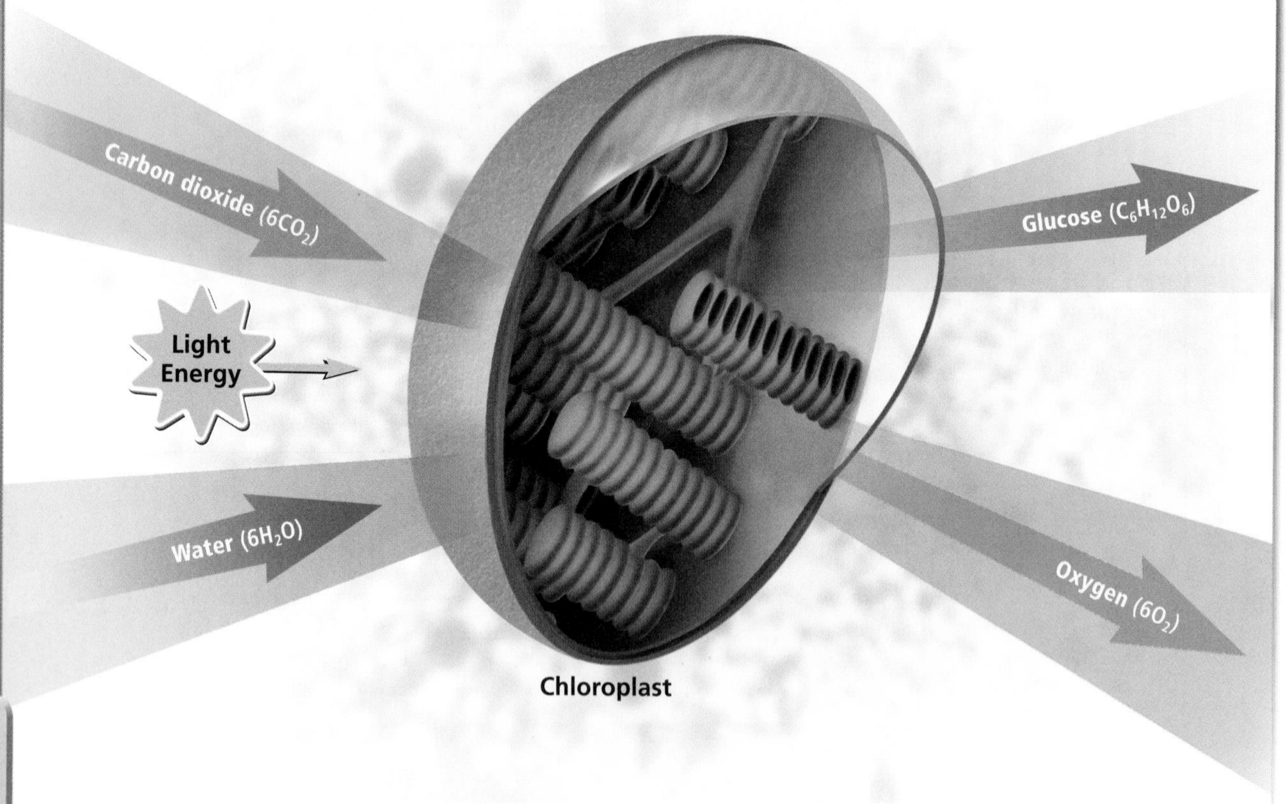

Water $6H_2O$ + Carbon dioxide $6CO_2$ + Light Energy ⟶ Glucose $C_6H_{12}O_6$ + Oxygen $6O_2$

APPENDIX

Chemical energy is stored in the bonds of the sugar and starch molecules. Cellular respiration is the process that releases energy from sugars. The process takes place in cellular structures called mitochondria, found in the cytoplasm. In cellular respiration, cells use oxygen to release energy as the molecules are broken down.

Compare the starting materials and products of photosynthesis with those of cellular respiration. The starting materials of one process are the products of the other process.

Cellular Respiration

1. **The starting materials** Glucose and oxygen enter the cell. Glucose is split into smaller molecules.
2. **The process** Inside the mitochondria more chemical bonds are broken in the smaller molecules. Oxygen is needed for this process.
3. **The products** Energy is released, and water and carbon dioxide are produced.

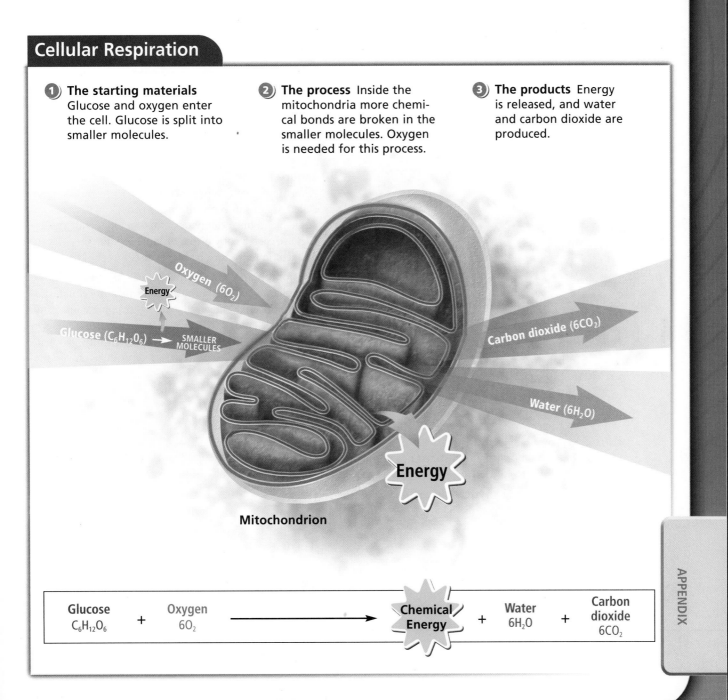

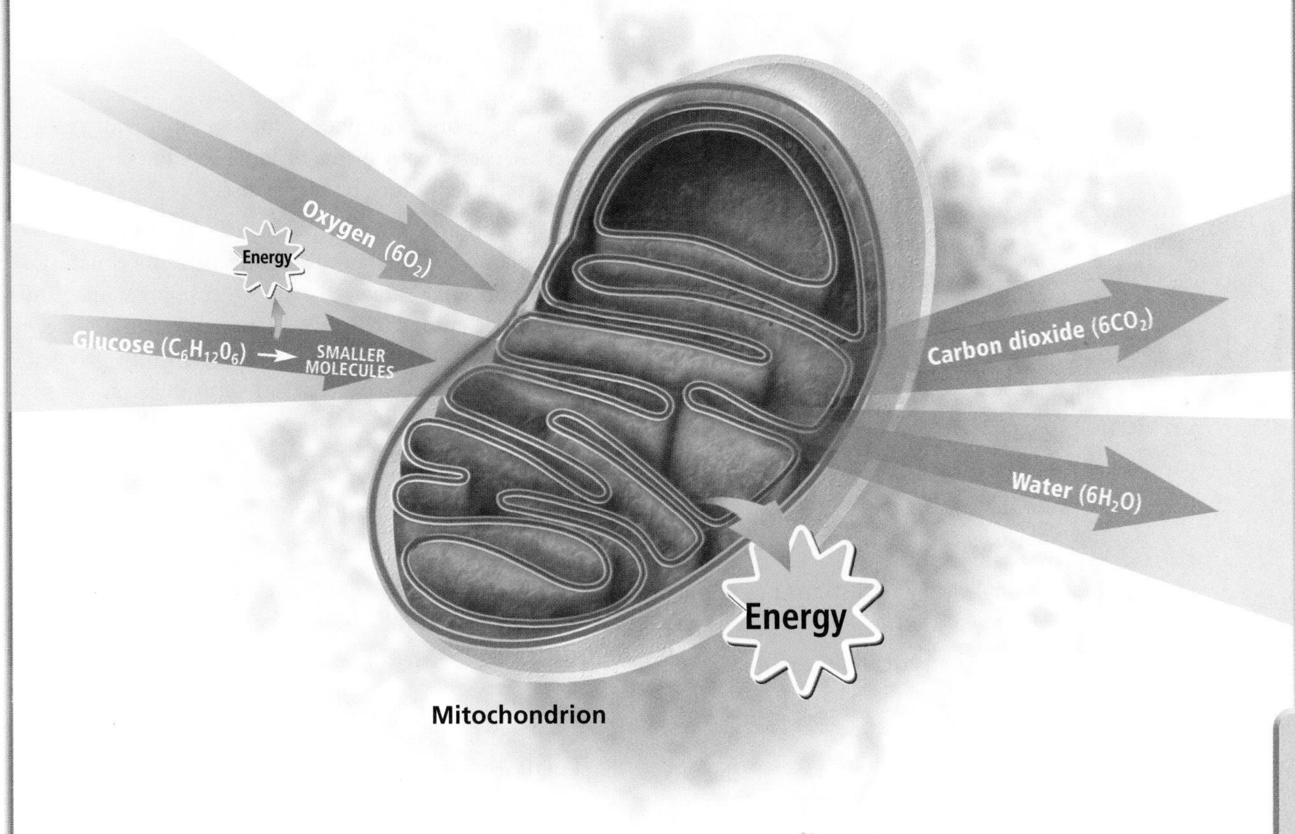

Glucose	+	Oxygen	→	Chemical Energy	+	Water	+	Carbon dioxide
$C_6H_{12}O_6$		$6O_2$				$6H_2O$		$6CO_2$

APPENDIX

Glossary

A

adaptation
A characteristic, a behavior, or any inherited trait that makes a species able to survive and reproduce in a particular environment. (p. xxi)

adaptación Una característica, un comportamiento o cualquier rasgo heredado que permite a una especie sobrevivir o reproducirse en un medio ambiente determinado.

algae
Protists that live mostly in water and use sunlight as a source of energy. *Algae* is a plural word; the singular is *alga.* (p. 31)

algas Protistas que viven principalmente en el agua y que usan la luz solar como fuente de energía.

amphibian
A cold-blooded vertebrate animal that lives in water and breathes with gills when it is young; as an adult, it moves onto land and breathes air with lungs. (p. 167)

anfibio Un vertebrado de sangre fría que vive en el agua y respira con branquias cuando es juvenil; cuando es adulto, se mueve a la tierra y respire aire con pulmones.

angiosperm (AN-jee-uh-SPURM)
A plant that has flowers and produces seeds enclosed in fruit. (p. 107)

Angiosperma Una planta que tiene flores y que produce semillas dentro de frutas.

arthropod
An invertebrate animal with an exoskeleton, a segmented body, and jointed legs. (p. 142)

artrópodo Un animal invertebrado con un exoesqueleto, un cuerpo segmentado y patas articuladas.

atom
The smallest particle of an element that has the chemical properties of that element.

átomo La partícula más pequeña de un elemento que tiene las propiedades químicas de ese elemento.

autotroph (AW-tuh-TRAHF)
An organism that captures energy from sunlight and uses it to produce energy-rich carbon compounds, usually through the process of photosynthesis. (p. 52)

autótrofo Un organismo que capta energía de la luz solar y la usa para producir compuestos de carbono ricos en energía, usualmente mediante el proceso de fotosíntesis.

B

behavior
An organism's action in response to a stimulus. (p. 62)

comportamiento La acción de un organismo en respuesta a un estímulo.

binary fission
A form of asexual reproduction by which some single-celled organisms reproduce. The genetic material is copied, and one cell divides into two independent cells that are each a copy of the original cell. Prokaryotes such as bacteria reproduce by binary fission. (p. 12)

fisión binaria Una forma asexual de reproducción mediante la cual algunos organismos unicelulares se reproducen. El material genético se copia y una célula se divide en dos células independientes las cuales son copias de la célula original. Los organismos procariotas, tales como las bacterias, se reproducen mediante fisión binaria.

biodiversity
The number and variety of living things found on Earth or within an ecosystem. (p. xxi)

biodiversidad La cantidad y variedad de organismos vivos que se encuentran en la Tierra o dentro de un ecosistema.

blubber
A layer of fat in some sea mammals that lies beneath the skin. It insulates the animal from cold and stores reserve energy. (p. 184)

grasa de ballena Una capa de tejido graso en algunos mamíferos marinos que yace bajo la piel. Aísla al animal del frió y almacena energía de reserva.

C

cell
The smallest unit that is able to perform the basic functions of life. (p. xv)

célula La unidad más pequeña capaz de realizar las funciones básicas de la vida.

cellular respiration
A process in which cells use oxygen to release energy stored in sugars. (p. 53)

respiración celular Un proceso en el cual las células usan oxígeno para liberar energía almacenada en las azúcares.

classification
The systematic grouping of different types of organisms by their shared characteristics.

clasificación La agrupación sistemática de diferentes tipos de organismos en base a las características que comparten.

cnidarian (ny-DAIR-ee-uhn)
An invertebrate animal such as a jellyfish that has a body with radial symmetry, tentacles with stinging cells, and a central internal cavity. (p. 128)

cnidario Un animal invertebrado tal como una medusa que tiene un cuerpo con simetría radial, tentáculos con células urticantes y una cavidad central interna.

compound
A substance made up of two or more different types of atoms bonded together.

compuesto Una sustancia formada por dos o más diferentes tipos de átomos enlazados.

consumer
A living thing that gets its energy by eating other living things in a food chain; consumers are also called heterotrophs. (p. 58)

consumidor Un organismo vivo que obtiene su energía alimentándose de otros organismos vivos en una cadena alimentaria; los consumidores también son llamados heterótrofos.

cycle
n. A series of events or actions that repeat themselves regularly; a physical and/or chemical process in which one material continually changes locations and/or forms. Examples include the water cycle, the carbon cycle, and the rock cycle.

v. To move through a repeating series of events or actions.

ciclo Una serie de eventos o acciones que se repiten regularmente; un proceso físico y/o químico en el cual un material cambia continuamente de lugar y/o forma. Ejemplos: el ciclo del agua, el ciclo del carbono y el ciclo de las rocas.

D

data
Information gathered by observation or experimentation that can be used in calculating or reasoning. *Data* is a plural word; the singular is *datum.*

datos Información reunida mediante observación o experimentación y que se puede usar para calcular o para razonar.

decomposer
An organism that feeds on and breaks down dead plant or animal matter. (p. 19)

descomponedor Un organismo que se alimenta de y degrada materia vegetal o animal.

density
A property of matter representing the mass per unit volume.

densidad Una propiedad de la materia que representa la masa por unidad de volumen.

DNA
The genetic material found in all living cells that contains the information needed for an organism to grow, maintain itself, and reproduce. Deoxyribonucleic acid (dee-AHK-see-RY-boh-noo-KLEE-ihk).

ADN El material genético que se encuentra en todas las céulas vivas y que contiene la información necesaria para que un organismo crezca, se mantenga a sí mismo y se reproduzca. Ácido desoxiribunucleico.

E

echinoderm
An invertebrate sea animal with a spiny skeleton, a water vascular system, and tube feet. (p. 139)

equinodermo Un animal invertebrado marino con esqueleto espinoso, sistema vascular acuífero y pies ambulacrales.

ectotherm
An animal whose body temperature changes with environmental conditions. (p. 170)

poiquilotermo o poiquilotérmico Un animal cuya temperatura corporal cambia con las condiciones del medio ambiente.

element
A substance that cannot be broken down into a simpler substance by ordinary chemical changes. An element consists of atoms of only one type.

elemento Una sustancia que no puede descomponerse en otra sustancia más simple por medio de cambios químicos normales. Un elemento consta de átomos de un solo tipo.

embryo (EHM-bree-OH)
A multicellular organism, plant or animal, in its earliest stages of development. (p. 98)

embrión Una planta o un animal en su estadio mas temprano de desarrollo.

endoskeleton
An internal support system; such a skeleton made of bone tissue is a distinguishing characteristic of vertebrate animals. (p. 157)

endoesqueleto Un sistema de soporte interno, como un esqueleto formado de tejido óseo es una característica distintiva de los animales vertebrados.

endotherm
An animal that maintains a constant body temperature. (p. 174)

homeotermo o endotermo Un animal que mantiene una temperatura corporal constante.

energy
The ability to do work or to cause a change. For example, the energy of a moving bowling ball knocks over pins; energy from food allows animals to move and to grow; and energy from the Sun heats Earth's surface and atmosphere, which causes air to move.

energía La capacidad para trabajar o causar un cambio. Por ejemplo, la energía de una bola de boliche en movimiento tumba los pinos; la energía proveniente de su alimento permite a los animales moverse y crecer; la energía del Sol calienta la superficie y la atmósfera de la Tierra, lo que ocasiona que el aire se mueva.

environment
Everything that surrounds a living thing. An environment is made up of both living and nonliving factors. (p. xix)

medio ambiente Todo lo que rodea a un organismo vivo. Un medio ambiente está compuesto de factores vivos y factores sin vida.

exoskeleton
The strong, flexible outer covering of some invertebrate animals, such as arthropods. (p. 143)

exoesqueleto La cubierta exterior fuerte y flexible de algunos animales invertebrados, como los artrópodos.

experiment
An organized procedure to study something under controlled conditions. (p. xxiv)

experimento Un procedimiento organizado para estudiar algo bajo condiciones controladas.

extinction
The permanent disappearance of a species. (p. xxi)

extinción La desaparición permanente de una especie.

F

fertilization
Part of the process of sexual reproduction in which a male reproductive cell and a female reproductive cell combine to make a new cell that can develop into a new organism. (p. 48)

fertilización El proceso mediante el cual una célula reproductiva masculina y una célula reproductiva femenina se combinan para formar una nueva célula que puede convertirse en un organismo nuevo.

flower
The reproductive structure of an angiosperm, containing male and female parts. (p. 108)

flor La estructura reproductiva de una angiosperma, contiene las partes masculinas y femeninas.

fruit
The ripened ovary of a flowering plant that contains the seeds. (p. 108)

fruta El ovario maduro de una planta floreciente que contiene las semillas.

G

genetic material
The nucleic acid DNA, which is present in all living cells and contains the information for a cell's growth, maintenance, and reproduction.

material genético El ácido nucleico ADN, ue esta presente en todas las células vivas y que contiene la información necesaria para el crecimiento, el mantenimiento y la reproducción celular.

germination (JUR-muh-NAY-shuhn)
The beginning of growth of a new plant from a spore or a seed. (p. 99)

germinación El inicio del crecimiento de una nueva planta a partir de una espora o una semilla.

gestation
In mammals, the period of time spent by a developing offspring inside the mother's body. (p. 186)

gestación En los mamíferos, el periodo de tiempo que pasa una cría en desarrollo dentro del cuerpo de la madre.

gill
A respiratory organ that filters oxgen dissolved in water. (p. 137)

branquia Un órgano respiratorio que filtra oxígeno disuelto en el agua.

gymnosperm (JIHM-nuh-SPURM)
A plant that produces seeds that are not enclosed in flowers or fruit. (p. 102)

Gimnosperma Una planta que produce semillas que no están dentro de las flores o las frutas.

H

heterotroph (HEHT-uhr-uh-TRAWF)
An organism that consumes other organisms to get energy. (p. 58)

heterótrofo Un organismo que consume a otros organismos para obtener energía.

hibernation
A sleeplike state in which certain animals spend the winter. Hibernation reduces an animal's need for food and helps protect it from cold. (p. 64)

hibernación Un estado parecido al de sueño en el cual ciertos animales pasan el invierno. La hibernación reduce la necesidad de alimento de un animal y le ayuda a protegerse del frío.

host cell
A cell that a virus infects and uses to make copies of itself. (p. 26)

célula hospedera Una célula que un virus infecta y usa para hacer copias de sí mismo.

hyphae
Threadlike tubes that form the structural parts of the body of a fungus. *Hyphae* is a plural word; the singular is *hypha*. (p. 67)

hifas Los tubos, similares a hilos, que forman las partes estructurales del cuerpo de un hongo.

hypothesis
A tentative explanation for an observation or phenomenon. A hypothesis is used to make testable predictions. (p. xxiv)

hipótesis Una explicación provisional de una observación o de un fenómeno. Una hipótesis se usa para hacer predicciones que se pueden probar.

I, J

incubation
The process of keeping eggs warm by bodily heat until they hatch. (p. 179)

incubación El proceso de mantener huevos cálidos por medio de calor corporal hasta que eclosionen.

insect
An arthropod with three body segments, six legs, two antennae, and compound eyes. (p. 145)

insecto Un artrópodo con tres segmentos corporales, seis patas, dos antenas y ojos compuestos.

interaction
The condition of acting or having an influence upon something. Living things in an ecosystem interact with both the living and nonliving parts of their environment. (p. xix)

interacción La condición de actuar o influir sobre algo. Los organismos vivos en un ecosistema interactúan con las partes vivas y las partes sin vida de su medio ambiente.

invertebrate
An animal that has no backbone. (p. 123)

invertebrado Un animal que no tiene columna vertebral.

K

kingdom
One of six large groupings of living things that have common characteristics. The kingdoms are Plantae, Animalia, Fungi, Protista, Archaea, and Bacteria. (p. 11)

reino Uno de los seis grandes grupos de organismos vivos que tienen características en común. Los reinos son Plantae, Animalia, Fungi, Protista, Archae y Bacteria.

L

larva
A free-living early form of a developing organism that is very different from its adult form. (p. 126)

larva Una etapa temprana de vida libre de un organismo en desarrollo que es muy diferente a su etapa adulta.

law
In science, a rule or principle describing a physical relationship that always works in the same way under the same conditions. The law of conservation of energy is an example.

ley En las ciencias, una regla o un principio que describe una relación física que siempre funciona de la misma manera bajo las mismas condiciones. La ley de la conservación de la energía es un ejemplo.

lichen (LY-kuhn)
An organism that results from a close association between single-celled algae and fungi. (p. 70)

liquen Un organismo que resulta de una asociación cercana entre algas unicelulares y hongos.

lung
A respiratory organ that absorbs oxygen from the air. (p. 137)

pulmón Un órgano respiratorio que absorbe oxígeno del aire.

M, N

mammal
A warm-blooded vertebrate animal whose young feed on milk produced by the mother's mammary glands. (p. 183)

mamífero Un animal vertebrado de sangre caliente cuyas crías se alimentan de leche producida por las glándulas mamarias de la madre.

mass
A measure of how much matter an object is made of.

masa Una medida de la cantidad de materia de la que está compuesto un objeto.

matter
Anything that has mass and volume. Matter exists ordinarily as a solid, a liquid, or a gas.

materia Todo lo que tiene masa y volumen. Generalmente la materia existe como sólido, líquido o gas.

meiosis (my-OH-sihs)
A part of sexual reproduction in which cells divide to form sperm cells in a male and egg cells in a female. Meiosis occurs only in reproductive cells. (p. 48)

meiosis Una parte de la reproducción sexual en la cual las células se dividen para formar espermatozoides en los machos y óvulos en las hembras. La meiosis sólo ocurre en las células reproductivas.

metamorphosis
The transformation of an animal from its larval form into its adult form. (p. 146)

metamorfosis La transformación de un animal de su forma larvaria a su forma adulta.

microorganism
A very small organism that can be seen only with a microscope. Bacteria are examples of microorganisms. (p. 10)

microorganismo Un organismo muy pequeño que solamente puede verse con un microscopio. Las bacterias son ejemplos de microorganismos.

migration
The movement of animals from one region to another in response to changes in the seasons or the environment. (p. 64)

migración El movimiento de animales de una región a otra en respuesta a cambios en las estaciones o en el medio ambiente.

R64

mobile
Able to move from place to place. (p. 130)

móvil Capaz de moverse de un lugar a otro.

molecule
A group of atoms that are held together by covalent bonds so that they move as a single unit.

molécula Un grupo de átomos que están unidos mediante enlaces covalentes de tal manera que se mueven como una sola unidad.

mollusk
An invertebrate animal with a soft body, a muscular foot, and a mantle. Many mollusks have a hard outer shell. (p. 136)

molusco Un animal invertebrado con cuerpo blando, un pie muscular y un manto. Muchos moluscos tienen una concha exterior dura.

molting
The process of an arthropod shedding its exoskeleton to allow for growth. (p. 143)

muda El proceso mediante el cual un artrópodo se despoja de su exoesqueleto para poder crecer.

O

organ
A structure in a plant or animal that is made up of different tissues working together to perform a particular function. (p. 44)

órgano Una estructura en un plato o en un animal compuesta de diferentes tejidos que trabajan juntos para realizar una función determinada.

organism
An individual living thing, made up of one or many cells, that is capable of growing and reproducing.

organismo Un individuo vivo, compuesto de una o muchas células, que es capaz de crecer y reproducirse.

P, Q

parasite
An organism that absorbs nutrients from the body of another organism, often harming it in the process. (p. 19)

parásito Un organismo que absorbe nutrientes del cuerpo de otro organismo, a menudo causándole daño en el proceso.

photosynthesis (FOH-toh-SIHN-thih-sihs)
The process by which green plants and other producers use simple compounds and energy from light to make sugar, an energy-rich compound. (p. 52)

fotosíntesis El proceso mediante el cual las plantas verdes y otros productores usan compuestos simples y energía de la luz para producir azúcares, compuestos ricos en energía.

placenta
An organ that transports materials between a pregnant female mammal and the offspring developing inside her body. (p. 186)

placenta Un órgano que transporta sustancias entre un mamífero hembra preñado y la cría que se está desarrollando dentro de su cuerpo.

plankton
Mostly microscopic organisms that drift in great numbers through bodies of water. (p. 33)

plancton Organismos, en su mayoría microscópicos, que se mueven a la deriva en grandes números por cuerpos de agua.

pollen
Tiny multicellular grains that contain the undeveloped sperm cells of a plant. (p. 100)

polen Los diminutos granos multicelulares que contienen las células espermáticas sin desarrollar de una planta.

predator
An animal that hunts other animals and eats them. (p. 63)

predador Un animal que caza otros animales y se los come.

prey
An animal that other animals hunt and eat. (p. 63)

presa Un animal que otros animales cazan y se comen.

producer
An organism that captures energy from sunlight and transforms it into chemical energy that is stored in energy-rich carbon compounds. Producers are a source of food for other organisms. (p. 19)

productor Un organismo que capta energía de la luz solar y la transforma a energía química que se almacena en compuestos de carbono ricos en energía. Los productores son una fuente de alimento para otros organismos.

R65

protozoa
Animal-like protists that eat other organisms or decaying parts of other organisms. *Protozoa* is a plural word; the singular is *protozoan.* (p. 34)

protozoarios Protistas parecidos a los animales que comen otros organismos o partes en descomposición de otros organismos.

R

reptile
A cold-blooded vertebrate that has skin covered with scales or horny plates and has lungs. (p. 168)

reptil Un vertebrado de sangre fría que tiene la piel cubierta de escamas o placas callosas y que tiene pulmones.

S

scale
One of the thin, small, overlapping plates that cover most fish and reptiles and some other animals. (p. 161)

escama Una de las pequeñas y delgadas placas traslapadas que cubren a la mayoría de los peces y reptiles y algunos otros animales.

seed
A plant embryo that is enclosed in a protective coating and has its own source of nutrients. (p. 100)

semilla El embrión de una planta que esta dentro de una cubierta protectora y que tiene su propia fuente de nutrientes.

sessile (SEHS-eel)
The quality of being attached to one spot; not free-moving. (p. 125)

sésil La cualidad de estar sujeto a un punto; sin libre movimiento.

sexual reproduction
A type of reproduction in which male and female reproductive cells combine to form offspring with genetic material from both cells. (p. 48)

reproducción sexual Un tipo de reproducción en el cual se combinan las células reproductivas femeninas y masculinas para formar una cría con material genético de ambas células.

species
A group of living things that are so closely related that they can breed with one another and produce offspring that can breed as well. (p. xxi)

especie Un grupo de organismos que están tan estrechamente relacionados que pueden aparearse entre sí y producir crías que también pueden aparearse.

sponge
A simple multicellular invertebrate animal that lives attached to one place and filters food from water. (p. 125)

esponja Un animal invertebrado multicelular simple que vive sujeto a un lugar y filtra su alimento del agua.

spore
A single reproductive cell that can grow into a multicellular organism. (p. 67)

espora Una célula reproductiva individual que puede convertirse en un organismo multicelular.

stimulus
Something that causes a response in an organism or a part of the body. (p. 55)

estímulo Algo que causa una respuesta en un organismo o en una parte del cuerpo.

system
A group of objects or phenomena that interact. A system can be as simple as a rope, a pulley, and a mass. It also can be as complex as the interaction of energy and matter in the four parts of the Earth system.

sistema Un grupo de objetos o fenómenos que interactúan. Un sistema puede ser algo tan sencillo como una cuerda, una polea y una masa. También puede ser algo tan complejo como la interacción de la energía y la materia en las cuatro partes del sistema de la Tierra.

T, U

technology
The use of scientific knowledge to solve problems or engineer new products, tools, or processes.

tecnología El uso de conocimientos científicos para resolver problemas o para diseñar nuevos productos, herramientas o procesos.

tentacle
A long, slender, flexible extension of the body of certain animals, such as jellyfish. Tentacles are used to touch, move, or hold. (p. 128)

tentáculo Una extensión larga, delgada y flexible del cuerpo de ciertos animales, como las medusas. Los tentáculos se usan para tocar, mover o sujetar.

theory
In science, a set of widely accepted explanations of observations and phenomena. A theory is a well-tested explanation that is consistent with all available evidence.

teoría En las ciencias, un conjunto de explicaciones de observaciones y fenómenos que es ampliamente aceptado. Una teoría es una explicación bien probada que es consecuente con la evidencia disponible.

tissue
A group of similar cells that together perform a specific function in an organism. (p. 44)

tejido Un grupo de células parecidas que juntas realizan una función específica en un organismo.

transpiration (TRAN-spuh-RAY-shuhn)
The movement of water vapor out of a plant and into the air. (p. 88)

transpiración El movimiento de vapor de agua hacia fuera de una planta y hacia el aire.

V, W, X, Y, Z

variable
Any factor that can change in a controlled experiment, observation, or model. (p. R30)

variable Cualquier factor que puede cambiar en un experimento controlado, en una observación o en un modelo.

vascular system (VAS-kyuh-lur)
Long tubelike tissues in plants through which water and nutrients move from one part of the plant to another. (p. 87)

sistema vascular Tejidos largos en forma de tubo en las plantas a través de los cuales se mueven agua y nutrientes de una parte de la planta a otra.

vertebrate
An animal with an internal backbone. (p. 157)

vertebrado Un animal que tiene columna vertebral interna.

virus
A nonliving disease-causing particle that uses the materials inside cells to make copies of itself. A virus consists of genetic material enclosed in a protein coat. (p. 14)

virus Una particular sin vida, que causa enfermedad y que usa los materiales dentro de las células para reproducirse. Un virus consiste de material genético encerrado en una cubierta proteica.

volume
An amount of three-dimensional space, often used to describe the space that an object takes up.

volumen Una cantidad de espacio tridimensional; a menudo se usa este término para describir el espacio que ocupa un objeto.

R67

Index

Page numbers for definitions are printed in **boldface** type.
Page numbers for illustrations, maps, and charts are printed in *italics*.

A

B

C

D

INDEX

R69

E

F

INDEX

G

H

I

INDEX

INDEX

N

O

INDEX

P, Q

R

INDEX

S

INDEX

INDEX

T, U

V

W

X, Y, Z

Acknowledgments

Photography

Cover © Buddy Mays/Corbis; **iii** Photograph of James Trefil by Evan Cantwell; Photograph of Rita Ann Calvo by Joseph Calvo; Photograph of Kenneth Cutler by Kenneth A. Cutler; Photograph of Douglas Carnine by McDougal Littell; Photograph of Linda Carnine by Amilcar Cifuentes; Photograph of Donald Steely by Marni Stamm; Photograph of Sam Miller by Samuel Miller; Photograph of Vicky Vachon by Redfern Photographics; **vi** © Orion Press/Corbis; **ix** *top* Photograph by Ken O'Donoghue; *bottom (both)* Photographs by Frank Siteman; **xiv, xv** © Mark Hamblin/Age Fotostock; **xvi, xvii** © Georgette Duowma/Taxi/Getty Images; **xviii, xix** © Ron Sanford/Corbis; **xx, xxi** © Nick Vedros & Assoc./Stone/Getty Images; **xxii** *left* © Michael Gadomski/Animals Animals; *right* © Shin Yoshino/Minden Pictures; **xxiii** © Laif Elleringmann/Aurora Photos; **xxiv** © Pascal Goetgheluck/Science Photo Library/Photo Researchers, Inc.; **xxv** *top left* © David Parker/Science Photo Library/Photo Researchers, Inc.; *top right* © James King-Holmes/Science Photo Library/Photo Researchers, Inc.; *bottom* Sinsheimer Labs/University of California, Santa Cruz; **xxvi, xxvii** *background* © Maximillian Stock/Photo Researchers, Inc.; **xxvi** Courtesy, John Lair, Jewish Hospital, University of Louisville; **xxvii** *top* © Brand X Pictures/Alamy; *center* Courtesy, AbioMed; **xxxii** © Chedd-Angier Production Company; **2, 3** © *background* Yva Momatiuk/John Eastcott/Minden Pictures; **3** *top* © Pat O'Hara/Corbis; *bottom* © Darrell Gulin/Corbis; **4** *top left* © Bruce Marlin/Cirrus Digital Imaging; *top right* © A.B. Sheldon; *bottom* © Chedd-Angier Production Company; **6, 7** © Science VU/Visuals Unlimited; **7** *top* Photograph by Frank Siteman; *center* Photograph by Ken O'Donoghue; *bottom* © Custom Medical Stock Photo; **9** Photograph by Ken O'Donoghue; **10** *background* © Lynda Richardson/Corbis; *inset* © Astrid & Hanns-Frieder Michler/Photo Researchers, Inc.; **12** © A.B. Dowsett/Science Photo Library/Photo Researchers, Inc.; **13** Photograph by Ken O'Donoghue; **14** *right* © Dr. Gopal Murti/Photo Researchers, Inc.; *left* © K.G. Murti/Visuals Unlimited; **15** © CNRI/ Photo Researchers, Inc.; **16** © Dennis Kunkel/Visuals Unlimited; **17** *left* © Tina Carvalho/Visuals Unlimited; *center* © D.M. Phillips/Visuals Unlimited; *right* © CNRI/Photo Researchers, Inc.; **18** *left* © Grant Heilman/Grant Heilman Photography; *left inset* © Dr. Kari Lounatmaa/Photo Researchers, Inc.; *center* © Roger Tidman/Corbis; *center inset* © Alfred Pasieka/Science Photo Library/Photo Researchers, Inc.; *right* © ML Sinbaldi/Corbis; *right inset* © Wolfgang Baumeister/Photo Researchers, Inc.; **19** *left* © Jack Novak/Photri-Microstock; *left inset* © Dr. Kari Lounatmaa/Photo Researchers, Inc.; *center* © Dennis Flaherty/Photo Researchers, Inc.; *center inset* © Microfield Scientific LTD/Science Photo Library/Photo Researchers, Inc.; *right* © Dr. P. Marazzi/Science Photo Library/Photo Researchers, Inc.; *right inset* © Dr. Kari Lounatmaa/Photo Researchers, Inc.; **20** © *background* Runk/Schoenberger/Grant Heilman Photography; *left inset* © Simko/Visuals Unlimited; *right inset* © Dwight R. Kuhn; **21** © T.A. Zitter/Cornell University; **22** *top* © Adam Hart-Davis, Leeds Public Health Laboraory/Photo Researchers, Inc. ; *bottom (both)* Photographs by Ken O'Donoghue; **24, 25** Photographs by Frank Siteman; **26** © Hans Gelderblom/Visuals Unlimited; **27** *(both)* © Lee D. Simon/Photo Researchers, Inc.; **28** © Bettmann/Corbis; **29** *left* © Judy White/GardenPhotos.com; *right* © Dennis Kunkel Microscopy, Inc.; **30** © Corbis; **31** Photograph by Ken O'Donoghue; **32** *left* © R. Kessel-G. Shih/Visuals Unlimited; *center* © Jan Hinsch/Photo Researchers, Inc.; *right* © Runk/Schoenberger/Grant Heilman Photography; **33** © R. Kessel-C.Y. Shih/Visuals Unlimited; **34** © Andrew Syred/Photo Researchers, Inc.; **35** © Ed Reschke/Peter Arnold, Inc.; **36** *top* © CNRI/Photo Researchers, Inc.; *bottom left* © Runk/Schoenberger/Grant Heilman Photography; *bottom center* © Ed Reschke/Peter Arnold, Inc.; *bottom right* © Andrew Syred/Photo Researchers, Inc.; **38** *left* © A.B. Dowsett/Science Photo Library/Photo Researchers, Inc.; *right* © Science VU/Visuals Unlimited; **40, 41** © Orion Press/Corbis; **41** *top, center* Photographs by Ken O'Donoghue; **43** © David G. Massey/AP Wide World Photos; **44** Photograph by Frank Siteman; **45** © Joe McDonald/Bruce Coleman, Inc.; **46** © Michael Doolittle/PictureQuest; **47** *background* © Photospin; *top* © A. Mercieca/Photo Researchers, Inc.; *center* © Jim Brandenburg/Minden Pictures; *center inset* © Yva Momatiuk/John Eastcott/Minden Pictures; *bottom* © Tim Fitzharris/Minden Pictures; **48** © Matt Brown/Corbis; **49** *left* © Joe McDonald/Corbis; *right* © David J. Wrobel/Visuals Unlimited; **50** © Frans Lanting/Minden Pictures; **51, 52** Photographs by Ken O'Donoghue; **53** *top* © Peter Dean/Grant Heilman Photography; *inset* © Pascal Goetgheluck/Science Photo Library/Photo Researchers, Inc.; **54** *(both)* © D. Suzio/Photo Researchers, Inc.; **55** *left* © Joel Arrington/Visuals Unlimited; *right* © Gary W. Carter/Visuals Unlimited; **57** © Josiah Davidson/Picturesque/PictureQuest; **58** © Yva Momatiuk/John Eastcott/Minden Pictures; **59** *left* © Brandon Cole; *right* © Rauschenbach/Premium Stock/PictureQuest ; **60** Photograph by Frank Siteman; **61** *left and left inset* © Dwight R.Kuhn; *center* © Steve Maslowski/Visuals Unlimited; *right* © Belinda Wright/DRK Photo; **62** © J. Sneesby/B. Wilkins/Getty Images; **63** *top* © Shin Yoshino/Minden Pictures; *bottom left* © Charles V. Angelo/Photo Researchers, Inc.; *bottom right* © Fred McConnaughey/Photo Researchers, Inc.; **64** *left* © Ron Austing/Photo Researchers, Inc.; *right* © Kevin Schafer; **65** *left, top right* © Steve Winter/National Geographic Image Collection; *bottom right* © Brian J. Skerry/National Geographic Image Collection; **66** Photograph by Ken O'Donoghue; **68** © IFA/eStock Photography/PictureQuest ; **69** *top* © Andrew Syred/Science Photo Library/Photo Researchers; *bottom* © Simko/Visuals Unlimited; **71** *(both)* © Lennart Nilsson/Albert Bonniers Forlag AB; **72** *top* © Bob Daemmrich/Stock Boston, Inc./PictureQuest; *bottom* Photograph by Ken O'Donoghue; **73** Photograph by Frank Siteman; **74** *top* © Matt Brown/Corbis; *center* © Rauschenbach/ Premium Stock/PictureQuest ; *bottom* © IFA/eStock Photography/PictureQuest; **78** *top* © Academy of Natural Sciences of Philadelphia/Corbis; *bottom* © The Granger Collection, New York; **79** *top left, top right* © The Natural History Museum, London; *bottom left* © Photo Researchers, Inc.; *bottom right* © The Granger Collection, New York; **80** *top left* © B. Boonyaratanakornit & D.S. Clark, G. Vrdoljak/EM Lab, UC Berkeley/Visuals Unlimited, Inc.; *top right* © OAR/National Undersea Research Program; *center* © Terry Erwin/Smithsonian Institution; *bottom* © Emory Kristoff/National Geographic Image Collection; **81** © Michael Bordelon/Smithsonian Institution; **82, 83** © David J. Job/AlaskaStock.com; **83** *(both)* Photographs by Ken O'Donoghue; **85** Photograph by Ken O'Donoghue; **86** *left* © Nick Garbutt/Nature Picture Library; *center* © D. Cavagnaro/Visuals Unlimited; *right* © Hal Horwitz/Corbis; **87** © Andrew Syred/Photo Researchers, Inc.; **88** *(both)* © Dr. Jeremy Burgess/Science Photo Library/Photo Researchers, Inc.; **89** *background* © Donna Disario/Corbis; *top left* © Photodisc/Getty Images; **90** © Craig K. Lorenz/Photo

Researchers, Inc.; **91** *top* © Delphoto/Premium Stock/PictureQuest; *inset* © Unicorn Stock Photo; **92** Photograph by Frank Siteman; **93** *top* © Ray Simmons/Photo Researchers, Inc.; *bottom* The Field Museum of Natural History, #GEO85637c; **94** Photograph by Frank Siteman; **95** © Dwight R. Kuhn; **96** *left* © Dr. Jeremy Burgess/Science Photo Library/Photo Researchers, Inc.; *inset* School Division, Houghton Mifflin Co.; *right* © Michael & Patricia Fogden/Corbis; **97** *left* © Dwight R. Kuhn; *right* © Sylvester Allred/Fundamental Photographs; **98** © Keren Su/Corbis; **99** © Dwight R. Kuhn; **100** © Martha Cooper/Peter Arnold, Inc.; **101** *background* © Photospin; *top* © Scott Barrow/ImageState; *bottom* © Bryan Mullenix/Getty Images; **102** Photograph by Ken O'Donoghue; **103** © Robert Gustafson/Visuals Unlimited; **104** *top* © Michael J. Doolittle/The Image Works; *bottom right* © James A. Sugar/Corbis; **105** Photograph by Frank Siteman; **106** *left* © Raymond Gehman/Corbis; *top right* © David Sieren/Visuals Unlimited, Inc.; *bottom right* © George Bernard/Science Photo Library; **107** Photograph by Ken O'Donoghue; **109** *background* © John Marshall; *left* © George D. Lepp/Corbis; *center* © Sergio Piumatti; *right* © Gary Braasch/Corbis; **110** © Ed Reschke; **111** *top* Photograph by Ken O'Donoghue; *bottom* © Custom Medical Stock Photo; **112** *left* © Frank Lane Picture Agency/Corbis; *right* © Eastcott/Momatiuc/Animals Animals; **113** *right* © Craig Aurness/Corbis; *inset* © Michael Newman/PhotoEdit; **114** *background* © Lance Nelson/Corbis; *top, left* National Cotton Council of America; *right* © Mary Kate Denny/PhotoEdit; **115** *left* © Patricia Agre/Photo Researchers, Inc.; *right* © Martin B. Withers, Frank Lane Picture Agency/Corbis; **116** *top* © Hal Horwitz/Corbis; *center left* © Martha Cooper/Peter Arnold, Inc.; *center right* © Dwight R. Kuhn; *bottom left* © Gary Braasch/Corbis; *bottom right* © George D. Lepp/Corbis; **117** *(both)* © Wolfgang Bayer/Bruce Coleman, Inc./PictureQuest ; **120, 121** © Norbert Wu; **121** *top* Photograph by Frank Siteman; *bottom* © Photospin; **123** © Andrew J. Martinez/Visuals Unlimited; **124** Photograph by Ken O'Donoghue; **125** *background* © Viola's Photo Visions, Inc./Animals Animals; **126** © Marty Snyderman/Visuals Unlimited; **127** *left* © Dwight R. Kuhn; *inset* © Thomas Kitchin/Tom Stack & Associates; **128** Photograph by Ken O'Donoghue; **129** © Photodisc/Getty Images; **130** *top left, bottom left* © John D. Cunningham/Visuals Unlimited; **131** *left* © Bob Evans/Peter Arnold, Inc.; *right* © Visuals Unlimited; **132** © Dwight R.Kuhn; **133** *top* © Larry Lipsky/DRK Photo; *center* © A. Flowers & L. Newman/Photo Researchers, Inc.; *bottom* © Richard Kessel/Visuals Unlimited; **134** *top* © Robert Pickett/Corbis; *bottom* Photograph by Ken O'Donoghue; **135** *top* Photograph by Ken O'Donoghue; *bottom* Photograph by Frank Siteman; **136** © Sinclair Stammers/Photo Researchers, Inc.; **137** *top* © Andrew J. Martinez/Photo Researchers, Inc.; *bottom* © Konrad Wothe/Minden Pictures; **138** *top* Photograph by Ken O'Donoghue; *bottom* © Fred Bavendam/Minden Pictures; **139** *top* © Andrew J. Martinez; *bottom* © David Wrobel/Visuals Unlimited; **140** *left* © Fred Winner/Jacana/Photo Researchers, Inc.; *right* © Gerald & Buff Corsi/Visuals Unlimited; **141** © Kevin Schafer/Getty Images; **142** Photograph by Frank Siteman; **143** *top* © Kelvin Aitken/Peter Arnold, Inc.; *bottom* © Barry Runk/Grant Heilman Photography; **144** *background* © Corbis-Royalty Free; *top* © Tim Davis/Corbis; *center* © Steve Wolper/DRK Photo; *bottom* © George Calef/DRK Photo; **146** Photograph by Ken O'Donoghue; **147** *top (all)* © Dwight R. Kuhn; *bottom* © Frans Lanting/Minden Pictures; **148** *top* © David Scharf/Peter Arnold, Inc.; **148** *bottom* © E.R. Degginger/Color-Pic, Inc.; **149** *left* © Science Photo Library/Photo Researchers, Inc.; *right* © Claus Meyer/Minden Pictures; **150** *top right* © Viola's Photo Visions, Inc./Animals Animals; *center left* © John D. Cunningham/Visuals Unlimited; *bottom left* © Konrad Wothe/Minden Pictures; *bottom right* © Gerald & Buff Corsi/Visuals Unlimited; **154, 155** © Paul A. Souders/Corbis; **155** *top* Photograph by Frank Siteman; *center left* © Steven Frame/Stock Boston Inc./PictureQuest; *center right* © Rod Planck/Photo Researchers, Inc.; **157** Photograph by Frank Siteman; **158** © Corbis-Royalty Free; **159** *background* © SeaLifeStyles Signature Series/Imagin; **160** *background* © Colla - V&W/Bruce Coleman, Inc.; *left* © Norbert Wu; *left inset* © Brandon Cole; *center* © Georgienne E. Bradley & Jay Ireland/Bradley Ireland Productions; *right* © Brandon Cole; **162** © David Doubilet; **163** *left* © Kennan Ward/Corbis; *right* © Frans Lanting/Minden Pictures; **164** *left* © Bianca Lavies/National Geographic Image Collection; *right* Photograph by Frank Siteman; **165** *left* © Dwight R. Kuhn; *right* © Joe McDonald/Bruce Coleman, Inc.; **166** *all* © Dwight R. Kuhn; **168** © Francois Gohier/Photo Researchers, Inc.; **169** © Carmela Leszczynski/Animals Animals; **170** *top* Photograph by Frank Siteman; *bottom* © Tui de Roy/Minden Pictures; **171** © Michael Fogden/Animals Animals; **172** *left* ©Frans Lanting/Minden Pictures; *center* © E.R. Degginger/Color-Pic, Inc.; *right* © Science Photo Library/Photo Researchers, Inc.; **173** Photograph by Ken O'Donoghue; **174** © Michael Quinton/Minden Pictures; **175** *left* © Julie Habel/Corbis; *center* © Randy Faris/Corbis; *right* © David-Young Wolff/PhotoEdit; **177** *background* © Tim Bird/Corbis; **178** *left* © Jim Brandenburg/Minden Pictures; *right* © Fritz Polking/Visuals Unlimited; **179** *top* © Ron Austing/Photo Researchers, Inc.; *bottom* © S. Nielsen/DRK Photo; **180** *top* © DigitalVision/PictureQuest; *bottom* Photograph by Frank Siteman **181** *top* Photograph by Frank Siteman; *bottom left* © Corbis-Royalty Free; *bottom center* © Frans Lemmens/Getty Images; *bottom right* © Arthur Morris/Corbis; **182** © Carleton Ray/Photo Researchers, Inc.; **183** *top* © Mitsuaki Iwago/Minden Pictures; *bottom* © Don Enger/Animals Animals; **184** Photograph by Frank Siteman; **185** *background* © Stephen Frink/Corbis; **186** *left* © Photodisc/Getty Images; *right* © Frans Lanting/Minden Pictures; **187** © Comstock; **188** *top left* © Dwight R. Kuhn; *top right* © Michael Fogden/Animals Animals; *center right* © Fritz Polking/Visuals Unlimited; *bottom left* © Frans Lanting/Minden Pictures; *bottom right* © Photodisc/Getty Images; **r28** © PhotoDisc/Getty Images.

Illustration and Maps

Robin Boutell/Wildlife Art Ltd. **132, 150** *(center right)*
Peter Bull/Wildlife Art Ltd. **26, 33, 34**
Myriam Kirkman-KO Studios **87**
Debbie Maizels **45, 56** *(inset)*, **89, 99, 101, 109, 118, 158**
MapQuest.com, Inc. **47**
Laurie O'Keefe **125, 130, 159, 185**
Mick Posen/Wildlife Art Ltd. **129**
Tony Randazzo/American Artists Rep. Inc. **56** *(background)*
Ian Jackson/Wildlife Art Ltd. **67, 70, 145, 150** *(bottom)*, **177**
Steve Oh/KO Studios **17, 27**
UNEP-WCMC, 2002 World Atlas of Biodiversity **81**
Dan Stuckenschneider/Uhl Studios **R11-R19, R22, R32**

ACKNOWLEDGMENTS

Content Standards: 5–8

A. Science as Inquiry

As a result of activities in grades 5–8, all students should develop

Abilities Necessary to do Scientific Inquiry

A.1 Identify questions that can be answered through scientific investigations. Students should develop the ability to refine and refocus broad and ill-defined questions. An important aspect of this ability consists of students' ability to clarify questions and inquiries and direct them toward objects and phenomena that can be described, explained, or predicted by scientific investigations. Students should develop the ability to identify their questions with scientific ideas, concepts, and quantitative relationships that guide investigation.

A.2 Design and conduct a scientific investigation. Students should develop general abilities, such as systematic observation, making accurate measurements, and identifying and controlling variables. They should also develop the ability to clarify their ideas that are influencing and guiding the inquiry, and to understand how those ideas compare with current scientific knowledge. Students can learn to formulate questions, design investigations, execute investigations, interpret data, use evidence to generate explanations, propose alternative explanations, and critique explanations and procedures.

A.3 Use appropriate tools and techniques to gather, analyze, and interpret data. The use of tools and techniques, including mathematics, will be guided by the question asked and the investigations students design. The use of computers for the collection, summary, and display of evidence is part of this standard. Students should be able to access, gather, store, retrieve, and organize data, using hardware and software designed for these purposes.

A.4 Develop descriptions, explanations, predictions, and models using evidence. Students should base their explanation on what they observed, and as they develop cognitive skills, they should be able to differentiate explanation from description—providing causes for effects and establishing relationships based on evidence and logical argument. This standard requires a subject matter knowledge base so the students can effectively conduct investigations, because developing explanations establishes connections between the content of science and the contexts within which students develop new knowledge.

A.5 Think critically and logically to make the relationships between evidence and explanations. Thinking critically about evidence includes deciding what evidence should be used and accounting for anomalous data. Specifically, students should be able to review data from a simple experiment, summarize the data, and form a logical argument about the cause-and-effect relationships in the experiment. Students should begin to state some explanations in terms of the relationship between two or more variables.

A.6 Recognize and analyze alternative explanations and predictions. Students should develop the ability to listen to and respect the explanations proposed by other students. They should remain open to and acknowledge different ideas and explanations, be able to accept the skepticism of others, and consider alternative explanations.

A.7 Communicate scientific procedures and explanations. With practice, students should become competent at communicating experimental methods, following instructions, describing observations, summarizing the results of other groups, and telling other students about investigations and explanations.

A.8 Use mathematics in all aspects of scientific inquiry. Mathematics is essential to asking and answering questions about the natural world. Mathematics can be used to ask questions; to gather, organize, and present data; and to structure convincing explanations.

Understandings about Scientific Inquiry

A.9.a Different kinds of questions suggest different kinds of scientific investigations. Some investigations involve observing and describing objects, organisms, or events; some involve collecting specimens; some involve experiments; some involve seeking more information; some involve discovery of new objects and phenomena; and some involve making models.

A.9.b Current scientific knowledge and understanding guide scientific investigations. Different scientific domains employ different methods, core theories, and standards to advance scientific knowledge and understanding.

A.9.c Mathematics is important in all aspects of scientific inquiry.

A.9.d Technology used to gather data enhances accuracy and allows scientists to analyze and quantify results of investigations.

A.9.e Scientific explanations emphasize evidence, have logically consistent arguments, and use scientific principles, models, and theories. The scientific community accepts and uses such explanations until displaced by better scientific ones. When such displacement occurs, science advances.

A.9.f Science advances through legitimate skepticism. Asking questions and querying other scientists' explanations is part of scientific inquiry. Scientists evaluate the explanations proposed by other scientists by examining evidence, comparing evidence, identifying faulty reasoning, pointing out statements that go beyond the evidence, and suggesting alternative explanations for the same observations.

A.9.g Scientific investigations sometimes result in new ideas and phenomena for study, generate new methods or procedures for an investigation, or develop new technologies to improve the collection of data. All of these results can lead to new investigations.

B. Physical Science

As a result of their activities in grades 5–8, all students should develop an understanding of

Properties and Changes of Properties in Matter

B.1.a A substance has characteristic properties, such as density, a boiling point, and solubility, all of which are independent of the amount of the sample. A mixture of substances often can be separated into the original substances using one or more of the characteristic properties.

B.1.b Substances react chemically in characteristic ways with other substances to form new substances (compounds) with different characteristic properties. In chemical reactions, the total mass is conserved. Substances often are placed in categories or groups if they react in similar ways; metals is an example of such a group.

B.1.c Chemical elements do not break down during normal laboratory reactions involving such treatments as heating, exposure to electric current, or reaction with acids. There are more than 100 known elements that combine in a multitude of ways to produce compounds, which account for the living and nonliving substances that we encounter.

Motions and Forces

B.2.a The motion of an object can be described by its position, direction of motion, and speed. That motion can be measured and represented on a graph.

B.2.b An object that is not being subjected to a force will continue to move at a constant speed and in a straight line.

B.2.c If more than one force acts on an object along a straight line, then the forces will reinforce or cancel one another, depending on their direction and magnitude. Unbalanced forces will cause changes in the speed or direction of an object's motion.

Transfer of Energy

B.3.a Energy is a property of many substances and is associated with heat, light, electricity, mechanical motion, sound, nuclei, and the nature of a chemical. Energy is transferred in many ways.

B.3.b Heat moves in predictable ways, flowing from warmer objects to cooler ones, until both reach the same temperature.

B.3.c Light interacts with matter by transmission (including refraction), absorption, or scattering (including reflection). To see an object, light from that object—emitted by or scattered from it—must enter the eye.

B.3.d Electrical circuits provide a means of transferring electrical energy when heat, light, sound, and chemical changes are produced.

B.3.e In most chemical and nuclear reactions, energy is transferred into or out of a system. Heat, light, mechanical motion, or electricity might all be involved in such transfers.

B.3.f The sun is a major source of energy for changes on the earth's surface. The sun loses energy by emitting light. A tiny fraction of that light reaches the earth, transferring energy from the sun to the earth. The sun's energy arrives as light with a range of wavelengths, consisting of visible light, infrared, and ultraviolet radiation.

C. Life Science

As a result of their activities in grades 5–8, all students should develop understanding of

Structure and Function in Living Systems

C.1.a Living systems at all levels of organization demonstrate the complementary nature of structure and function. Important levels of organization for structure and function include cells, organs, tissues, organ systems, whole organisms, and ecosystems.

C.1.b All organisms are composed of cells—the fundamental unit of life. Most organisms are single cells; other organisms, including humans, are multicellular.

C.1.c Cells carry on the many functions needed to sustain life. They grow and divide, thereby producing more cells. This requires that they take in nutrients, which they use to provide energy for the work that cells do and to make the materials that a cell or an organism needs.

C.1.d Specialized cells perform specialized functions in multicellular organisms. Groups of specialized cells cooperate to form a tissue, such as a muscle. Different tissues are in turn grouped together to form larger functional units, called organs. Each type of cell, tissue, and organ has a distinct structure and set of functions that serve the organism as a whole.

C.1.e The human organism has systems for digestion, respiration, reproduction, circulation, excretion, movement, control, and coordination, and for protection from disease. These systems interact with one another.

C.1.f Disease is a breakdown in structures or functions of an organism. Some diseases are the result of intrinsic failures of the system. Others are the result of damage by infection by other organisms.

Reproduction and Heredity

C.2.a Reproduction is a characteristic of all living systems; because no individual organism lives forever, reproduction is essential to the continuation of every species. Some organisms reproduce asexually. Other organisms reproduce sexually.

C.2.b In many species, including humans, females produce eggs and males produce sperm. Plants also reproduce sexually—the egg and sperm are produced in the flowers of flowering plants. An egg and sperm unite to begin development of a new individual. That new individual receives genetic information from its mother (via the egg) and its father (via the sperm). Sexually produced offspring never are identical to either of their parents.

C.2.c Every organism requires a set of instructions for specifying its traits. Heredity is the passage of these instructions from one generation to another.

C.2.d Hereditary information is contained in genes, located in the chromosomes of each cell. Each gene carries a single unit of information. An inherited trait of an individual can be determined by one or by many genes, and a single gene can influence more than one trait. A human cell contains many thousands of different genes.

C.2.e The characteristics of an organism can be described in terms of a combination of traits. Some traits are inherited and others result from interactions with the environment.

National Science Education Standards

Regulation and Behavior

C.3.a All organisms must be able to obtain and use resources, grow, reproduce, and maintain stable internal conditions while living in a constantly changing external environment.

C.3.b Regulation of an organism's internal environment involves sensing the internal environment and changing physiological activities to keep conditions within the range required to survive.

C.3.c Behavior is one kind of response an organism can make to an internal or environmental stimulus. A behavioral response requires coordination and communication at many levels, including cells, organ systems, and whole organisms. Behavioral response is a set of actions determined in part by heredity and in part from experience.

C.3.d An organism's behavior evolves through adaptation to its environment. How a species moves, obtains food, reproduces, and responds to danger are based in the species' evolutionary history.

Populations and Ecosystems

C.4.a A population consists of all individuals of a species that occur together at a given place and time. All populations living together and the physical factors with which they interact compose an ecosystem.

C.4.b Populations of organisms can be categorized by the function they serve in an ecosystem. Plants and some microorganisms are producers—they make their own food. All animals, including humans, are consumers, which obtain food by eating other organisms. Decomposers, primarily bacteria and fungi, are consumers that use waste materials and dead organisms for food. Food webs identify the relationships among producers, consumers, and decomposers in an ecosystem.

C.4.c For ecosystems, the major source of energy is sunlight. Energy entering ecosystems as sunlight is transferred by producers into chemical energy through photosynthesis. That energy then passes from organism to organism in food webs.

C.4.d The number of organisms an ecosystem can support depends on the resources available and abiotic factors, such as quantity of light and water, range of temperatures, and soil composition. Given adequate biotic and abiotic resources and no disease or predators, populations (including humans) increase at rapid rates. Lack of resources and other factors, such as predation and climate, limit the growth of populations in specific niches in the ecosystem.

Diversity and Adaptations of Organisms

C.5.a Millions of species of animals, plants, and microorganisms are alive today. Although different species might look dissimilar, the unity among organisms becomes apparent from an analysis of internal structures, the similarity of their chemical processes, and the evidence of common ancestry.

C.5.b Biological evolution accounts for the diversity of species developed through gradual processes over many generations. Species acquire many of their unique characteristics through biological adaptation, which involves the selection of naturally occurring variations in populations. Biological adaptations include changes in structures, behaviors, or physiology that enhance survival and reproductive success in a particular environment.

C.5.c Extinction of a species occurs when the environment changes and the adaptive characteristics of a species are insufficient to allow its survival. Fossils indicate that many organisms that lived long ago are extinct. Extinction of species is common; most of the species that have lived on the earth no longer exist.

D. Earth and Space Science

As a result of their activities in grades 5–8, all students should develop an understanding of

Structure of the Earth System

D.1.a The solid earth is layered with a lithosphere; hot, convecting mantle; and dense, metallic core.

D.1.b Lithospheric plates on the scales of continents and oceans constantly move at rates of centimeters per year in response to movements in the mantle. Major geological events, such as earthquakes, volcanic eruptions, and mountain building, result from these plate motions.

D.1.c Land forms are the result of a combination of constructive and destructive forces. Constructive forces include crustal deformation, volcanic eruption, and deposition of sediment, while destructive forces include weathering and erosion.

D.1.d Some changes in the solid earth can be described as the "rock cycle." Old rocks at the earth's surface weather, forming sediments that are buried, then compacted, heated, and often recrystallized into new rock. Eventually, those new rocks may be brought to the surface by the forces that drive plate motions, and the rock cycle continues.

D.1.e Soil consists of weathered rocks and decomposed organic material from dead plants, animals, and bacteria. Soils are often found in layers, with each having a different chemical composition and texture.

D.1.f Water, which covers the majority of the earth's surface, circulates through the crust, oceans, and atmosphere in what is known as the "water cycle." Water evaporates from the earth's surface, rises and cools as it moves to higher elevations, condenses as rain or snow, and falls to the surface where it collects in lakes, oceans, soil, and in rocks underground.

D.1.g Water is a solvent. As it passes through the water cycle it dissolves minerals and gases and carries them to the oceans.

D.1.h The atmosphere is a mixture of nitrogen, oxygen, and trace gases that include water vapor. The atmosphere has different properties at different elevations.

D.1.i Clouds, formed by the condensation of water vapor, affect weather and climate.

D.1.j Global patterns of atmospheric movement influence local weather. Oceans have a major effect on climate, because water in the oceans holds a large amount of heat.

D.1.k Living organisms have played many roles in the earth system, including affecting the composition of the atmosphere, producing some types of rocks, and contributing to the weathering of rocks.

Earth's History

D.2.a The earth processes we see today, including erosion, movement of lithospheric plates, and changes in atmospheric composition, are similar to those that occurred in the past. Earth history is also influenced by occasional catastrophes, such as the impact of an asteroid or comet.

D.2.b Fossils provide important evidence of how life and environmental conditions have changed.

Earth in the Solar System

D.3.a The earth is the third planet from the sun in a system that includes the moon, the sun, eight other planets and their moons, and smaller objects, such as asteroids and comets. The sun, an average star, is the central and largest body in the solar system.

D.3.b Most objects in the solar system are in regular and predictable motion. Those motions explain such phenomena as the day, the year, phases of the moon, and eclipses.

D.3.c Gravity is the force that keeps planets in orbit around the sun and governs the rest of the motion in the solar system. Gravity alone holds us to the earth's surface and explains the phenomena of the tides.

D.3.d The sun is the major source of energy for phenomena on the earth's surface, such as growth of plants, winds, ocean currents, and the water cycle. Seasons result from variations in the amount of the sun's energy hitting the surface, due to the tilt of the earth's rotation on its axis and the length of the day.

E. Science and Technology

As a result of activities in grades 5–8, all students should develop

Abilities of Technological Design

E.1 Identify appropriate problems for technological design. Students should develop their abilities by identifying a specified need, considering its various aspects, and talking to different potential users or beneficiaries. They should appreciate that for some needs, the cultural backgrounds and beliefs of different groups can affect the criteria for a suitable product.

E.2 Design a solution or product. Students should make and compare different proposals in the light of the criteria they have selected. They must consider constraints—such as cost, time, trade-offs, and materials needed—and communicate ideas with drawings and simple models.

E.3 Implement a proposed design. Students should organize materials and other resources, plan their work, make good use of group collaboration where appropriate, choose suitable tools and techniques, and work with appropriate measurement methods to ensure adequate accuracy.

E.4 Evaluate completed technological designs or products. Students should use criteria relevant to the original purpose or need, consider a variety of factors that might affect acceptability and suitability for intended users or beneficiaries, and develop measures of quality with respect to such criteria and factors; they should also suggest improvements and, for their own products, try proposed modifications.

E.5 Communicate the process of technological design. Students should review and describe any completed piece of work and identify the stages of problem identification, solution design, implementation, and evaluation.

Understandings about Science and Technology

E.6.a Scientific inquiry and technological design have similarities and differences. Scientists propose explanations for questions about the natural world, and engineers propose solutions relating to human problems, needs, and aspirations. Technological solutions are temporary; technologies exist within nature and so they cannot contravene physical or biological principles; technological solutions have side effects; and technologies cost, carry risks, and provide benefits.

E.6.b Many different people in different cultures have made and continue to make contributions to science and technology.

E.6.c Science and technology are reciprocal. Science helps drive technology, as it addresses questions that demand more sophisticated instruments and provides principles for better instrumentation and technique. Technology is essential to science, because it provides instruments and techniques that enable observations of objects and phenomena that are otherwise unobservable due to factors such as quantity, distance, location, size, and speed. Technology also provides tools for investigations, inquiry, and analysis.

E.6.d Perfectly designed solutions do not exist. All technological solutions have trade-offs, such as safety, cost, efficiency, and appearance. Engineers often build in back-up systems to provide safety. Risk is part of living in a highly technological world. Reducing risk often results in new technology.

E.6.e Technological designs have constraints. Some constraints are unavoidable, for example, properties of materials, or effects of weather and friction; other constraints limit choices in the design, for example, environmental protection, human safety, and aesthetics.

E.6.f Technological solutions have intended benefits and unintended consequences. Some consequences can be predicted, others cannot.

F. Science in Personal and Social Perspectives

As a result of activities in grades 5–8, all students should develop understanding of

Personal Health

F.1.a Regular exercise is important to the maintenance and improvement of health. The benefits of physical fitness include maintaining healthy weight, having energy and strength for routine activities, good muscle tone, bone strength, strong heart/lung systems, and improved mental health. Personal exercise, especially developing cardiovascular endurance, is the foundation of physical fitness.

F.1.b The potential for accidents and the existence of hazards imposes the need for injury prevention. Safe living involves the development and use of safety precautions and the recognition of risk in personal decisions. Injury prevention has personal and social dimensions.

F.1.c The use of tobacco increases the risk of illness. Students should understand the influence of short-term social and psychological factors that lead to tobacco use, and the possible long-term detrimental effects of smoking and chewing tobacco.

F.1.d Alcohol and other drugs are often abused substances. Such drugs change how the body functions and can lead to addiction.

F.1.e Food provides energy and nutrients for growth and development. Nutrition requirements vary with body weight, age, sex, activity, and body functioning.

F.1.f Sex drive is a natural human function that requires understanding. Sex is also a prominent means of transmitting diseases. The diseases can be prevented through a variety of precautions.

F.1.g Natural environments may contain substances (for example, radon and lead) that are harmful to human beings. Maintaining environmental health involves establishing or monitoring quality standards related to use of soil, water, and air.

Populations, Resources, and Environments

F.2.a When an area becomes overpopulated, the environment will become degraded due to the increased use of resources.

F.2.b Causes of environmental degradation and resource depletion vary from region to region and from country to country.

Natural Hazards

F.3.a Internal and external processes of the earth system cause natural hazards, events that change or destroy human and wildlife habitats, damage property, and harm or kill humans. Natural hazards include earthquakes, landslides, wildfires, volcanic eruptions, floods, storms, and even possible impacts of asteroids.

F.3.b Human activities also can induce hazards through resource acquisition, urban growth, land-use decisions, and waste disposal. Such activities can accelerate many natural changes.

F.3.c Natural hazards can present personal and societal challenges because misidentifying the change or incorrectly estimating the rate and scale of change may result in either too little attention and significant human costs or too much cost for unneeded preventive measures.

Risks and Benefits

F.4.a Risk analysis considers the type of hazard and estimates the number of people that might be exposed and the number likely to suffer consequences. The results are used to determine the options for reducing or eliminating risks.

F.4.b Students should understand the risks associated with natural hazards (fires, floods, tornadoes, hurricanes, earthquakes, and volcanic eruptions), with chemical hazards (pollutants in air, water, soil, and food), with biological hazards (pollen, viruses, bacterial, and parasites), social hazards (occupational safety and transportation), and with personal hazards (smoking, dieting, and drinking).

F.4.c Individuals can use a systematic approach to thinking critically about risks and benefits. Examples include applying probability estimates to risks and comparing them to estimated personal and social benefits.

F.4.d Important personal and social decisions are made based on perceptions of benefits and risks.

Science and Technology in Society

F.5.a Science influences society through its knowledge and world view. Scientific knowledge and the procedures used by scientists influence the way many individuals in society think about themselves, others, and the environment. The effect of science on society is neither entirely beneficial nor entirely detrimental.

F.5.b Societal challenges often inspire questions for scientific research, and social priorities often influence research priorities through the availability of funding for research.

F.5.c Technology influences society through its products and processes. Technology influences the quality of life and the ways people act and interact. Technological changes are often accompanied by social, political, and economic changes that can be beneficial or detrimental to individuals and to society. Social needs, attitudes, and values influence the direction of technological development.

F.5.d Science and technology have advanced through contributions of many different people, in different cultures, at different times in history. Science and technology have contributed enormously to economic growth and productivity among societies and groups within societies.

F.5.e Scientists and engineers work in many different settings, including colleges and universities, businesses and industries, specific research institutes, and government agencies.

F.5.f Scientists and engineers have ethical codes requiring that human subjects involved with research be fully informed about risks and benefits associated with the research before the individuals choose to participate. This ethic extends to potential risks to communities and property. In short, prior knowledge and consent are required for research involving human subjects or potential damage to property.

F.5.g Science cannot answer all questions and technology cannot solve all human problems or meet all human needs. Students should understand the difference between scientific and other questions. They should appreciate what science and technology can reasonably contribute to society and what they cannot do. For example, new technologies often will decrease some risks and increase others.

G. History and Nature of Science

As a result of activities in grades 5–8, all students should develop understanding of

Science as a Human Endeavor

G.1.a Women and men of various social and ethnic backgrounds—and with diverse interests, talents, qualities, and motivations—engage in the activities of science, engineering, and related fields such as the health professions. Some scientists work in teams, and some work alone, but all communicate extensively with others.

G.1.b Science requires different abilities, depending on such factors as the field of study and type of inquiry. Science is very much a human endeavor, and the work of science relies on basic human qualities, such as reasoning, insight, energy, skill, and creativity—as well as on scientific habits of mind, such as intellectual honesty, tolerance of ambiguity, skepticism, and openness to new ideas.

National Science Education Standards

Nature of Science

G.2.a Scientists formulate and test their explanations of nature using observation, experiments, and theoretical and mathematical models. Although all scientific ideas are tentative and subject to change and improvement in principle, for most major ideas in science, there is much experimental and observational confirmation. Those ideas are not likely to change greatly in the future. Scientists do and have changed their ideas about nature when they encounter new experimental evidence that does not match their existing explanations.

G.2.b In areas where active research is being pursued and in which there is not a great deal of experimental or observational evidence and understanding, it is normal for scientists to differ with one another about the interpretation of the evidence or theory being considered. Different scientists might publish conflicting experimental results or might draw different conclusions from the same data. Ideally, scientists acknowledge such conflict and work towards finding evidence that will resolve their disagreement.

G.2.c It is part of scientific inquiry to evaluate the results of scientific investigations, experiments, observations, theoretical models, and the explanations proposed by other scientists. Evaluation includes reviewing the experimental procedures, examining the evidence, identifying faulty reasoning, pointing out statements that go beyond the evidence, and suggesting alternative explanations for the same observations. Although scientists may disagree about explanations of phenomena, about interpretations of data, or about the value of rival theories, they do agree that questioning, response to criticism, and open communication are integral to the process of science. As scientific knowledge evolves, major disagreements are eventually resolved through such interactions between scientists.

History of Science

G.3.a Many individuals have contributed to the traditions of science. Studying some of these individuals provides further understanding of scientific inquiry, science as a human endeavor, the nature of science, and the relationships between science and society.

G.3.b In historical perspective, science has been practiced by different individuals in different cultures. In looking at the history of many peoples, one finds that scientists and engineers of high achievement are considered to be among the most valued contributors to their culture.

G.3.c Tracing the history of science can show how difficult it was for scientific innovators to break through the accepted ideas of their time to reach the conclusions that we currently take for granted.

Project 2061
Benchmarks for Science Literacy

1. The Nature of Science

By the end of the 8th grade, students should know that

1.A The Scientific World View

1.A.1 When similar investigations give different results, the scientific challenge is to judge whether the differences are trivial or significant, and it often takes further studies to decide. Even with similar results, scientists may wait until an investigation has been repeated many times before accepting the results as correct.

1.A.2 Scientific knowledge is subject to modification as new information challenges prevailing theories and as a new theory leads to looking at old observations in a new way.

1.A.3 Some scientific knowledge is very old and yet is still applicable today.

1.A.4 Some matters cannot be examined usefully in a scientific way. Among them are matters that by their nature cannot be tested objectively and those that are essentially matters of morality. Science can sometimes be used to inform ethical decisions by identifying the likely consequences of particular actions but cannot be used to establish that some action is either moral or immoral.

1.B Scientific Inquiry

1.B.1 Scientists differ greatly in what phenomena they study and how they go about their work. Although there is no fixed set of steps that all scientists follow, scientific investigations usually involve the collection of relevant evidence, the use of logical reasoning, and the application of imagination in devising hypotheses and explanations to make sense of the collected evidence.

1.B.2 If more than one variable changes at the same time in an experiment, the outcome of the experiment may not be clearly attributable to any one of the variables. It may not always be possible to prevent outside variables from influencing the outcome of an investigation (or even to identify all of the variables), but collaboration among investigators can often lead to research designs that are able to deal with such situations.

1.B.3 What people expect to observe often affects what they actually do observe. Strong beliefs about what should happen in particular circumstances can prevent them from detecting other results. Scientists know about this danger to objectivity and take steps to try and avoid it when designing investigations and examining data. One safeguard is to have different investigators conduct independent studies of the same questions.

1.C The Scientific Enterprise

1.C.1 Important contributions to the advancement of science, mathematics, and technology have been made by different kinds of people, in different cultures, at different times.

1.C.2 Until recently, women and racial minorities, because of restrictions on their education and employment opportunities, were essentially left out of much of the formal work of the science establishment; the remarkable few who overcame those obstacles were even then likely to have their work disregarded by the science establishment.

1.C.3 No matter who does science and mathematics or invents things, or when or where they do it, the knowledge and technology that result can eventually become available to everyone in the world.

1.C.4 Scientists are employed by colleges and universities, business and industry, hospitals, and many government agencies. Their places of work include offices, classrooms, laboratories, farms, factories, and natural field settings ranging from space to the ocean floor.

1.C.5 In research involving human subjects, the ethics of science require that potential subjects be fully informed about the risks and benefits associated with the research and of their right to refuse to participate. Science ethics also demand that scientists must not knowingly subject coworkers, students, the neighborhood, or the community to health or property risks without their prior knowledge and consent. Because animals cannot make informed choices, special care must be taken in using them in scientific research.

1.C.6 Computers have become invaluable in science because they speed up and extend people's ability to collect, store, compile, and analyze data, prepare research reports, and share data and ideas with investigators all over the world.

1.C.7 Accurate record-keeping, openness, and replication are essential for maintaining an investigator's credibility with other scientists and society.

3. The Nature of Technology

By the end of the 8th grade, students should know that

3.A Technology and Science

3.A.1 In earlier times, the accumulated information and techniques of each generation of workers were taught on the job directly to the next generation of workers. Today, the knowledge base for technology can be found as well in libraries of print and electronic resources and is often taught in the classroom.

3.A.2 Technology is essential to science for such purposes as access to outer space and other remote locations, sample collection and treatment, measurement, data collection and storage, computation, and communication of information.

3.A.3 Engineers, architects, and others who engage in design and technology use scientific knowledge to solve practical problems. But they usually have to take human values and limitations into account as well.

3.B Design and Systems

3.B.1 Design usually requires taking constraints into account. Some constraints, such as gravity or the properties of the materials to be used, are unavoidable. Other constraints, including economic, political, social, ethical, and aesthetic ones, limit choices.

3.B.2 All technologies have effects other than those intended by the design, some of which may have been predictable and some not. In either case, these side effects may turn out to be unacceptable to some of the population and therefore lead to conflict between groups.

3.B.3 Almost all control systems have inputs, outputs, and feedback. The essence of control is comparing information about what is happening to what people want to happen and then making appropriate adjustments. This procedure requires sensing information, processing it, and making changes. In almost all modern machines, microprocessors serve as centers of performance control.

3.B.4 Systems fail because they have faulty or poorly matched parts, are used in ways that exceed what was intended by the design, or were poorly designed to begin with. The most common ways to prevent failure are pretesting parts and procedures, overdesign, and redundancy.

3.C Issues in Technology

3.C.1 The human ability to shape the future comes from a capacity for generating knowledge and developing new technologies—and for communicating ideas to others.

3.C.2 Technology cannot always provide successful solutions for problems or fulfill every human need.

3.C.3 Throughout history, people have carried out impressive technological feats, some of which would be hard to duplicate today even with modern tools. The purposes served by these achievements have sometimes been practical, sometimes ceremonial.

3.C.4 Technology has strongly influenced the course of history and continues to do so. It is largely responsible for the great revolutions in agriculture, manufacturing, sanitation and medicine, warfare, transportation, information processing, and communications that have radically changed how people live.

3.C.5 New technologies increase some risks and decrease others. Some of the same technologies that have improved the length and quality of life for many people have also brought new risks.

3.C.6 Rarely are technology issues simple and one-sided. Relevant facts alone, even when known and available, usually do not settle matters entirely in favor of one side or another. That is because the contending groups may have different values and priorities. They may stand to gain or lose in different degrees, or may make very different predictions about what the future consequences of the proposed action will be.

3.C.7 Societies influence what aspects of technology are developed and how these are used. People control technology (as well as science) and are responsible for its effects.

4. The Physical Setting

By the end of the 8th grade, students should know that

4.A The Universe

4.A.1 The sun is a medium-sized star located near the edge of a disk-shaped galaxy of stars, part of which can be seen as a glowing band of light that spans the sky on a very clear night. The universe contains many billions of galaxies, and each galaxy contains many billions of stars. To the naked eye, even the closest of these galaxies is no more than a dim, fuzzy spot.

4.A.2 The sun is many thousands of times closer to the earth than any other star. Light from the sun takes a few minutes to reach the earth, but light from the next nearest star takes a few years to arrive. The trip to that star would take the fastest rocket thousands of years. Some distant galaxies are so far away that their light takes several billion years to reach the earth. People on earth, therefore, see them as they were that long ago in the past.

4.A.3 Nine planets of very different size, composition, and surface features move around the sun in nearly circular orbits. Some planets have a great variety of moons and even flat rings of rock and ice particles orbiting around them. Some of these planets and moons show evidence of geologic activity. The earth is orbited by one moon, many artificial satellites, and debris.

4.A.4 Large numbers of chunks of rock orbit the sun. Some of those that the earth meets in its yearly orbit around the sun glow and disintegrate from friction as they plunge through the atmosphere—and sometimes impact the ground. Other chunks of rocks mixed with ice have long, off-center orbits that carry them close to the sun, where the sun's radiation (of light and particles) boils off frozen material from their surfaces and pushes it into a long, illuminated tail.

4.B The Earth

4.B.1 We live on a relatively small planet, the third from the sun in the only system of planets definitely known to exist (although other, similar systems may be discovered in the universe).

4.B.2 The earth is mostly rock. Three-fourths of its surface is covered by a relatively thin layer of water (some of it frozen), and the entire planet is surrounded by a relatively thin blanket of air. It is the only body in the solar system that appears able to support life. The other planets have compositions and conditions very different from the earth's.

4.B.3 Everything on or anywhere near the earth is pulled toward the earth's center by gravitational force.

4.B.4 Because the earth turns daily on an axis that is tilted relative to the plane of the earth's yearly orbit around the sun, sunlight falls more intensely on different parts of the earth during the year. The difference in heating of the earth's surface produces the planet's seasons and weather patterns.

4.B.5 The moon's orbit around the earth once in about 28 days changes what part of the moon is lighted by the sun and how much of that part can be seen from the earth—the phases of the moon.

4.B.6 Climates have sometimes changed abruptly in the past as a result of changes in the earth's crust, such as volcanic eruptions or impacts of huge rocks from space. Even relatively small changes in atmospheric or ocean content can have widespread effects on climate if the change lasts long enough.

4.B.7 The cycling of water in and out of the atmosphere plays an important role in determining climatic patterns. Water evaporates from the surface of the earth, rises and cools, condenses into rain or snow, and falls again to the surface. The water falling on land collects in rivers and lakes, soil, and porous layers of rock, and much of it flows back into the ocean.

4.B.8 Fresh water, limited in supply, is essential for life and also for most industrial processes. Rivers, lakes, and groundwater can be depleted or polluted, becoming unavailable or unsuitable for life.

4.B.9 Heat energy carried by ocean currents has a strong influence on climate around the world.

4.B.10 Some minerals are very rare and some exist in great quantities, but—for practical purposes—the ability to recover them is just as important as their abundance. As minerals are depleted, obtaining them becomes more difficult. Recycling and the development of substitutes can reduce the rate of depletion but may also be costly.

4.B.11 The benefits of the earth's resources—such as fresh water, air, soil, and trees—can be reduced by using them wastefully or by deliberately or inadvertently destroying them. The atmosphere and the oceans have a limited capacity to absorb wastes and recycle materials naturally. Cleaning up polluted air, water, or soil or restoring depleted soil, forests, or fishing grounds can be very difficult and costly.

4.C Processes that Shape the Earth

4.C.1 The interior of the earth is hot. Heat flow and movement of material within the earth cause earthquakes and volcanic eruptions and create mountains and ocean basins. Gas and dust from large volcanoes can change the atmosphere.

4.C.2 Some changes in the earth's surface are abrupt (such as earthquakes and volcanic eruptions) while other changes happen very slowly (such as uplift and wearing down of mountains). The earth's surface is shaped in part by the motion of water and wind over very long times, which act to level mountain ranges.

4.C.3 Sediments of sand and smaller particles (sometimes containing the remains of organisms) are gradually buried and are cemented together by dissolved minerals to form solid rock again.

4.C.4 Sedimentary rock buried deep enough may be reformed by pressure and heat, perhaps melting and recrystallizing into different kinds of rock. These re-formed rock layers may be forced up again to become land surface and even mountains. Subsequently, this new rock too will erode. Rock bears evidence of the minerals, temperatures, and forces that created it.

4.C.5 Thousands of layers of sedimentary rock confirm the long history of the changing surface of the earth and the changing life forms whose remains are found in successive layers. The youngest layers are not always found on top, because of folding, breaking, and uplift of layers.

4.C.6 Although weathered rock is the basic component of soil, the composition and texture of soil and its fertility and resistance to erosion are greatly influenced by plant roots and debris, bacteria, fungi, worms, insects, rodents, and other organisms.

4.C.7 Human activities, such as reducing the amount of forest cover, increasing the amount and variety of chemicals released into the atmosphere, and intensive farming, have changed the earth's land, oceans, and atmosphere. Some of these changes have decreased the capacity of the environment to support some life forms.

4.D Structure of Matter

4.D.1 All matter is made up of atoms, which are far too small to see directly through a microscope. The atoms of any element are alike but are different from atoms of other elements. Atoms may stick together in well-defined molecules or may be packed together in large arrays. Different arrangements of atoms into groups compose all substances.

4.D.2 Equal volumes of different substances usually have different weights.

4.D.3 Atoms and molecules are perpetually in motion. Increased temperature means greater average energy, so most substances expand when heated. In solids, the atoms are closely locked in position and can only vibrate. In liquids, the atoms or molecules have higher energy, are more loosely connected, and can slide past one another; some molecules may get enough energy to escape into a gas. In gases, the atoms or molecules have still more energy and are free of one another except during occasional collisions.

4.D.4 The temperature and acidity of a solution influence reaction rates. Many substances dissolve in water, which may greatly facilitate reactions between them.

4.D.5 Scientific ideas about elements were borrowed from some Greek philosophers of 2,000 years earlier, who believed that everything was made from four basic substances: air, earth, fire, and water. It was the combinations of these "elements" in different proportions that gave other substances their observable properties. The Greeks were wrong about those four, but now over 100 different elements have been identified, some rare and some plentiful, out of which everything is made. Because most elements tend to combine with others, few elements are found in their pure form.

4.D.6 There are groups of elements that have similar properties, including highly reactive metals, less-reactive metals, highly reactive nonmetals (such as chlorine, fluorine, and oxygen), and some almost completely nonreactive gases (such as helium and neon). An especially important kind of reaction between substances involves combination of oxygen with something else—as in burning or rusting. Some elements don't fit into any of the categories; among them are carbon and hydrogen, essential elements of living matter.

4.D.7 No matter how substances within a closed system interact with one another, or how they combine or break apart, the total weight of the system remains the same. The idea of atoms explains the conservation of matter: If the number of atoms stays the same no matter how they are rearranged, then their total mass stays the same.

4.E Energy Transformations

4.E.1 Energy cannot be created or destroyed, but only changed from one form into another.

4.E.2 Most of what goes on in the universe—from exploding stars and biological growth to the operation of machines and the motion of people—involves some form of energy being transformed into another. Energy in the form of heat is almost always one of the products of an energy transformation.

4.E.3 Heat can be transferred through materials by the collisions of atoms or across space by radiation. If the material is fluid, currents will be set up in it that aid the transfer of heat.

4.E.4 Energy appears in different forms. Heat energy is in the disorderly motion of molecules; chemical energy is in the arrangement of atoms; mechanical energy is in moving bodies or in elastically distorted shapes; gravitational energy is in the separation of mutually attracting masses.

4.F Motion

4.F.1 Light from the sun is made up of a mixture of many different colors of light, even though to the eye the light looks almost white. Other things that give off or reflect light have a different mix of colors.

4.F.2 Something can be "seen" when light waves emitted or reflected by it enter the eye—just as something can be "heard" when sound waves from it enter the ear.

4.F.3 An unbalanced force acting on an object changes its speed or direction of motion, or both. If the force acts toward a single center, the object's path may curve into an orbit around the center.

4.F.4 Vibrations in materials set up wavelike disturbances that spread away from the source. Sound and earthquake waves are examples. These and other waves move at different speeds in different materials.

4.F.5 Human eyes respond to only a narrow range of wavelengths of electromagnetic radiation—visible light. Differences of wavelength within that range are perceived as differences in color.

4.G Forces of Nature

4.G.1 Every object exerts gravitational force on every other object. The force depends on how much mass the objects have and on how far apart they are. The force is hard to detect unless at least one of the objects has a lot of mass.

4.G.2 The sun's gravitational pull holds the earth and other planets in their orbits, just as the planets' gravitational pull keeps their moons in orbit around them.

4.G.3 Electric currents and magnets can exert a force on each other.

5. The Living Environment

By the end of the 8th grade, students should know that

5.A Diversity of Life

5.A.1 One of the most general distinctions among organisms is between plants, which use sunlight to make their own food, and animals, which consume energy-rich foods. Some kinds of organisms, many of them microscopic, cannot be neatly classified as either plants or animals.

5.A.2 Animals and plants have a great variety of body plans and internal structures that contribute to their being able to make or find food and reproduce.

5.A.3 Similarities among organisms are found in internal anatomical features, which can be used to infer the degree of relatedness among organisms. In classifying organisms, biologists consider details of internal and external structures to be more important than behavior or general appearance.

5.A.4 For sexually reproducing organisms, a species comprises all organisms that can mate with one another to produce fertile offspring.

5.A.5 All organisms, including the human species, are part of and depend on two main interconnected global food webs. One includes microscopic ocean plants, the animals that feed on them, and finally the animals that feed on those animals. The other web includes land plants, the animals that feed on them, and so forth. The cycles continue indefinitely because organisms decompose after death to return food material to the environment.

5.B Heredity

5.B.1 In some kinds of organisms, all the genes come from a single parent, whereas in organisms that have sexes, typically half of the genes come from each parent.

5.B.2 In sexual reproduction, a single specialized cell from a female merges with a specialized cell from a male. As the fertilized egg, carrying genetic information from each parent, multiplies to form the complete organism with about a trillion cells, the same genetic information is copied in each cell.

5.B.3 New varieties of cultivated plants and domestic animals have resulted from selective breeding for particular traits.

5.C Cells

5.C.1 All living things are composed of cells, from just one to many millions, whose details usually are visible only through a microscope. Different body tissues and organs are made up of different kinds of cells. The cells in similar tissues and organs in other animals are similar to those in human beings but differ somewhat from cells found in plants.

5.C.2 Cells repeatedly divide to make more cells for growth and repair. Various organs and tissues function to serve the needs of cells for food, air, and waste removal.

5.C.3 Within cells, many of the basic functions of organisms—such as extracting energy from food and getting rid of waste—are carried out. The way in which cells function is similar in all living organisms.

5.C.4 About two-thirds of the weight of cells is accounted for by water, which gives cells many of their properties.

5.D Interdependence of Life

5.D.1 In all environments—freshwater, marine, forest, desert, grassland, mountain, and others—organisms with similar needs may compete with one another for resources, including food, space, water, air, and shelter. In any particular environment, the growth and survival of organisms depend on the physical conditions.

5.D.2 Two types of organisms may interact with one another in several ways: They may be in a producer/consumer, predator/prey, or parasite/host relationship. Or one organism may scavenge or decompose another. Relationships may be competitive or mutually beneficial. Some species have become so adapted to each other that neither could survive without the other.

5.E Flow of Matter and Energy

5.E.1 Food provides molecules that serve as fuel and building material for all organisms. Plants use the energy in light to make sugars out of carbon dioxide and water. This food can be used immediately for fuel or materials or it may be stored for later use. Organisms that eat plants break down the plant structures to produce the materials and energy they need to survive. Then they are consumed by other organisms.

5.E.2 Over a long time, matter is transferred from one organism to another repeatedly and between organisms and their physical environment. As in all material systems, the total amount of matter remains constant, even though its form and location change.

5.E.3 Energy can change from one form to another in living things. Animals get energy from oxidizing their food, releasing some of its energy as heat. Almost all food energy comes originally from sunlight.

5.F Evolution of Life

5.F.1 Small differences between parents and offspring can accumulate (through selective breeding) in successive generations so that descendants are very different from their ancestors.

5.F.2 Individual organisms with certain traits are more likely than others to survive and have offspring. Changes in environmental conditions can affect the survival of individual organisms and entire species.

5.F.3 Many thousands of layers of sedimentary rock provide evidence for the long history of the earth and for the long history of changing life forms whose remains are found in the rocks. More recently deposited rock layers are more likely to contain fossils resembling existing species.

6. The Human Organism

By the end of the 8th grade, students should know that

6.A Human Identity

6.A.1 Like other animals, human beings have body systems for obtaining and providing energy, defense, reproduction, and the coordination of body functions.

6.A.2 Human beings have many similarities and differences. The similarities make it possible for human beings to reproduce and to donate blood and organs to one another throughout the world. Their differences enable them to create diverse social and cultural arrangements and to solve problems in a variety of ways.

6.A.3 Fossil evidence is consistent with the idea that human beings evolved from earlier species.

6.A.4 Specialized roles of individuals within other species are genetically programmed, whereas human beings are able to invent and modify a wider range of social behavior.

6.A.5 Human beings use technology to match or excel many of the abilities of other species. Technology has helped people with disabilities survive and live more conventional lives.

6.A.6 Technologies having to do with food production, sanitation, and disease prevention have dramatically changed how people live and work and have resulted in rapid increases in the human population.

6.B Human Development

6.B.1 Fertilization occurs when sperm cells from a male's testes are deposited near an egg cell from the female ovary, and one of the sperm cells enters the egg cell. Most of the time, by chance or design, a sperm never arrives or an egg isn't available.

6.B.2 Contraception measures may incapacitate sperm, block their way to the egg, prevent the release of eggs, or prevent the fertilized egg from implanting successfully.

6.B.3 Following fertilization, cell division produces a small cluster of cells that then differentiate by appearance and function to form the basic tissues of an embryo. During the first three months of pregnancy, organs begin to form. During the second three months, all organs and body features develop. During the last three months, the organs and features mature enough to function well after birth. Patterns of human development are similar to those of other vertebrates.

6.B.4 The developing embryo—and later the newborn infant—encounters many risks from faults in its genes, its mother's inadequate diet, her cigarette smoking or use of alcohol or other drugs, or from infection. Inadequate child care may lead to lower physical and mental ability.

6.B.5 Various body changes occur as adults age. Muscles and joints become less flexible, bones and muscles lose mass, energy levels diminish, and the senses become less acute. Women stop releasing eggs and hence can no longer reproduce. The length and quality of human life are influenced by many factors, including sanitation, diet, medical care, sex, genes, environmental conditions, and personal health behaviors.

6.C Basic Functions

6.C.1 Organs and organ systems are composed of cells and help to provide all cells with basic needs.

6.C.2 For the body to use food for energy and building materials, the food must first be digested into molecules that are absorbed and transported to cells.

6.C.3 To burn food for the release of energy stored in it, oxygen must be supplied to cells, and carbon dioxide removed. Lungs take in oxygen for the combustion of food and they eliminate the carbon dioxide produced. The urinary system disposes of dissolved waste molecules, the intestinal tract removes solid wastes, and the skin and lungs rid the body of heat energy. The circulatory system moves all these substances to or from cells where they are needed or produced, responding to changing demands.

6.C.4 Specialized cells and the molecules they produce identify and destroy microbes that get inside the body.

6.C.5 Hormones are chemicals from glands that affect other body parts. They are involved in helping the body respond to danger and in regulating human growth, development, and reproduction.

6.C.6 Interactions among the senses, nerves, and brain make possible the learning that enables human beings to cope with changes in their environment.

6.D Learning

6.D.1 Some animal species are limited to a repertoire of genetically determined behaviors; others have more complex brains and can learn a wide variety of behaviors. All behavior is affected by both inheritance and experience.

6.D.2 The level of skill a person can reach in any particular activity depends on innate abilities, the amount of practice, and the use of appropriate learning technologies.

6.D.3 Human beings can detect a tremendous range of visual and olfactory stimuli. The strongest stimulus they can tolerate may be more than a trillion times as intense as the weakest they can detect. Still, there are many kinds of signals in the world that people cannot detect directly.

6.D.4 Attending closely to any one input of information usually reduces the ability to attend to others at the same time.

6.D.5 Learning often results from two perceptions or actions occurring at about the same time. The more often the same combination occurs, the stronger the mental connection between them is likely to be. Occasionally a single vivid experience will connect two things permanently in people's minds.

6.D.6 Language and tools enable human beings to learn complicated and varied things from others.

6.E Physical Health

6.E.1 The amount of food energy (calories) a person requires varies with body weight, age, sex, activity level, and natural body efficiency. Regular exercise is important to maintain a healthy heart/lung system, good muscle tone, and bone strength.

6.E.2 Toxic substances, some dietary habits, and personal behavior may be bad for one's health. Some effects show up right away, others may not show up for many years. Avoiding toxic substances, such as tobacco, and changing dietary habits to reduce the intake of such things as animal fat increases the chances of living longer.

6.E.3 Viruses, bacteria, fungi, and parasites may infect the human body and interfere with normal body functions. A person can catch a cold many times because there are many varieties of cold viruses that cause similar symptoms.

6.E.4 White blood cells engulf invaders or produce antibodies that attack them or mark them for killing by other white cells. The antibodies produced will remain and can fight off subsequent invaders of the same kind.

6.E.5 The environment may contain dangerous levels of substances that are harmful to human beings. Therefore, the good health of individuals requires monitoring the soil, air, and water and taking steps to keep them safe.

6.F Mental Health

6.F.1 Individuals differ greatly in their ability to cope with stressful situations. Both external and internal conditions (chemistry, personal history, values) influence how people behave.

6.F.2 Often people react to mental distress by denying that they have any problem. Sometimes they don't know why they feel the way they do, but with help they can sometimes uncover the reasons.

8. The Designed World

By the end of the 8th grade, students should know that

8.A Agriculture

8.A.1 Early in human history, there was an agricultural revolution in which people changed from hunting and gathering to farming. This allowed changes in the division of labor between men and women and between children and adults, and the development of new patterns of government.

8.A.2 People control the characteristics of plants and animals they raise by selective breeding and by preserving varieties of seeds (old and new) to use if growing conditions change.

8.A.3 In agriculture, as in all technologies, there are always trade-offs to be made. Getting food from many different places makes people less dependent on weather in any one place, yet more dependent on transportation and communication among far-flung markets. Specializing in one crop may risk disaster if changes in weather or increases in pest populations wipe out that crop. Also, the soil may be exhausted of some nutrients, which can be replenished by rotating the right crops.

8.A.4 Many people work to bring food, fiber, and fuel to U.S. markets. With improved technology, only a small fraction of workers in the United States actually plant and harvest the products that people use. Most workers are engaged in processing, packaging, transporting, and selling what is produced.

8.B Materials and Manufacturing

8.B.1 The choice of materials for a job depends on their properties and on how they interact with other materials. Similarly, the usefulness of some manufactured parts of an object depends on how well they fit together with the other parts.

8.B.2 Manufacturing usually involves a series of steps, such as designing a product, obtaining and preparing raw materials, processing the materials mechanically or chemically, and assembling, testing, inspecting, and packaging. The sequence of these steps is also often important.

8.B.3 Modern technology reduces manufacturing costs, produces more uniform products, and creates new synthetic materials that can help reduce the depletion of some natural resources.

8.B.4 Automation, including the use of robots, has changed the nature of work in most fields, including manufacturing. As a result, high-skill, high-knowledge jobs in engineering, computer programming, quality control, supervision, and maintenance are replacing many routine, manual-labor jobs. Workers therefore need better learning skills and flexibility to take on new and rapidly changing jobs.

8.C Energy Sources and Use

8.C.1 Energy can change from one form to another, although in the process some energy is always converted to heat. Some systems transform energy with less loss of heat than others.

8.C.2 Different ways of obtaining, transforming, and distributing energy have different environmental consequences.

8.C.3 In many instances, manufacturing and other technological activities are performed at a site close to an energy source. Some forms of energy are transported easily, others are not.

8.C.4 Electrical energy can be produced from a variety of energy sources and can be transformed into almost any other form of energy. Moreover, electricity is used to distribute energy quickly and conveniently to distant locations.

8.C.5 Energy from the sun (and the wind and water energy derived from it) is available indefinitely. Because the flow of energy is weak and variable, very large collection systems are needed. Other sources don't renew or renew only slowly.

8.C.6 Different parts of the world have different amounts and kinds of energy resources to use and use them for different purposes.

8.D Communication

8.D.1 Errors can occur in coding, transmitting, or decoding information, and some means of checking for accuracy is needed. Repeating the message is a frequently used method.

8.D.2 Information can be carried by many media, including sound, light, and objects. In this century, the ability to code information as electric currents in wires, electromagnetic waves in space, and light in glass fibers has made communication millions of times faster than is possible by mail or sound.

8.E Information Processing

8.E.1 Most computers use digital codes containing only two symbols, 0 and 1, to perform all operations. Continuous signals (analog) must be transformed into digital codes before they can be processed by a computer.

8.E.2 What use can be made of a large collection of information depends upon how it is organized. One of the values of computers is that they are able, on command, to reorganize information in a variety of ways, thereby enabling people to make more and better uses of the collection.

8.E.3 Computer control of mechanical systems can be much quicker than human control. In situations where events happen faster than people can react, there is little choice but to rely on computers. Most complex systems still require human oversight, however, to make certain kinds of judgments about the readiness of the parts of the system (including the computers) and the system as a whole to operate properly, to react to unexpected failures, and to evaluate how well the system is serving its intended purposes.

8.E.4 An increasing number of people work at jobs that involve processing or distributing information. Because computers can do these tasks faster and more reliably, they have become standard tools both in the workplace and at home.

8.F Health Technology

8.F.1 Sanitation measures such as the use of sewers, landfills, quarantines, and safe food handling are important in controlling the spread of organisms that cause disease. Improving sanitation to prevent disease has contributed more to saving human life than any advance in medical treatment.

8.F.2 The ability to measure the level of substances in body fluids has made it possible for physicians to make comparisons with normal levels, make very sophisticated diagnoses, and monitor the effects of the treatments they prescribe.

8.F.3 It is becoming increasingly possible to manufacture chemical substances such as insulin and hormones that are normally found in the body. They can be used by individuals whose own bodies cannot produce the amounts required for good health.

9. The Mathematical World

By the end of the 8th grade, students should know that

9.A Numbers

9.A.1 There have been systems for writing numbers other than the Arabic system of place values based on tens. The very old Roman numerals are now used only for dates, clock faces, or ordering chapters in a book. Numbers based on 60 are still used for describing time and angles.

9.A.2 A number line can be extended on the other side of zero to represent negative numbers. Negative numbers allow subtraction of a bigger number from a smaller number to make sense, and are often used when something can be measured on either side of some reference point (time, ground level, temperature, budget).

9.A.3 Numbers can be written in different forms, depending on how they are being used. How fractions or decimals based on measured quantities should be written depends on how precise the measurements are and how precise an answer is needed.

9.A.4 The operations + and – are inverses of each other—one undoes what the other does; likewise x and ÷ .

9.A.5 The expression *a/b* can mean different things: *a* parts of size *1/b* each, *a* divided by *b,* or *a* compared to *b.*

9.A.6 Numbers can be represented by using sequences of only two symbols (such as 1 and 0, on and off); computers work this way.

9.A.7 Computations (as on calculators) can give more digits than make sense or are useful.

9.B Symbolic Relationships

9.B.1 An equation containing a variable may be true for just one value of the variable.

9.B.2 Mathematical statements can be used to describe how one quantity changes when another changes. Rates of change can be computed from differences in magnitudes and vice versa.

9.B.3 Graphs can show a variety of possible relationships between two variables. As one variable increases uniformly, the other may do one of the following: increase or decrease steadily, increase or decrease faster and faster, get closer and closer to some limiting value, reach some intermediate maximum or minimum, alternately increase and decrease indefinitely, increase or decrease in steps, or do something different from any of these.

9.C Shapes

9.C.1 Some shapes have special properties: triangular shapes tend to make structures rigid, and round shapes give the least possible boundary for a given amount of interior area. Shapes can match exactly or have the same shape in different sizes.

9.C.2 Lines can be parallel, perpendicular, or oblique.

9.C.3 Shapes on a sphere like the earth cannot be depicted on a flat surface without some distortion.

9.C.4 The graphic display of numbers may help to show patterns such as trends, varying rates of change, gaps, or clusters. Such patterns sometimes can be used to make predictions about the phenomena being graphed.

9.C.5 It takes two numbers to locate a point on a map or any other flat surface. The numbers may be two perpendicular distances from a point, or an angle and a distance from a point.

9.C.6 The scale chosen for a graph or drawing makes a big difference in how useful it is.

9.D Uncertainty

9.D.1 How probability is estimated depends on what is known about the situation. Estimates can be based on data from similar conditions in the past or on the assumption that all the possibilities are known.

9.D.2 Probabilities are ratios and can be expressed as fractions, percentages, or odds.

9.D.3 The mean, median, and mode tell different things about the middle of a data set.

9.D.4 Comparison of data from two groups should involve comparing both their middles and the spreads around them.

9.D.5 The larger a well-chosen sample is, the more accurately it is likely to represent the whole. But there are many ways of choosing a sample that can make it unrepresentative of the whole.

9.D.6 Events can be described in terms of being more or less likely, impossible, or certain.

9.E Reasoning

9.E.1 Some aspects of reasoning have fairly rigid rules for what makes sense; other aspects don't. If people have rules that always hold, and good information about a particular situation, then logic can help them to figure out what is true about it. This kind of reasoning requires care in the use of key words such as if, and, not, or, all, and some. Reasoning by similarities can suggest ideas but can't prove them one way or the other.

9.E.2 Practical reasoning, such as diagnosing or troubleshooting almost anything, may require many-step, branching logic. Because computers can keep track of complicated logic, as well as a lot of information, they are useful in a lot of problem-solving situations.

9.E.3 Sometimes people invent a general rule to explain how something works by summarizing observations. But people tend to overgeneralize, imagining general rules on the basis of only a few observations.

9.E.4 People are using incorrect logic when they make a statement such as "If *A* is true, then *B* is true; but *A* isn't true, therefore *B* isn't true either."

9.E.5 A single example can never prove that something is always true, but sometimes a single example can prove that something is not always true.

9.E.6 An analogy has some likenesses to but also some differences from the real thing.

10. Historical Perspectives

By the end of the 8th grade, students should know that

10.A Displacing the Earth from the Center of the Universe

10.A.1 The motion of an object is always judged with respect to some other object or point and so the idea of absolute motion or rest is misleading.

10.A.2 Telescopes reveal that there are many more stars in the night sky than are evident to the unaided eye, the surface of the moon has many craters and mountains, the sun has dark spots, and Jupiter and some other planets have their own moons.

10.F Understanding Fire

10.F.1 From the earliest times until now, people have believed that even though millions of different kinds of material seem to exist in the world, most things must be made up of combinations of just a few basic kinds of things. There has not always been agreement, however, on what those basic kinds of things are. One theory long ago was that the basic substances were earth, water, air, and fire. Scientists now know that these are not the basic substances. But the old theory seemed to explain many observations about the world.

10.F.2 Today, scientists are still working out the details of what the basic kinds of matter are and of how they combine, or can be made to combine, to make other substances.

10.F.3 Experimental and theoretical work done by French scientist Antoine Lavoisier in the decade between the American and French revolutions led to the modern science of chemistry.

10.F.4 Lavoisier's work was based on the idea that when materials react with each other many changes can take place but that in every case the total amount of matter afterward is the same as before. He successfully tested the concept of conservation of matter by conducting a series of experiments in which he carefully measured all the substances involved in burning, including the gases used and those given off.

10.F.5 Alchemy was chiefly an effort to change base metals like lead into gold and to produce an elixir that would enable people to live forever. It failed to do that or to create much knowledge of how substances react with each other. The more scientific study of chemistry that began in Lavoisier's time has gone far beyond alchemy in understanding reactions and producing new materials.

10.G Splitting the Atom

10.G.1 The accidental discovery that minerals containing uranium darken photographic film, as light does, led to the idea of radioactivity.

10.G.2 In their laboratory in France, Marie Curie and her husband, Pierre Curie, isolated two new elements that caused most of the radioactivity of the uranium mineral. They named one radium because it gave off powerful, invisible rays, and the other polonium in honor of Madame Curie's country of birth. Marie Curie was the first scientist ever to win the Nobel prize in two different fields—in physics, shared with her husband, and later in chemistry.

10.I Discovering Germs

10.I.1 Throughout history, people have created explanations for disease. Some have held that disease has spiritual causes, but the most persistent biological theory over the centuries was that illness resulted from an imbalance in the body fluids. The introduction of germ theory by Louis Pasteur and others in the 19th century led to the modern belief that many diseases are caused by microorganisms—bacteria, viruses, yeasts, and parasites.

10.I.2 Pasteur wanted to find out what causes milk and wine to spoil. He demonstrated that spoilage and fermentation occur when microorganisms enter from the air, multiply rapidly, and produce waste products. After showing that spoilage could be avoided by keeping germs out or by destroying them with heat, he investigated animal diseases and showed that microorganisms were involved. Other investigators later showed that specific kinds of germs caused specific diseases.

10.I.3 Pasteur found that infection by disease organisms—germs—caused the body to build up an immunity against subsequent infection by the same organisms. He then demonstrated that it was possible to produce vaccines that would induce the body to build immunity to a disease without actually causing the disease itself.

10.I.4 Changes in health practices have resulted from the acceptance of the germ theory of disease. Before germ theory, illness was treated by appeals to supernatural powers or by trying to adjust body fluids through induced vomiting, bleeding, or purging. The modern approach emphasizes sanitation, the safe handling of food and water, the pasteurization of milk, quarantine, and aseptic surgical techniques to keep germs out of the body; vaccinations to strengthen the body's immune system against subsequent infection by the same kind of microorganisms; and antibiotics and other chemicals and processes to destroy microorganisms.

10.I.5 In medicine, as in other fields of science, discoveries are sometimes made unexpectedly, even by accident. But knowledge and creative insight are usually required to recognize the meaning of the unexpected.

10.J Harnessing Power

10.J.1 Until the 1800s, most manufacturing was done in homes, using small, handmade machines that were powered by muscle, wind, or running water. New machinery and steam engines to drive them made it possible to replace craftsmanship with factories, using fuels as a source of energy. In the factory system, workers, materials, and energy could be brought together efficiently.

10.J.2 The invention of the steam engine was at the center of the Industrial Revolution. It converted the chemical energy stored in wood and coal, which were plentiful, into mechanical work. The steam engine was invented to solve the urgent problem of pumping water out of coal mines. As improved by James Watt, it was soon used to move coal, drive manufacturing machinery, and power locomotives, ships, and even the first automobiles.

11. Common Themes

By the end of the 8th grade, students should know that

11.A Systems

11.A.1 A system can include processes as well as things.

11.A.2 Thinking about things as systems means looking for how every part relates to others. The output from one part of a system (which can include material, energy, or information) can become the input to other parts. Such feedback can serve to control what goes on in the system as a whole.

11.A.3 Any system is usually connected to other systems, both internally and externally. Thus a system may be thought of as containing subsystems and as being a subsystem of a larger system.

11.B Models

11.B.1 Models are often used to think about processes that happen too slowly, too quickly, or on too small a scale to observe directly, or that are too vast to be changed deliberately, or that are potentially dangerous.

11.B.2 Mathematical models can be displayed on a computer and then modified to see what happens.

11.B.3 Different models can be used to represent the same thing. What kind of a model to use and how complex it should be depends on its purpose. The usefulness of a model may be limited if it is too simple or if it is needlessly complicated. Choosing a useful model is one of the instances in which intuition and creativity come into play in science, mathematics, and engineering.

11.C Constancy and Change

11.C.1 Physical and biological systems tend to change until they become stable and then remain that way unless their surroundings change.

11.C.2 A system may stay the same because nothing is happening or because things are happening but exactly counterbalance one another.

11.C.3 Many systems contain feedback mechanisms that serve to keep changes within specified limits.

11.C.4 Symbolic equations can be used to summarize how the quantity of something changes over time or in response to other changes.

11.C.5 Symmetry (or the lack of it) may determine properties of many objects, from molecules and crystals to organisms and designed structures.

11.C.6 Cycles, such as the seasons or body temperature, can be described by their cycle length or frequency, what their highest and lowest values are, and when these values occur. Different cycles range from many thousands of years down to less than a billionth of a second.

11.D Scale

11.D.1 Properties of systems that depend on volume, such as capacity and weight, change out of proportion to properties that depend on area, such as strength or surface processes.

11.D.2 As the complexity of any system increases, gaining an understanding of it depends increasingly on summaries, such as averages and ranges, and on descriptions of typical examples of that system.

12. Habits of Mind

By the end of the 8th grade, students should know that

12.A Values and Attitudes

12.A.1 Know why it is important in science to keep honest, clear, and accurate records.

12.A.2 Know that hypotheses are valuable, even if they turn out not to be true, if they lead to fruitful investigations.

12.A.3 Know that often different explanations can be given for the same evidence, and it is not always possible to tell which one is correct.

12.B Computation and Estimation

12.B.1 Find what percentage one number is of another and figure any percentage of any number.

12.B.2 Use, interpret, and compare numbers in several equivalent forms such as integers, fractions, decimals, and percents.

12.B.3 Calculate the circumferences and areas of rectangles, triangles, and circles, and the volumes of rectangular solids.

12.B.4 Find the mean and median of a set of data.

12.B.5 Estimate distances and travel times from maps and the actual size of objects from scale drawings.

12.B.6 Insert instructions into computer spreadsheet cells to program arithmetic calculations.

12.B.7 Determine what unit (such as seconds, square inches, or dollars per tankful) an answer should be expressed in from the units of the inputs to the calculation, and be able to convert compound units (such as yen per dollar into dollar per yen, or miles per hour into feet per second).

12.B.8 Decide what degree of precision is adequate and round off the result of calculator operations to enough significant figures to reasonably reflect those of the inputs.

12.B.9 Express numbers like 100, 1,000, and 1,000,000 as powers of 10.

12.B.10 Estimate probabilities of outcomes in familiar situations, on the basis of history or the number of possible outcomes.